Ethics of Emerging Technologies

Ethics of Emerging Technologies
Scientific Facts and Moral Challenges

Thomas F. Budinger, M.D., Ph.D.
Miriam D. Budinger, M.D.

WILEY

JOHN WILEY & SONS, INC.

This book is printed on acid-free paper. ∞

Published by John Wiley & Sons, Inc., Hoboken, New Jersey

Published simultaneously in Canada

For general information on our other products and services or for technical support, please contact our Customer Care Department within the United States at (800) 762-2974, outside the United States at (317) 572-3993 or fax (317) 572-4002.

Wiley also publishes its books in a variety of electronic formats. Some content that appears in print may not be available in electronic books. For more information about Wiley products, visit our web site at www.wiley.com.

Library of Congress Cataloging-in-Publication Data:

Budinger, Thomas F. (Thomas Francis), 1932-
  Ethics of emerging technologies / by Thomas F. Budinger and Miriam D. Budinger.
     p. cm.
  ISBN 13: 978-0-471-69212-6 (cloth)
  ISBN 10: 0-471-69212-3 (cloth)
  1. Technology--Philosophy. 2. Technology--Moral and ethical aspects. 3. Science--Moral and ethical aspects. I. Budinger, Miriam D. II. Title.
  T14.B784 2006
  174'.96--dc22

                                                                                              2005024174

Printed in the United States of America
10 9 8 7 6 5 4 3 2 1

To Your Grandchildren and Ours

Contents

6 Ethics of Genetically Modified Organisms   186

7 Medical Ethics   223

8 Ethics of Human and Animal Experimentation   277

Preface

This text is written for students, scientists, engineers, managers, administrators, and anyone who is pursuing or will pursue a career in various areas of the engineering, medical, chemical, and biological sciences. The material is also relevant to those in social sciences and public policy, as well as anyone interested in learning more about these contemporary ethical issues. This book expands upon a series of lectures from a course in science, engineering, and biotechnology ethics for undergraduate and graduate students at the University of California at Berkeley.

The topics selected for discussion are those currently emerging and likely to have great societal and intergenerational implications. Scientists of today are involved in interdisciplinary approaches to complex problems that require adaptation to multinational cultures and talents. The interdisciplinary nature of modern technology and commercial involvement has motivated the inclusion in this text of a broad range of topics in order to give the reader a breadth of technical and philosophical perspectives. The fields range from the Internet to stem cell research and human cloning. In each instance, sufficient information is provided about the technical aspects of each topic so that the reader can better make informed decisions about the ethical consequences of each.

Tools and strategies are introduced to help the reader systematically evaluate an ethical dilemma, and these strategies are used repeatedly to show how they can be applied to a variety of situations. Cases, problem sets, and debates are used to gain experience with ethical problem solving. The intent is also to demonstrate that ethical dilemmas may have more than one "right" solution, and to encourage readers to remain open-minded when evaluating ethical issues, since perceptions about the "right" decision may shift over time.

The emerging technologies discussed in this text are those that have major ethical issues associated with them. We list some of them here:

- Privacy, pirating, and plagiarism in the information technologies.
- Reproductive and therapeutic cloning.
- Gene therapy as treatment or enhancement.
- Genetic modification of plants and organisms.
- Consequences of synthetic biology.
- Alternative energy sources.
- Neuroengineering as therapy or enhancement.
- Sustainability of the resources of the planet.
- Space exploration and terraforming Mars.

Each of these topics is highly controversial, and each raises multiple ethical questions. The intent is not to bring ready answers to these problems but to encourage openness to others' ways of thinking about them. Because we are now a global community, it is increasingly necessary to respect the beliefs of others and recognize how differing opinions can influence the thinking about a dilemma as well as become a major part of their solution.

ACKNOWLEDGMENTS

This text benefited from U.C. Berkeley faculty and others who gave lectures on the following topics: William Kastenberg, Ph.D. on ethics of nuclear energy and risk analysis; Jane Mauldon, Ph.D. on experimentation on human subjects; Donald McQuade, Ph.D. on authorship; Kent Udell, Ph.D. on environmental ethics; Mr. Daniel Herbert on business ethics. Lectures on medical ethics were given by Peter Dietrick, M.D., Peter Lichty, M.D., Louis McNabb, M.D., and Scott Taylor, Ph.D. The text has benefited from the input and classroom involvement of students and teaching assistants during the five times the course Bioengineering 100 has been taught at the University of California, Berkeley from 2000 to 2005. Particular acknowledgement is given to the following students, teaching assistants, and graduate student instructors: Robert Blazej, Morgan Carlson, Cecilia Chang, Jeffery Comreau, Jesse Engelberg, Camron Gorguinpour, Pamela Jackson, Brian McNabb, Krishna Parthasarathy, and Matthew Schoenecker. Amanda Herbert Bilby, Ph.D. historian, documented many of the facts surrounding cases of ethical transgressions. Bibliographic assistance was provided by Robert Smith and Dr. Kathleen Brennan, who also helped with the illustrations.

Rosa Lynn Pincus, Ph.D. provided incisive and helpful suggestions during the early stages of the writing of this text. We are grateful to those who reviewed and made valuable suggestions to improve various aspects of the book. These included Gary Baldwin, Ph.D, William Jagust, M.D., Robin Bhaerman, M.A, CIP, Paul Citron, Ph. D., Daniel Garry, M.D., Ph.D., Peggy Lemaux, Ph.D., Joseph Loscalzo, M.D., Ph.D., Stephen Paik, M.B.A., Martin Pera, Ph.D., Malcolm Potts, M.B., B.Chir., Ph.D., Scott Taylor, Ph.D., Michael West, Ph.D., and William Wulf, Ph.D.

Support for videotaping of selected lectures that contributed to the text was from the Lawrence Berkeley National Laboratory. Support for the preparation of this book was provided by a grant from the Whitaker Foundation. The members of the foundation encouraged the preparation of this text as part of their mission to educate bioengineers. We also acknowledge the generosity of our children and grandchildren who gave us the gift of time to complete this work.

1

Ethical Principles, Reasoning, and Decision Making

INTRODUCTION

Over the last 50 years, new technologies have emerged that have affected the lives of most people on Earth, and along with the benefits have come increasingly complex moral challenges. Each of these technologies has been associated with ethical issues that are new to humanity, have not been considered before, and have solutions that might have uncertain consequences that are difficult to predict. The issues are varied and the moral consequences are wide-reaching. Examples include the conflicts the new technologies present to basic rights of the individual when the technologies can benefit many, privacy rights in the face of the potential of widespread electronic dissemination of personal information, the conflict of the autonomy of the embryo vs. the rights of the parents, and conflicts over the use of stem cells. Other examples are the thirst for more energy vs. the duty to maintain an uncontaminated and sustainable environment, conflicts over the use of genetically modified foods and organisms, the increasing need for organ transplants vs. limited resources, and the conflict between a corporate goal of increasing profits and a company's responsibilities toward all of its stakeholders.

Globalization has led to the need for ethical decisions to be multicultural because decisions affecting one country or culture might have consequences contrary to the policies or doctrines of another country. Thus, both the unanticipated consequences of emerging technologies and globalization present new ethical challenges that must address differing scientific, political, religious, ethnic, and other beliefs and precedents. To address these challenges requires knowledge of the scientific facts. Moreover, it requires solutions that take into account the moral beliefs and theories that have been a part of decision making for hundreds of years but that often tend to be forgotten in today's fast-paced world. A third element is to learn and practice strategies or tools for assessing and prioritizing solutions to ethical problems. This book is designed to address all of these needs by reviewing technical information about the major new technologies so that the ethical problems can be discussed with the facts in mind by revisiting the major moral, spiritual, and ideological theories and by providing a means of approaching the evaluation and decision making that can be applied to each ethical dilemma.

We begin by presenting a case that exemplifies many aspects of this book; that is, how to identify, address, and work through solutions for the kinds of problems the scientist, engineer, corporation, or government body might encounter.

Case 1.1 Cellclone: Multiple Ethical Problems in Biotechnology

At a press conference held at Cellclone headquarters, it is announced that the company is offering a new service to produce customized tissues and organ replacements that are an identical genetic match for each patient. In addition, the company is offering better engineered viral vectors for the insertion of new genes into cells to restore normal cell function (e.g., gene therapy). This announcement is also made in newspapers as an advertisement, and by e-mail to a list of addresses purchased from Spam, Inc.

Cellclone is a private U.S. company with some collaboration with a university and an associated medical school. To produce the genetically matched tissues, the cell nucleus from a human egg is removed by microsurgery and replaced with the nucleus from a cell of the person for whom the tissues are being produced. Cellclone acquires eggs from female employees and also buys eggs from students with a reimbursement of $10,000. Cellclone replaces the nucleus of these eggs with the nucleus of a cell taken from the individual for whom the tissues are to be generated. This method is known as somatic cell nuclear transfer, or more popularly, cloning (in this case, not cloning to produce living individuals but cloning to produce cells that can replace diseased tissues in an individual). The particular method used for separating the various possible tissues from the growing "artificial embryo" was patented by Chinese scientists, and neither their patent nor the reference to their authorship appears in published technical reports or reports to the investment bankers. Cellclone's technical reports of the success of transplanted tissues do not include the 15% incidence of allergic reactions seen in safety studies in experimental animals. Some of these reactions occurred in nonhuman primates, where it was conjectured they were due to an incompatibility between some material in the unmatched donor egg and the recipient.

The first trials sponsored by Cellclone in human subjects involved muscle replacement. They were conducted by a medical school professor who proceeded without having the protocol methods reviewed by his institution because he was working at a Cellclone facility. He did, however, have a federal grant from the National Institutes of Health to investigate cloning using established and federally approved cell lines previously grown from embryos. The subsequent technical reports and scientific papers were authored by large numbers of individuals, including some who participated in but did not make an essential contribution to the experiments (e.g., a summer research intern, a technician who made substantial measurements).

This case presents a host of potential ethical problems that include

- The ethics of human cloning, rights of an embryo
- Theft of intellectual property (patent infringement)
- Human experimentation
- Animal experimentation
- Responsibilities of a corporation to stockholders and the public
- Professional responsibilities
- Authorship
- The risks of emerging technologies

The case represents many of the conflicts that practitioners in biotechnology face as they go about their daily work. The situations surrounding each aspect of this case have choices behind them that present ethical quandaries for the decision maker:

- There is the dilemma faced by the student who may be improperly induced by a high monetary reward to undergo surgery to remove her eggs.
- There is the unacceptable use of employees for human experimentation when some risk is involved.
- There is the problem of the medical school professor who is receiving federal funding for cloning research but is now interested in participating in research not allowed by the federal government.
- There is the question of the safety of the product being offered, given the allergic reactions that have not been thoroughly evaluated.
- There is the issue of the misrepresentation by the corporation that treats the procedure as its own without giving credit to the Chinese originators.
- There is the situation in which authors of the scientific and engineering publications have given credit inappropriately to nonessential participants.

The situations described in the case are examples of the types of conflicts and moral problems that can arise with emerging technologies. We shall discuss these issues more thoroughly in relevant chapters covering authorship, business ethics, human experimentation, and biotechnology. Before proceeding further, we introduce a strategy used throughout this text for designing solutions to such dilemmas, and review the moral theories and principles that provide guidance to analyzing the right thing to do when encountering dilemmas.

DESIGNING SOLUTIONS TO ETHICAL DILEMMAS

To help in resolving dilemmas such as those described in the above case, we introduce here the Four A's strategy, a systematic approach that can be used by individuals or institutions to design solutions to ethical problems, or to make decisions intended to prevent their occurrence. This strategy assists in organizing one's thinking about a dilemma and in coming to a decision about what action to take when there are two or more possible solutions. Even if all possible actions appear to be morally equivalent, the task of the professional is to determine the optimal solution. The strategy given here was derived from a number of sources, among them Swazey and Bird (1995), Weil (1993), Velasquez (2001), and Slossburger (1993). The Four A's strategy is simply a way to systematically apply guidelines when assessing the various aspects of an ethical dilemma so that alternate solutions become apparent and their consequences can be evaluated. The assessment includes the application of moral theories, the prioritization of stakeholders (e.g., those directly and indirectly affected by a decision), the prioritization of duties, and a formal analysis of risk where appropriate.

The four steps below form a foundation for more specific elaboration on a range of ethical problems, whether business ethics, stem cell research ethics, or considerations of risks and benefits of emerging technologies. The strategy is applicable to problems that might arise for individuals, universities, corporations, governments, or other entities.

Table 1.1   Basic Strategy of the Four A's

Approach a dilemma by a systematic consideration of the four A's (i.e., Acquire facts, Alternatives, Assessment, and Action):

- **Acquire Facts**
  - Define uncertainties, clarify ambiguities
  - Get the facts
- **Alternatives**
  - List alternate solutions
  - Develop alternate plans in parallel
- **Assessment**
  - Assess possible solutions according to the moral theories of virtue, justice, duty, rights, and utilitarianism
  - Identify and prioritize stakeholders affected by the decision
  - Perform risk analysis when appropriate
- **Action**
  - Decide on a plan or plans for action
  - Keep alternate action plans under consideration should they be needed
  - Adjust and adapt, recognizing that the process is a dynamic one and that an initial solution may require revision
  - Be open to new options

Frequently, ethical problems can be characterized by referring to conflicting moral theories. The codification of ethical behavior seldom guides us to a clear universal solution. However, through examples and case studies, the application of the Four A's strategy is a useful decision-making guide, even when no clear "best action" appears evident at the outset.

Case 1.1 contains a number of actions whose ethical basis should have been questioned. We have selected one example of the many ethical questions apparent in this case to demonstrate the applicability of the Four A's strategy outlined in Table 1.1, that is, the act of buying eggs from students.

The ethical question here is whether buying eggs violated the autonomy or rights of the students. One problem is that, because the large monetary offering of $10,000 might overly tempt students to donate their eggs, this could be considered a form of coercion. Another ethical question that might be raised is whether the donor has violated a duty not to sell body parts, but for the present, we will address the dilemma of improper coercion and the violation of the individual's autonomy. We proceed as follows:

Acquire facts:
The corporation scientists can learn from asking others what precedents have been established for egg buying. What is the experience of others in terms of price and method of acquisition? What prices are being offered in local university newspaper

advertisements? What rules have been established regarding this type of human research by the National Institutes of Health (NIH) or by the Federal Drug Administration (FDA)? Does it appear that the FDA will allow this project to proceed in the absence of its approval to proceed? Does the university have any rules barring this type of activity since it represents a form of selling of body parts that may be illegal? What will the university's Institutional Review Board (IRB, an ethics board that reviews protocols and ensures that study participants are fully informed of all aspects of the study) require in order to clear the protocol to begin? Could non-human primate eggs be used? Are there medical risks to the product that result from the manner of acquiring the eggs?

Alternatives:
There being no other alternative but to use human eggs, the corporation might consider establishing its own ethics board. One of its functions might be to establish a policy on the method of advertising and the amount of remuneration. What method of seeking volunteers for egg donation would be most feasible? Are there other procedures that need to be followed to perform the proposed research activities? Has this project been sufficiently evaluated for the company to allow it to proceed? Are there sufficient company quality assurance and clinical guidelines in place to carefully evaluate the project for early warning signs of possible problems?

Assessment:
Assess the possibilities with regard to the moral theories: Is the student egg donor being unduly coerced by the high fee you are willing to pay? Is the donor being informed of the risks of ovarian stimulation and egg retrieval? The bioethics board might be seen as a rubber stamp committee, paid by the corporation and a target for criticism. Would this board be objective in determining whether the risks to student egg donors were fully disclosed and that there was not undue monetary inducement or coercion violating the autonomy of students? Thus, in order to reduce bias in the ethics board's decision making, the board could decide to reduce reimbursement to the board members to that normally paid scientific review panels, as was done by another company doing the first therapeutic cloning (Green et al. 2002).

Reevaluate the safety of the project and whether everything has been done to assure the best interests of the egg donors and those to receive the product.

Action:
Establish the bioethics board and heed its advice. Contact the NIH to determine the accepted procedure. In evaluating the safety of the procedure to the egg donor, the bioethics board also should investigate the safety of the product itself. The bioethics board should monitor animal experiments that must be done to detect unanticipated consequences of the procedures. In the actual initial experiments, it was the bioethics board's vigilance that led to methods to avoid allergic reactions (Green et al. 2002). Thinking through the strategy of the Four A's systematically has the advantage of identifying other areas of concern, and perhaps preventing events that could be disastrous for the company.

Other ethical issues in this text are approached through similar examples and solutions. The Four A's strategy brings decision makers' thought processes to bear on a seemingly unsolvable dilemma by providing a framework for a systematic, thorough analysis. Case studies, discussions, debates, and the problem sets will help hone these problem-solving skills.

Another approach to ethical engineering practices, described by Pinkus and coauthors (1997a), can also be applied to scientific innovations as discussed in Chapter 12. The framework for evaluation of ethical aspects of new projects utilizes three core concepts described in Table 1.2. These core concepts were the common themes that evolved from analyses of large engineering projects. The ethics of good engineering goes beyond the individual and includes the responsibilities of the organization and responsibilities to the public.

EVOLUTION OF ETHICAL PRINCIPLES

To design a solution to an ethical problem requires knowledge of the principal moral theories. The major ethical theories (i.e., virtue, justice, duty, rights, and utilitarianism), as well as the concepts they have in common (autonomy, beneficence, non-malfeasance, and justice), are fundamental to evaluating alternative approaches to ethical problems. Selecting the appropriate alternative solutions is basic to selecting the right course of action. While these topics may seem abstract and even dry, they enter into our everyday decision making along with other sources of guidance. These other sources (such as legal, religious, scientific, cultural, ideological, and historical precedents) are sometimes helpful but often have less relevance than the basic moral theories when resolving difficult ethical problems. Frequently, the moral theories are in direct conflict with one another, and no simple codification can show the way to a solution.

Historically, the ethical thinking of the Judeo-Christian/Islamic religions contained religious and moral philosophies similar to Buddhism, such as the dictum of doing to others as you would want done to you (Golden Rule). The concepts of virtue and the creation of personal values, which led to doing the "right thing," underpin the philosophy of Plato and Aristotle. However, cultures around the world developed quite different definitions of acceptable and unacceptable individual and governmental behavior. As we shall see, a

Table 1.2   Engineering Ethics Framework

1. Competence — Recognition that the engineer is a knowledge expert within an organization of other knowledge experts. Together, they must be competent to estimate the risks and failure potentials of a new technology.

2. Responsibility — Recognition that knowledge has power and must be used wisely and safely by both the engineer and the organization.

3. Cicero's Creed II — Recognition that it is the responsibility of the individual and the organization to insure the safety of the public.

virtuous person from the East might have different ideas about political, medical, environmental, and business ethics than a virtuous person from the West. The theories and principles of ethics can be understood based on their definitions, but engineers and scientists in the modern world must also understand the conflicts between these principles and the differences in allowed ethical behavior for individuals from different social, religious, and geographical cultures. Personal, religious, and cultural values can be the determining factors underlying individual or corporate behavior in a situation demanding an ethical decision.

The material below is divided into sections showing how various moral theories arose and how they present useful ideas for many ethical dilemmas. It is imperative to keep in mind the importance of global differences in cultures and value systems while we categorize the five major moral theories of virtue, justice, rights, duty, and utilitarianism. We will also outline some prevailing ideologies and modifications of moral theories (such as rules ethics, rights ethics, and cultural relativism), as they also play some role in the logical consideration of problems associated with the new technologies. Table 1.3 gives a listing of the moral theories, principles, religions, and ideologies, one or several of which may arise in the consideration of a particular ethical dilemma.

Virtue Ethics

Aristotle advocated the ethics of virtue, defining a good person as someone with qualities such as courage, wisdom, loyalty, and fairness. However, the Greeks were also elitist, seeing themselves as superior to others. They saw eugenics (from the Greek meaning good birth) as a virtue, and even Plato wanted to improve society by creating better children through arranged breeding (Pence 2004a).

Table 1.3   Summary of Moral Theories, Principles, Religions, and Ideologies

THEORIES	PRINCIPLES	RELIGIONS	IDEOLOGIES
Virtue	Autonomy	Judaism	Anthropocentrism
Justice	Beneficence	Hinduism	Biocentrism
Rights	Non-malfeasance	Islam	Ecocentrism
Duty	Justice (medical)	Buddhism	Deep Ecology
Utilitarianism		Daoism	Ecofeminism
(Consequentialism)		Confucianism	Pantheism
Pragmatism		Christianity	Pluralism
Cultural Relativism			Transcendentalism
Subjectivism			
Pluralism			
Minimum Conception of Morality			

In Athens in about 350 B.C., when there were only around 300,000 citizens and no competing religious doctrines or theologies, Aristotle taught the importance of attaining moral virtues. These virtues were attributes to be acquired through habit formation in order to reach a proper balance between extremes in conduct, emotion, desire, and attitude (Pence 2004b). For example, Aristotle considered truthfulness a virtue, not in the context of either telling or not telling a lie, but in terms of finding a medium between the extremes of excessive truth and insufficient truth. Thus, the virtue of truthfulness according to Aristotle is the Golden Mean between revealing too much information and giving inadequate information about a situation. For example, it is not virtuous to make negative comments about the appearance of an individual even though the comments might be truthful. Nor is it virtuous to remain silent when there is some information a person deserves to have for his or her success, or to hide fraudulent activities of a corporation.

Another example of virtue ethics dictates that it is better to achieve happiness or self-fulfillment through a life of internal good (e.g., good behavior and virtuous acts) rather than through a life of pleasure or mere contentment. The good in engineering is exemplified by the creation of useful products or technologies while respecting the autonomy of clients and the public. The internal good of science is found in discovering the truths of the physical, biological, and cognitive world. The internal good of teaching is learning, and the internal good of medicine is the promotion of health while respecting the autonomy of others.

Judeo-Christian philosophies emphasize virtue and justice under the just authority of a god or supreme being responsible for the welfare of the faithful. The associated moral thinking resisted ideas of eugenics, insisting that human beings are made in the image of God and should evolve naturally. Over time, the ideas of the early philosophers became intertwined with religious teachings. The influence of early Christian and Jewish teachings spread through the Bible and Torah to the far reaches of the Roman Empire. Teachings of Islam also emphasized the importance of virtue in the life of the individual.

As we learn about the other moral theories that have evolved since virtue was espoused by Greek, Chinese, and other philosophers, it is important to appreciate that virtue is a theory about self and the consequences of actions relative to an individual's personal values. Other theories we discuss below address how one's actions affect others rather than self. As virtue includes the personal attribute of loyalty to other persons, institutions, and corporations, we will see that virtue may be in conflict with duty, particularly in business ethics (Chapter 4). The moral theories of virtue, duty, and utilitarianism come into conflict as a result of the evolution of institutions in our modern society. These conflicts will be pointed out and discussed frequently in this text. The Four A's strategy will be used as the basic tool for evaluating and resolving these conflicts.

Justice Ethics

Justice ethics addresses the need to treat everyone equally without favoritism or bias. Justice dictates the need for equal burden or equal benefit, an impartiality that must be maintained so that multiple individuals are treated fairly. For example, justice becomes an issue in assigning tax benefits or burdens, and when distributing limited resources such as flu vaccines. However, as we will appreciate throughout this text, treating everyone equally becomes a challenge. In many cases a dilemma arises because treating one person based on what is due or owed will often come in conflict with what is due or owed to another person. That is, execution of the minimum principle, which requires that similar people be treated alike, requires a definition of "similar" for each situation. How are the relevant characteristics of a diverse population to be evaluated in terms of equality? When relevant dissimilarities are found to exist, how should the treatment differ? Additional definitions of various aspects of justice are provided below to show the range of situations to which the basic tenets of justice are applied.

- *Procedural justice* refers to social processes (most familiarly, in the judicial system).
- *Retributive justice* applies to issues of correction and punishment.
- *Distributive justice* refers to the fair distribution of social benefits and burdens.
- *Egalitarian justice* refers to theories of justice that stress equal access to primary social goods; libertarian theories of justice give primacy to social and economic freedom; Marxist theories emphasize need ("to each according to his needs; from each according to his abilities").

Justice often sits at the intersection of the moral theory of duty to respect the rights of each individual and the moral theory of doing that which will make the most people happy. We will discuss each of these moral theories later, but for the present, it is helpful to see how justice can be a valuable tool for making decisions about the rights and duties of the individual. The six points listed in Table 1.4 are part of the method used to design solutions to the ethical dilemma of distributing benefits or burdens. It is helpful to consider these six points in reference to real life situations such as awarding salary increases, assigning grades in school, and awarding research or engineering contracts. Consider these points also in the context of assigning burdens, such as in punishment for plagiarizing. Philosophers and others (e.g., the Belmont Report discussed in Chapter 8) have proposed these axioms.

Table 1.4   Justice Axioms

---

1. Give an equal share.
2. Give a share according to need.
3. Give according to effort.
4. Give according to merit.
5. Give according to free market value, e.g., according to the market value of the individual, the competition, and/or how much another would pay for the same service.
6. Give according to contribution of the individuals.

Case 1.2 Justice in a Workplace Situation

Mr. A, who works at a plant that uses a toxic material, develops signs of toxicity. The company doctor confirms his toxic exposure and tells Mr. A's manager of the findings. The manager decides not to tell Mr. A but moves him to another job. Ms. C and Mr. B are upset because they feel they have not been treated equally, and come to talk to the manager. What should he tell them? What is just here? It is not a solution to dodge the issue, nor can he tell them Mr. A has a bad disease. Giving them a bribe to let things be is not the best of solutions. In deciding a solution, all six of the above points about justice may apply. For example, Mr. A is in need of keeping his job, but Mr. B and Ms. C feel their efforts and contributions should also be rewarded.

The intent here is not to present a solution to this case, but to lead the reader to the conclusion that the manager should do something to equalize the situation for Mr. B and Ms. C and to not harm Mr. A further. It is also to point out that ethical decisions can go beyond the individual to the institution, to groups of individuals, to the society. All stakeholders are involved in the interpretation of justice.

Rules Ethics

In the sixteenth century, there was a thrust for the codification of ethics through rules and laws. Thomas Hobbes (1588–1679) was one who saw life in a natural state as ". . . solitary, poor, nasty, brutish and short" (Hobbes 1962a). He believed that individual liberties led to selfish decisions. From his perspective, a society governed by rules that regulated conduct was preferred to a society where individual rights and liberties were of prime importance. He felt that civilization was based on the fear of death and a desire for power such that a society could not be trusted to govern itself; instead, there must be the construction of a commonwealth with a sovereign who would have absolute rule and keep peace by having the final word on matters of law, morality, and religion (Hobbes 1962b). This moral theory is deemed archaic and unacceptable by many cultures. However, the concept of sovereign rule is still viewed as acceptable governance in many countries as well as families, and in some situations (such as the need for martial law after a natural disaster).

Rights Ethics

In contrast to Hobbes, John Locke (1632–1704) believed that people are essentially good, rational, and able to act for the common good. He insisted that

> . . . men being, by nature, all free, equal, and independent, no one can be put out of this estate, and subjected to the political power of another, without his own consent. The only way whereby any one divests himself of his natural liberty, and puts on the bonds of civil society, is by agreeing with other men to join and unite into a community for their comfortable, safe, and peaceable living one amongst another, in a secure enjoyment of their properties, and a greater security against any, that are not of it (Locke 1690a).

In short, citizens join in a social contract whereby they exchange some of their rights in order to protect society. He further believed that individual liberty and other basic rights should prevail over government, which should not intervene unless by consent of the people (Locke 1690b).

Locke's influence on the formation of a modern ethical basis for liberty and democracy can be seen in the parallelisms between his writings and the U.S. Declaration of Independence and Bill of Rights. Universal rules of behavior became codified as laws, as exemplified in these documents. Locke's thinking along these lines is evident in the passage below:

> . . . all men equally have the right to punish transgressors: civil society originates when, for the better administration of the law, men agree to delegate this function to certain officers. Thus . . . government is instituted by a "social contract"; its powers are limited, and they involve reciprocal obligations; moreover, they can be modified or rescinded by the authority which conferred them. It is a rendering of the facts of constitutional government in terms of thought, and it served its purpose as a justification of the Revolution settlement in accordance with the ideas of the time (*Internet Encyclopedia of Philosophy* 2005, no date, www.iep.utm.edu/l/locke.htm, accessed January 2, 2005).

The constructive doctrines elaborated in his *Second Treatise* became the basis of social and political philosophy for generations. Rights ethics provides the basis for the moral thinking behind many of the contemporary ethical debates we will discuss later, including the right to life of the fetus vs. the rights of the mother, the contested right of physician-assisted suicide, and the rights of astronauts to be told all the facts that might affect their survival. The rights of the individual are also of paramount importance in the Kantian theory of duty presented below.

Duty Ethics

While the rights of the individual and the concepts of equality were being asserted and tested in the American colonies, the German philosopher Immanuel Kant (1724–1804) was espousing another doctrine or theory known as the ethics of duty, or deontological ethics. He held that the motivation for an action should not come from feelings or desires, because these emotions may be irrational due to the influence of previous experiences. Rather, an action should be thought through according to one's duty, or according to the universal maxim for doing good: what is right, what is appropriate, what is obligatory, what is moral in and of itself. In his *Foundations of the Metaphysics of Morals,* written in 1785, Kant states: "Act only according to that maxim by which you can at the same time will that it should become a universal law" (Kant 1785a).

This statement defines what is known as the Categorical Imperative. According to Kant, the worth of an act does not lie in the effect of the act but rather in the moral motive. As we shall see below, duty ethics does not look to the consequences of actions as the determinants for what is right or wrong, but instead emphasizes the duty of the individual or institution to act based on a maxim of doing good. Humans should not be treated as means to an end but as ends unto themselves (for example, a patient's well-being must never be jeopardized in a clinical trial, even though the new treatment might benefit society as a whole). To act rationally and freely, to seek what is right and to act accordingly, this is to act morally. Kant calls this capacity "autonomy," meaning the freedom of an individual to decide rationally on the right act (Pence 2004c; also Shannon 1993).

Three major imperatives embody the moral thinking of Kant, paraphrased below from an English translation of his German text (Kant 1785b):

1. To have genuine moral worth, an action must be done out of duty or for the sake of duty. (A distinction must be drawn between duty meaning an action one is obligated to perform and duty referring to the motive for the act, which might best be called dutifulness.) If one chooses a course of action based on what one feels is the *right thing to do,* the act is of moral worth. However, Kant asserts that the same act, motivated solely by the need to comply with a law or a rule, is a morally worthless act. We note here that this imperative is closely related to the moral theory of virtue.

2. An action performed out of duty accrues moral worth from the motive of dutifulness and not from the consequences of the action. As already noted, Kant's ethics is an ethics of motive, and *not an ethics of consequences.* Though this ethics of duty is contrary to consequentialism or utilitarianism ethics, as we will see below, it does not mean that acts in accord with utilitarianism cannot be acts motivated by an individual's sense of the moral worth of the action.

3. Duty (in the sense of dutifulness) is the binding obligation to perform an act (a duty) out of respect for a moral law. The individual motivated by a desire to be seen as a good person acts out of a sense of what one is compelled to do and not necessarily out of a sense of what is a right action. The third imperative of Kant is that the act must be motivated by a duty to moral acts and *not by an obligation.*

The imperatives presume that the individual must know what is morally right and does not act merely in obedience to rules or laws but through the individual's own sense of "the right thing." But how can Kant assume that individuals have an innate sense of what is the right act? How does the individual know what is an act of moral worth, unless the values of acts are learned through nurturing, taught principles, observation, and experience? Can an individual from one culture be expected to know which acts are accepted in another culture as acts in accord with a moral law if the two cultures have differing notions of duty? The universal notion of "doing to others as you would have them do to you," known as the Golden Rule, and the duty to respect the rights of others seem to have a common morality, yet they are not always practiced between families, tribes, societies, or governments.

Recall that about 100 years before Kant, the English philosopher John Locke challenged the concept of a monarchy on the basis that people have fundamental rights that others must respect. Both duty ethics and rights ethics achieve the same end: individuals must be respected and actions that are ethical maintain this respect for the individual. Difficulties arise when attempting to prioritize the basic rights of one person that are in conflict with the basic rights of another, and in balancing the good of an individual vs. the good of a society. These difficulties will become more apparent when we present the theory of utilitarianism below.

The philosophies of virtue, justice, duty, and rights have many common threads, especially with regard to loyalty and the dictum to respect the rights of others. Loyalty dictated by virtue ethics has a natural alignment with duty to respect the rights of others. The importance of duty ethics in professional activities will be seen throughout this book. But duties and loyalties of the professional to respect the autonomy of others are frequently in conflict with the theory of utilitarianism, where consequences of acts rather that the inherent worth of an act become important. Indeed, it is extremely difficult to resolve problems when the rights of one individual are at odds with those of another individual or a group of individuals, or when one duty to perform an act is in direct conflict with another duty. We will attempt to put these theories into a coherent framework and present methods and guidance that allow a rational approach to making the right ethical decision.

Before going further, let us consider a case where two duties are in conflict: the professional duty to preserve life and the duty to respect the autonomy or right of a patient to die.

Case 1.3 Professional Duty vs. Rights

Consider the dilemma of a physician when the duty to preserve the life of a patient is in conflict with the wishes of the patient to be left alone to die. The case of a severely burned patient illustrates this ethical dilemma, as well as the ethical dilemmas regarding issues of informed consent, competency, and autonomy vs. medical paternalism (Kliever 1989).

In 1973, Dax Cowart was severely burned following a propane-gas explosion that killed his father (Burton 1989). Sixty-seven percent of his body sustained severe burns, including his ears, eyes, and hands. As an experienced pilot, he recognized the severity of his situation at the scene of the accident, and asked the farmer who first arrived at the scene for a gun to shoot himself. The farmer refused, and he was transported to the burn center at the University of Texas Medical Center, where he refused treatment. However, despite his insistence, and despite the fact that he was fully competent to make this decision, the staff instead received consent from his mother to proceed. He endured excruciatingly painful treatment that kept him alive, but in the end, when discharged after 232 days in the hospital, he was horribly disfigured, blind, and had only limited use of one hand.

He is now married, a 1986 graduate of Baylor law school, and is a successful attorney. After winning a lawsuit against the oil company responsible for the gas leak, he is financially comfortable. Nevertheless, he insists that he had the right to refuse treatment and to die. He does not argue that every burn patient should be allowed to die, but that every competent burn patient should be allowed to make that decision.

Rachels considers his desire to die a rational reaction, justified because his concept of self was transfigured by the accident:

> What his injury had done, from his point of view, was to destroy his ability to lead the life that made him the distinctive individual that he was. There could be no more rodeos, no more aeroplanes, no more dancing with the ladies, and a lot more. [His] position was that if he could not lead that life, he didn't want to live (Rachels 1986).

The procedures and problems surrounding issues of the right to die, autonomy, paternalism, and futile medical care are subjects further discussed in Chapter 7.

The next case emphasizes the conflict between three duties: duty to assure safety, the duty to accomplish an engineering goal, and the duty to respect the rights of individuals.

Case 1.4 The Challenger Catastrophe

This case underscores the serious decisions facing engineers who work on projects where human lives are at risk. On January 28, 1986, the launch of the space shuttle Challenger went ahead despite concerns voiced by project engineers that it was not safe. At issue was the possibility that the extremely cold temperature at the time of launch might interfere with the function of the O-rings in the solid rocket booster, causing the escape of hot gases formed by the combustion of solid rocket propellant. The O-rings had been problematic in the past. During testing in January 1985 at 53°F, there were signs that hot gases had escaped through the O-ring seals. Because of concerns that the cold temperature increased the rigidity of the seals and made them ineffective, the engineers redesigned them, but they were not ready at the time of the launch in January. On the scheduled launch day in January 1986, temperatures were in the low 20s. Since this was the coldest temperature for any launch and no data existed to ensure the function of the O-rings at this temperature, the engineers most familiar with the problem recommended that the launch be delayed until warmer conditions prevailed. Overruled by management, the launch proceeded and the resulting explosion killed all on board. The post-explosion investigation indicated the failure of the O-rings in the rocket booster, which allowed the escape of hot gases formed by combustion of the solid rocket propellant. The resulting flames from these gases burned through the external tank, ignited the liquid propellant, and caused the spacecraft to explode.

What are the proper procedures to be followed in such a situation? Who should make the go/no go decision? What if there are insufficient data to make a decision? Should the astronauts have been informed and allowed to participate in the go/no go decision? What was the duty of the concerned engineers in this situation and how far should they have gone to stop the launch? At what point do budgetary concerns or political implications undermine decision making and exacerbate the pressure to "go"?

The types of issues raised here are similar to those that may arise during other projects. Ethical decisions may be required on a day-to-day basis and may have far-reaching consequences. Those making the decisions have the responsibility to make the best possible choices. This catastrophe highlights the importance of recognizing that modern engineering involves multiple competencies, an understanding of respective responsibilities, and an overriding acceptance of Cicero's Creed II (i.e., the priority of public safety). Pinkus et al. (1997b) provide further discussion of the ethical issues raised by this catastrophe.

Modern Duty Ethics and Ross's Prima Facie Duties

A major problem with Kantianism involves the resolution of problems when there are conflicting duties. That is, if one duty is performed, an equally important duty would be neglected.

The major problem with the Kantian philosophy is that it does not provide a method to resolve dilemmas wherein the subjective principles of action (called maxims) come into conflict with the categorical imperatives. The test of a maxim as a genuinely universal moral principle involves a test against three criteria:

1. Is the maxim one that should become a universal law?
2. Does the maxim involve treating yourself or another as *an end* in contrast to *a means to an end*?
3. Does the maxim respect the autonomy of others?

A simple example often mentioned in the literature on ethical dilemmas is that a maxim about lying to save another cannot pass all three categories. For example, if you have hidden a child in a house that is being invaded by thieves, and you are asked if there are others in the house, you might lie to protect your child. The maxim might be: Lying is acceptable if it is done to protect a child. Many other similar dilemmas can be cited.

However, Kant's moral theory does not allow a means of deciding between two possible actions that are both one's duty and in conflict with one another. To resolve these problems, W. D. Ross, a twentieth century philosopher, proposed a modification of Kant's moral theory. He suggests that there are seven prima facie duties (see Table 1.5), each of which is an obligatory duty unless there is a conflict with a greater prima facie duty (Ross 1930). Over the past eight decades, during which the development of technologies, including global communications, has brought about new ethical dilemmas, it can be shown that these seven prima facie duties cover practically all situations. For a given situation, the moral action depends on which of the seven duties apply, and of those that apply, which has priority over the others.

According to Ross, the selection of which prima facie duty overrides another comes from an individual's intuition. Thus, a soldier's duty to prevent harm to others might override the duty to prevent harm to himself. The duty to increase general happiness by giving to the poor might be overridden by the duty of fidelity. The duty to prevent harm might override other prima facie duties when considering the benefits and risks of many new technologies. An example would be the duty to define as far as possible the risks of genetic

Table 1.5   Prima Facie Priorities

1. **Fidelity:** Duty to keep commitments. This includes the duty to honor one's professional code.
2. **Reparation:** Duty to correct past wrongs.
3. **Gratitude:** Duty to repay.
4. **Justice:** Duty to prevent unfair distribution of benefits. This includes limited resources such as organs for transplantation.
5. **Beneficence:** Duty to increase general happiness.
6. **Self-improvement:** Duty to better oneself. This includes a professional's duty to continuing education.
7. **Non-malfeasance:** Duty to prevent harm. This includes duty to make risk assessments of the consequences of emerging technologies.

modification of plants and animals for future generations. Another example is the duty to evaluate the long-term consequences of deep brain stimulation.

One can argue that the intuitive weighing of our various duties is how we practically work out choices of action in everyday life. A summary statement by Feldman (1978a) rephrases the Ross philosophy, which can help in deriving a strategy for each case:

> An act is morally right if and only if it is a prima facie duty and no alternative is a more stringent prima facie duty.

But Feldman's statement raises the question: what is meant by stringent? The concept of an obligatory act rises above others in the attempts to define stringent. If two conflicting acts are prima facie duties, then the one which is more obligatory should be that on which the act is based. As will be shown throughout this text, in most cases the confusion and debate regarding Kantianism and Ross's theory can be lifted if one applies the design strategy to each ethical dilemma (i.e., engaging a problem by resolving ambiguities, getting advice, developing alternative solutions, and taking action while recognizing the dynamics of the selected acts). One duty that seems universal and encompasses or underlies most of Ross's seven priorities is the duty to be kind, which is a priority of two great philosophers:

> There is no duty more obligatory than the repayment of kindness (Cicero 106–43 B.C.).
> Forget injuries, never forget kindnesses (Confucius 551–479 B.C.).

The next case exemplifies a situation involving conflicting duties. The dilemma for FBI agent Mark Felt was between doing the virtuous or right thing regarding truths of importance to the American people, or adhering to the responsibilities of an FBI agent to keep secrets. While it may never be known what reasoning he used when he decided to leak the FBI's most sensitive information to the press during the Watergate scandal, he may have used a strategy along the lines of the Four A's in his decision making.

Case 1.5 Duty vs. Loyalty: The Morality of Deep Throat

In 1972, a felonious break-in occurred at the Democratic National Committee headquarters in the Washington, DC, apartment complex known as Watergate. An anonymous tipster known only as Deep Throat provided secret insider information to investigative reporters Bob Woodward and Carl Bernstein of the *Washington Post*, information that enabled them to trace the break-in directly to top officials in the White House, and ultimately to then-President Nixon himself. The *Washington Post* story, along with other damning information, including Senate hearings and secret White House tape recordings, led to the threat of impeachment and Nixon's resignation in 1974.

Over the years, there has been much speculation about the identity of Deep Throat. Only three people knew his name and steadfastly kept his name a secret for over 30 years until May 2005 when Felt, at age 91, revealed his identity. At the time of the Watergate scandal, Felt was a staunch, loyal member of the FBI and was considered to be next in line to take over the FBI after J. Edgar Hoover, but the job went to outsider (and Nixon supporter) L. Patrick Gray III instead, who became Felt's supervisor. Gray had agreed to keep the White House informed of all FBI findings in the Watergate case, which infuriated Felt, who felt the FBI's integrity had been compromised. When Felt began to learn of information that led to Nixon and others in the administration, he could not go to Gray with his concerns. He turned instead to *Washington Post* reporter Bob Woodward, someone who could reveal to the public the truth and at the same time minimize his own vulnerability as the source of the information. Felt's ongoing supply of information and encouragement allowed Woodward and his co-investigator, reporter Carl Bernstein, to stick to their position despite vicious and desperate attacks from the White House to discredit them. Felt's decision to tell the truth forced him to choose between his oath of loyalty to the FBI and his duty to protect the country from the criminal activities at the highest levels of the White House.

The above case represents the conflicts between virtue and the duty to keep a secret or to be loyal to the professional organization's policy. Next, we introduce the moral theory of doing good for the most. This moral theory also comes into conflict with both virtue and duty.

UTILITARIANISM

Originating in the late eighteenth and early nineteenth centuries, the theory of utilitarianism was espoused by Jeremy Bentham (1748–1832) and later by James Mill (1773–1836) and his son, John Stuart Mill (1806–1873). An often quoted or paraphrased statement of the fundamental principle or morality proposed by J. S. Mill is the basis for discussion of utilitarianism:

> . . . actions are right in proportion as they tend to promote happiness, wrong as they tend to produce the reverse of happiness. By happiness is intended pleasure and the absence of pain; by unhappiness, pain and the privation of pleasure (Mill 2001).

Happiness might be hedonistic (e.g., physical pleasure) or intellectual, thus the categories of hedonistic utilitarianism or ideal utilitarianism, respectively. Utilitarianism originally was a movement away from outmoded Christian ethics and the injustices of English aristocracy that favored the few, toward changes that favored the majority. The reforms encouraged by utilitarianism were precipitated by slavery, poor factory conditions, child labor abuse, the poor treatment of prisoners, and other inequalities.

Consequentialism is defined as ". . . the doctrine that the moral value of any action always lies in its consequences and that it is by reference to the consequences that actions, and indeed such things as institutions, laws and practices, are to be justified if they can be justified at all" (Smart and Williams 1998a). The specification of a happy consequence is the major connector between a theory of consequences and utilitarianism.

As summarized by Pence (2004d), the four basic tenets of utilitarianism are:

- **Consequentialism:** Consequences count, not motives or intentions.
- **The maximization principle:** The number of people affected by consequences matters; the more people affected, the more important the effect.
- **A theory of value (or of good):** Good consequences are defined by pleasure (hedonic utilitarianism) or what people prefer (preference utilitarianism).
- **A scope-of-morality premise:** Each being's happiness is to count as one and no more.

Two important parts of utilitarianism known as act utilitarianism and rule utilitarianism are introduced here to help the reader understand that the consequentialism explanation of utilitarianism is only part of the theory. That is, utilitarianism goes beyond the doctrine of preferring an act or decision that produces the greatest happiness for the greatest number. As we will see through our examples, this simplification leads to an erroneous analysis when considering alternative solutions. An additional problem is that utilitarianism can lead to conflicts with Kantian philosophy of the moral worth of an act based on duty. An action in accordance with utilitarianism can still have as its motivation moral worth in addition to duty. For example, an engineer decides to check and recheck the calculations to ensure the safety of a structure. His sense of duty motivates this extensive rechecking. The consequences of this rechecking to prevent harm would also lead the engineer, bound by duty, to follow the professional code of engineers, to go to these extremes of rechecking, but if the professional code is the motivation, then the action of the engineer has not the same moral value, according to Kant.

Case 1.6 Taxation

Suppose a population taxation goal is to be achieved by either taxing all citizens equally or by taxation in proportion to the wealth of each citizen. In the first alternative, the happiness of the wealthy would be enhanced and this would be the greatest happiness for those individuals, but the poor would suffer. The second alternative would achieve some happiness for the poor whose tax bill would be reduced but would achieve less happiness for the wealthy. Clearly, without a quantification of the value of the happiness times the number of wealthy people minus the quantification of the unhappiness times the number of poor citizens, we cannot know how well taxation according to wealth supersedes the alternative of equal taxation where we also have no measure of the value of happiness for all citizens. The notion of "greatest happiness for the greatest number of people" is difficult to apply in a situation when there are no data to evaluate the balance between benefits to all citizens. In dealing with this problem, Feldman (1978b) proposes a definition of act utilitarianism that brings to the fore the notion of utility:

> An act is right if and only if there is no other act the agent could have done instead that has higher utility than it has.

This definition emphasizes an equivalence between act utilitarianism and consequentialism. It helps us understand how to make decisions when consideration of the six points of justice does not lead to a clear selection of the right decision in a particular dilemma. From this, we can conclude that utilitarianism is understood by evaluating the consequences of one's actions. However, the examples below show why utilitarianism ethics required a modification, and why rule utilitarianism is needed for the defense of utilitarianism as a moral theory. The first is the application of utilitarianism (consequentialism) to fulfilling promises. This is a category of human behavior that presents serious problems to act utilitarianism. The second is the conflict between punishment and simple application of consequentialism (act utilitarianism).

Case 1.7 Promise to a Dying Person

A father and his son sailing around the world find themselves stranded on an island in the South Pacific following a storm. The father has been seriously injured, is near death, and asks his son to promise to bury him with his dead wife's wedding ring, which the father wears around his neck in a gold locket. The son promises. After his father's death, the utilitarian son is faced with two decisions about what he should do. What is the utility of burying his father? It is difficult to dig a grave in the rocky soil and much easier to cast him into the ocean on the outgoing tide. Secondly, he might need the gold locket, chain, and ring to trade for his safe return to civilization. What is the utility of keeping the promise to bury the father's keepsake?

One can argue that the consequence of not burying his father will be feelings of having treated his father callously, but the utilitarian son concludes the difficulty of burial is a consequence worse than the seeming callousness of casting his father into the sea. With respect to the jewelry, the son applies act utilitarianism and concludes the act of keeping his promise is of less moral value that that of taking the jewelry, which will have the consequence of securing the son's safe passage and, therefore, his happiness. How can there be any pain in this act? The decision between keeping the promise and breaking the promise is clearly dictated by the moral theory of consequentialism. But would we conclude the decision to break the promise was ethically correct? Thus, the category of human behavior known as keeping promises presents serious problems to utilitarianism because breaking the promise in this case would be consistent with utilitarianism, but to most, this is the wrong thing to do. A form of utilitarianism known as rule utilitarianism is needed for the defense of utilitarianism as a moral theory.

Examples of the type described in Case 1.7 led to rule utilitarianism wherein the correct moral action is *to keep promises*. Rule utilitarianism deems that rules of behavior are derived by a determination of whether particular acts are right or wrong. Rules are justified by their general utilitarian consequences, and the overall utility is maximized by a system of rules, compared to the alternative of having individuals decide on conduct in particular circumstances based on their own judgment. Human beings may find it impossible to always be impartial about utility in matters involving their self-interest. Rules have the value of imposing a degree of "objectivity." Consistency requires that rules be applied in the same way to relevantly similar circumstances. Rule utilitarianism would guide the son to decisions regarding the act of burial and the separate act of keeping the promise to bury the father with the jewelry. The next example emphasizes further problems with utilitarianism.

Case 1.8 Punishment Conflicts with Utilitarianism

A student writes a project paper and uses material copied electronically from the Web. His act of plagiarism is discovered by the scanning systems used by many university professors. His punishment is to receive an F in the course. What utility does this act have and to what extent is there any happiness for anyone in the act? Have the basic tenets of utilitarianism been refuted by this example?

The subject of punishing a wrongdoer has been a major problem for the theory of utilitarianism because it is an act which causes pain and does not necessarily give happiness or pleasure. As Feldman (1978c) states: "How can we be justified in producing something intrinsically bad as a response to those whose main fault is that they produced something intrinsically bad? Can two evils make a good?" The utility of some punishments can be rehabilitation or possibly a deterrent to potential future offenders, but the more general concept of retributive punishment seems the norm and this concept is at odds with act utilitarianism.

In answer to this conflict, as well as the previously mentioned cases where the moral theory of justice and act utilitarianism oppose each other, rule utilitarianism is used to guide decisions regarding the preferred acts. Arguments for and against utilitarianism are found in the analysis by Feldman (1978d) and a debate between Smart and Williams (Smart and Williams 1998b).

Analysis of Duty Ethics vs. Utilitarianism

Kant's moral theory of duty (deontology) prescribes the correctness of an act is based on moral duty and not on a judgment of its consequences. Religious revelation ("divine command") provides the historically common foundation for deontological moral principles: things are right or wrong if, and only if, commanded or forbidden by God. The concepts of natural law and human reason are cited as sources for what is right or wrong, but in modern societies throughout the world, these same concepts seldom lead to similar conclusions by individuals dealing with ethical dilemmas. This is because these concepts are too general and too distant for consistent use in formulating methods for approaching ethical problems in various cultures and everyday situations.

It is hard to imagine that an abandoned child raised by wolves, a tribeswoman from New Zealand, and a youth from an affluent European family would all have the same moral basis for action under a given circumstance, or that they would act similarly in a situation requiring acts that require consideration of the rights of other individuals. The Problem Set at the end of this chapter will allow practice with how some moral principles can override others. But for the present, we consider some reconciliation of differences between duty ethics and utilitarianism.

Some philosophers (and many sociobiologists) take the position that deontological principles are simply those shown to have consistently favorable consequences over a long period of experience; that is, they have evolved as accepted duties through the test of time. Accordingly, these have been sanctioned by custom, religious practice, and in many cases, have led to laws.

Though situations that seem to pit duty ethics against utilitarian ethics arise throughout the world, the two theories do share the aspect of valuing individual rights. Kant argues for the supremacy of individual rights, and in fact his philosophy, as well as Locke's, strongly influenced the Declaration of Independence and the Bill of Rights of the United States. John Stuart Mill, a major proponent of act utilitarianism, also felt that individuality was an important component of happiness and necessary for individual and social progress. He believed in the freedom to have and express one's own opinions without reserve, and to act upon them at one's own risk or peril, as long as it does not harm another.

As long as ". . . he refrains from molesting others in what concerns them, and merely acts according to his own inclination and judgment in things which concern himself, the same reasons which show that opinion should be free, prove also that he should be allowed, without molestation, to carry his opinions into practice at his own cost" (Mill 1859). But the utilitarianism theory gives far less priority to the individual's rights than does the Kantian theory of duty.

As we have seen, utilitarianism is related to consequentialism, where the correctness of an act or ethical decision is based on whether the consequences will lead to the greatest happiness. The consequence usually sought is a state or situation wherein happiness or maximum utility of the individuals is reached, but not necessarily all of the individuals will benefit and some might be harmed. That leads to a major conflict between the theory of utilitarianism and individual autonomy, as doing the best for the most will often violate the rights or autonomy of a few. We present the arguments between utilitarianism and duty through examples of situations that are perplexing to the utilitarian. That is, we consider the role of injustice in utilitarianism, and the role of doing the best for the most in deontology. The first situation is presented is a case against utilitarianism.

Case 1.9 Individual Rights vs. Utilitarianism

A sheriff concludes that if he falsifies some documents he will cause the death of an innocent man but thereby save the lives of 100 citizens who are facing a horrible and painful death.

The issue here is that although the sheriff's act is unjust, by committing it he will minimize pain (McClosky 1963). J. C. C. Smart, in arguing for utilitarianism, admits that the utilitarian is forced by logic to accept the unjust act (Smart and Williams 1998c). This would be a situation of the lesser of two evils, and an example of the end justifying the means. This case exemplifies negative utilitarianism, wherein one is considering minimization of pain rather than maximization of happiness. The reader will note that the act here is judged solely on the end consequence and not on the virtue of the act itself. Recall that duty ethics would deem this act outside the rules of behavior commonly accepted by society.

These examples lead to the chief persuasive argument for utilitarianism: ". . . the dictates of any deontological ethics (duty ethics) will always, on some occasions, lead to the existence of misery that could, on utilitarian principles, have been prevented" (Smart and Williams 1998d).

Autonomy vs. Utilitarianism

The conclusion that action based on consequence might violate our duty to respect the rights or autonomy of a particular individual is overruled by the utilitarian rule of the greatest happiness for the most. But as we formulated earlier, the rule that should be considered is that of the act that gives the greatest *utility*. This rule is applied when a decision is made to improve the quality of life (such as water supply, hydroelectric power, or roads) for many individuals at the expense or violation of the autonomy of one or a few individuals. The

Golden Rule of doing to others as you would have them do to you takes a lower priority than the rule of doing that which gives the greatest utility, generally interpreted as the greatest happiness for the majority.

This theory is clearly different from Kantian theory because Kant's major principles ignore the consequences of an act as determinants of right or wrong, and instead emphasize the value of the act in and of itself. However, the right vs. wrong decision for both the moral theories of duty and utilitarianism is guided by whether an action maximizes happiness or does not cause pain. When confronted by dilemmas regarding duty to the rights of a few vs. the happiness of the majority, an analysis of the alternative actions and their dynamics, as guided by the Four A's, will be useful.

Case 1.10 Eminent Domain vs. Rights of the Land Owner

Governments have the power to acquire land (real property) against the will of the owner if the acquisition is necessary for the completion of a road, dam, or other important public project. In the United States this power is known as *eminent domain* and in other countries that follow the principles of English law, the term *compulsory purchase* is used. The Fifth Amendment to the U.S. Constitution stipulates, in addition to just compensation being made to the owner, that "public use" of the property be demonstrated. However, over the years since 1930 the definition of "public use" has expanded to include *economic development.* Thus, the power of eminent domain has allowed property seizures to enable commercial development. A public purpose used to justify this seizure is that the commercial development will generate more jobs and more tax revenue for the local government. In 2005, the issue of eminent domain for commercial interests was argued in the U.S. Supreme Court (*Kelo et al. v. City of New London, CT*, No. 04-108, Decided June 23, 2005). The suit was brought by owners of 15 homes on a desirable 1.54 acres of waterfront property against the town of New London, Connecticut, which planned to turn the area into offices, new housing, and a marina and to allow a $300 million research facility to be built by a pharmaceutical company. In this case of eminent domain vs. Fifth Amendment rights, the Court ruled by a 5 to 4 vote in favor of commercial interests, the majority citing land redevelopment in an economically depressed city qualifying as "public use."

This case establishes a precedent for similar actions on the part of communities around the United States and leaves small property owners vulnerable. The justification for eminent domain intended by the Fifth Amendment was pubic usage and not commercial interests that indirectly might be in the interests of the local community. Clearly, any developer could make a case to the local and state government that they can improve the appearance and utility of a sector of land and in addition bring more tax monies to the local government for the benefit of the public. But is this just? As we shall see in Chapter 5, the global initiatives to improve the environment might use this precedent to seize lands whose current use is judged not to be consistent with initiatives to maintain a sustainable environment. The Supreme Court decision gives local governments wide latitude to decide when a seizure for "public use" is justified.

Utilitarianism vs. Justice: Allocation of Limited Resources

The issue of limited medical resources becomes especially significant when the allocation means life or death for the patient. Important examples are the distribution of penicillin to wounded soldiers in World War II and, more recently, the allocation of organs (e.g., heart, liver, kidney) when the supply of organs or the distribution of kidney dialysis machines is insufficient to meet the need. Further discussion of organ transplantation will be made later in this chapter and in Chapter 7. For the present, we will present the moral considerations behind selecting a method of allocation, whether it be for a drug, organ, or physician's services.

Perhaps the most famous modern philosophical treatment of justice relevant to utilitarianism and other moral theories is that of Rawls (Rawls 1999a). The rationalization that the rights of the individual can be extended to an argument for the rights of the social system is not accepted by Rawls. The essence of the superiority of justice is contained in his words:

> Each person possesses an inviolability founded on justice that even the welfare of society as a whole cannot override. For this reason justice denies that the loss of freedom for some is made right by a greater good shared by others. It does not allow that the sacrifices imposed on a few are outweighed by the larger sum of advantages enjoyed by many (Rawls 1999b).

The two competing moral theories surrounding allocation methods are utilitarianism and justice (as in equity of opportunity and the Rawls concept of equity of liberty). These two opposing doctrines came into focus in 1841 when survivors of a collision with an iceberg found themselves in an overloaded boat.

Case 1.11 Case of the Sinking Lifeboat

After a ship collided with an iceberg and sank in the North Atlantic in March of 1841, a severely overloaded lifeboat was in jeopardy of capsizing in the icy waters, thereby ending the lives of all on board. In order to save the most, 14 unmarried men who were not essential to rowing the boat were forced overboard by the sailors. Shortly thereafter, the survivors in the lifeboat were rescued. One of the seamen, the only one who had not left town after arriving ashore, was arrested and brought to trial (*United States v. Holmes 1842*). His legal defense was that he was doing that which would be best for the most, a utilitarian argument. Justice Oliver Holmes argued that the most just method of deciding who should die was to draw lots. This case is the focus of the argument that it is far better to decide by a method by which all have the same chance to survive than to decide by a method of selection that may be unjust and biased.

How do we reconcile the court's decision with the Rawls theory of justice? What would a student of Rawls say about this situation? We find an answer in the following:

> . . . injustice is tolerable only when it is necessary to avoid an even greater injustice (Rawls 1999c).

The argument for random selection made by Childress below is compelling by comparison to arguments for deliberate selection. Quoting from Ramsey, Childress explains this concept as follows:

> The individual's personal and transcendent dignity, which on the utilitarian approach would be submerged in his social role and function, can be protected and witnessed to by a recognition of his equal right to be saved. Such a right is best preserved by procedures which establish equality of opportunity. Thus, selection by chance more closely approximates the requirements established by human dignity than does utilitarian calculation. It is not infallibly just, but it is preferable to the alternatives of letting all die or saving only those who have the greatest social responsibilities and potential contribution (Childress 1970, Childress 1997).

But the firm rule of random selection cannot be operable in all circumstances. We give some examples, which are somewhat contrary to the general assertion of Childress that exceptions are rare and unlikely in any case. In some situations, the loss of one or a few individuals would be disastrous for the others (e.g., loss of the president in a grave national emergency).

First consider the situation in which penicillin was in short supply in a non-combat-zone hospital during World War II: the two groups needing the drug were soldiers with syphilis and soldiers with wound infections. Should lots have been drawn to give all individuals equal opportunity for a cure regardless of their diagnosis? In another situation, when physician and medical staff personnel are limited at a time of massive causalities, the method of triage is employed to assess whether the wounded can be saved based on the severity and character of their wounds. Those selected for treatment are given the best opportunity for life, while those whom a physician decides would receive only futile care are given comforting care. True, they are not left to die (thus, the situation is different from throwing people out of a boat). Nevertheless, it is far from the doctrine of equal opportunity. Perhaps one could argue it is equal opportunity to life for those who have some chance of living.

OTHER MORAL THEORIES

Cultural Relativism

Cultural relativism takes the position that moral beliefs and principles are relative to individual cultures or, in the extreme, individual persons. Thus, the rightness or wrongness of a situation may vary from place to place (even person to person), and not all absolute or universal moral standards can apply to all at all times. Consequently, concepts of rightness and wrongness are meaningless apart from the specific contexts in which they arise. This theory arose in the twentieth century to justify the correctness of the morality of one culture when rules and accepted behavior of that culture were not acceptable to another culture. For example, Eskimos sometimes kill infants at birth as part of their culture, but this is not acceptable to other cultures. Certain foods accepted by some cultures are considered unethical to consume by other cultures.

This theory has strong objections, not the least of which is the fact that it does not recognize the moral theories of rights, duty, utilitarianism, virtue, and justice. However, in modern business ethics, cultural relativism does become important (e.g., methods of negotiations and bribery are the culture for some but not all). The scientist, engineer, and corporation leadership should study the moral codes of the societies with which they have collaborations and business negotiations.

Pragmatism

Pragmatism is a theory of ethics that developed because of the absence of any one theory or code of behavior that was applicable to the majority of situations. Pragmatism is defined by Schinzinger and Martin (2000) as follows:

> [It is] . . . a theory about morality that emphasizes the limitations of abstract rules. . . . Pragmatists emphasize the importance of particular contexts in which facts and values must be weighed and balanced, including values such as rights, duties, and virtues. Pragmatists also emphasize flexibility in integrating and harmonizing competing values. Rather than applying fixed rules or ideals, moral decision making is essentially a matter of extending moral reasons into new and often uncharted situations. [It] is best understood as a methodological emphasis on sensitivity to moral complexity, reasonable compromise, and close attention to the manifold dimension of cases in their complete context . . . it is not so much an alternative to the search for valid moral principles as it is an insistence on greater flexibility and tolerance in how principles are applied.

The philosophers William James (1842–1910) and John Dewey (1859–1952) espoused this theory as a rejection of the search for general ethical theories. Pragmatism emphasizes maximum good by impartially considering the interests of everyone affected. The concept is to be flexible in integrating competing values rather than applying fixed rules or ideals. Solving an ethical dilemma is a matter of extending moral reasons into new and sometimes unprecedented situations. In later chapters, particularly in Chapter 4 on Business Ethics, we shall give examples of how the professional engineer, scientist, or manager may make a pragmatic decision where the end result would seem to justify the means. What is important is to ensure that the action taken is, in the end, ethical and does not lead to the slippery slope of unwanted consequences.

Pluralism

As we have emphasized, the reader should be aware that there is no single theory or single method for resolving all disagreements. Moral principles can collide and there can be disagreement about how to resolve the collision (e.g., differences between individual rights vs. utilitarianism). The pluralist position is that in a heterogeneous culture there may be many sources of moral value, and consequently, there may exist a multitude of moral points of view on many issues (consider abortion). This applies all the more strongly across cultures, each of which may have significantly different views of principles, or different principles entirely.

Subjectivism

The theory of ethical subjectivism started with the idea that morality is a matter of sentiment rather than fact. In other words, when a person says that something is morally good or bad, this means that he or she approves of that thing, or disapproves of it, and nothing more (Rachels 2003a). Thus, in a classroom debate on abortion, euthanasia, human enhancement, capital punishment, or some other ethics issue, the opponents might argue one side or another based on their feelings, emotions, or personal values and not a higher set of moral rules.

Many arguments launched against simple subjectivism show that this theory is flawed. However, without some knowledge of moral theories to guide the student and professional, there may be little else to guide their reasoning in even rather simple situations requiring ethical decisions.

Even without philosophical training or the minimum exposure to the theories and principles of this chapter, there is one guiding concept, known as the minimum conception of morality, that calls for reasoning based on facts. This concept is developed further below.

Minimum Conception of Morality

One seeks some guidance or underlying concept that might unite the theories and principles described above. Is there some fundamental construct that would provide a rationalization for moral actions independent of the various theories which, as we have seen, are sometimes in conflict? In this, we are guided by a moral philosophy that defines morality as: ". . . at the very least, the effort to guide one's conduct by reason—that is, to do what there are the best reasons for doing—while giving equal weight to the interests of each individual who will be affected by what one does" (Rachels 2003b).

Deontologists and utilitarianism philosophers would agree with the incorporation of this minimum conception in some form. In fact, this minimum conception of morality can be seen as a statement of justice and also conforms in general with the Aristotelian concept of virtue. Note that the Four A's strategy is a method whereby this principle can be put into practice.

Other Concepts and Ideologies

Above we have summarized the major ethical theories, but there are other concepts around which actions have been justified. We list here seven religious or spiritual philosophies and will use these philosophies in an examination of how human beings have considered their relationships to the environment in Chapter 5. Though there are additional spiritual philosophies, seven of the principal ones are Confucianism, Hinduism, Buddhism, Daoism, Islam, Judaism, and Christianity. Modern thinking in ethics has been influenced by Darwin (1809–1882, evolutionary ethics, survival of the fittest), Freud (1856–1939, behaviorism, decision making defined as the struggle between the ego, superego, and the id), and Pavlov (1869–1936, actions as conditioned responses). Other ideologies include biocentrism (everything in nature has value, nature has its own intrinsic worth independent of its utility for humans), anthropocentrism (human beings have rights superior to the rights of plants, animals, and the environment), right to life (an embryo and fetus have the full rights of an individual human being), and animal rights (animals have autonomy and we have a duty to respect them as we do human beings). Communism is an ideology whose proponents might be called utilitarians, and democracy is an ideology whose proponents might be called Kantians.

More recently, four additional ethical concepts have been formulated by a board of philosophers, ethicists, religious leaders, theologians, scientists, physicians, and lawyers to help sort out ethical issues when medical experimentation involves human beings. While designed for, and, therefore, especially applicable to, ethical issues involving human experimentation, they have come to be applied to other situations as well. The four principles are autonomy, beneficence, non-malfeasance, and justice.

PRINCIPLES THAT GUIDE HUMAN EXPERIMENTATION

In the paragraphs above, we have given a synopsis of the major moral theories and illustrated how they can be in conflict. Below we give a synopsis of perhaps the most important set of principles of moral philosophy in the twentieth century. This synopsis, expanded upon in Chapter 8, demonstrates how four principles derived from the major ethical theories were brought together in a codification of accepted methods for human experimentation. The four principles, which are at the basis for evaluating whether a human subject should participate in a research study, are also essential to evaluating and making conclusions about many other moral questions, including those in the practice of medicine (to be discussed in Chapter 7). They are listed with definitions below:

1. **Autonomy:** This principle refers to self-rule; that is, people have the right to make decisions about matters that affect them as individuals, and this right should be respected. The question should be asked whether the issue at hand pertains only to the individual in such a way that the individual's rights should prevail or whether there are consequences to the action that might affect or harm other individuals or society. Often cited examples of violations of autonomy include deception and therapeutic misconception, such as an individual under the false notion that a procedure to which he or she submits will have some personal benefit when it will only benefit a research objective.

2. **Beneficence:** This is the principle of doing good, namely, helping others by protecting them and working to ensure their well-being. Beneficence demands that no subject of research be exposed to risks incommensurate with the potential for benefits to be gained by others. This term is used in medical research and is related to utilitarianism. Research on human subjects with their consent can be done only if the benefits to the patient or others outweigh the risks, an evaluation that is done by an institutional review board as will be discussed in Chapter 8.

3. **Non-malfeasance:** This is the principle of doing no harm, of doing no wrong, of avoiding injury, of not putting anyone at unnecessary risk, and of not intruding on lives unnecessarily. Research designed to obtain scientific information from human subjects often involves medical procedures that can be painful or even put the subject at risk of physical harm. The principle of non-malfeasance is to avoid harm but not necessarily in every situation. If the benefit of the research is great relative to the risks, as determined by a committee of peers, and the subject agrees to the experiment after being fully informed, then some discomfort or risk is allowed.

4. **Justice:** The application of justice as a moral theory and as a principle for guidance in human experimentation requires that individuals be selected as subjects fairly without bias to a particular class or group. This requirement is not only to assure equality of treatment but for scientific reasons. For example, experiments done on one ethnic population might not give results applicable to another ethnic group due to a difference in the genetic make-up between the two groups, and experimental results in adults may not be applicable to children.

In considering any experimental situation involving human volunteers, the principles of autonomy, beneficence, non-malfeasance, and justice may provide differing viewpoints and may lead to conflicting conclusions. The person proposing the experiment must show how each principle is served. Key to the evaluation of the ethics of a particular experiment is an evaluation of the risks and benefits. This is so because often there is some risk to the subject, a violation of the principle of non-malfeasance. This risk of harm to the individual must be seen as outweighed by the benefits if the experiment is to be allowed. All other conditions must be met as well.

It seemed at the outset an impossible task to codify methods for ensuring that the four principles described above would be upheld in all human experimentation, yet it was possible to derive a set of guidelines and procedures for the protection of human subjects using the autonomy of the experimental subject as the first principle. The same principles apply to the medical treatment of patients. Chapter 7 on medical ethics and Chapter 8 on human and animal experimentation will expand on these principles.

The example below emphasizes the priority of individual rights (autonomy) whenever medical research or medical procedures are to be performed on humans. Duty ethics, non-malfeasance, beneficence, and utilitarianism are some of the other principles and moral theories that come to bear on almost all research involving humans as well as decisions on the medical care of individuals.

Case 1.12 Promise to Donate an Organ

A patient dying of kidney disease has a cousin who is a good match for a kidney transplant. The cousin volunteered to donate his kidney, a beneficent thing to do. The initial tests were good, but when the cousin found out that he was also to undergo genetic testing, he refused and declined to donate his kidney after all. The patient who had been given false hope sued the cousin. The issues in question have to do with autonomy (the donor's right to make a decision regarding himself), beneficence, and duty. To what extent does the cousin have a duty to proceed to donate the kidney, given he had undergone the initial testing implying he would proceed, then backed out, perhaps harming the patient who had interrupted a further search for a donor? The principle of non-malfeasance with respect to harm to the recipient and to the donor is an issue in this case. The recipient may have been harmed because after the first testing he was no longer continuing the search for a donor. Should the cousin follow through based on virtue ethics? The cousin cannot be required to put himself at risk, yet by first agreeing to the donation he has put the recipient at some risk.

The Four A's strategy can be applied to this dilemma. For example, the acquisition of facts and **advice** might lead to important new perspectives. The **alternatives** do include a continuing search for a donor. The **assessment** should include all of the stakeholders including the family of the donor and the doctors' responsibilities. **Action** includes a respect for the autonomy of the donor as well as the requirement of the responsible caretakers to find a donor.

Case 1.13 Physician-Assisted Suicide

In this case, the medical dictum of "to do no harm" (i.e., non-malfeasance) and the societal abhorrence toward suicide are in direct conflict with autonomy and the dictum to relieve pain and suffering in a dying patient. A 65-year-old woman with metastatic breast cancer is near death, as evidenced from anemia, weight loss, heart failure, and breathing difficulties due to lung metastases. She asks the physician to help her die by giving her a lethal injection. He is unable to do this, but knows that an increase in the dose of a narcotic to relieve pain will result in death due to respiratory failure. He asks the nurse to increase the dose of morphine, but she notes that the patient's respirations are failing and her personal ethics will not allow her to follow these orders. The method of dealing with this situation is discussed in Chapter 7.

The principles of non-malfeasance, justice, beneficence, and justice came to bear on matters of human experimentation in a study done in a Third World country. As demonstrated in the next case, there was an epidemic of meningitis in children in which the international community tried to help by providing antibiotics. Justice is often at issue in countries lacking the resources to respond in such cases: also at issue is the question of duty when treating a dire disease in children with a high mortality rate when the appropriate dose of a medication, which had been studied only in adults, cannot be predicted.

Case 1.14 Medical Experimentation in Children

Pfizer, a large U.S. pharmaceutical company, was sued in August 2001 by parents of children who died or were injured by participating in a study in Africa of Pfizer's experimental drug for meningitis.

- The parents alleged that they were never told that the treatment was experimental (it had never been tested in children for this indication).

- They were not told that the drug proven to treat the condition and used in the other arm of the study was being given at a lower dose than recommended, or that full doses were available elsewhere (http://www2.warwick.ac.uk/fac/soc/law/elj/lgd/2004_1/ ford accessed June 14, 2005).

- They contended that the study was not the humanitarian effort Pfizer claimed, but rather an effort to obtain study data quickly and cheaply in order to license the drug for children.

- It was also alleged that Pfizer never obtained consent from the hospital to conduct the study and later fabricated a letter on letterhead from the hospital Institutional Review Board. However, the hospital did not have an ethics committee or the letterhead on the date shown on the letter (www.washingtonpost.com/ac2/wp-dyn?pagename= article&contentId=A63515-2001Jan15&notFound=true).

Did Pfizer act in its own best interest here? Were these actions driven by profit motives or by utilitarianism? Did they follow the guidelines for informed consent for a clinical trial? Were they breaking international law? Is there justice in what they did? If fewer children died with the new therapy, perhaps it was good. However, there was an epidemic of meningitis underway at the time, and the drug dose had not been properly evaluated in children, whose metabolic pathways may significantly differ from those of adult patients. There may have been earlier data to support the decision to use a lower dose than that stated for use in adults, but how was this decided? Who decided? Did some of the deaths relate to this lower dose? What were the circumstances under which the trial was reviewed and approved? Was there falsification concerning approval for the study by the authorities? Did Pfizer do any harm? They could have. The full facts are not known at the present time, but the case stimulates possible ways to think about an ethical situation involving society (utilitarianism), individual rights (autonomy), justice, and duty (corporate duties to shareholders vs. duty to the children, as is discussed further in Chapter 4). In the meantime, it appears that the case will not come to trial. The U.S. court has declined to hear it, and the African parents have no faith in the corrupt Nigerian courts where the trial has been stalled since 2001 and was adjourned more than 14 times by 2005 (www2.warwick.ac.uk/fac/soc/law/elj/lgd/2004_1/ford/, accessed June 14, 2005).

However, although the situation depicted here gives the plaintiffs' objections and information in published reports, any implication that Pfizer did not act morally is unwarranted without all of the facts. Those involved at Pfizer say that because there was an epidemic underway and the children needed rapid treatment to save lives, they did move rapidly but also they did obtain review and approval of the protocol before proceeding. Without a fair trial, the truth will probably never be known.

To some, the ethics of animal experimentation have become even more controversial than human experimentation. This is generally because of conflicting ideologies about animal rights. The major argument against animal experimentation is that the harm to animals is not justified by the benefits to society of the proposed experiments. The principal ideological conflicts revolve around the status of animals in the moral community and, perhaps more importantly, around virtue issues regarding infliction of suffering. Utilitarianism, with emphasis on minimizing discomfort, provides the arguments that led to rules for ethical behavior in animal experimentation (further discussed in Chapter 8).

Dangers of Expediency and the Slippery Slope Theory

The slippery slope concept refers to situations where an act, decision, or allowance might lead to unwanted consequences or events that cannot be controlled. An example is the Supreme Court decision not to allow physician-assisted suicide, even though it was argued that the individual patient had a right to decide on his or her life. The Supreme Court anticipated that the varied consequences of allowing physician-assisted suicide were grave, and that such a decision was a step onto a slippery slope, the bottom of which would be far worse than granting individual rights in this case. Where would the line be drawn regarding others who might be assisted in suicide? The insane? The disabled? The elderly who could not pay for medical or other care? In everyday decisions, the consequences of doing what might be seen as the right thing could be disastrous. The avoidance of the slippery slope, although not identified with utilitarianism in general, should be high on the list of considerations when faced with ethical dilemmas in personal and professional life choices. Case 1.10, the Supreme Court case of Eminent Domain vs. Rights of the Land Owner that allowed property seizures for "public purpose" to promote economic development in a depressed city, is an example of a slippery slope situation.

As will be discussed in more detail in the chapters on business ethics and medical ethics, there are some common prodromal circumstances, seemingly innocent acts that become gateways leading to major ethical errors. The concept of the slippery slope applies to situations where one purposely limits the right of an individual because societal wisdom perceives that the allowance of rights (e.g., physician-assisted suicide) could lead to dire legal and ethical consequences. One of the major goals of this book is to provide information that will lead the scientist and engineer to alternate paths in order to avoid entering a series of actions that, though seemingly not morally wrong at the time, nevertheless can lead to unethical and even illegal activities.

The most common gateway to unethical behavior is the temptation to resolve a problem or perform an action because it is convenient or expedient. This temptation to adopt an expedient solution occurs frequently in the corporate environment where there are multiple stakeholders. As we shall see, many of the stakeholders' interests may not be consistent with the corporate goal of making a maximum profit. Examples in Chapter 4 will show how one corporation became mired in a disastrous situation because of a few expedient acts on the part of seemingly innocent engineers, and how another corporation avoided possible ruin by avoiding the temptation to make an expedient decision.

Successful decision making and principled ethical behavior are closely related to vigilance concerning the consequences of expedient acts. Throughout this book, many cases will be presented wherein ethical theories and principles were violated. The intent is to emphasize how easily some early decisions and reckless approaches to problems have led individuals and entire corporations into dire situations. Protection from the negative consequences of a pragmatic solution to a problem is afforded by the consideration of how alternative solutions will affect the stakeholders, and by seeking advice.

Learning Values and Cheating in Life

Though we might like to think that all cognitive human beings were born with an innate set of values and a fundamental set of virtues, this is not the case. Clearly, some behaviors are innate to survival in animals of all species, such as avoiding some dangers not necessarily taught by parents, seeking food, and protecting the young. But behaviors such as kindness, loyalty, sharing, telling the truth, and not stealing or cheating are learned. Few would argue that cheating is wrong, but because of its widespread occurrence (Callahan 2004), it seems that the value of honesty is not learned, or if it is learned, is not held as an important virtue. However, this value can be learned and reinforced by the many rewards gleaned from success without cheating (George 2003; see also Chapter 4). Guidance in behavior that is considered virtuous by the majority appears in the form of codes of professional conduct, credos of business corporations, policies of foundations and universities, religious rules, and codes of laws. While these codes and laws may fall short of guiding behavior in all circumstances, generally, when considered along with the ethical principles emphasized throughout this text, they are valuable guides that form a basis for working through both personal and professional ethical dilemmas.

SYNOPSIS OF TOOLS FOR RESOLVING ETHICAL DILEMMAS

Here we review the tools for designing solutions to ethical problems or in making decisions to avoid ethical situations that can lead an individual, institution, or corporation into ruin. First recall the Four A's presented in Table 1.1: 1) acquire facts, 2) consider alternatives, 3) assess alternatives relative to moral theories and stakeholders, and 4) act but be prepared to modify the action plan.

Frequently during steps 2 and 3, two or more actions come into conflict where duty is involved. In these cases, the selection of the prima facie duty (Table 1.5) that overrides others involves a logical thought process. Frequently, a decision for the correct action is based on the intuition of the individual faced with the need to select one duty over another. During this stage, it is important to enumerate the stakeholders, particularly in business ethics (see Chapter 4), and to analyze how equity or justice (Table 1.4) to the stakeholders is affected by the consequences of alternative actions.

In step 2 of the design strategy, the decision maker evaluates a number of alternative solutions or courses of action. Selection of the optimum course of action involves not only attention to the conflicting moral theories but a prioritization or ranking of the alternatives using some or all of the following tools:

Consequences to stakeholders: The listing and prioritization of stakeholders involved in an ethical dilemma becomes particularly important in business ethics and this subject will be discussed in Chapter 5 when the design strategy is expanded to facilitate decision making in business or corporate ethical problems.

Justice and equity (Table 1.4): A consideration of an action with respect to the fairness to each affected individual arises in distributing rewards and in situations of limited resources. These situations occur in corporate and medical ethics discussed in Chapters 4 and 7, respectively.

Prima facie priorities (Table 1.5): This philosophical modernization of Kant's deontology when there are conflicting duties will usually apply to actions with respect to one's personal duties.

Risk analysis: The consequences of various alternative solutions can in some cases be quantified using theories of risk assessment. This is a complex area of engineering, public policy, and government regulation settings, which will be discussed from the engineering and public policy viewpoints when we consider the ethics of emerging technologies in Chapter 12. Risk analysis involves an attempt to quantitatively prioritize alternative solutions. Risk assessment asks three questions: What can go wrong? How likely is it to happen? What are the consequences? These three questions will also help guide the analysis of alternative solutions when at step 2 in the Four A's strategy. We will return to the uses of risk analysis in ethical problems in Chapter 12.

CHAPTER SUMMARY

The theories of virtue, justice, rights, duty, and utilitarianism, together with the principles of autonomy, non-malfeasance, and beneficence, represent guiding concepts when thinking about ethical problems. Strategies for analysis and solutions are presented as the Four A's Strategy for Dilemmas and the Ethical Engineer Framework in Tables 1.1 and 1.2. But without consideration of all ideologies applicable to an ethical problem, one might choose a solution that is not optimal for the situation. Table 1.3 gives a list of the theories and principles defined in this chapter along with many contemporary ideologies that may influence the selection of a solution for major and minor dilemmas. The cases presented in this chapter have illustrated the applications of different moral theories as well as conflicts between the theories of justice, duty, rights, and utilitarianism. Also emphasized is the importance of thinking through the consequences of actions relative to stakeholders and the prioritization of duties. Many of these concepts can come into play within a single case with several "right" or "good" solutions, making it necessary to decide on a solution when there is no "best" solution. Further, ethical concepts may change over time, making it important to keep an open mind about both the solution and the consequences.

No single moral philosophy can be applied to all situations, and even proponents of one philosophy disagree on important details. For example, the maxim of the utilitarian to select that action leading to the greatest happiness (act utilitarianism) will not always lead to an ethically correct act. As a result, it was necessary to incorporate into the theory of utilitarianism a branch called rule utilitarianism. The rules are those that, if followed by society, will maximize happiness. Three examples of these rules are: keep promises, do not lie, and punish individuals for wrong behavior.

The moral theory of duty as espoused by Kant gave moral value for acts based on the motive of doing the right thing and not on motives for the consequences of benefits of the act. This theory required the introduction of the prima facie priorities, a more modern philosophy that allows the selection of an action when multiple duties are in conflict.

Along with knowledge of the major moral theories, the professional engineer or scientist needs a systematic approach to resolving dilemmas. An overview of the approach espoused in this text is given in Tables 1.1–1.5. The use of the Four A's is an effective way to think through a problem. Further protection from making a slippery slope decision is gained if the decision maker also takes into consideration the consequences of an action to all stakeholders, the order of priorities, and the axioms of justice and other moral theories. In some complex situations, the initial fact finding, as well as the evaluation of alternatives, should also include the use of risk analysis (see Chapter 12). But perhaps of greatest importance to the art of decision making when faced with ethical problems is to resolve the facts and ambiguities before considering alternative solutions, to seek help from others, and to avoid expedient decisions.

REFERENCES

Bernstein, C. and B. Woodward. 1974. *All the President's Men*. New York, NY: Simon & Schuster, Inc., New York, NY.

Burton, K. 1989. "A chronicle: Dax's case as it happened." In: *Dax's Case: Essays in Medical Ethics and Human Meaning*. L.D. Kliever (ed.). Dallas, TX: Southern Methodist University Press, pp. 1–12.

Callahan, D. 2004. *The Cheating Culture: Why More Americans Are Doing Wrong to Get Ahead*. Orlando, FL: Harcourt, Inc.

Childress, J. F. 1970. "Who shall live when not all can live?" *Soundings* **53**:339–355.

Childress, J. F. 1997. "Who shall live when not all can live?" In: *Bioethics: An Introduction to the History, Methods and Practice*. N. S. Jecker, A. R. Jonsen, and R. A. Pearlman (eds.). Sudbury, MA: Jones and Bartlett Publishers, p. 24.

Feldman, F. 1978a. *Introductory Ethics*. Upper Saddle River, NJ: Prentice-Hall, Inc., p. 156.

Feldman, F. 1978b, p. 26.

Feldman, F. 1978c, p. 55.

Feldman, F. 1978d, pp. 16–79.

George, W. W. 2003. *Authentic Leadership: Rediscovering the Secrets to Creating Lasting Value*. Hoboken, NJ: Jossey Bass/John Wiley & Sons.

Green, R. M., K. O. DeVries, J. Bernstein, et al. 2002. "Overseeing research on therapeutic cloning: A private ethics board responds to its critics." *Hastings Center Report* **32**(3):27–33.

Hobbes, T. 1962a. *Leviathan*. Revised Collier Books Edition, Michalel Oakeshott (ed.). New York: Macmillan Publishing Co., Inc., p. 11.

Hobbes, T. 1962b, pp. 12–13.

Internet Encyclopedia of Philosophy. 2005. "John Locke." www.utm.edu/research/iep/l/locke.htm, accessed January 3, 2005.

Johnson, D. G. 2001. *Computer Ethics*. Upper Saddle River, NJ: Prentice-Hall, Inc., p. 156.

Kant I. 1785a. *Foundations of the Metaphysics of Morals*. Second Edition (revised, 1997), translated by L. W. Beck. Upper Saddle River, NJ: Prentice-Hall Publishers, Inc., p. 39.

Kant, I. 1785b, page ix.

Kliever, L. D. (ed.). 1989. *Dax's Case: Essays in Medical Ethics and Human Meaning*. Dallas, TX: Southern Methodist University Press.

Locke, J. 1690a. "Two Treatises of Government. The Second Treatise of Government." Chapter VIII: "Of the Beginning of Political Societies," Section 95. In: John Locke. *The Second Treatise of Government and a Letter Concerning Toleration,* J. W. Gough (ed.). Oxford, England: Basil Blackwell, 1966, p. 49.

Locke J. 1690b. Chapter IX: "Of the Ends of Political Society and Government," Section 131. In: John Locke. *The Second Treatise of Government and a Letter Concerning Toleration*, J. W. Gough (ed.). Oxford, England: Basil Blackwell, 1966, pages 65–66.

Martin M. W. and R. Schinzinger. 2005. *Ethics in Engineering*. Boston, MA: McGraw-Hill Publishers.

McCloskey, H. J. 1963. "A note on utilitarian punishment." *Mind* **72**:599.

Mill, J. S. 2001. *Utilitarianism*. Second Edition, George Sher (ed.). Indianapolis, IN: Hackett Publishing Company, Inc., p. 7.

Mill, J. S. 1859. "On Individuality, As One of the Elements of Wellbeing." In: John Stuart Mill. *On Liberty*, David Bromwich and George Kateb (eds.). New Haven, CT: Yale University Press, New Haven, CT, 2003, p. 121.

Pence, G. E. 2004a. *Classic Cases in Medical Ethics*. Fourth Edition, New York, NY: McGraw-Hill Companies, Inc., p. 12.

Pence, G. E. 2004b, p. 10.

Pence, G. E. 2004c, pp. 16–18.

Pence, G. E. 2004d, p. 19.

Pence, G. E. 2004e, p. 78.

Pinkus, R. L. B, L. J. Shuman, N. P. Hummon, and H. Wolfe. 1997a. *Engineering Ethics: Balancing Cost, Schedule, and Risk—Lessons Learned from the Space Shuttle*. New York: Cambridge Press, p. 33.

Pinkus, R. L. B, L. J. Shuman, N. P. Hummon, and H. Wolfe. 1997b, p. 277.

Rachels, J. 2003a. *The Elements of Moral Philosophy*. Fourth Edition. New York: McGraw-Hill Publishers, Inc., p. 34.

Rachels, J. 2003b, p. 14.

Rachels, J. 1986. *The End of Life: Euthanasia and Morality*. Oxford, UK: Oxford University Press, p. 54.

Rawls, J. 1999a. *A Theory of Justice*. Cambridge, MA: Belknap Press of Harvard University Press.

Rawls, J. 1999b, p. 3.

Rawls, J. 1999c, p. 4.

Ross, W.D. 1930. *The Right and the Good*. Oxford, UK: Oxford University Press, pp. 20–22.

Schinzinger, R. and M. W. Martin. 2000. *Introduction to Engineering Ethics*. New York: McGraw-Hill Companies, Inc., pp. 55–57.

Shannon, T. (ed.). 1993. *Bioethics*. Fourth Edition. Mahwah, NJ: Paulist Press, p. 5.

Slossburger, E. 1993. *The Ethical Engineer*. Philadelphia: Temple University Press, pp. 29–37.

Smart J. C. C. and B. Williams. 1998a. *Utilitarianism For and Against*. Cambridge, UK: Cambridge University Press, p. 79.

Smart and Williams. 1998b, pp. 1–155.

Smart and Williams. 1998c, pp. 67–71.

Smart and Williams. 1998d, p. 62.

Swazey, J. P. and S. J. Bird. 1995. "Teaching and learning research ethics." *Professional Ethics* **4**:155–178.

United States v. Holmes, 1842. Circuit Court, E.D. Pennsylvania. 26 F.Cas. 360.

Velasquez, M. G. 2001. *Business Ethics: Concepts and Cases*. Fifth Edition. Englewood Cliffs, NJ: Prentice-Hall Publishers, Inc., p. 38.

Weil, V. 1993. "Teaching Ethics in Science." In: *Ethics, Values and the Promise of Science*, Forum Proceedings of Sigma Xi, the Scientific Research Society. Research Triangle Park, NC, pp. 243–248.

ADDITIONAL READING

Below are listed a number of books on engineering ethics and moral philosophies which are particularly relevant to the issues of decision making by the professional scientist and engineer. Additional listings of books relevant to business ethics, medical ethics, computer and Internet ethics, stem cell research, and germline engineering are also provided at the end of the respective chapters.

Callahan, J. (ed.). 1988. *Ethical Issues in Professional Life*. New York: Oxford University Press, Inc., 470 pp.

Feldman, F. 1978. *Introductory Ethics*. Upper Saddle River, NJ: Prentice Hall, Inc., 255 pp.

Fleddermann, C. 1999. *Engineering Ethics*. Upper Saddle River, NJ: Prentice-Hall, Inc., 135 pp.

Harris, C. E., M. S. Pritchard, and M. J. Rabins. 2000. *Engineering Ethics: Concepts and Cases*. Second Edition. Stamford, CT: Wadsworth/Thomson Learning Publishers, 300 pp.

Herkert, J. (ed.). 2000. *Social, Ethical and Policy Implications of Engineering*. New York: IEEE Press, 339 pp.

Jecker, N. S., A. R. Jonsen, and R. A. Pearlman (eds). 1997. *Bioethics: An Introduction to the History, Methods and Practice*. Sudbury, MA: Jones and Bartlett Publishers, 416 pp.

Johnson, D. G. (ed.). *Ethical Issues in Engineering*. Upper Saddle River, NJ: Prentice-Hall, Inc., 392 pp.

Kant, Immanuel. 1997. *Foundations of the Metaphysics of Morals and What Is Enlightenment?* Second Edition (revised), translated by L. W. Beck. Upper Saddle River, NJ: Prentice-Hall, Inc., 332 pp.

Nietzsche, Friedrich. 1989. *Beyond Good and Evil: Prelude to a Philosophy of the Future.* Translated by Walter Kaufmann. Vintage Books Edition. New York: Random House, Inc., 256 pp.

Pinkus, R. L. B., L. J. Shuman, N. P. Hummon, and H. Wolfe. 1997. *Engineering Ethics. Balancing Cost, Schedule, and Risk—Lessons Learned from the Space Shuttle.* New York: Cambridge University Press, 377 pp.

Rachels, J. 2003. *The Elements of Moral Philosophy.* Fourth Edition. New York: McGraw-Hill Publishers, Inc., 218 pp.

Rawls, John. *Lectures on the History of Moral Philosophy.* 2000. Herman, B. (ed.). Cambridge, MA: Harvard University Press, 384 pp.

Rawls, J. A. 2000. *Theory of Justice.* Cambridge, MA: Belknap Press of Harvard University Press, 538 pp.

Schinzinger R. and M. W. Martin. 2000. *Introduction to Engineering Ethics.* New York: McGraw-Hill Companies, Inc., 260 pp.

Schlossberger, E. 1993. *The Ethical Engineer.* Philadelphia: Temple University Press, 284 pp.

Smart, J. J. C. and B. Williams. 1998. *Utilitarianism For and Against.* Cambridge, UK: Cambridge University Press, First published 1973, reprinted 1998, 155 pp.

Thurman, R. 1998. *Inner Revolution: Life, Liberty and the Pursuit of Real Happiness.* New York: Riverhead Books, 322 pp.

Whitbeck, C. 1998. *Ethics in Engineering Practice and Research.* Cambridge UK: Cambridge University Press, 330 pp.

PROBLEM SET

1.1 Prepare a diagram showing how the theories of virtue, justice, duty, rights, and utilitarianism are or are not related to each other. (Hint: consider how consequences do or do not play a role in each theory.)

1.2 Following the Challenger incident there was another disaster that took the lives of all the astronauts on board the Columbia spacecraft. From the Internet, learn some details of the Columbia incident and then state what ethical violation(s) you might argue for the way the Columbia mission was handled.

1.3 Using the design method presented in this chapter, show how you would approach the following problem:

As the engineering manager of air conditioning units for Fargo West Bus Company, you receive a notice that there might be some flaw in the units because a virus-like upper respiratory infection has spread through the school children who were bussed from Fresno to Modesto, California, for a special summer session. The suspicion is reinforced early in the fall school year when only a few of 80 students showed up at the pick-up points two weeks after the bus runs started. These are new buses put into service at the beginning of the summer, thus there is no previous experience with them. You go to your boss to advise an immediate withdrawal of the buses, but he refuses on the grounds that there is no evidence. You realize you have taken the first step on a slippery slope to an ethical disaster because you had not followed part of the first step in solving a problem—getting the facts. How should you have proceeded?

1.4 Match the following with a moral philosophy:

a. Ten Commandments

b. Right to life

c. Declaration of Independence

d. Avoidance of a drinking party

e. War

f. Competition (sport)

g. Eminent domain

1.5 Mary is asked to give her kidney to a sibling who has renal failure due to diabetes and cannot find a donor. She becomes conflicted because on the one hand, she feels she has a duty to help her sibling, yet on the other hand she is fearful of surgery and also believes she might develop diabetes and need both of her kidneys. How should Mary approach this dilemma? Use the design strategy to outline the possible considerations in this dilemma. Use 80 words or less.

1.6 There is evidence that violent behavior, including willingness to murder, is associated with a chemical imbalance in the human brain. A serial killer is brought to trial and the defense states that brain images taken by modern positron emission tomography methods show abnormalities that might be associated with the defendant's uncontrolled behavior. The defense argues that to give the death sentence would be analogous to giving the death sentence to people with psychiatric disorders such as schizophrenia or bipolar disorder because they also have mood swings that could be a threat to society. With what you have learned regarding moral thinking in this chapter, discuss in 100 words or less the arguments for and against the death penalty in the context of the circumstances described above.

1.7 After 360 years, the Roman Catholic Church, which had threatened Galileo with excommunication, admitted that Galileo was correct in espousing a sun-centric galaxy. Today, the Pope and some government leaders have declared stem cell research and its applications immoral based on the argument that they are related to the purposeful death of an embryo or fetus. In Chapter 10, you will learn about multiple methods of stem cell use, some of which do not involve use of embryos. However, for this problem, consider the argument that the embryo tissue is already excess tissue in a test tube that will be discarded, as is the case for about 0.4 million excess embryos currently in fertilization clinics in the United States. Is it better to discard these embryos, as is currently being done, or to put the embryo cells to a useful purpose? Use the design strategy and 100 words or less.

1.8 Add at least one prima facie duty to the list in Table 1.5. Explain a situation wherein this new duty would override other duties in the list. If you cannot think of another prima facie duty, then give an example possibly from your own experience wherein a few of the listed duties were in conflict and describe how the one action taken was justified by the logic of one duty overriding another. You could also describe a situation where the wrong action or prima facie duty was chosen.

1.9 The following problem involves rights and autonomy vs. utilitarianism.

A city proposes to build a hospital in order to provide better health care for its residents. The plans involve the destruction of a house whose owner is adamantly opposed to leaving. What would a pure rights ethicist say about the proposal? What about a pure consequentialist or utilitarian? Use no more than 30 words for each answer.

1.10 Essential to the moral philosophy of Kant and the philosophy of John Rawls is the concept that there is an innate moral good that guides our decisions to do the right thing. In the early years of the twenty-first century, a child raised by wolves was found. This child did have non-wolf innate talents because the child could learn language. Can you find evidence that this child had some innate notions of how to behave based on one or more of the moral philosophies of this chapter? Use the Internet to search for evidence that some innate behavior exists to guide day-to-day decisions about how to behave.

1.11 A local government is approached by developers of a new shopping center to seize 20 homes on either side of a small road in order to improve access. The justification for this action against land-owner rights is the Fifth Amendment. Present an argument for not allowing this eminent domain action by the local government using the slippery slope theory.

2

Ethics in Scientific Research

"Let's put Professor Gift's name on our paper so the reviewers think more of it."

"If I include in my C.V. the paper I plan to write, I can get it done and submitted before the recruiter asks to see it."

"Dr. Smith's genetic lab has a recent paper that is in disagreement with ours. They are probably wrong, and ours is already in the mail anyway, so we need not cite it."

INTRODUCTION

This chapter focuses on the ethics of a scientist or engineer in the conduct of research to discover new knowledge or develop a new technology. The main moral theories that apply to individual behavior in the conduct of scientific and engineering research include virtue, duty, non-malfeasance, and autonomy. While it is expected that all scientists and engineers will conduct their professional lives with the utmost attention to ethical principles, there are those who do not follow the rules of conduct. Because of continuing problems of data falsification in publications of research supported by the federal government, Congress demanded action by the scientific community and agencies supporting research. Guidelines and a formal definition of scientific misconduct resulted. The guidelines were originally intended for government agencies investigating publicly funded scientific research institutions to ensure that public funds were being spent appropriately, and to investigate misconduct if it was found. As the policy evolved, it received intensive review, input, and revision based on comments not only from individual researchers but also from other government agencies, educational institutions, and a committee of the National Academy of Sciences. The final guidelines have come to be applied to everyone carrying out scientific research of any kind. The policy went into effect in December of 2000 and is printed in its entirety in Appendix 2.1 (www.ostp.gov/html/001207_3.html, accessed February 16, 2005).

Table 2.1   Violations of Accepted Practices in Scientific Publications

Faking research data - 0.3%
Plagiarism - 1.4%
Removing data - 6%
Multiple publications of the same data - 4.7%
Inappropriate inclusion of authors - 10%
Changed a study design - 15%
Inadequate record keeping - 27.5%

Whereas one can estimate that at least one in every ten students is guilty of at least minor acts of plagiarism, the prevalence of more serious acts of falsification and the extent of fabrication of data can be appreciated from a survey of over 8,000 grantees of funding for scientific research reported in 2005 (Martinson et al. 2005). The anonymous questionnaire, returned by 3,247 respondents, revealed the following statistics shown in Table 2.1.

This is a serious situation, not only because of the occurrence of misconduct, but also because scientifically incorrect information cannot easily be removed from the scientific literature. In addition, there are detrimental effects for those who innocently or through their own negligence become involved in a fraudulent misrepresentation. Not every scientist and engineer realizes the scope of the definition of scientific misconduct, or the extent to which seemingly minor acts can decimate a career. Below is the generally accepted definition.

DEFINITION OF SCIENTIFIC MISCONDUCT

Research misconduct is defined as fabrication, falsification, or plagiarism in proposing, performing, or reviewing research, or in reporting research results.

- **Fabrication is making up data or results and recording or reporting them.**
- **Falsification is manipulating research materials, equipment, or processes, or changing or omitting data or results such that the research is not accurately represented in the research record.**
- **Plagiarism is the appropriation of another person's ideas, processes, results, or words, without giving appropriate credit.**
- **Research misconduct does not include honest error or differences of opinion.**

A finding of research misconduct requires that:

- **There be a significant departure from accepted practices of the relevant research community.**
- **The misconduct be committed intentionally or knowingly or recklessly.**
- **The allegation be proven by a preponderance of evidence.**

However, there is not unanimous acceptance of these basic definitions by the scientific community. One area of contention is the definition of plagiarism because it does not recognize the difference between one who conveys information without attribution to oneself or the actual author and one who presents information as if it were authored by the presenter. Conveyance of published ideas, puzzles, or teaching materials as exercises wherein attribution is not given to the creator (but also not taken by the conveyor) cannot practically be called plagiarism. This is an area of concern for universities in particular because conveyance of ideas without citation is often used in teaching.

A second area of contention, and a more universally held disagreement about the definition of scientific misconduct, involves the statement "There be a significant departure from accepted practices of the relevant research community." This is problematic because the terms "significant departure" and "accepted practices" are subject to interpretation. Research methods frequently represent a departure from accepted practice, and there is no practical scale by which to judge objectively what a "significant departure" is. Is the production of a blastocyst using a human egg and human sperm or rabbit egg and human sperm and subsequent implantation into a rabbit uterus to be considered scientific misconduct? How would the scientific community judge one who is responsible for repeated cloning attempts on nonhuman primates? What about the research done before the first liver transplant in 1963 when pioneer surgeon Dr. Thomas Starzl performed 50 experiments in dogs, most of which were failures, but each gave clues for improving the technique that allowed the success of the first liver transplant in humans? These are only a few examples to illustrate how the definition of scientific misconduct needs to be carefully applied so as not to discourage or penalize innovative scientific research.

CATEGORIES OF SCIENTIFIC MISCONDUCT

We separate types of scientific misconduct according to the professional category of the perpetrator (e.g., senior experimenter, research technician, graduate student, professor). A further separation can be made based on whether the behavior is a serious fraudulent activity or a less serious/minor departure from prudent or acceptable behavior. However, the latter distinction does not convey the severity of the consequences of a reported misconduct, because it is not uncommon for seemingly small transgressions to have serious repercussions, not only for those involved from the sidelines but also for the accusers as well. Even a whistleblower (see below) is frequently ridiculed or put at a career disadvantage for attempts to expose the misconduct of another.

Unethical Conduct by Researchers

Accusations of dishonesty may result in ruined careers and lifelong negative consequences. Even if an allegation is disproved, hours of time and effort can be wasted reacting to it and attempting to regain a credible reputation. The serious consequences of scientific misconduct are reiterated in a publication issued by the National Academies Press (www.nap.edu/readingroom/books/obas/contents/misconduct.html, accessed January 7, 2006):

The ethical transgressions discussed in earlier sections—such as misallocation of credit or errors arising from negligence—are matters that generally remain internal to the scientific community. Usually they are dealt with locally through the mechanisms of peer review, administrative action, and the system of appointments and evaluations in the research environment. But misconduct in science is unlikely to remain internal to the scientific community. Its consequences are too extreme: it can harm individuals outside of science (as when falsified results become the basis of a medical treatment), it squanders public funds, and it attracts the attention of those who would seek to criticize science. As a result, federal agencies, Congress, the media, and the courts can all get involved.

Within the scientific community, the effects of misconduct in terms of lost time, forfeited recognition to others, and feelings of personal betrayal can be devastating. Individuals, institutions, and even entire research fields can suffer grievous setbacks from instances of fabrication, falsification, or plagiarism - even if they are only tangentially associated with the case.

When individuals have been accused of scientific misconduct in the past, the institutions responsible for responding to those accusations have taken a number of different approaches. In general, the most successful responses are those that clearly separate a preliminary investigation to gather information from a subsequent adjudication to judge guilt or innocence and issue sanctions if necessary. During the adjudication stage, the individual accused of misconduct has the right to various due process protections, such as reviewing the evidence gathered during the investigation and cross-examining witnesses.

In addition to falsification, fabrication, and plagiarism, other ethical transgressions directly associated with research can cause serious harm to individuals and institutions. Examples include cover-ups of misconduct in science, reprisals against whistleblowers, malicious allegations of misconduct in science, and violations of due process in handling complaints of misconduct in science.

Perhaps the most devastating consequence of all is that each instance of scientific misconduct heightens mistrust, not only in the view of the scientific community but also for the public (Kennedy 2000). Such mistrust can be difficult to overcome, and setbacks in progress may affect individual researchers and program funding for years to come. Efforts are increasing on the part of the Office for Research Integrity to find patterns of causality and overall solutions. The goal is to ensure that scientists using novel or unorthodox research methods are not falsely accused, and that autonomy in research is not unduly curtailed.

Table 2.2 summarizes the types of unethical behaviors of experimenters.

Ethical Transgressions by Authors

The publication of a manuscript is the primary vehicle by which a scientific or other document is put into the hands of the public, thereby opening the subject for reanalysis, scrutiny, and debate beyond the immediate peer reviewers and the editor.

Table 2.2 Summary of Unethical Conduct by Researchers

- Cover-up of errors
- Misuse of funds
- Fabrication of data
- Deletion of data without justification
- Falsification of data
- Making major protocol deviations
- Unorthodox manipulation of data during data analysis
- Performance of inappropriate statistical analyses
- Performance of unauthorized human or animal experiments
- Knowingly participating in unauthorized human or animal experiments
- Failing to report wrongs when there is a responsibility to do so
- Misrepresentation or purposeful exclusion of relevant data from others
- Misrepresentation of originality of ideas, writings, software, and hardware—plagiarism

The importance of publications to the documentation of original ideas, to give evidence of progress in grants and contracted projects, and to establish the progress of an individual's career leads to the fact that authorship of publications is one of the three most valuable currencies of a scientist and engineer. (The other two are patented inventions and named identification with successful creations [e.g., Coulter Counter, Faraday effect, the tesla].)

The ethical principles and moral theories that underpin authorship in scientific and engineering technical contributions include:

Duty: There is an ethical requirement to return to the community the product of the scientists' efforts from the investment the public may have made in the efforts of the scientist.

Justice: Credit to those who contributed in the work of a publication or patent must be given in terms of joint authorship, appropriate citations, and acknowledgments.

Virtue: Virtue applies to every aspect of the conduct of research, including the analysis, presentation of data, and review of work by others. Appropriate sophistication in criticism of another's work or conclusions will convey the truth without impugning the character of the author(s) being criticized.

Autonomy: The rights of others are manifested in many aspects of authorship, including proper acknowledgments of others' contributions, ideas, previous work, and sponsorship.

Non-Malfeasance: The responsibilities of authorship extend from strict avoidance of falsification of data to the requirement to report all of the pertinent data and facts. There are many acceptable ethical approaches to authorship, but situations still occur in which authors make

Table 2.3 Categories of Misconduct by Authors

Major Ethical Transgressions by Authors

- Describing data or artifacts that do not exist.
- Describing documents or objects that have been forged.
- Misrepresenting real data or deliberately distorting evidence or data.
- Presenting another's ideas, text. or work without attribution (plagiarism), including deliberate violation of copyright.
- Presenting statistical analyses in which required conditions are not met or discussed.
- Omitting negative results from corollary experiments.

Minor Ethical Transgressions by Authors

- Failing to assume the null hypothesis.
- Misrepresenting authorship by omitting an author.
- Misrepresenting authorship by including a noncontributing author.
- Misrepresenting publication status.
- Failure to mention equally likely interpretations or hypotheses not tested or not testable.
- Citing work irrelevant or unsupportive to a point being made in the text.
- Citing work as proving a point that it does not.
- Citing literature without verifying the legitimate source of the citation.
- Omitting citing the work of competitors whose work has been represented.

unnecessary mistakes and fail to recognize the seriousness of a transgression, sometimes with career-threatening repercussions. One example is the graduate student who was dropped from the academic department in which he worked after he cited as already published a manuscript that had only been submitted for publication. The important thing is that the authors (junior and senior alike) are clear as to the procedures, because even a seemingly small erroneous decision can be a serious transgression. A summary of the major and minor types of scientific misconduct by authors is given in Table 2.3 (expanded from Chubin 1990).

Ethical Transgressions by Reviewers

Most journals are dedicated to as rapid publication as possible so that the results can be disseminated, openly judged, and confirmed in repeat experimentation. Prior to publication, however, confidentiality is mandatory whenever one is privy to information such as preliminary experimental results, protocols intended for study, or final results, because these may change based upon peer or editorial review for publication. This type of break in confidentiality by preliminary disclosure, communication, or even theft of an idea or information is a serious breach of ethical behavior deserving harsh censorship of the perpetrator. Some of the transgressions occurring in this area are summarized in Table 2.4 (Chubin 1990 in part).

Table 2.4   Ethical Transgressions by Reviewers, Editors, or Staff

Ethical Transgressions by Reviewers

- Misrepresenting facts or lying in communications with authors or editors.
- Unreasonably delaying review in order to achieve personal gain.
- Stealing ideas or text from a manuscript under review.
- Allowing conflicts of interest (an explanation of this term follows later in this chapter) to bias recommendations for publication.

Ethical Transgressions by Editors or Staff

- Forging or fabricating a referee's report.
- Lying to the author about the review process.
- Stealing ideas or text from a manuscript under review.
- Allowing conflicts of interest to bias acceptability of a manuscript.

Examples of Scientific Misconduct

Below we give 10 true examples of scientific misconduct and one example (Case 2.11) of a situation that was, and continues to be, misconstrued as scientific misconduct.

Case 2.1 Piltdown Man

An older example of scientific misconduct involving fabrication, which resulted in repercussions that extended over decades, was that of the Piltdown man. Fragments of an unusual skull were found in a gravel pit near Piltdown, England, in 1908, and additional fragments were found over the next several years. Also found in the same gravel pit was an unusual jawbone with ape-like characteristics not usually associated with a human skull. Some questioned whether the skull was authentic and believed the skull and the jawbone belonged to two different species, but the controversial theory won out that it was real and represented the skull of a man. For a number of years the discovery changed the thinking about the theory of evolution of man. It was named the Piltdown man and was thought to date back to the Lower Pleistocene period. However, when tests became available with which to date the specimen, it was found that while the skull was around 50,000 years old, the jawbone was only a few decades old. The perpetrator of the hoax was thought to be the original discoverer of the skull, who had since died, but was found to have also misrepresented information about other antiquities.

Case 2.2 Data Fabrication during an NIH-Sponsored Investigation

An investigator in Baltimore received a large NIH grant to assess the epidemiology of AIDS and other sexually transmitted diseases. A large research institute was hired to manage the project. Eleven months into the study, the investigator received a call from a data collection manager who had noted an unusually high number of interviews and inaccuracies, raising concerns that a study interviewer was fabricating the data. This allegation was subsequently proven to be true, and further checks showed that other interviewers were faking data as well. The entire study was cast into disarray. Since there was no way of knowing which data were good and which had been fabricated, the entire database had to be viewed as tainted. Sorting things out took months, which wasted money and caused great frustration (Marshall 2000).

Case 2.3 Scientific Misconduct by a Nanotechnology Physicist

Jan Hendrik Schön, a physicist working at Bell Laboratories, was highly regarded for his work in high-temperature superconductivity and for creating field-effect transistors from small molecules. His work was thought to be so innovative that it had the potential to earn him a Nobel Prize. However, when others were unable to reproduce his results and closely scrutinized his publications, it was noted that similar graphs appeared in three unrelated papers. This finding shocked all who knew him. The subsequent investigation by an independent panel resulted in 16 counts of scientific misconduct, including the duplication, falsification, and destruction of data.

The Executive Summary of the investigative committee found the following (http://lucent.com/news_events/pdf/summary.pdf, accessed February 17, 2005):

Proper laboratory records were not systematically maintained by Hendrik Schön in the course of the work in question. In addition, virtually all primary (raw) electronic data files were deleted by Hendrik Schön, reportedly because the old computer available to him lacked sufficient memory. No working devices with which one might confirm claimed results are presently available, having been damaged in measurement, damaged in transit, or simply discarded. Finally, key processing equipment no longer produces the unparalleled results that enabled many of the key experiments. Hence, it is not possible to confirm or refute directly the validity of the claims in the work in question.

The evidence that manipulation and misrepresentation of data occurred is compelling. In its mildest form, whole data sets were substituted to represent different materials or devices. Hendrik Schön acknowledges that the data are incorrect in many of these instances. He states that these substitutions could have occurred by honest mistake. The recurrent nature of such mistakes suggests a deeper problem. At a minimum, Hendrik Schön showed reckless disregard for the sanctity of data in the value system of science. His failure to retain primary data files compounds the problem.

More troublesome are the substitutions of single curves, or even parts of single curves, in multiple figures representing different materials or devices, and the use of mathematical functions to represent real data. Hendrik Schön acknowledges these practices in many instances, but states that they were done to achieve a more convincing representation of behavior that was nonetheless observed. Such practices are completely unacceptable and represent scientific misconduct.

Schön was fired and his career ruined, but the findings also raised serious questions about the validity of the scientific research process, as well as the responsibilities of coauthors, mentors, and department heads, not only in his own department but also in research laboratories across the country.

Case 2.4 Scientific Misconduct by a Physicist at a U.S. National Laboratory

Noted nuclear physicist Victor Ninov and his 14 collaborators received acclaim when they announced the discovery of element 118 and its immediate decay product, element 116. The discovery was made by bombarding targets of lead with an intense beam of high-energy krypton ions in the 88-inch cyclotron and analyzing the high energy particle collision data. However, others were unable to verify their claims despite a number of confirmation attempts, including reanalysis of the original data. As a result, the scientific team except for Ninov submitted a brief statement of retraction to *Physical Review Letters,* the same publication in which the original results were announced:

> In 1999, we reported the synthesis of element 118 in the (lead-krypton) reaction based upon the observation of three decay chains, each consisting of an implanted heavy atom and six sequential high-energy alpha decays, correlated in time and position. Prompted by the absence of similar decay chains in subsequent experiments, we (along with independent experts) re-analyzed the primary data files from our 1999 experiments. Based on these re-analyses, we conclude that the three reported chains are not in the 1999 data. We retract our published claim for the synthesis of element 118 (www.lbl.gov/Science-Articles/Archive/118-retraction.html, accessed Nov. 18, 2002).

Ninov was fired from the laboratory.

The accusations of scientific misconduct raised questions about earlier data Ninov had analyzed in connection with the discovery of elements 110 and 112, although the existence of these elements has now been confirmed. Another act of misconduct occurred during the retraction procedure. There were graduate students assigned authorship on the original 1999 paper who were told they did not need to sign the retraction letter. The ethics of this advice and of their acceptance are inconsistent with responsible behavior.

Case 2.5 Misrepresentation of Data in an NIH Grant Application

The following is a copy of the public listing of the scientific misconduct of an investigator found to have plagiarized the work of others in applying for an NIH grant, and the consequences (http://grants2.nih.gov/grants/guide/notice-files/NOT-OD-02-008.html):

> NOTICE: NOT-OD-02-008 Department of Health and Human Services

> Notice is hereby given that the Office of Research Integrity (ORI) and the Assistant Secretary for Health have taken final action in the following case:

David A. Padgett, Ph.D., The Ohio State University: Based on the report of an investigation conducted by The Ohio State University (OSU), the Respondent's admissions, and additional analysis conducted by ORI in its oversight review, the U.S. Public Health Service (PHS) found that Dr. Padgett, an Assistant Professor at the OSU College of Dentistry, engaged in scientific misconduct in grant application 1 R01 AG20102-01 submitted to the National Institute of Aging, National Institutes of Health (NIH).

Specifically, PHS finds that Dr. Padgett plagiarized and misrepresented as his own research data for Figures 1 and 2 of this NIH grant application, data which represented unpublished experiments originally conducted by a researcher at another institution for a private company. The plagiarism was a significant misrepresentation because the data appeared in the preliminary results section of the NIH grant application. The respondent used these experiments, which were relevant to the proposed research, to support the request for funding.

Dr. Padgett has entered into a Voluntary Exclusion Agreement (Agreement) in which he has voluntarily agreed for a period of three (3) years, beginning on October 4, 2001:

(1) to exclude himself from serving in any advisory capacity to PHS, including but not limited to service on any PHS advisory committee, board, and/or peer review committee, or as a consultant; and

(2) that Dr. Padgett and any institution employing him are required to certify, in every PHS application or report in which he is involved, that all persons who contribute original sources of ideas, data, or research results to the applications or reports are properly cited or otherwise acknowledged. This requires Dr. Padgett and the institution, with respect to Dr. Padgett's contributions to the application or report, to certify that all individuals (both within and outside the institution) who contributed to the application or report are acknowledged. The institution must also send a copy of the certification to ORI.

Case 2.6 Darsee Case

John Darsee, M.D., was a young cardiologist who was a rising star in cardiovascular research. His career was marked by a large number of publications, a fellowship in the well-known laboratory of Eugene Braunwald, M.D., at Brigham and Women's Hospital at Harvard University, and the offer of a faculty position at Harvard. His accomplishments came under scrutiny in 1981 when others at the laboratory suspected that an abstract he was writing contained a description of research he had never done, a concern they voiced to the director of the laboratory, Dr. Robert Kloner. Reportedly, when asked for the raw data he went to the lab and ". . . [made] recordings from a single dog, charting them with dates and entries as if they were from several experiments, i.e., he made up the data while the fellows and lab tech watched!" (www.unmc.edu/ethics/data/darsee.htm, accessed November 29, 2005). When confronted, he admitted to falsifying these data but denied other instances of wrongdoing. While stripped of his NIH fellowship and his faculty appointment, he continued to work in the lab and continued publishing.

A review of his other work during two separate internal reviews at Harvard (one by his supervisors Drs. Kloner and Braunwald, and the second by the dean of the School of Medicine) absolved Darsee of other misconduct. Several months later, NIH questioned Harvard-generated data that was at variance with data submitted from other centers as part of a multicenter study. An NIH-sponsored investigation that followed found Darsee guilty of multiple acts of scientific misconduct that occurred while he was an undergraduate and later at Emory and at Harvard. They found multiple acts of manipulation or fabrication of data affecting 8 papers and 32 abstracts while at Emory, with coauthors listed who did not even know their names were on the abstracts. The misconduct while at Harvard resulted in the retraction of 9 papers and 21 abstracts. Darsee was barred from NIH funding and from sitting on advisory boards for 10 years (www.unmc.edu/ethics/data/darsee.htm, accessed November 29, 2005).

There was much discussion as to how he could have perpetrated fraud over such an extended period of time, and how two committees failed to spot evidence of fraud when investigating him. It became clear that investigations of misconduct need to be delegated to respected individuals outside the home institution of the accused to avoid bias. It also pointed out the need to tighten the definitions of what constitutes scientific misconduct, the responsibilities of coauthors, and the responsibilities of peer review (Engler et al. 1987).

Case 2.7 Slutsky: Data Fabrication and Authorship

The issues here are similar to those of the Darsee case. Robert Slutsky, M.D., was a cardiologist at the University of California, San Diego (UCSD). He was also a highly prolific writer who published 137 articles in seven years, the majority of which were published while he was a trainee at UCSD in cardiology, nuclear medicine, and then radiology. Early in 1985, a colleague on the faculty raised questions about apparent duplications in two of Slutsky's publications. An investigating committee requested the retraction of two fraudulent articles in April 1985 (www.ama-assn.org/public/peer/7_13_94/pv3111x.htm, accessed February 17, 2005). Soon thereafter, he resigned his appointments as a radiology resident (trainee) and a nonsalaried associate clinical professor in the Department of Radiology.

During the following year, faculty committees investigated Slutsky's entire bibliography of 137 articles published in seven years. Of these, 77 (including reviews) were classified as valid, 48 were judged questionable, and 12 were deemed fraudulent (Engler et al. 1987). Between 1983 and 1984, he had published one paper every 10 days (Locke 1986). His misconduct was also widely published in the press, including information on his resignation, his attempt to join a group medical practice in New York, and his release from that practice on September 20, 1985 (*Los Angeles Times* 1985).

The Slutsky case was among the most thoroughly investigated by administrators at UCSD, newspaper reporters, and scientists who analyzed the impact of his publications. Whitely et al. (1994) studied the rate of citation of Slutsky's publications to assess how long the scientific community continued to report his data. The results are shown in Figure 2.1.

During the period 1979 to 1985, before Slutsky's fraudulent activities were exposed, the average rate of citation was similar for Slutsky's articles and the controls; however, after the Slutsky fraudulent articles were exposed in the period 1986 to 1990, the Slutsky articles received one less citation per article per year on average (Whitely et al. 1994). As this was one of the most publicized cases of misconduct, it would seem that the fall of literature citations should have been more precipitous than shown in Figure 2.1. Thus, there is concern that that even though fraud is widely publicized, the scientific community remains uninformed. Moreover, publicity about fraud is not sufficient, either for purifying the literature or for discovering false information in the literature. Below, a more quantitative analysis was done on a series of fraudulent publications by Breuning.

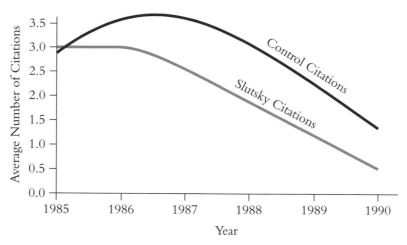

Figure 2.1 The two curves compare the average rate of publication citation for 86 articles by Slutsky (lighter line) compared to the average for the controls (176 articles taken from the same journals, darker line). The rate of rise in average number of citations from 1979 was about 0.5 per year per paper for both Slutsky articles and controls. The peak in citations shows an accummulation of citations for the nonfraudulent publications, but not much difference in the rate of decline in citations between for the two groups thereafter. Redrawn from data by Whitely et al. *JAMA* **272**:171, 1994 (July 13), © 1994, with permission from the AMA.

Case 2.8 Breuning Case: Faked Clinical Trials on Retarded Children

The Breuning case is useful to discuss because it formed the basis for an investigation of the scientific and medical impact of false medical information. While at the University of Pittsburgh, Stephen Breuning published a number of studies from 1980 to 1984 on the use of drugs to control hyperactive retarded children. The studies claimed that stimulant drugs were more effective and had fewer side effects than tranquilizers in the treatment of retarded children. Breuning's work probably led some states to change their policies on treating retarded children (Garfield and Welljams-Dorof 1990). NIH investigated Breuning's work and concluded that "None of the described studies of psychopharmacologic treatment had been carried out. . . . Only a few of experimental subjects were ever studied" (Holden 1987). Breuning pleaded guilty to making false statements on federal grant applications, for which he was jailed and put on probation for five years. An analysis of the citations of Breuning's publications for the eight years from 1980 to 1988 was done by Garfield and Welljams-Dorof (1990). Of 200 citations, 80 were self-citations; of the remaining publications studied, which represented a relatively high number for the period, they concluded that Breuning's fraudulent data had little influence on the findings or conclusions of others, and had little impact on science once exposed.

Case 2.9 Baltimore Case

Referred to as the Baltimore Case, the accusations were leveled at Thereza Imanishi-Kari, who was the supervisor of Margot O'Toole, a graduate student in her laboratory. David Baltimore was the overall supervisor of the laboratory. This is a case of scientific misconduct that dragged on over a 10-year period, during which time the case was reviewed by Tufts University, MIT, NIH, the U.S. Congress, the Office of Research Integrity, NIH Appeals Board of the Department of Health and Human Services, and even the Secret Service (http://rcr.ucsd.edu/content/resources/misconduct.html, accessed December 23, 2002).

In 1986, Margot O'Toole, a postdoctoral fellow working in the laboratory of Thereza Imanishi-Kari, questioned data in a paper by Imanishi-Kari submitted to the journal *Cell,* and alleged that she had falsified and possibly fabricated data. O'Toole was fired. In 1989, an MIT review found no fault with the paper and David Baltimore, a coauthor, defended his colleague's paper. Later in 1989, the Secret Service presented some evidence of foul play. Finally, in 1991, NIH investigators concluded that Imanishi-Kiri forged entire sets of data during the period 1986–1988 in order to support her *Cell* paper. In 1991, her coauthor requested that the paper be retracted.

Representative John Dingell, who chaired some of the congressional hearings surrounding this case, gives the following summary that shows how complex the investigation became:

> In 1986, Dr. David Baltimore, with the assistance of Dr. Imanishi-Kari, published a paper in the journal *Cell*. Margot O'Toole, a young postdoctoral student working in Imanishi-Kari's lab (not . . . in Baltimore's lab), attempted to replicate research conducted by Imanishi-Kari. She could not. Her subsequent questions about the research were summarily dismissed by the individuals and institutions involved. At this point, the matter came to the attention of the Oversight Subcommittee, which held bipartisan public hearings (as it had on several other cases of alleged scientific misconduct). The Subcommittee held its last hearing on this case in 1990. From there, it went to the Office of Scientific (now Research) Integrity within the Department of Health and Human Services, and then to the most recent appeals panel.

> Bullying? If there was any bullying, it was of Margot O'Toole, who was driven from science and found herself answering phones at a relative's moving company.

> Vindication? Hardly. The HHS panel's opinion confirms that many of Dr. O'Toole's concerns were justified. The research she questioned was "rife with errors of all sorts," said the panel's decision. Moreover, the panel cited findings of "significant inaccuracies or misstatements" by previous scientific reviews of the *Cell* paper. Had Imanishi-Kari dealt openly and appropriately with O'Toole's criticisms in the privacy of their laboratory, rather than letting the problem fester and retaliating against O'Toole, the controversy would never have erupted.

> Is David Baltimore a victim and martyr in this cause? Dr. Baltimore, not the Subcommittee, forced the scope of the inquiry to be expanded. Imanishi-Kari did not conduct her work in Dr. Baltimore's laboratory (again contrary to the *Post*'s June 30 editorial). Had she done so, his direct supervision might have prevented the publication of the *Cell* paper with the flawed data. Nonetheless, Baltimore sought to defend

his friend and collaborator by turning the subcommittee's investigation into a referendum on his own honor and reputation—despite the fact that research he conducted was never, at any time, under question.

Baltimore's understandable impulse was misguided. The Subcommittee set out only to examine the process by which scientists receiving vast sums of taxpayer money resolve allegations of misconduct leveled by other scientists. Baltimore insisted instead on pitting the competence and veracity of O'Toole against that of Imanishi-Kari. He argued that the public's representatives had no right and no cause to examine how the public's money is spent on scientific endeavors. Would anyone of sane mind accept a similar argument that elected officials have no right to question public expenditures on weapons systems, the space station, solar energy, or for that matter, food stamps?

The HHS appeal panel did not find a preponderance of the evidence to support a charge of outright fraud or intentional misconduct by Imanishi-Kari. But nor did their opinion rebut much of the Secret Service's sworn testimony that portions of her laboratory records had apparently been altered. "We recognize," said the panel, "that some of the apparent anomalies and peculiarities in the records may be impossible to explain fully at this point." The panel acknowledged that data supporting a key table in the *Cell* paper were simply "inexplicable," and thus "evidence that it is unlikely that Dr. Imanishi-Kari would fabricate such results." (This is most curious logic: if the data supporting the table were "explicable," would that have been evidence of fabrication?) The panel simply accepted Imanishi-Kari's claim of haphazard record-keeping.

Some scientists familiar with the Secret Service's analysis remain convinced that the forensic, statistical, and scientific evidence of falsification is overwhelming (www.house. gov/commerce_democrats/comdem/press/104pr12.htm, accessed November 22, 2002; Web site no longer available).

Case 2.10 Summerlin Case of Patchwork Mice

Summerlin was a biology researcher in the area of tissue transplantation immunology, whose interest was in tissue transplantation between genetically unrelated animals. A major impediment to transplantation has been the rejection of tissues or organs as a result of immunologic incompatibility, and many researchers have focused their careers on the study of how to prevent, treat, or control this phenomenon. Summerlin, along with coauthors of indisputable reputations, published original work reporting the successful transplantation of skin between mice with genetic differences (Summerlin et al. 1973). The paper attracted immediate national and international attention because of its far-reaching implications for improved success for many types of transplantation. However, when the findings could not be replicated, he came under the scrutiny of his peers and superiors at Memorial Sloan-Kettering Cancer Center in New York.

When his superior and coauthor at Sloan Kettering threatened to publish a retraction of their previously reported success, Summerlin brought his transplanted mice to a meeting to demonstrate to him the black patches and the integrity of his work. After the mice were returned to the laboratory, a technician found that the patches could easily be removed with alcohol. It was then discovered that Summerlin had used a black felt pen to create the illusion of a successful transplantation between two separate species. Summerlin argued that he had painted the patches because of the pressure he was under to succeed (Broad 1982, Steneck 1994).

Case 2.11 Cold Fusion: Scientific Misconduct or Journalism Malfeasance?

In March of 1989, scientists Dr. B. Stanley Pons and Dr. Martin Fleischmann of the University of Utah announced at a press conference they had achieved cold fusion, purportedly a highly efficient means by which to generate power. However, disturbing factors in the early months of 1989 were:

- The announcement of their findings at a press conference was highly irregular, violating the standard procedure of first publishing data as a scientific manuscript through a peer-reviewed scientific journal. While preliminary information may be presented at meetings as an oral presentation or a poster, presenting research results to the press risks sensationalism and improper reporting. In the absence of a formal description of their methods, those who tried to duplicate their work could not do so, leading to a rejection of their claims. Peer review might have spared them their damaged reputations.

- The claims were not consistent. They originally indicated they were able to generate four times the amount of power over the amount used to produce it, but later said the amount of power generated was 10 times that needed to produce it.

- Scientists were not only unable to reproduce their results, but those who came to observe the experiments of Pons and Fleischmann were unable to verify the results they claimed. Since 1989, some laboratories have been able to reproduce their work occasionally. Though research has continued in this area, the data remain inconsistent.

Pons and Fleischmann left the University of Utah for Texas A&M, but later left the country with their reputations tarnished and their work dismissed as fraudulent by the majority of scientists and the public. This case is often cited as an example of scientific misconduct. However, as the facts emerged over time, this case is now deemed a case of malfeasance due to the publication of an unsubstantiated accusation by a journalist, as described below.

First, it is important to examine the claim by the majority of physicists that cold fusion is physically impossible. This claim is based on the observation that two charged nuclei cannot get close enough to fuse because of Coulomb repulsive forces. The closer the charges get to each other, the greater the repulsive force; to effect a fusion, tremendous energy must be applied. But could there be a phenomenon in nature that would allow the nuclei to approach each other through neutralization of one or both of the nuclei of, say, hydrogen or deuterium? If so, then cold fusion should be physically possible. In fact, there is such a mechanism. It involves an interaction of a negative muon (a particle about 207 times the mass of an electron) with one of the hydrogen isotopes (hydrogen or deuterium). If the muon-deuteron assemblage is close in space to another deuterium nucleus, fusion could occur. This reaction, first observed by Luis Alvarez and coworkers (Alvarez et al. 1956), seems to have been ignored as proof of the feasibility of cold fusion by the majority of physicists. Physicists following Alvarez recognized the phenomenon of muon catalyzed cold fusion (Ashmore et al. 1958, Ishida et al. 2003). However, most scientists have discarded this approach as impractical or are unaware of experiments that have shown that low-energy nuclear reactions are real. As the Pons-Fleischmann experiment is very simply the observation of heat or tritium emitted from a bath or cell containing platinum and palladium electrodes, it is not surprising that the phenomenon would occur at unpredictable times if cosmic particles such as muons were involved.

A second important aspect of this case is that there were claims of fraudulent activities at Texas A&M at a time a colleague of Pons and Fleischmann was attempting to verify their experiments. A non-scientist reporter interviewed students and faculty at Texas A&M as part of his research for a book on cold fusion. He published an unsubstantiated article in *Science*, claiming that a graduate student helping with these experiments had spiked the experimental electrolytic cells with tritium to give the result of a successful fusion experiment (Taubes 1990). The graduate student who was implicated by this report was exonerated by a university investigation. Additional investigations at Texas A&M found that what was primarily at fault was Taubes's publication of an unsubstantiated accusation (Mallove 1999).

As a result of the embroilment surrounding the subject of cold fusion, further research in this area has suffered substantial setbacks. While in 2005 there were scientists in some 50 laboratories in the world working on cold fusion, most do so in secrecy because of concerns that their scientific judgment, or the use of their funding, might be questioned if the truth were known.

The most concise statement of the potential problems and principles of behavior for journalists, including newspaper reporters, campus paper reporters, foundation reports to trustees, and employees of an institution's public affairs organization, is the code of ethics adopted in 1996 by professional journalists. This code defines the duties of journalists in achieving their goals of public enlightenment such that justice and democracy are served. Their overall mission is the quest for truth and the provision of a fair and comprehensive account of events and issues. The reporting in the cold fusion case did not live up to these standards, two of which were violated in this case:

- Test the accuracy of information from all sources and exercise care to avoid inadvertent error. Deliberate distortion is never permissible.
- Diligently seek out subjects of news stories to give them the opportunity to respond to allegations of wrongdoing.

Scientists, engineers, students, educators, managers, corporation board members, and the public in general rely on journalists for accurate reporting on technology and everyday events. The culture of news reporting is part of scientific and engineering enterprises ranging from local news to carefully prepared brochures and advertisements. The ethics of virtue and justice demand fair and honest reporting. There are few safeguards to prevent biased reporting in view of the First Amendment rights except the right to rebut erroneous and misleading articles, such as was done by Mallove (1999) following the investigation of the cold fusion case.

RESPONSIBLE AUTHORSHIP

The number of authors contributing to a scientific body of work is increasing, the number of disciplines contributing is increasing, and different universities and laboratories around the world might contribute to the same work. As a result, the act of putting together a written work has grown more complex, and the issue of authorship itself has come under increasing scrutiny. The two main issues are: Who should be authors and how should they be listed?

In addition, as will be detailed below, there is now an effort to require all authors to take responsibility for the accuracy of the scientific article, as well as to have substantially participated in the development of the science or engineering represented in the article. The importance of this was underscored with the retraction in 2005 of an important paper on DNA repair published in 1997, after it was proven that falsified data had been supplied by one of the authors without the knowledge of the lead author (Cooper et al. 2005). Prior to publication, it appears that none of the other authors of the paper had validated, verified, or reviewed the data, even though the data were critical to the conclusions they reported. Their conclusions may eventually prove to be correct, but the incident has had a negative effect on the careers of all the authors and may result in sanctions by the agency that granted support for the study.

In the early 2000s, many biomedical journals established their own rules of authorship. These rules prescribed criteria for authorship far more restrictive than has been the practice in life sciences. The National Academy of Sciences has set less stringent guidelines for authorship. Both sets of guidelines are presented below. Rather than feeling confused, the reader should use these lists to glean the nuances of authorship considered important by these groups. Later we suggest a practical approach, drawn from each set of guidelines and based on ethical principles, to aid in the decision about who should be an author and why.

Guidelines for Authorship for Bioscience Journals

The set of guidelines now used by many life science journals evolved because the editors of medical and biological journals saw the need for and took responsibility for formulating the strict guidelines published in 1997 in the *Journal of the American Medical Association (JAMA)*. Below is a synopsis of these guidelines for authorship, updated in 2004 by the Office of Research Integrity, Department of Health and Human Services.

- Authorship credit should be based on 1) substantial contributions to conception and design, or acquisition of data, or analysis and interpretation of data; 2) drafting the article or revising it critically for important intellectual content; and 3) final approval of the version to be published. Authors should meet conditions 1, 2, and 3.

- When a large, multicenter group has conducted the work, the group should identify the individuals who accept direct responsibility for the manuscript. These individuals should fully meet the criteria for authorship defined above, and editors will ask these individuals to complete journal-specific author and conflict of interest disclosure forms. When submitting a group-authored manuscript, the corresponding author should clearly indicate the preferred citation and should clearly identify all individual authors as well as the group name. Journals will generally list other members of the group in the acknowlegments. The National Library of Medicine indexes the group name and the names of individuals the group has identified as being directly responsible for the manuscript.

- Acquisition of funding, collection of data, or general supervision of the research group, alone, does not justify authorship. (Discussed below and in Problem 2.12.in the Problem Set.)
- All persons designated as authors should qualify for authorship, and all those who qualify should be listed.
- Each author should have participated sufficiently in the work to take public responsibility for portions of the content.

Some journals now require publication of the names of one or more author(s) identified as "guarantors" who assume responsibility for the integrity of the work as a whole from inception to published article. Those who have contributed by acquiring funding or assisting in manuscript preparation, as well as those who have merely served as technical advisors and data collectors, should be described only in the acknowledgments. These requirements are particularly relevant to multicenter trials in which scores of authors might otherwise be listed. Indeed, in some physics experiments, over 100 authors have been listed, and it is not surprising that these requirements have not yet been adopted by those in the physical sciences. An example of how problematic this can be is the landmark paper on the human genome, published in *Science* in 2001, in which over 275 authors are listed (Venter et al. 2001). It seems impossible that all listed authors read the manuscript, or if so, that their changes were incorporated before publication as evidenced by errors in the text and footnotes.

The guidelines cited above are generally viewed as too stringent and impractical to be followed by most research groups. A strict adherence to these guidelines would result in a conclusion that many papers have no qualified authors. The above guidelines were developed to avoid the problems listed as minor or major (fraudulent) unethical practices in Table 2.2. But fulfilling each of the three requirements in the JAMA guidelines in order to earn authorship would result in injustices to many who contributed ideas and personal effort to a scientific endeavor. Recognizing this problem, the National Academy of Sciences developed guidelines for responsible authorship.

National Academy of Sciences Guidance on Responsible Authorship

The National Academy of Sciences provided guidelines for determining authorship (www.nap.edu/readingroom/books/obas/contents/authorship.html, accessed February 17, 2005):

- The division of credit within research groups and best practices for authorship among local researchers as well as among all collaborators should be discussed at regular group meetings.
- The allocation of credit and ranking of authors involves consideration of the differences in the roles and status of the team members as well as the collaborators.
- The inclusion of the name of someone as author who had little or nothing to do with the content of a paper dilutes the credit due those who actually did the work. It also inflates the credentials of those so "honored," and makes the proper attribution of credit ambiguous. This is also known as gift authorship.
- The allocation of authorship denotes that an author has taken responsibility for the paper and its contents. Thus, unless a footnote or the text of the paper explicitly

assigns responsibility for different parts of the paper to others, the authors whose names appear on a paper must share responsibility for all of it. It is the practice of some journals to require all authors to sign a document attesting to their acceptance of responsibility for the content of the paper.

The following case is an example of the problem of assigning credit (www.nap.edu/reading room/books/obas/contents/aauthorship.html, accessed February 17, 2005).

Case 2.12 Who Should Get Credit for the Discovery of Pulsars?

A much-discussed example of the difficulties associated with allocating credit between junior and senior researchers was the 1967 discovery by Jocelyn Bell, then a 24-year-old graduate student, of pulsars. Over the previous two years, Bell and several other students, under the supervision of Bell's thesis advisor, Anthony Hewish, had built a 4.5-acre radiotelescope to investigate scintillating radio sources in the sky. After the telescope began functioning, Bell was in charge of operating it and analyzing its data under Hewish's direction.

One day, Bell noticed a "bit of scruff" on the data chart (McGrayne 1993):

The first thing I noticed was that sometimes within the record there were signals that I could not quite classify. They weren't twinkling or manmade interference. I began to remember that I had seen this particular bit of scruff before and from the same part of the sky.

By measuring the period of its recurrence, she determined that it had to be coming from an extraterrestrial source. Together Bell and Hewish analyzed the signal and found several similar examples elsewhere in the sky. After discarding the idea that the signals were coming from an extraterrestrial intelligence, Hewish, Bell, and three other people involved in the project published a paper announcing the discovery, which was given the name "pulsar" by a British science reporter.

Many argued that Bell should have shared the Nobel Prize awarded to Hewish for the discovery, saying that her recognition of the signal was the crucial act of discovery. Others, including Bell herself, said that she received adequate recognition in other ways and should not have been so lavishly rewarded for doing what a graduate student is expected to do in a project conceived and set up by others.

Fairness in crediting the originator of the work is a basic tenet to publication and is especially important since an individual's promotion, funding, reputation, and career advancements may depend on it. Failure to do so can have lasting consequences to the offender, since word gets around and may result in peer censorship.

Practical Guidelines for Assigning Authorship

With all these guidelines, what is a practical approach to the authorship dilemma? Below is a list of statements that should be addressed early in every multiple authorship situation.

Rather than judge specific guidelines, each group needs to discuss, develop, and adopt policies it will follow regarding authorship. The policy should be agreed to by all members of the group, given to all newcomers to the group, and reaffirmed on a regular basis. This should also be reassessed in the context of the journal requirements for an intended publication, in case adjustments are needed. These elements are:

- All authors shall be responsible for the integrity of the data.
- All of the authors shall read and agree to the text of the publication.
- No honorary or gift authorship should ever be offered or accepted. At many centers, it is the custom to automatically list the director of a laboratory as an author, but this should not be allowed.
- The technical support staff should be acknowledged in the listings but not included as authors unless they are an integral part of the conception, design, performance, or data analysis of the study. If one is included as an author, all of the duties of authorship must be followed.
- The group should decide on what level of acknowledgment should be given to those outside the laboratory for contributions of material or analyses.
- The group should decide how authors should be listed, recognizing the precedents for first author, last author (frequently called lead author), who is frequently the senior member of the scientific team, and precedents for alphabetizing the remaining authors.

A written policy that includes answers to these questions will serve as a guide to avoid misunderstanding, resentment, and disappointments.

Ranking of Authors

Recognition of the responsible persons who have contributed to a scientific or technological contribution is done through authorship. The ethical issue in the selection of authors and deciding on the ranking of names is mainly justice. The most common method for listing a series of authors when three or more are involved is to put them in alphabetical order. Frequently, the most senior investigator or laboratory director is listed last on a multi-authored paper. This last-place listing usually signals the leadership role of that author but is ambiguous with respect to how much scientific innovation and work should be attributed to the last-place author. It has become a custom to list the laboratory director last on any paper, no matter whether he or she had anything to do with the paper. That practice has led to ambiguities regarding who did the work. It is also a dangerous practice because it may lead to publications without critical reading and analysis by the most senior scientist. A statistical analysis of authorship listing patterns by Zuckerman (1968) shows the more senior and recognized the author, the less likely he or she would be listed first. Contemporary standards do not exist, and perceptions of the meaning of the ranking of authors have junior investigators confused (Tarnow 1999). Though the practice of recognizing the senior investigator last is prevalent in many life science institutions, publications from medical schools and some physical sciences laboratories are not consistent in how the senior investigator is listed. One not familiar with the field and the identity of the lead investigators will not know whether the author list denotes a hierarchy of credit or not.

STEALING AND COPYRIGHT

The next section discusses aspects of misrepresenting the work of others as one's own work. There are clear cases of publishing ideas and data belonging to others without giving a proper attribution or without permission from the true author or owner. But there are also many cases where the use of others' data or ideas might be acceptable. The majority of cases in this gray zone are those where attribution is given, but permission is not obtained. It is in these cases that issues of copyright infringement arise when authors fail to recognize the requirements to obtain permission to use original data, artwork, or other copyrighted material. A gate keeper for the latter situation is the publisher. One of the most common types of copyright infringement is the act of producing a copy of an entire book to avoid purchasing that book, or making multiple copies of a chapter of a book to hand out at a scientific meeting or in a class.

Case 2.13 Stealing the Work of Others

Professor Jones is brought before the committee on tenure as a result of a complaint from a student who did 90% of the work that Jones is taking credit for. The committee reviews the case and discovers that Jones did the same thing in his previous place of employment. They decide to suspend Jones for a year without pay because he has plagiarized. Jones says everyone understands this is how it is done and appeals to the grievance committee. They send the case back to the tenure committee, saying Jones should be dismissed. As president of the university what do you do? In the end, the case went to civil litigation. The judge upheld the university's decision. (One of the later exercises in the Problem Set will be for the reader to act as the president of the institution in deciding this situation.)

Plagiarism

Plagiarism is the appropriation of another person's ideas, processes, results, or words without giving appropriate credit. It is one of the most common transgressions of ethical behavior, particularly among students. It is one of the major articles in the federal definition of scientific misconduct, and it is an aspect of responsible authorship that remains a major concern of educators and editors.

Plagiarism takes many forms in scientific writing. Paraphrasing without giving attribution is plagiarism just as much as the outright reproduction of an observation or interpretation of another word-for-word copy without attribution. The definition needs some clarification, because plagiarism does not include use of commonly known axioms, theories, equations, and diagrams. For example, it is not necessary to cite Einstein every time one uses $E = mc^2$. Frequently, we name entities to ease communication, such as declaring how the Green's function appears in a derivation, but to not mention Green is not plagiarism. But suppose that a colleague mentions at a scientific meeting the concept of an intraoral touch pad for quadriplegics to enable communication, and another scientist takes the idea without giving attribution. This would be considered plagiarism. Suppose that a research paper postulates the idea that the originator cells for the lining of the capillaries in the brain come from bone marrow, and that neurodegeneration is due to dysfunctional progenitor cells from the individual's bone marrow cells. If another researcher picks up this idea and performs experiments to demonstrate the truth without giving attribution, that researcher has plagiarized. The simple rule for responsible authorship is to give citations for every important idea that is not that of the authors writing the paper.

The simple rule above is not always easy to practice, particularly in the environment of educational programs at the undergraduate and graduate levels. Suppose that one is given an assignment in creative writing. One frequently learns about an idea or concept from a remote source and then elaborates on this idea without full attribution. Suppose that one of the assignments in a university ethics course leads one to ask fellow classmates for ideas on approaching some of our ethical dilemmas. Is this wrong? Of course not. The principal method of approaching ethical problems is fact finding, discussion, and understanding the viewpoints of others. However, copying the creative writing of others (i.e., homework), or even paraphrasing without attribution, is wrong. It is very difficult to be original. This culture puts an enormous value on originality and cleverness. How does one deal with the issue of originality? How does one establish the integrity of one's accomplishments in the context of these pressures? A start is to disclose that part of the writing that is original and that part that was derived from others.

A special kind of violation is that of auto-plagiarism, where an author fails to acknowledge the publisher who has copyrighted the author's material and now is the rightful owner of the work. If a body of work by an author has been copyrighted and released as a publication by a publisher, that author must obtain permission from the publisher to use those materials. Once the copyright is assigned to a publisher, the author loses the right to use his or her material without permission but remains the true author. This might present a conflict, and in some cases might be seen as unjust, but the publisher, not the author, controls the distribution of such information. It is a copyright violation, not plagiarism.

Case 2.14 Plagiarism by the Innocent

A student submits a paper representing the work done for a bioengineering class. He has spent the semester assisting in the data analysis of data collected by a graduate student on human subjects in accord with institutional procedures of human subject research. The work of approximately eight hours per week involves creating some improved software for analysis of the response of subjects to nonharmful stimuli. The paper describes the reason for the research and the principles of the instrument that collects the data by abstracting sentences from a paper in progress not written by the student. The figures in the text of the paper are taken from slides used by the graduate student and from figures out of technical reports. There are no citations and no references to the source of material in the paper. The student's professor did not supervise the work but is expert in the type of work performed. Through questioning, the professor learns the student does not know the definition of the abbreviations or the terms used in the paper. This student is a senior and has not taken any courses on ethics. The professor learns that the student is generally very reliable, accurate, and a great asset to the research being performed. In fact, other than the paper, prepared during exam week of the semester during which the course was given, the student has been of substantial help to the graduate student he was assisting.

The question before the professor is how to deal with this case, using the principles of justice and non-malfeasance. Upon learning that the student did not understand the word *plagiarism* or how to avoid plagiarism in practice, a career-threatening punishment would be unfair or unjust. The professor requires only that the student study the course reader on ethics that preceded this text and rewrite the paper.

The above case shows leniency on the part of the professor and a possible violation of the policy at most universities and colleges. The presumption of innocence based on a claim the student did not know what constituted plagiarism does not exempt the professor from reporting the incident. The reason for reporting cases of plagiarism is that if a student who claims innocence plagiarizes a second or third time, the violations lead to the conclusion of unacceptable academic behavior, which will result in penalties ranging from loss of course credit to expulsion from school. If professors do not report even minor violations, the offending student will not be detected. The ethical principal is justice to the equitable credit to other students who did not plagiarize.

Clearly, education regarding plagiarism needs to be emphasized early in a student's academic life. Plagiarism prevention programs can be implemented using Web-based tutorials for students (e.g., www.umuch.edu/vailtutor/index.html) or a Web site called TURNITIN (www.turnitin.com), which allows professors to check submitted papers for plagiarism. In 2005, the cost to utilize this Web site, which will search the Internet for phrases similar to those in the submitted article or paper, was $150 for up to 150 students or 150 reports in 365 days.

Motives for Scientific Misconduct and Countermeasures

Countermeasures for fraud in science cannot be deployed without a better understanding of the principal causes of scientific misconduct. The motivations for scientific misconduct are not fully understood, and specific information as to its etiology is lacking. The National Academy of Sciences (1992) lists six possible causes:

- Funding and career pressures of the contemporary research environment.
- Inadequate institutional oversight.
- Inappropriate forms of collaborative arrangements between academic scientists and commercial firms.
- Inadequate training in the methods and traditions of science.
- The increasing scale and complexity of the research environment, leading to the erosion of peer review, mentorship, and educational processes in science.
- The possibility that misconduct in science is an expression of a broader social pattern of deviation from traditional norms.

But the majority opinion is that the pressure to publish scientific papers, to obtain grants, and to achieve recognition by one's peers is the major temptation that leads to scientific fraud. These are classified as external pressures. Frequently, these external pressures tempt the junior and senior scientist to transgress from morally accepted practices (List 1985). The internal rewards of the joy of finding truth and sharing it with others can, if learned, be the major countermeasure for scientific misconduct.

A countermeasure might be to remove some of the incentives for fraud, such as competition for prizes or monetary rewards (List 1985), but this is impractical. It is more practical to instill in senior researchers the need to actively mentor junior investigators on avoiding major digressions in virtuous behavior. However, this mentoring regarding the obvious must also include instructions and examples of the ethical issues regarding the minor departures such as equity in referencing the work of others and justice in authorship credits (Alberts and Shine 1994).

Another explanation is that an individual perpetrating misconduct has a personality disorder for which there may be a medical or psychological explanation. Others believe that the perpetrator lapses into misconduct due to alienation from colleagues, thereby becoming isolated from the usual social interactions. Mounting work, career pressures, and disappointment with achievements can lessen moral resolve and lead to the temptation to cheat (Hackett 1999).

Scientific misconduct is different from the tort law concept of fraud because in an assertion of fraud, there is a plaintiff who is alleged to have been harmed. In scientific misconduct, the scientific community might be harmed in that scientists are put to the task of conducting time-consuming experiments in an attempt to duplicate the data. Fraud is difficult to detect, there may be multiple offenses by a single scientist, and there is no standard or reliable method for retraction of fraudulent papers in the scientific literature. Eventually, the truth is documented, but it could take as long as 50 years. Society is harmed by the diversion of resources away from other more productive research.

The previous reliance on the scientific community to discover errors and unwarranted conclusions, and to prevent the propagation of inaccurate and false data, has all been disappointing. Codes of conduct and guidelines for responsible conduct of research have been issued by many universities, foundations, and institutions. An informative federal organization known as the Office for Research Integrity provides guidance and analyses of all facets of the responsible conduct of research for the individual student, investigator, author, reviewer, mentor, and institution (Steneck 2004). It can be accessed at http://ori.dhhs.gov.

Discovery of Scientific Misconduct and Whistleblowing

Whistleblowing occurs when an employee or someone in academia, industry, or some other professional organization becomes aware of and reports unethical, inappropriate, or illegal activity on the part of another. The perpetrator may be an individual employee, a supervisor, or someone in upper management of a company, a public agency, or an organization. It is the responsibility of students, scientists, engineers, professionals, and employees to report transgressions or wrongdoing they observe. Deficiencies in the practice of research may not be uncovered without the actions and cooperation of a whistleblower. The importance of taking on the responsibility of reporting wrong doing by others or whole corporations cannot be minimized, particularly in scientific inquiry as many scientific and academic endeavors are self-regulated. As such, a high level of vigilance and shared responsibilities are required to maintain integrity within the respective communities.

The term *whistleblower* is used to denote an individual who alerts others to the fact that some error, departure from normal behavior, or fraudulent activity is occurring. It usually implies that there is a perpetrator of wrongful activity and the whistleblower is accusing the individual or institution of wrongdoing. Whistleblowing is a hard subject to approach with enthusiasm. The public generally looks with distain on the concept of whistleblowing, and in particular, the whistleblower is usually looked upon as a "snitch" deserving of some form of retaliation. This is also the perception of those in business, who often see whistleblowing ". . . as an act of extreme disloyalty to the company and to coworkers" (Fleddermann 2003a). Nevertheless, the concept of whistleblowing has fundamental merit and, in fact, is a responsibility of any individual whenever five criteria presented below are met.

But who should blow the whistle, to whom and when? What are the consequences of being found guilty of misconduct and how are they imposed? Should they be anonymous? Should the reporter be protected? The most senior person aware of the misconduct should be the one to go to an immediate supervisor, if this is deemed appropriate, or to whatever level of management that has the power to intervene and make a decision about the allegations. The two most important factors determining when whistleblowing is appropriate is first, that the misconduct must be sufficiently serious to warrant it, and second, that if safety is an issue, the whistle should be blown as early and as soon as possible, with specific identification of the misconduct to those in charge.

When it is an illegal or unethical behavior by an individual involving misconduct of a nature not involving safety per se, the charge should be leveled according to the misdeed. Perhaps the charge is the result of an honest error that might be settled by a frank discussion with the errant individual. If more than that, e.g., if the intent has been to mislead or deceive, or if the charge involves a moral issue, then the whistleblower must proceed in a timely manner to make the charge known, initially on as small a scale as possible (to the individual) or to the person's immediate supervisor. Higher management should be brought in only if this fails.

Criteria for Whistleblowing
(Adopted from Fleddermann 2003b, Harris et al. 2000, De George 1999.)

1. Need: There must be an impending harm or a dangerous situation that can be avoided by blowing the whistle. The situation about which one blows the whistle must be important. This might include a pattern of small things that can add up to an issue of major importance. For example, a pattern of "tweaking" or changing the ingredients or components of products without the proper studies carries with it an ever-increasing potential for harm.

2. Proximity: The whistleblower must have firsthand knowledge of the wrongdoing. In addition the whistleblower must have the expertise to make a realistic assessment of the situation. For example, hearsay is not an adequate basis for whistleblowing.

3. Capability: By the action of the whistleblower, there must be a high probability of stopping the activity. All whistleblowing activities carry with them some risk to the career and future financial security of the whistleblower. This risk must be balanced by confidence that the whistleblower has access to the proper channels to ensure that the situation is resolved.

4. Last Resort: Whistleblowing should be attempted only if there is no one else more capable or more proximate to blow the whistle, and all other lines of action have been explored.

5. Adequate Documentation: An allegation of misconduct must be based on adequate documentation. An allegation of misconduct based solely on verbal accounts will harm both the reporter and the scientist, and is unlikely to result in a beneficial or remedial outcome for anyone. The legal tenet of reasonableness is used to determine whether to blow the whistle: it should be done when any reasonable person with similar skills and scientific knowledge working under the same ethical precepts would do the same.

Not all concerns about research conduct should result in an allegation of research misconduct. As for any other ethical decision, it is appropriate to first balance the risks and benefits of the proposed action. If justified, the decision may be to report an allegation of misconduct, but circumstances might instead warrant other approaches. Making an allegation of research misconduct should not be a first step to remedy questions about the practice of research because impressions should be validated before making serious charges; perceived problems may consist only of disagreements or misunderstandings that can be resolved by other means.

Whistleblowing is never to be undertaken lightly or for the wrong reasons, and not until the truth of the accusation(s) can be documented as fully as possible. Because such accusations are serious, the magnitude of the wrongdoing should be sufficiently important to justify putting one's reputation or career on the line, since the ramifications of accusing another of wrongdoing are likely to be large. For the same reasons, the whistleblower should have first-hand information about the wrongdoing and should have the facts and evidence of the wrongdoing before proceeding. Other options besides whistleblowing should have been considered and exhausted before proceeding.

Whenever whistleblowing occurs, it is a generally difficult and career-threatening occurrence for the perpetrator. Sometimes, there are unexpectedly negative consequences for the whistleblower as well. It is also an exceedingly difficult time for those having to investigate the allegations and an emotional time for coworkers of the accused who, even though tangentially involved, may also come to have their reputations tarnished and their work interrupted. It is also extremely time-consuming to track down and write up the details, to say nothing of the time wasted in behind the scenes discussions by everyone even remotely aware of the incident. The work of an entire laboratory can come to a standstill.

The integrity of research depends in part on self-policing. One should not be silent about perceived wrongdoing, particularly if consequences may include harm to patients or to human or animal subjects, or if there is a waste of scarce resources or publication of misleading findings. But there are deterrents to reporting misconduct because the report might implicate the reporter, and one might find that risk greater than the risk that some harm would occur if the misconduct is not reported. However, not reporting misconduct that is discovered by others could implicate one for complicity, or could at least lead to questions as to why nothing was said or done earlier. The prevention of the need for whistleblowing within a corporation can be facilitated by fair corporate practices that include a culture that values ethical behavior, open communication without fear of retaliation, and a willingness to admit and rectify mistakes (Fleddermann 2003c).

CONSEQUENCES OF BEING ACCUSED OF SCIENTIFIC MISCONDUCT, AND PROTECTION WHEN ACCUSED OF SCIENTIFIC MISCONDUCT

What protection is afforded the accused? This is a major issue that needs to be addressed by federal policy makers as well as by individual institutions. The current system violates the moral theory of justice, as we argue below. Currently, the accused researcher's institution performs the investigation. Most institutional policies do not require the presumption of innocence and do not allow scientists to see the evidence against them. Further, investigations into research misconduct use a low standard of proof, requiring only "preponderance of evidence" (that is, that the evidence points to a finding of guilt of the accused more likely than not). But many agree that accused scientists should be guaranteed more procedural safeguards during investigations of alleged misconduct (Mello and Brennan 2003). During the period 1992 to 2001, among 883 allegations of misconduct reported to the Office of Research Integrity from 248 institutions, there were 703 investigations, with misconduct documented in 16% (Rhoades 2004).

The guidelines of 2000 also provide a broader, more comprehensive, and uniform misconduct policy. The language "fabrication, falsification, plagiarism" is retained, as proposed in 1995, but now, proof of the researcher's intent to present false or plagiarized information is required. The new guidelines of 2000 also suggest the following safeguards for accused scientists:

- Institutions should inform and interview the researchers about the allegations when the inquiry begins.
- The accused scientists should be allowed to confront witnesses, to have full access to evidence against them, and to present their own evidence at both the inquiry stage and the full-investigation stage.
- They should be granted the right to be represented by counsel throughout the entire process.

The Office of Research Integrity recommended, but did not require, institutional compliance with the 2000 guidelines. The lack of requirements to follow the guidelines might lead to serious problems for the stakeholders, the institution, and the falsely accused, as will be seen in Problem 2.3 of the Problem Set of this chapter. The major protection for those accused of scientific misconduct is the guideline that the institution must provide sufficient evidence to convince peers that the accused intended to deceive. This recommendation does not mean that a particular institution will adhere strictly to this guideline. The ethical issue here is justice.

Whistleblower Protection Act (WPA)

In 1989, Congress unanimously passed the original WPA, and in 1994 voted unanimously to strengthen it with new legislation. At the time, these revisions were put in place to broaden the ability of federal employees to blow the whistle in situations where previous rules surrounding national security had prevented them from speaking out about possible terrorism threats, but they also offer protection and advice for anyone involved in whistleblowing. See Chapter 4, "Business Ethics," for a further discussion of this subject.

Approaches to Scientific Misconduct: What Should Be Done?

As early as 1989, the Institute of Medicine recognized that difficulties arise when scientists try to police their own:

> Until very recently, governmental and private institutions assumed that the principle of self-regulation was sufficient to promote and maintain the integrity and quality of federally funded research. Recent cases of scientific fraud, however, have challenged the wisdom of relying entirely upon investigator honesty and unwritten collegial standards as the sole means of assuring integrity and quality in research. These cases have fostered a perception that existing institutional and professional controls have been inadequate to assure integrity in federally funded research (Institute of Medicine 1989).

The approach now is to encourage integrity in scientific research at every level. The responsibility for change rests with the individual researcher, with government agencies, and with the institutions carrying on scientific research.

Responsibilities of Individual Researchers

An editorial written by members of the National Academies panel evaluating scientific fraud emphasizes that the individual researcher must be at the forefront of maintaining scientific credibility. This begins with upholding the standards for the responsible approach to a scientific project, including, for example, a scrupulous literature search, a thoughtful protocol, a truthful consent form, an examination of conscience for an absence of bias regarding outcomes or other inducements that might bias interpretation of data, and the straightforward use of funding for the project. The accuracy of the data is imperative, the analysis must be truthful, and the report timely (including careful attention to the rules of authorship), the identification of and reporting of any misconduct occurring during the study, as well as a willingness to help others who may want to duplicate the work. Because teams of investigators are now often involved in a study, the leader of the project must be directly involved not only in ensuring the accuracy of data, but also in the subsequent rigorous review of the data to ensure that there are checks and balances in place to spot discrepancies. The individual researcher also has responsibilities to address possible conflicts of interest in negotiating participation in a research study, or in negotiating any contract relating to a study. In particular, the presence of any bias that could affect the performance of services or the outcome of the study must be addressed. "Research universities are concerned about financial conflict of interest (individual and institutional) because it strikes to the heart of the integrity of the institution and the public's confidence in that integrity" (Association of American Universities 2001).

Responsibilities of Government, Universities, and Research Institutions

The responsibilities of federal agencies are clearly defined in Section III of the Federal Policy on Research Misconduct reproduced at the end of this chapter (see also www.ostp.gov/html/001207_3.html, accessed February 17, 2005). As stated, the federal agencies have ultimate oversight authority for federally funded research, but research institutions bear primary responsibility for prevention and detection of research misconduct and for the inquiry, investigation, and adjudication of research misconduct alleged to have occurred in association with their own institution. In most cases, agencies will rely on the researcher's home institution to make the initial response to allegations of research misconduct and to inform the agency of the findings.

Agencies will usually refer allegations of research misconduct made directly to them to the appropriate research institution. However, a federal agency may proceed with its own investigation if there is a need to protect public health, safety, or other interests, or if the home institution is not able to reasonably conduct the investigation. An example is the FDA evaluation in 2005 of the serious side effects reported for certain over-the-counter anti-inflammatory drugs, the result of which was the withdrawal of several from the market.

Scientists who have successfully obtained very large grants are viewed with favor by the institution's central administration because as much as 50% of the funds for these grants go directly to the administration as "indirect costs." These funds increase the monies available to administrators to run the university as well as fund other projects. As a result, there has been a tendency to exploit scientists to generate these funds, even affecting the ethical handling of charges of misconduct. For example, a scientist who is believed to be involved in some questionable practices and who is a major grantee of federal funds could be favored in an investigation of scientific misconduct.

CONFLICT OF INTEREST AS A CATEGORY OF SCIENTIFIC MISCONDUCT

We usually consider a conflict of interest to exist when there is some financial arrangement that might bias, prejudice, or influence the actions of a professional engineer or scientist. There are a number of categories in which a conflict of interest is of major importance as an ingredient of ethical transgressions. The omission of reporting a potential conflict of interest is considered a violation of accepted practices in most situations involving organized scientific meetings, publications, and the conduct of research at the major universities around the world. Generally, even if it *appears* that a conflict of interest exists, even if one does not, professional duty requires that it be reported. Once the potential conflict is reported, the institution's officials make a judgment as to how to manage the situation.

Table 2.5   Examples of Conflicts of Interest

- Scientist who holds shares in a vendor.
- Scientist who presents a paper at a scientific meeting on a product from a company providing financial support for the presenter.
- University researcher who consults with a company supporting her graduate students.
- Professor who owns a company whose products are similar to the subject of current research using university resources and students.
- Researcher who receives first class travel and an honorarium from a drug company to present results of a drug study.

Managing Conflicts of Interest

An essential feature of managing conflicts of interest is to follow a professional course that avoids all conflicts of interest. It is equally important to avoid any situation that has the appearance of a conflict of interest. To avoid both requires ongoing vigilance. This duty is not merely to maintain one's own professional reputation but also to protect the reputations of colleagues one represents. Scientific and professional societies as well as universities have developed methods of mitigating or removing the stigma of conflicts of interest. Scientific and professional societies usually require a declaration by speakers at meetings and by authors regarding their potential conflicts of interest. Currently, some technical journals and all medical journals require a signed document in which the authors declare any financial or administrative relationship with the commercial entity whose product is the subject of the publication.

One example of how a university is managing potential conflict of interests is at the University of California at Berkeley (UCB). As a result of the 1980 Political Reform Act that requires state institutions to form conflict of interest committees, UCB established a Sponsored Projects Office (SPO) to monitor and manage compliance in all research undertaken on campus. These responsibilities are in addition to the oversight of all funding for research and other activities such as grants and contracts. All investigators, including students who might work on a project, must sign a form disclosing any potential or known conflict of interest. A committee, chaired by a faculty member and co-chaired by a representative from the SPO, evaluates all cases of potential conflict. The mission of that committee is not to police faculty and staff but to "manage conflict of interest." Thus, the majority of cases it examines are allowed to go forward, sometimes with some limitations. One resolution might be to further define how finances or resources are used or shared so as to avoid actual conflict between the mission of the university, the sponsor of the research, federal or state law, and the individual. Another resolution might be to ask the professional to take a leave of absence for a limited period required to pursue the enterprise that is in conflict with current employment. It is much easier to provide information up front about a potential conflict of interest than to have to react to an accusation of wrongdoing. Such accusations can lead to serious consequences for the individual and compromise the integrity of the project. Willful withholding of information regarding activities that are prohibited can result in penalties as great as discharge from the institution or fines in terms of withheld pay.

The situations outlined in Table 2.5 can be generalized into categories of potential conflicts of interest described below.

Individual financial conflict: This occurs when an individual's personal and professional lives are in competition. The concern is that a professor's interests away from the university might compromise the interests of the university or compromise an academic career. For example, if a university professor owns or is being paid by a company to do research with a product being developed by that company, there is a risk that the research may be biased in terms of data collection, analysis, and outcome. Similarly, if a professor's wife owns a company and the professor contracts with the company for services or products, this would represent a financial conflict of interest because the professor's wife (and he) would be seen to gain from the arrangement. It gives rise to the appearance of doing something wrong and must be avoided.

Institutional conflict of interest: There may be situations in which the campus or the university has a conflict with a professor's activities. The institution must be protected against a conflict of interest or even the appearance of a conflict of interest. This is tricky because the university encourages filing patents on inventions and encourages start-up companies, but sometimes these can be problematic for the university.

University officer conflict of interest: For example, say the president of University A is a large stockowner in QualCom. If University A were to use Eudora, a product of QualCom, as its only Internet company, that would represent a conflict of interest.

Consultant conflict of interest: Acting as a consultant for a company can also be problematic. Although the university allows time for this, part-time employment is a concern because it is time spent away from the university. Again, there may be an *appearance* of wrongdoing.

Nonfinancial conflicts of interest: These conflicts of interest might involve scientific integrity (such as publishing data too early in an attempt to advance one's career), conflict of conscience (such as if a university policy is in conflict with one's own belief about what is right), or a conflict of commitment. For example, a professor might have a conflict if deflected from his duties to teach, do research, and perform service to the institution because of outside obligations (e.g., consulting, starting a company, membership on a board of directors).

The reason that universities have methods to cope with conflicts of interest, even when in fact no conflict exists, is based on the special role a university has with respect to public trust. There is a need to protect the reputation of the institution as well as the reputation of the professors and students. If a conflict of interest arises and the situation is mismanaged, the negative publicity can have an impact on the funding of the university by foundations and alumni, resulting in a less attractive institution for future students. The university must be seen as caring about key values and having a steadfast commitment to students and to academic freedom. It must ensure that research results are accurate and obtained without bias so that there is confidence in the results and so that the university's commitment to excellence is evident. Otherwise, media coverage may be brought to focus on the university, and there can be legal repercussions as well. In addition to teaching and community service, the university's goal is to promote the excellence of the research generated by the faculty and students, and not to focus on whether a partnership with outside interests will accrue to the wealth of the university. In this respect, justice requires the university to prohibit unfair advantages given to one corporation over another in the use of the university resources for business purposes.

The central administration at UCB is proactive by requiring everyone planning to be involved in research to sign a disclosure listing all associations that might be seen as a potential conflict of interest, and manages such disclosures through peer review by a faculty committee.

Conflicts between Goals of Commercialism and the Tenets of Academia
Conflicts of interest in the modern university now extend beyond the activities of individual scientists and students. In addition to the university's traditional roles in the quest for

knowledge, teaching, and service toward social ends, the new culture of a university now includes an active participation in these areas with industry. There is now an active presence of industry on campus that fosters the exchange of ideas in laboratories and in seminars, and forges relationships that involve funding, student internships, cooperation in problem solving, and an interfacing of the tenets of academia with those of profit-oriented corporate cultures. To many, it would appear to be a win-win situation. The university gains through funding of the research and potential patent rights, and the students gain through experience. The corporation gains by avoiding personnel recruiting, capital investment in resources, and human relations costs for shutting down one project and starting another because these are borne by the university.

The concern is that the juxtaposition of two such diverse entities is bound to result in conflicts of interest. For example, the quest for corporate funding may unduly influence the priorities of professors and the leadership of the universities. It may also result in commercial control of research directions and ownership of intellectual property. These concerns gain credibility when comparing federal funding of university research in the United States in 2000 ($17.5 billion) with funding by commercial organizations in the same year ($2.3 billion). Though federal funding is currently about eight times larger, commercial funding of university research has increased by a factor of 100 in 20 years.

Professors at universities are concerned that the central administration is being taken over by businesspersons, whose pursuit of financial success threatens the freedom of the faculty to decide what to teach, whom to recruit, and whom to promote. These concerns are odds with the advantages of partnerships between academia and the corporate world in the scholarly work, as described in *Universities in the Market Place: The Commercialization of Higher Education* (Bok 2003a). The recognition of the advantages and disadvantages of the partnership with corporations, and even the university's commercialization of a service or an invention, requires the establishment of guidelines. However, Derek Bok, a former president of Harvard University, concludes (2003b):

> Unfortunately, the structure of governance in most universities is not equal to the challenge of resisting the excesses of commercialization. Presidents and deans are ultimately responsible for upholding basic academic values, but they are exposed to strong conflicting pressures that make it hard for them to carry out this duty effectively.

CHAPTER SUMMARY

Scientific misconduct encompasses acts of data falsification, fabrication, and plagiarism, as well as dishonest representations in scientific writings and engineering documents. Misconduct may also consist of inappropriate claims by authors, biased evaluations of scientific grants and manuscripts by reviewers, and conflicts of interest in the activities of scientists, professors, and even institutions. The intentional stealing of ideas of others without proper attribution is perhaps is the most serious example of scientific misconduct. Virtue and justice dictate a level of professional behavior when involved in the enterprises of scientific publications, ranging from the just selection of authors to the virtuous behavior of reviewers and editors.

Though the problems of data fabrication, data falsification, and data omission result in consequences to the scientific community at large, the prevalence is not nearly as great as misconduct in scientific authorship through plagiarism. Minor offenses include deviations from fair practices in crediting authors and lifting references from the Internet without consulting the original paper in reviews and scientific paper discussion sections. More serious offenses by the scientist or engineer include a disregarding of the fundamental duty to represent the whole truth. Editors and reviewers may also err through their disrespect for the autonomy of authors by unfair management of their manuscripts.

Within the academic setting, the two most prevalent categories of misconduct by students are plagiarism and cheating on exams and assignments. In most universities, plagiarism is labeled as scientific misconduct whether innocent or not; however, the innocent plagiarism case is protected from a charge of misconduct by the definition of 2000. The guidelines for how to proceed in cases of observed misconduct involve five well-defined steps, in addition to application of the Four A's strategy for resolving dilemmas.

The subject of scientific authorship is reviewed in detail because the ethics are often misunderstood and serious repercussions can result. The final topic is the increasing involvement of corporations that use the university as a research facility. The ethical issues here involve conflicts of interest between the university management's obligations to expand income to finance the growth of the university and justly serving the needs of students and society.

REFERENCES

Alberts, B. and K. Shine. 1994. "Scientists and the integrity of research." *Science* **266**:1660–1661.

Alvarez, L. W., M. Bradner, F. S. Crawford, et al. 1956. (December 10). "The catalysis of nuclear reactions by μ mesons." Printed for the U.S. Atomic Energy Commission, University of California, Radiation Laboratory, Berkeley, CA (contract No. W-7405-eng-48).

Ashmore, A., R. Nordhagen, K. Strauch, and B. M. Townes. 1958. "The gamma-rays from muon catalysed fusion of hydrogen and deuterium." *Proc Phil Soc* **71**:161–172.

Association of American Universities. 2001. *Report on Individual and Institutional Financial Conflict of Interest.* www.aau.edu/research.

Bloom, M. 1988. "Article embroils JAMA in ethical controversy." *Science* **239**:135–136.

Bok, D. 2003a. *Universities in the Marketplace: The Commercialization of Higher Education.* Princeton, NJ: Princeton University Press.

Bok, D. 2003b, p. 185.

Broad, W. 1982. "Coping with fraud." *Science* **215**(4532):479.

Chubin, D. E. 1990. "Scientific Malpractice and Contemporary Politics of Knowledge." In: *Theories of Science in Society.* S. E. Cozzens and T. F. Gieryn (eds.). Bloomington, IN: Indiana University Press, pp. 144–163.

Cooper, P. K., T. Nouspikel, and S. G. Clarkson. 2005. "Retraction." *Science* **308**:1740

Dale, M. 2000. *Ethical Challenges and Practical Solutions for Managers in Research: Workshop Proceedings.* Sigma Xi Scientific Research Society, p. 57 (was available from Office of Research Integrity at: http://ori.dhhs.gov/html/publications/studies.asp. Information at this site, or at www.sigmaxi.org/programs/overview/history.shtml, is no longer available).

Davidson, K. 2005 (June 9). "Study charts science's ethical swamplands." *San Francisco Chronicle*, pp. A1 and A12.

De George, R. T. 1999. *Business Ethics.* Fifth Edition. Upper Saddle River, NJ: Prentice-Hall, Inc., pp. 250–259.

Engler, R. L., J. W. Covell, P. J. Friedman, et al. 1987. "Misrepresentation and responsibility in medical research." *N Engl J Med* **317**:1383–1389.

Fleddermann, C. B. 2003a. "Whistleblowing" (in Chapter 6, Section 6.5, "The Rights and Responsibilities of Engineers"). In: *Engineering Ethics.* Second Edition. Upper Saddle River, NJ: Pearson Prentice Hall, Pearson Education, Inc., p. 92.

Fleddermann, C. B. 2003b, p. 93.

Fleddermann, C. B. 2000c, p. 94.

Friedman, P. J. 1990. "Correcting the literature following fraudulent publication." *JAMA* **263**:1416–1419.

Garfield, E. and A. Welljams-Dorof. 1990. "The impact of fraudulent research on the scientific literature: the Stephen E. Breuning case." *JAMA* **263**:1424–1426.

Hackett, E. J. 1999. "A social control perspective on scientific misconduct." In: *Perspectives on the Scholarly Misconduct in the Sciences.* J. M. Braxton (ed). Columbus, OH: Ohio State University Press, pp. 103–105.

Hamilton, D. 1991 (March 8). "Verdict in sight in the 'Baltimore case.'" *Science* **251**(4998):1168–1172.

Harris, C. E. Jr., M. S. Pritchard, and M. J. Rabins. 2000. *Engineering Ethics, Cases and Concepts.* Second Edition. Belmont, CA: Wadsworth/Thomson Publishers, pp.198–199.

Holden, C. 1987. "NIMH finds a case of serious misconduct." *Science* **235**:1567–1577.

Institute of Medicine 1989. *The Responsible Conduct of Research in the Health Sciences.* Report of a Study by a Committee on the Responsible Conduct of Research (Pub. IOM–89–01). Washington, DC: The National Academies Press, p 18.

Ishida, K., K. Nagamine, T. Matsuzaki, and N. Kawamura. 2003. "Muon catalyzed fusion." *J Phys G: Nucl Part Phys* **29**:2043–2045.

Kennedy, D. 2000. "Reflections on a retraction." *Science* **289**:1137.

List, C. J. 1985. "Scientific fraud: social deviance or the failure of virtue?" *Science, Technology, and Human Values* **10**(4):27–36.

Locke, R. 1986 (Dec 4). "Another damned by publications." *Nature* **324**:401.

Los Angeles Times, 1985. San Diego Editions, September 17, September 21, and September 25, respectively.

Mallove, E. F. 1999. "MIT and cold fusion: A special report." *Infinite Energy* **4**:64–118.

Mallove, E. F. 2000. "The triumph of alchemy: Professor John Bockris and the transmutation crisis at Texas A&M." *Infinite Energy* **6**:32.

Marshall, E. 1986. "San Diego's tough stand on research fraud." *Science* **234**:534–535.

Marshall, E. 2000 (December 1). "Scientific misconduct. How prevalent is fraud? That's a million-dollar question." *Science* **290**:1662–1663.

Martinson, B. C., M. S. Anderson, and R. de Vries. 2005 (June 9). "Scientists behaving badly." *Nature* **435**:737–738.

McGrayne, S. B. 1993. *Nobel Prize Women in Science: Their Lives, Struggles, and Monumental Discoveries*. Second Edition. Washington, DC: Joseph Henry Press, p. 367.

Mello, M. and T. A. Brennan. 2003. "Due process in investigations of research misconduct." *N Engl J Med* **329**:1280–1286.

National Academy of Science. 1992. *Responsible Science: Ensuring the Integrity of the Research Process* (Volume I). Washington, DC: National Academies Press, p. 69. Available online at www.nap.edu/openbook/0309047315/html/69.html.

Pfeifer, M. P. and G. L. Snodgrass. 1990. "The continued use of retracted, invalid scientific literature." *JAMA* **263**:1420–1423.

Rhoades, L. J. 2004. "New Institutional Research Misconduct Activity: 1992–2001." Office of Research Integrity, Department of Health and Human Services. Available at http://ori.dhhs.gov, accessed November 23, 2005.

Steneck, J. 1994 (May–June). "Research universities and scientific misconduct—history, policies, and the future." *J Higher Educ* **65**(3):310–330.

Steneck, M. H. 2004. *Introduction to the Responsible Conduct of Research*. Washington, DC: Health and Human Services Department, Office of Research Integrity, 178 pp.

Summerlin, W. T., C. Broutbar, R. B. Foanes, et al. 1973 (March). "Acceptance of phenotypically differing cultured skin in man and mice." *Transplantation Proceedings* **5**(1):707–710.

Tarnow, E. 1999. "The authorship list in science: Junior physicists' perceptions of who appears and why." *Science and Engineering Ethics* **5**:73–88.

Taubes, G. 1990. "Cold fusion conundrum at Texas A&M." *Science* **248**:1299–1304.

U.S. Department of Health and Human Services. 1995. *Integrity and Misconduct in Research: Report of the Commission on Research Integrity*, Washington, DC, p. 22, www.faseb.org/opar/cri/html, accessed March 1, 2005.

Venter, J. C., M. D. Adams, E. W. Myers, et al. 2001. "The sequence of the human genome." *Science* **291**(5507):1304–1351.

Whitely, W., D. Rennie, and A. Hafner. 1994 (July 13). "The scientific community's response to evidence of fraudulent publication." *JAMA* **272**:171.

Zuckerman, H. 1968 (November). "Patterns of name ordering among authors of scientific papers: a study of social symbolism and its ambiguity." *American Journal of Sociology* **74**:276–291.

ADDITIONAL READING

Broad, W. and N. Wade. 1982. *Betrayers of the Truth*. New York: Simon & Schuster, pp. 153–157.

Department of Health and Human Services, Office of Research Integrity. 2005 (May). "Handling Misconduct." Available at ori.ddhs.gov/misconduct/index.shtml, accessed November 26, 2005.

Djerassi, C. 1989. *Cantor's Dilemma*. New York: Doubleday Publishing, 229 pp.

Glazer, M. 1983. "Ten whistleblowers and how they fared." *Hastings Center Report* **13**(6):33–41.

Gunsalus, C. K. 1998. "How to blow the whistle and still have a career afterwards." *Science and Engineering Ethics* **4**(1):51–64. (This is an excellent summary for whistle-blowers about what to expect and how to proceed. It is essential reading for anyone involved in an allegation of research misconduct.)

Hixson, J. 1976. *The Patchwork Mouse*. New York: Anchor Press, 228 pp.

Kohn, A. 1988. *False Prophets: Fraud and Error in Science and Medicine*. Cambridge, MA: Basil Blackwell, 248 pp.

LaFollette, M. G. 1996. *Stealing into Print*. Berkeley, CA: University of California Press, 293 pp.

Shannon, T. 1981 (August–September). "Whistle-blowing and countersuits: the President's Commission and fraudulent research." *IRB* **3**(7):6–7.

Woolf, P. K. 1986. "Pressure to publish and fraud in science." *Ann Int Med* **104**(2):254–256.

Zuckerman, H. A. 1988. "Introduction: Intellectual property and diverse rights of ownership in science." *Science, Technology, and Human Values* **13**(1&2):7–16.

APPENDIX

Appendix 2.1: Federal Policy on Research Misconduct

(www.ostp.gov/html/001207_3.html, accessed May 25, 2005)

The Office of Science and Technology Policy (OSTP) published a request for public comment on a proposed federal research misconduct policy in the October 14, 1999, Federal Register (pp. 55722–55725). OSTP received 237 sets of comments before the public comment period closed on December 13, 1999. After consideration of the public comments, the policy was revised and has now been finalized. This notice provides background information about the development of the policy, explains how the policy has been modified, and discusses plans for its implementation. Effective date: December 6, 2000.

FEDERAL POLICY ON RESEARCH MISCONDUCT[1]

I.   Research[2] Misconduct Defined

Research misconduct is defined as fabrication, falsification, or plagiarism in proposing, performing, or reviewing research, or in reporting research results.

- *Fabrication* is making up data or results and recording or reporting them.
- *Falsification* is manipulating research materials, equipment, or processes, or changing or omitting data or results such that the research is not accurately represented in the research record.[3]
- *Plagiarism* is the appropriation of another person's ideas, processes, results, or words without giving appropriate credit.
- Research misconduct does not include honest error or differences of opinion.

II.   Findings of Research Misconduct

A finding of research misconduct requires that:

- There be a significant departure from accepted practices of the relevant research community; and
- The misconduct be committed intentionally, or knowingly, or recklessly; and
- The allegation be proven by a preponderance of evidence.

III. Responsibilities of Federal Agencies and Research Institutions[4]

Agencies and research institutions are partners who share responsibility for the research process. Federal agencies have ultimate oversight authority for Federally funded research, but research

---

[1]No rights, privileges, benefits or obligations are created or abridged by issuance of this policy alone. The creation or abridgment of rights, privileges, benefits or obligations, if any, shall occur only upon implementation of this policy by the Federal agencies.

[2]Research, as used herein, includes all basic, applied, and demonstration research in all fields of science, engineering, and mathematics. This includes, but is not limited to, research in economics, education, linguistics, medicine, psychology, social sciences, statistics, and research involving human subjects or animals.

[3]The research record is the record of data or results that embody the facts resulting from scientific inquiry, and includes, but is not limited to, research proposals, laboratory records, both physical and electronic, progress reports, abstracts, theses, oral presentations, internal reports, and journal articles.

[4]The term "research institutions" is defined to include all organizations using federal funds for research, including, for example, colleges and universities, intramural federal research laboratories, federally funded research and development centers, national user facilities, industrial laboratories, or other research institutes. Independent researchers and small research institutions are covered by this policy.

institutions bear primary responsibility for prevention and detection of research misconduct and for the inquiry, investigation, and adjudication of research misconduct alleged to have occurred in association with their own institution.

- ○ Agency Policies and Procedures. Agency policies and procedures with regard to intramural as well as extramural programs must conform to the policy described in this document.

- ○ Agency Referral to Research Institution. In most cases, agencies will rely on the researcher's home institution to make the initial response to allegations of research misconduct. Agencies will usually refer allegations of research misconduct made directly to them to the appropriate research institution. However, at any time, the Federal agency may proceed with its own inquiry or investigation. Circumstances in which agencies may elect not to defer to the research institution include, but are not limited to, the following: the agency determines the institution is not prepared to handle the allegation in a manner consistent with this policy; agency involvement is needed to protect the public interest, including public health and safety; the allegation involves an entity of sufficiently small size (or an individual) that it cannot reasonably conduct the investigation itself.

- ○ Multiple Phases of the Response to an Allegation of Research Misconduct. A response to an allegation of research misconduct will usually consist of several phases, including: (1) an inquiry – the assessment of whether the allegation has substance and if an investigation is warranted; (2) an investigation – the formal development of a factual record, and the examination of that record leading to dismissal of the case or to a recommendation for a finding of research misconduct or other appropriate remedies; (3) adjudication – during which recommendations are reviewed and appropriate corrective actions determined.

- ○ Agency Follow-up to Institutional Action. After reviewing the record of the investigation, the institution's recommendations to the institution's adjudicating official, and any corrective actions taken by the research institution, the agency will take additional oversight or investigative steps if necessary. Upon completion of its review, the agency will take appropriate administrative action in accordance with applicable laws, regulations, or policies. When the agency has made a final determination, it will notify the subject of the allegation of the outcome and inform the institution regarding its disposition of the case. The agency finding of research misconduct and agency administrative actions can be appealed pursuant to the agency's applicable procedures.

- ○ Separation of Phases. Adjudication is separated organizationally from inquiry and investigation. Likewise, appeals are separated organizationally from inquiry and investigation.

- ○ Institutional Notification of the Agency. Research institutions will notify the funding agency (or agencies in some cases) of an allegation of research misconduct if (1) the allegation involves Federally funded research (or an application for Federal funding) and meets the Federal definition of research misconduct given above, and (2) if the institution's inquiry into the allegation determines there is sufficient evidence to proceed to an investigation. When an investigation is complete, the research institution will forward to the agency a copy of the evidentiary record, the investigative report, recommendations made to the institution's adjudicating official, and the subject's written response to the recommendations (if any). When a research institution completes the adjudication phase, it will forward the adjudicating official's decision and notify the agency of any corrective actions taken or planned.

- ○ Other Reasons to Notify the Agency. At any time during an inquiry or investigation, the institution will immediately notify the Federal agency if public health or safety is at risk; if agency resources or interests are threatened; if research activities should be suspended; if there is reasonable indication of possible violations of civil or criminal law; if Federal action is required to protect the interests of those involved in the investigation; if the research institution believes the inquiry or investigation may be made public prematurely so that appropriate steps can be taken to safeguard evidence and protect the rights of those involved; or if the research community or public should be informed.

○ <u>When More Than One Agency Is Involved</u>. A lead agency should be designated to coordinate responses to allegations of research misconduct when more than one agency is involved in funding activities relevant to the allegation. Each agency may implement administrative actions in accordance with applicable laws, regulations, policies, or contractual procedures.

IV. Guidelines for Fair and Timely Procedures

The following guidelines are provided to assist agencies and research institutions in developing fair and timely procedures for responding to allegations of research misconduct. They are designed to provide safeguards for subjects of allegations as well as for informants. Fair and timely procedures include the following:

○ <u>Safeguards for Informants</u>. Safeguards for informants give individuals the confidence that they can bring allegations of research misconduct made in good faith to the attention of appropriate authorities or serve as informants to an inquiry or an investigation without suffering retribution. Safeguards include protection against retaliation for informants who make good faith allegations, fair and objective procedures for the examination and resolution of allegations of research misconduct, and diligence in protecting the positions and reputations of those persons who make allegations of research misconduct in good faith.

○ <u>Safeguards for Subjects of Allegations</u>. Safeguards for subjects give individuals the confidence that their rights are protected and that the mere filing of an allegation of research misconduct against them will not bring their research to a halt or be the basis for other disciplinary or adverse action absent other compelling reasons. Other safeguards include timely written notification of subjects regarding substantive allegations made against them; a description of all such allegations; reasonable access to the data and other evidence supporting the allegations; and the opportunity to respond to allegations, the supporting evidence and the proposed findings of research misconduct (if any).

○ <u>Objectivity and Expertise</u>. The selection of individuals to review allegations and conduct investigations that have appropriate expertise and have no unresolved conflicts of interests help to ensure fairness throughout all phases of the process.

○ <u>Timeliness</u>. Reasonable time limits for the conduct of the inquiry, investigation, adjudication, and appeal phases (if any), with allowances for extensions where appropriate, provide confidence that the process will be well managed.

○ <u>Confidentiality During the Inquiry, Investigation, and Decision-Making Processes</u>. To the extent possible consistent with a fair and thorough investigation and as allowed by law, knowledge about the identity of subjects and informants is limited to those who need to know. Records maintained by the agency during the course of responding to an allegation of research misconduct are exempt from disclosure under the Freedom of Information Act to the extent permitted by law and regulation.

V. Agency Administrative Actions

○ <u>Seriousness of the Misconduct</u>. In deciding what administrative actions are appropriate, the agency should consider the seriousness of the misconduct, including, but not limited to, the degree to which the misconduct was knowing, intentional, or reckless; was an isolated event or part of a pattern; or had significant impact on the research record, research subjects, other researchers, institutions, or the public welfare.

○ <u>Possible Administrative Actions</u>. Administrative actions available include, but are not limited to, appropriate steps to correct the research record; letters of reprimand; the imposition of special certification or assurance requirements to ensure compliance with applicable regulations or terms of an award; suspension or termination of an active award; or suspension and debarment in accordance with applicable government-wide rules on suspension and debarment. In the event of suspension or debarment, the information is made publicly available through the List of Parties Excluded from Federal Procurement and Nonprocurement Programs maintained by the U.S. General Services Administration. With respect to administrative actions imposed upon government employees, the agencies must comply with all relevant federal personnel policies and laws.

○ In Case of Criminal or Civil Fraud Violations. If the funding agency believes that criminal or civil fraud violations may have occurred, the agency shall promptly refer the matter to the Department of Justice, the Inspector General for the agency, or other appropriate investigative body.

VI. Roles of Other Organizations

This Federal policy does not limit the authority of research institutions, or other entities, to promulgate additional research misconduct policies or guidelines or more specific ethical guidance.

PROBLEM SET

2.1. A professor, who is responsible for a $40 million grant for research over a period of the next five years, is the object of a graduate student's complaint of fraudulent activity regarding the insertion of experimental results in a manuscript. As an advisor in the university's office of the provost, you are assigned the case for further investigation and appropriate action. How would you proceed? What are your resources and what should be the guiding policies? Do not dwell on the prerequisites of the whistleblower because this act whether done correctly or not is already complete. (200 words in outline or bullet form, please.)

2.2. How would one approach the following problem: A student reports to a faculty member the fact that two problems on a four-problem assignment at Stanford were the same as the text on the Internet from Purdue. What would you advise the faculty member to do as the responsible mentor for the professor whose course generated the assignment at Stanford?

2.3. The following data are generated in an observation of the number of individuals per million with adult leukemia vs. time over the last 90 years. Your mission is to show there is no increase during the same period that electromagnetic noise increased a factor of 100 in most frequency bands.

   1910, 2; 1920, 2; 1930 4; 1940, 4;1950, 3; 1960, 3;
   1970, 5; 1980, 4; 1990,10; 2000, 5.

You plot these data and note a trend that seems to show an increase but on reflection consider the 1990 statistic too much an outlier. Therefore, you decide to write the conclusion that since there is no increase in leukemia in 10 decades of increase in electromagnetic field intensity, there is no conclusive evidence that electromagnetic fields cause leukemia. A graduate student in public health sees your report and goes to the source of data, only to discover you left out the data point from 1990. That student reports misconduct, and you are brought to disciplinary action.

A. How could this situation have been prevented and how would your advocates argue that you did no wrong to justify the disciplinary action?

B. How should the graduate student whistleblower be advised regarding her actions?

2.4. An experiment was done to show the possible neurological effects on experimental animals fed beef originating from England and Canada. Based on the differences in the average incidence of animals falling from a balance beam during the experiment whose numbers are shown below, a scientist reports Canadian beef is safer than English beef. You think this Canadian scientist is a fraud. Given the data below, calculate the significance, if any, of the difference between the two populations. The numbers represent the number of balance beam falls in 10 trials for 20 individual animals in each category.

Population A received beef from England over the last 10 years:

1, 0, 2, 3, 2, 4, 1, 0, 5, 8, 3, 2, 0, 1, 5, 4, 0, 0, 1, 5

Population B received beef from Canada over the last 10 years:

0, 0, 4, 3, 5, 3, 3, 0, 2, 0, 4, 6, 2, 0, 0, 1, 0, 0, 1, 0

What would you do with the conclusion from this calculation:

If you are in public health?

If you are the owner of a hamburger factory in Great Britain?

If you are a student in an upper division bioengineering course?

(Hint: Consult texts on statistics and the use of the Student's *t* distribution.)

2.5. An experiment in the genetics of tomatoes was performed using the university's botanical garden, which is supervised by Ms. A. The originator of the idea for the experiment was Professor B, who died before the experiment started. The experimenters were two graduate students, Ms. C and Mr. D, as well as a summer undergraduate student Mr. E. Assistant Professor F supervised the students but did not actually do the experiments. Administrative Assistant G obtained the funding from the old chair endowment of Professor B, helped write the manuscript, and dealt with the requested revisions from the journal editor; she has a degree in law and is not a scientist.

Who should be listed as an author on the manuscript, considering the fact that Mr. D. did the major work which led to the discovery described in the manuscript, but the general idea for starting the experiment was that of a dead scientist?

2.6. Referring to 2.5 above, suppose that Assistant Professor F is the lead author, and all of the others, except Assistant G and Lab Assistant E, are on the paper, with Professor B as last author. Assistant G and Lab Assistant E go to the department chair and claim injustice. Why is G claiming injustice? How should this situation be handled?

2.7. A professor in upper division engineering presents a problem set for a course in magnetic resonance imaging techniques. The problem set contains a problem that the professor found on the Internet and copied verbatim as it was a good problem very appropriate to the course. No acknowlegment was given to the originator of the problem. A student who is doing poorly finds the problem on the Internet and reports to the department chair that there has been plagiarism. How should the department chair approach this situation? (This is not strictly an authorship problem, but it does present an aspect of plagiarism seldom encountered in academia. Make a clear argument that it is not plagiarism—the answer is simple when an analysis is made.)

2.8. The properties of a particular biologic cell line might help a researcher in University A determine the mechanism of aging. The cell line is a specialty of a professor at University B. The University B professor agrees to give University A researcher the cells if A agrees to include B and B's technician in any papers written by A. How should A deal with this problem before writing the paper?

How should A deal with the laboratory leader C, whose policy it is to have her name on all manuscripts using the resources of her laboratory?

2.9. Scientist A wishes to demonstrate whether there is abnormal isotope uptake in a part of the brain of a series of imaging studies done for other purposes, and asks a technician, whom he pays a regular salary, to investigate this possibility. The technician, during normal working hours, analyzes over 100 patient studies and carefully records the information. A scientific paper is written by A as the sole author. A acknowledges the technician in a footnote, which also acknowledges the director of the laboratory and the doctors who referred the patients for the other study to which the current study became an afterthought.

Give arguments why the referring doctors, the lab director, and the technician should or should not be coauthors.

2.10. Refer to Case 2.13 in the text. As the president of the university, write the letter to the professor in 70 words or less. Give a compelling reason(s) for the action taken.

2.11. This problem involves an ethical dilemma faced by the *Journal of the American Medical Association* (*JAMA*) editor in Chicago, Illinois. It is a true story (Bloom 1988). The *JAMA* ran a series of editorials and discussions on the ethics of "pulling the plug" on the respirators or life support systems for patients who were in coma for long periods of time or had some evidence of not being revivable. (Later in the course you will encounter the dilemmas related to medical and moral ethics here, but this case is about the responsibilities of authors and editors.) One physician who was involved in long-term care of patients wrote a considered opinion about her philosophy and her experiences in physician-assisted suicide, and sent the paper to the *JAMA* to be published under conditions of anonymity. After the *JAMA* article appeared and the newspaper the *Chicago Tribune* cited the article, the Chicago District Attorney approached the *JAMA* editor and demanded the identity of the author on the grounds that the author had committed a crime or a number of crimes, and these were felonies. He argued that the editor was protecting a criminal, who might be committing more murders. The editor was advised by legal counsel that he was protected by the First Amendment and the Illinois Reporter's Privilege Act, arguing that the author had requested anonymity as a condition of publication. But the editor, a physician himself, had some long-standing deep feelings against pulling the plug and physician-assisted suicide.

Blue Team: Make arguments for revealing the author's identity.

Gold Team: Make arguments for not revealing the author's identity.

Use the Four A's in an argument that leads to the Blue or the Gold decision. Make an outline or use a bullet approach to keep within 80 carefully thought-through words. You will need these words and other notes you make to participate in a debate on this topic of "Authorship and Confidentiality."

2.12. This is not a design problem; from the list below (A–H) you are to choose the order of authors of a scientific paper based on the description of the participation of each.

A—Manager of the laboratory who handled personnel and budgets as well as travel and special tasks for the group. Went to all the meetings of the team and edited the manuscript.

B—Laboratory leader who set up the laboratory and hired the staff, including the   scientists. She encouraged the experiments, criticized the interpretation of some of the results and was principal investigator on the NIH grant that supported the investigators. She did not write the paper.

C—Undergraduate student who came up with the idea for the particular experiment whose results are the subject of the paper in question. He had the idea but could not participate because he had to attend a class on Wednesdays.

D—Laboratory technician who performed all the immunoassays and recorded the data.

E—Laboratory technician who did all the electron microscopy and thin sectioning of tissues.

F—Grant writer who did a PubMed Internet search of the literature and wrote the paper.

G—Graduate student who coordinated the activities of D and E and argued frequently with F about the paper content. He prepared the letter of transmittal to the journal for B to sign.

H—German scientist who sent for use in this experiment a sample of a rare and otherwise unavailable antibody. He insisted that he and his assistant be authors when the experiment was reported but did not need to see the manuscript.

Give your rationale for the authorship in 100 words or less. This problem requires some thought.

2.13. Using Google, find two systems for tracking whether a manuscript submitted by a student contains plagiarism.

2.14. Journalism ethics was touched upon in Case 2.11 in which the actions of journalist Gary Taubes were claimed to be unethical due to omissions and erroneous commissions. Can you verify that this journalist has a pattern of unethical conduct relative to the professional code of ethics? Using the Internet or any other source, are there other examples of his unethical conduct? Write an evaluation of Gary Taubes's conduct and include in the 100 words (bulleted phrases are allowed) a section in his defense.

2.15. Reproductive cloning of mammalian animals (sheep, cows, and rodents) has been successful since 1996, but the reproductive cloning of humans has been banned. Attempts at therapeutic cloning to establish human embryonic stem cell lines for individual patients (Chapter 10) was not possible as of 2006. However, in 2004 and again in 2005, a South Korean team led by S. Hwang reported success. Those data have been judged to be fraudulent by Hwang's peers and an investigative panel at his university. Give a concise chronology of this case (40 words using the Internet) and then analyze the possible reasons this fraud occurred (80 words).

3

Information Technology Ethics

INTRODUCTION

We define the information enterprises as those new technologies that have revolutionized the way in which information is developed, accessed, made available, stored, and transmitted. Information technology (IT) is a fast-growing market, with spending in 2001 of over $800 billion (www.itaa.org/about/index.cfm, accessed Dec. 26, 2004). Although the advantages of the Information Technology Revolution are countless, as with all emerging technologies, there are ethical issues. The purpose of this chapter is to describe these issues and consider methods to cope with the invasion of the rights of individuals, the loss of autonomy, and malfeasance resulting from trespassing, theft, spying, sabotage, and other illegal activities. The stakeholders and methodologies include:

Stakeholders	Enabling Technologies
Authors	Computers and software
Inventors of IT	Communication devices
Consumers	Surveillance cameras
Public at large	Internet and World Wide Web
Business	Telecommunications systems
Government	Satellites

The Agricultural and Industrial Revolutions brought great advances to our lives. Just as then, the Information Technology Revolution is impacting human lives in profound and complex ways, but at a much faster pace. Just as then, there are and will continue to be both positive and negative effects as this area continues to evolve.

Case 3.1 Invasion of Privacy

Customer: "Hi, Pizza Hut? I'd like to order."

Operator: "May I have your NIDN first, sir?"

Customer: "My National ID Number, yeah, hold on, it's 6102049998-45-54610."

Operator: "Thank you, Mr. Sheehan. I see you live at 1742 Meadowland Drive, and the phone number's 494-2366. Your office number over at Lincoln insurance is 745-2302, and your cell is 266-2566. Which number are you calling from, sir?"

Customer: "Huh? I'm at home. Where d'ya get all this information?"

Operator: "We're wired into the system, sir."

Customer: "Oh. I'd like to order a couple of your All-Meat Special pizzas."

Operator: "I don't think that's a good idea, sir."

Customer: "Whaddya mean?"

Operator: "Sir, your medical records indicate that you've got high blood pressure and extremely high cholesterol. Your National Health Care provider won't allow such an unhealthy choice."

Customer: "Damn. What do you recommend, then?"

Operator: "You might try our low-fat Soybean Yogurt pizza. I'm sure you'll like it."

Customer: "What makes you think I'd like something like that?"

Operator: "Well, you checked out 'Gourmet Soybean Recipes' from your local library last week, sir. That's why I made the suggestion."

Customer: "All right, all right. Give me two family-sized ones, then."

Operator: "That should be plenty for you, your wife, and your four kids, sir. That comes to $49.99."

Customer: "Lemme give you my credit card number."

Operator: "Sorry sir, you'll have to pay in cash. Your credit card is over its limit."

Customer: "I'll run over to the ATM for some cash before your driver gets here."

Operator: "That won't work either, sir. Your checking account's overdrawn."

Customer: "Just send the pizzas. I'll have the cash ready. How long will it take?"

Operator: "About 45 minutes, sir. You might want to pick 'em up while you're out getting the cash, but carrying pizzas on a motorcycle can be a little awkward."

Customer: "How the hell do you know I'm riding a bike?"

Operator: "It says here you're in arrears on your car payments, so your car got repo'ed. But your Harley's paid up, so I just assumed that you'd be using it."

Customer: "@#%/$@&?#!"

Operator: "I'd advise watching your language, sir. You've already got a July 2006 conviction for cussing out a cop."

Customer: (Speechless)

Operator: "Will there be anything else, sir?"

Customer: "No, nothing. Oh, yeah, don't forget the two free liters of Coke your ad says I get with the pizzas."

Operator: "I'm sorry sir, but our ad's exclusionary clause prevents us from offering free soda to diabetics."

~ Anonymous e-mail

The positive and negative aspects of the computer age, along with the fight against terrorism, have forced institutions and governments to recognize new responsibilities for the protection of the rights of citizens. As a result, a new ethics called information technology ethics has emerged as we grapple with the new technology and how it fits into our lives.

Those involved in the development of information technology must be responsible for thinking through all aspects of its use. In particular, it will be important to analyze not only the benefits of increased ease, speed, and breadth of capabilities, but also what possible negative consequences might evolve from its use in the future. Could we have anticipated such problems as hacking, hoaxes, viruses, invasion of privacy, cookie spies, copyright violations, plagiarism, identity theft, and other crimes? We might respond NO, except perhaps with the exception of plagiarism by stealing words or ideas that are protected by copyright laws. But how can those responsible for developing this technology better anticipate future problems? It is not easy to define just how this should be done. Later in the chapter we develop a case in order to debate and exercise mentally how problematic issues in this area might be thought through. This chapter will also examine some areas in which computers have brought the greatest changes to our lives, and those in which this technology is challenging our values. We will examine how solutions to these challenges might evolve. We will also present case examples of IT ethical dilemmas and use the Four A's strategy to evaluate solutions to these dilemmas.

Some of these issues to keep in mind as we proceed through the chapter include:

- Should the Internet be regulated?
- To what extent should the government be given access to private information as a means of curbing terrorism?
- Is it ethical to monitor the workplace?
- How can plagiarism be thwarted?
- Is spam considered trespassing in the same way as telemarketing?

We will return to discuss these and other issues after presenting some of the areas of the information technologies that are both beneficial and problematic.

There was a span of only 50 years from the first computer development for weather forecasting and code breaking during the Second World War to the invention of the Internet and the development of the World Wide Web, which allowed global communications. The history of this development is summarized in Table 3.1 (www.comoputerhistory.org/exhibits/internet_history, accessed February 17, 2005).

Table 3.1 Landmarks in Computer Technology Development

1940s	First computer patched together by graduate students at the University of California at Berkeley and Stanford. Uses in World War II included communications code breaking and weapon design.
1950s	The UNIVAC I made by Remington Rand (later merged with Siemens) was the first commercial computer to attract widespread public attention. Cost: more than $1M each, plus $185,000 for the high-speed printer. Computer-assisted manufacturing was initiated. Computer databases and computations were facilitated.
1960s	Online transaction processing began with IBM's SABRE airline reservation system using telephone lines to link 2,000 terminals in 65 cities to a pair of IBM 7090 computers, delivering data on any flight in less than three seconds.
	The modem allowed computers to connect to the telephone network by means of standard telephone handsets.

continued

Table 3.1 (continued)

1970s	First ATM installed. The first computer-to-computer network, known as ARPANET, expanded when the Department of Defense established links between the University of California, Santa Barbara; the University of California, Los Angeles; SRI International; and the University of Utah. The first advertisement for a microprocessor, the Intel 4004 with 2250 transistors, could perform up to 90,000 operations per second in 4-bit words. The 8-inch floppy diskette was invented at IBM.
1980s	The National Science Foundation formed the NSFNET, linking five supercomputer centers at Princeton University, Pittsburgh, the University of California at San Diego, the University of Illinois at Urbana–Champaign, and Cornell University. Soon, several regional networks developed; eventually, the government reassigned pieces of the ARPANET to the NSFNET. The NSF allowed commercial use of the Internet for the first time in 1991, and in 1995, it decommissioned the backbone, leaving the Internet a self-supporting industry.
1990s	First computer worm launched by a bored teenager.
	World Wide Web formed when Tim Berners-Lee developed HyperText Markup Language and other specifications and links to servers that allowed a user using a browser to have access to a site. The first World Wide Web server and browser became available to the general public in 1991.

Other technologies that underpin the effectiveness of computer technologies include the manufacture of electronic components, and the development of satellite communications, wireless telecommunications, and fiber-optic information transmission.

THE BENEFITS AND PROBLEMS GENERATED BY THE INFORMATION TECHNOLOGIES

Benefits

Computers have a multitude of uses. They run our cars and appliances more efficiently, allow almost instantaneous access to a world of information, allow us to pay bills, to do research, to study, to vote, to keep in touch with friends and family. Anyone can become an instant publisher on the World Wide Web and make his or her views known across the globe to anyone wanting to read them without having to go through the cost or editing involved in a publication. Customers are using this technology to shop on the Internet in increasing numbers for needed items, supplies, and gifts. Businesses rely on it to record sales, track inventory, communicate with customers and suppliers, store customer information, and perform robotic manufacturing procedures and other tasks previously given to salaried employees. Physicians use it for telemedicine to visualize the patient, perform physical examinations, access laboratory work, and even perform remotely controlled medical and surgical procedures. It is also used in combat situations to pinpoint troops on the ground and monitor their status.

This technology is also being used to record and store detailed information about virtually every aspect of our daily lives, including our purchases, the Internet sites we access, the books we borrow from the library, our telephone calls, and our movements in public places.

Entities such as the government, universities, financial institutions, businesses, health care institutions, organ transplant registries, communications, airlines and the entire travel industry, ground transportation, and shipping all rely on computerized information to organize and access information in a timely way. Even our entertainment, everything from games to music to movies to Internet chat rooms, is either generated by computers or available on line. Since the 1990s, computer technologies have been adapted for use in the composition of music, movies, videos, games, and scientific or entertainment exercises in virtual reality. The technology of digitization has revolutionized the way pictures, sound, and language are published and communicated.

Problems Associated with IT

When computers are "down," there can be a widespread ripple effect for the individual, the local economy, the country, and even the world (as, for example, when worldwide air travel was shut down when computers at a major airport crashed). New innovations surrounding their use are emerging so fast there are gaps in our ability to keep up with them. The computers and software we used only a few years ago are already so obsolete that we are no longer able to retrieve the information from them with our newer computers and software. While computers have increased our ability to more rapidly find, acquire, store, and transmit information electronically, the same technologies have been responsible for major problems such that local, state, and federal governments have been forced to intervene. The same is true in other countries around the world, and attempts at harmonization to bring international customs, practices, and laws into better alignment with one another have not been altogether successful.

The major problems are security requirements to protect individual autonomy and privacy on the Internet. There are regulatory issues regarding taxation, data retention, outsourcing, freedom of speech, to what extent certain topics should be banned (e.g., child and adult pornography), and how to punish cybercrimes. There has been an increasing invasion of personal privacy made possible by cookies, hacking, identity theft, cell phone photography, and the access to personal information granted by the U.S. Patriot Act. Individuals who misuse computers have found devious ways to spy, steal, sell, and commit fraud and other crimes. The theft of intellectual property, made easy by virtue of the very way in which computers function, is a growing problem that has yet to be fully addressed. We find ourselves in a reflex mode of trying to fix problems such as spam (unsolicited advertising by e-mail) that jams our in boxes. The situation is so problematic that the U.S. Congress and other lawmakers have had to consider ways to address questions such as "Who should take action to protect the public and what action should this be?" Some of these problems were not anticipated even as recently as a few years ago.

Invasion of Privacy Due to Faulty Engineering of Computer Systems and Software Programs

One reason that individual privacy has been compromised is flawed engineering of computer systems and software programs. These flaws and weaknesses can have serious consequences for the user because they allow break-ins into personal or other computers. There is an increasing demand that companies releasing programs to the public take the time to ensure that the programs are bug-free and secure, because the time spent in trying to recover from errors can be significant and the loss of privacy and theft of information can be devastating. An example is the surreptitious use of a program known as cookies. By means of cookies attached to one's computer at the time of a visit to a Web site, an individual's interests and preferences can be recorded and stored in a data base so that when the site is revisited, a list of options that might interest the visitor based on previous information can be offered. E-mail ads from Web sites can also be deployed to the individual's e-mail address that are specific to interests indicated while perusing the Web site.

Further, it is becoming increasingly difficult to maintain the security of personal information stored on computers that are accessed through Web sites. Sophisticated hackers have successfully accessed personal information on thousands of individuals by breaking into databanks at schools, universities, corporations, banks and other financial institutions, military bases, and hospitals and doctors' offices. Recent examples involve the breaching of college and university databanks that store personal data on college applicants, students, alumni, and staff at a number of large colleges and universities. Confidential data that were accessed included names, addresses, e-mail addresses, phone numbers, birth dates, Social Security numbers (SSNs), test scores, financial aid information, and credit card numbers. Human error has contributed to other instances where personal data have been inadvertently placed on an insecure server, or inadvertently posted on the Internet. In other instances, class lists of grades using names and SSNs have been posted publicly by universities. As a result, the individual might be a victim of identity theft or credit card fraud or both.

Case 3.2 Exposure of 1.4 Million Names and Social Security Numbers

In 2004, the California Department of Social Services (DSS) collaborated with an investigator from the University of Connecticut in a research project involving the activities of in-home care workers and care recipients. The DSS provided the names and Social Security numbers (SSNs) for over 1,000,000 individuals. The Connecticut investigator came to University of California as a visiting faculty member, and on arrival, asked to have her computer connected into the university's computer network. The university technicians initially declined because the required safeguards and proper firewall for her computer were not installed. The Connecticut investigator took some measures to provide the required protection, and her computer was linked to the university system and the Internet. A paper form was used by the technician to document the content of the computer. Among the questions asked was whether the computer contained personal or sensitive information. The technician, who has since left the university, claimed at the time that no sensitive information was contained in the files.

Sometime after this computer was connected to the Internet, there was a hacking incident and it was found that there were Social Security numbers and associated names available to this hacker. This incident led to a review by the administrations of two universities, the Institutional Review Board (IRB) of these two universities, the National Institutes of Health, the California Department of Social Services, and California legislators who were concerned about preventing identity theft. The major stakeholders at risk were the many individuals whose privacy was invaded. The ethical duty of the state of California was to inform all of the individuals that they might be at risk, and the duty of the IRBs was to stop the research and take measures to protect subjects of social and behavioral science research from exposure of private information in the future. The method for this protection is in Appendix 3.1.

This case can lead to more severe consequences, including long-term negative consequences to public policy of benefit to society. There is currently a debate among legislators regarding whether public agencies should ban sharing information with researchers unless it is required by law, a move that would make a wealth of important public health data unavailable to researchers in the future. The alternative is to create regulations whereby Social Security numbers, names, or other personal data could not be entered on computer databases unless confidentiality is controlled by strict methods to prevent future incidents.

On the one hand, we want to protect the privacy of individuals; on the other hand, we do not want to promote regulations that restrict public health research. The utilitarian would argue that the controlled distribution of SSNs and names should be allowed in the interests of scientific research, and that the benefits of research outweigh the risks of identity theft and other negative consequences if proper safeguards are put in place. One would think that legislation should be directed to implementing safeguards rather than banning access to this important public health information.

Efforts to ensure the cessation of these errors must include the reevaluation of software to make it even more resistant to hacking or illegal entry. Other safeguards include limiting the use of SSNs, providing more training and ongoing retraining for all personnel who handle these data, and ongoing reevaluation of Web sites to ascertain the presence of weak links in the system. To prevent a replication of the situation in Case 3.2, the following options were considered by the university in 2005:

1. Never connect a computer having personal or sensitive information to the Internet.
2. Establish a system of encryption and deencryption for all originators and recipients of sensitive data.
3. Deidentify all sensitive data stored on electronic files and store key codes in a file that cannot be accessed from the Internet.
4. Prevent the release of any sensitive data by the state of California to any non-official investigator.

Case 3.3 Colleges Leaking Confidential Data

Sometime during the summer of 2002, computers at Princeton University were used to hack into the computerized admissions system to read confidential student files at Yale University. The Web site went live in April so that undergraduate applicants could find out whether they had been admitted to Yale by using their federal identification numbers and birth dates. Yale admissions discovered that there were several unauthorized logins to Yale's files on its admissions Web site, and these were found to have originated from computers in Princeton's admissions offices. Princeton has since suspended its associate dean, who was also director of admissions. He admitted to accessing Yale's files using the information he took from the files of students who had applied to both universities. He claimed he was merely checking the site's security.

This case is clearly a violation of the autonomy of the students' rights to privacy. Since then, there have been reports of additional failures of computer security at colleges and universities across the country. Some of the schools include the University of California, San Diego State University, Georgia Institute of Technology, the University of Texas at Austin, and New York University (Shevitz 2004). In addition to attacks by hackers, another part of the problem is blamed on human error and the problems are expected to continue. The equipment is changing, training for personnel is inadequate (partly due to changing equipment upgrades), hackers are becoming more sophisticated, and not all Web sites have been tested thoroughly to ascertain safety from unauthorized entry. Since identity theft increased almost sixfold in the three years from 2000 to 2003 (Shevitz, 2004), and can cause years of monetary and personal anguish for those defrauded before the perpetrator is caught, these breaches are especially unsettling. The U.S. Congress made identity theft with intent to commit a felony a federal crime in 1998 (Baase 2003a).

Risks to Privacy by Outsourcing Private Information to Other Countries

Two situations that occurred in October 2003 and April 2004, described below, are examples of how the privacy of medical records outsourced for transcription were jeopardized from afar by disgruntled workers in another country.

Case 3.4 Risks to Privacy by Outsourcing Electronic Medical Records

The records of a patient at a university medical center in northern California were sent to Pakistan for medical transcription. A Pakistani transcriber demanded money she believed she was owed from previous work and accompanied her demand with the patient's medical records and a threat to post them on the Internet. She was subsequently paid by one of the subcontractors handling the hospital records and withdrew her threat, but it shows the vulnerability of the confidentiality of outsourced documents.

Another situation in April 2004 involved an Ohio company that outsources the transcription of medical files from dozens of U.S. hospitals to India. There was an extortion attempt from Indian workers who threatened to reveal confidential medical information unless they received a cash payoff. Although the workers had broken into a manager's office to obtain patient records, they found only nonconfidential training documents.

What is alarming about these examples is the intent to obtain monetary gain by leaking sensitive confidential information by someone on the other side of the world in another country where the cultural ethics may be different than in the United States.

Collection, Storage, Selling, and Unauthorized Use of Personal Information

People visiting a Web site may have information about themselves monitored and encoded. It is now possible to monitor an individual's pattern of information gathering, e-commerce business practices, Web sites visited, and other private information. Also monitored is the previous Web site that provided the link to the currently visited site, and the parts of the Web site you visited. Information regarding your preferences, your purchases, your e-mail address, and other private information becomes part of a folio on each individual. These data could also be sold or shared with others, which in turn may result in even more unsolicited advertisements by e-mail (spam). Because of the anonymity of the providers on Web sites, their legitimacy or professionalism cannot be verified, and they may be fraudulent. E-mails designed to look exactly like those from a reputable company requesting confirmation and sensitive information may be scams.

Further, personal medical records previously held confidential in a physician's office are now increasingly stored on a computer and are often forwarded electronically, either to a transcriber, consultant, or a new health care provider. As such, they are open to hacking and invasion of privacy. The Health Information and Privacy Protection Act (HIPPA, discussed further in Chapter 7, "Medical Ethics") will not protect the patient who consents to have the data transmitted. The methods for treating electronic data are seldom secure: encryption is seldom used, there are as yet no standards for the protection of routine clinical data, and research data are generally kept on computers whose access is limited only through passwords.

Loss of Privacy by Monitoring

There has been an increasing use of surveillance methodologies to track the movement of individuals through surveillance cameras in public places (parking lots, malls, airports) and in the workplace. Home surveillance cameras can monitor the outside perimeter of the home itself, as well as the neighborhood, neighbors, would-be visitors, personnel working in the home, and children, pets, or the elderly who are home alone. Global position system (GPS) monitoring allows guidance when lost, but also allows the secret tracking of individuals by anyone attaching a simple device. The technology can locate a vehicle with an accuracy of a few meters, and can detect its speed—all of which can be done anonymously and from afar by use of a satellite. Phones can be tapped, and small cameras, microphones, and telephones hidden on a person or in the environment allow surreptitious photography and video and/or audio monitoring of a person's actions and words. Bridge toll plazas, stoplights, and other "smart" road devices can record the license plate numbers of cars. The use of surveillance to confirm behaviors that are unscrupulous (for example, a cheating spouse, abuse by a nanny, a teenager being led astray by unsavory acquaintances) raises other ethical issues.

Further, the U.S. Patriot Act, an anti-terror law adopted in 2001 following the 9/11 attacks, gave the federal government the power to obtain private information on citizens to assist in national security investigations. Among other things, the act legalized secret surveillance and the searching of citizens, their homes, and cars and gave the government the power to subpoena from Internet service companies or other providers personal data on their subscribers (among them name, address, credit card data, and details of their Internet use) without their knowledge. A suit brought by the American Civil Liberties Union argued to no avail that the surveillance actions ". . . broadly violated the Constitution by giving federal authorities unchecked powers to obtain private information" and that the subpoena power ". . . violated the Fourth Amendment because it did not allow for review by a court" (Preston 2004).

There are fears that the increasing use of surveillance, even for purposes of law enforcement, the social good, or one's own good, might prove to be detrimental to democracy itself. As pointed out by Deborah Johnson, Jeffrey Reiman, and others, people who know they are being watched behave differently than they might otherwise: they become more inhibited, are less likely to act with abandon, are more likely to curtail any new or overt actions that might draw interest or attention, and feel less freedom to be one's self. Over time, these inhibitions may become more ingrained to the point that they feel they must act conventionally all the time and are unwilling or unable to exercise their autonomy. When few will take risks or step forward with new ideas, the heart of democracy and freedom is affected (Johnson 2001, Reiman 1995).

Many feel there is an increasing need for such monitoring and its use. Who among us would not hesitate to use this technology if one suspected abuse of a child or loved one? But there are now means to monitor which are invasive (heat-monitoring devices can detect the exact whereabouts of a person in his own house), and as smaller and more efficient devices for monitoring become available, there is an increasing sense throughout society of being watched. Monitoring in the public realm is now widespread. Parking lots, airports, train stations, and shopping malls are fully covered by surveillance devices, some of which can be remotely controlled to follow the movements of selected individuals. The expense of these activities in the public realm is enormous. But rather than catching the behavior, it is more important to come up with ways to prevent it. We shall discuss these issues later in the chapter.

Many individuals are employed in a workplace where monitoring of the work site itself and employee performance are either in place or deemed necessary. As summarized from Baase (2003b), with new technology it is now possible to track the exact location of individual employees and company vehicles, eavesdrop on telephone calls and other conversations, monitor employee e-mails, visualize computer screen activities, track keystrokes, follow Internet activities, count the number of orders processed, and tally the amounts rung up on cash registers. Surveillance cameras are widely used to prevent theft from shoplifting by both employees and customers. In 2000, shoplifting accounted for the loss of $12.85 billion from employees and $10.15 billion from customers (Baase 2003c). Monitoring activities on the part of employers have become increasingly secretive, and have become increasingly utilized as new technology has become available.

One of the main arguments in favor of monitoring has been to safeguard the company physical and intellectual property from access for illegal purposes and to increase productivity. But as an employer, do the benefits outweigh employee disgruntlement with the process? Do the procedures invite micromanagement? From the viewpoint of an employee, is surveillance justified? Fair? Is this an invasion of privacy and autonomy? The subject is controversial. Employers are concerned about employees who might breach the security of the organization or illegally access the company's data base to mine for company secrets or intellectual property that could be easily downloaded and made public or sold illegally. Other problems employers cite involve the use of company computers for illicit purposes such as illegal downloads, selling drugs, betting on professional sports, or harassing another employee. Evidence that employees have overstepped the boundaries of trust supports the need for countermeasures of monitoring. But those against this monitoring are concerned that monitoring may be extended to include monitoring of personal information or activities. They argue that taking a break to surf the Internet, play computer games, or handle a personal piece of business reduces stress and makes for better morale.

Ideally, employees and management together should develop the principles and guidelines for employee monitoring. The procedures should be reviewed periodically so all employees are aware of the expectations and concerns of management, what is being monitored, and the consequences of misconduct.

Ethics of Camera Surveillance in Public Places

One might question whether there is any greater invasion of privacy than by the unannounced observation of people in a public place. This can be done by means of a television camera or by a wireless handheld device with zoom or audio capabilities that allow unfettered peering, eavesdropping, and photographing of someone without permission. We have come to accept TV cameras at a shopping mall, tourist attraction, or airport for security purposes, but this does not render permission for some random person to have our private personal or other information to be observed, collected, or used in any fashion the collector desires. Webcam systems (small cameras with wireless connections to the Internet that are accessed by computer) can be purchased by individuals and used to remotely monitor one's property and home environment. However, these systems can be positioned to spy on neighbors and the neighborhood as well. In fact, a commercial enterprise could advertise an Internet service whereby a customer could access any number of locations to view persons or events using cameras associated with the Web site. These devices could be on building tops, on street corners, in athletic clubs, and in other public or even private environments. Through cursor control on a remote computer any user can point the camera and acquire magnified pictures of any scene in range of the Webcam. The result is that anyone who can access a particular Web site will be able to collect image data at a magnification that would allow inspection of what people look like, what book they have in their hands, or with whom they are talking.

Unauthorized Use of Intellectual Property (IP)

Another problem area that has developed with the advent of computers has to do with the unauthorized and/or illegal use of intellectual property. The term *intellectual property* has been defined as that property that is novel and unique and results from the work of authors and inventors. In the United States, such work is protected by patent, license, copyright, trademark, and trade secret law. Thus, the types of intellectual property that are protected under these laws include ideas, patents, written materials, recorded materials, films, computer codes, and manufacturing processes—anything that is unique and original. However, protection of intellectual property has become more complex because the use of computers has radically changed the way that intellectual property is digitized, transmitted, accessed, and used.

Up until the 1990s, when someone read a book, listened to a song, or saw a movie, the intellectual property rights associated with these uses were under reasonable control of the originator. But with the proliferation of computers, the rapid growth of computer networks and the World Wide Web, and the increased digitalization of information, it is exceedingly easy to copy and distribute original works on the Web. Whenever digital information is accessed by computer, it is accessed by making a copy, which is transferred to the computer's memory and displayed on the monitor. Reading a book, hearing a song, or viewing a movie on your computer automatically involves making a copy of the work. Making a permanent copy of that work takes only a click of the mouse. With a copy in one's computer, it is a simple matter to forward it to anyone anywhere in the world. The effect is that the originator of the work loses control over the property rights to the work: control over distribution, reproduction, and reimbursement for the work is lost (Committee on Intellectual Property Rights and the Emerging Information Infrastructure 2000a). In addition, the work can be substantially altered without knowledge of the originator, and claims can be made by others to ownership of this altered work.

Another problem is that piracy of IP is occurring in all parts of the world. IP is being sent from one part of the world to another, where piracy laws may differ despite attempts at harmonization, a world in which stakeholders may have very different concerns, interests, and laws. When laws differ, whose are to prevail? Cultural attitudes and values regarding intellectual property also differ. In China for example, ideas are thought to be ". . . rooted in the contributions of ancestors" and is "in the air." As a consequence, one who expresses an idea has no right to it. . . ."it is a social expression, not an individual one" (Committee on Intellectual Property Rights and the Emerging Information Infrastructure 2000b). The concept of intellectual property in creative expression is completely foreign, not only in China but also in most Asian countries, though credit for authorship is universally recognized.

The preservation of property rights is similarly problematic in all forms of digital information whether in book, music, or movie form. To understand the complexity of the situation, it is important to understand how difficult the preservation of IP rights has become. Imagine, for example, that you have spent millions of dollars making a movie, which upon release is pirated and within minutes copied to a CD. Subsequently, if it is distributed for sale or given to friends, who also distribute it to others, it is not long before revenues for the movie are severely impacted. Or imagine the consequences of authoring a book only to find that it has been scanned into a home computer and displayed on the Internet for all to read and copy without your knowledge or permission. Other affected areas include the piracy of music, videos, games, and software.

Some software and IP are protected by patent; music, books, articles, pieces of artwork, and movies are protected by copyright and patent laws. The reader is referred to two excellent sources for in-depth reviews of the types of legal protections available for IP: *The Digital Dilemma: Intellectual Property in the Information Age* by the Committee on Intellectual Property Rights and the Emerging Information Infrastructure, et al. and *Computer Ethics* by D. G. Johnson.

One of the main problems is that the perpetration of piracy of IP in cyberspace can be done cheaply and almost instantaneously, and can be sent immediately to someone on the other side of the world. Further, copies of digital information are of the same quality as the original so each and every reproduction can be made without sacrificing quality. They are also easily altered, raising still further problems due to fraud, plagiarism, and theft. When digital information from one source is combined with information from another source, the interpretation of intellectual property infringement becomes complex, exceedingly difficult, and a growing problem (Committee on Intellectual Property Rights and the Emerging Information Infrastructure 2000a). An industry trade group estimated in 2002 that the piracy of software alone costs $12 billion a year worldwide (Forester and Morrison

2a). The infringement constitutes theft because the ille-
bility to reimburse the originator of the product.
e are many stakeholders who are affected. These may
aterial or, in the case of a video or an album of songs,
ng into the production of the work (the vocalist, production staff, postproduction staff, distributors, stock-
stakeholders may be affected by a single act of piracy.
al property has impacted concepts of how public access
istorically, copyrights and other laws have been devel-
t intervals to give authors and inventors exclusive rights
ations. Also, in the past, the public had wide access to
formation placed in public places such as archives,
ents, without cost or payment by the interested party.
orary could be signed out repeatedly; books bought by
n to a circle of friends. However, new concerns about
ave resulted in changing ideas about public access to
m on a Web site ("born digital," in colloquial terms).
l the world, modified, and claimed as one's own with
mation is a high priority for those who own it.
scussions and decisions about IP issues takes into
resolution (Committee on Intellectual and Property
Infrastructure 2000c):

perty in the context of digital music.

nd archiving of the social and cultural heritage and
structure has on IP and society.

ehavior with respect to IP, and whether a "copy" is
protection of digital IP.

IP.

The infringement of intellectual property rights needs to be viewed as an ethical dilemma.
The music industry has approached the problematic downloading of songs from the
Internet with a vengeance. The providers of Internet access are forced to identify their customers who have pirated, and these customers are sued. It is the responsibility of the industries to set boundaries, to work with the U.S. Congress to pass necessary laws, to initiate
punishments for those who pirate, and to set up a system of surveillance or other protections to stop unfair distribution.

Hacking and Hoaxes

Hacking, hoaxes, scams, viruses, plagiarism, theft, deception, and fraud are examples of
possible harassments propagated through the Internet. Infringement of privacy is especially decried as more and more issues come to the fore. The ethical issues here are virtue
and non-malfeasance. But when controls or regulations are introduced as countermeasures,
the utilitarian consequences can violate individual rights.

Hacking originally began as a game to challenge technology and to identify the weaknesses or vagaries of a computer program. Later, those skilled in the art of computer use found that they could also gain access to other computers and wreak havoc in a variety of ways (everything from changing data, altering financial and school records, accessing sales or marketing plans of rival companies, destroying Web sites, obtaining highly secret documents and information, and even sabotaging the system). Stolen information has made identity theft a significant problem for the victim. Others found they were able to track online purchases and sell this information for merchandising use. Major concerns about an individual's right to privacy became more apparent as these activities became more and more ubiquitous. These activities also violate ethical principles, including the right to autonomy and anonymity. Is hacking ever ethical? Perhaps, if national security were an issue we could argue that the end justifies the means. But if terrorists make it a priority to wreak havoc on the United States by hacking into and disrupting computers on Wall Street, in federal and state governments, and in flight and airport tracking, the consequences are potentially so harmful that extreme measures of protection would seem to be warranted.

Another entity is the so-called benevolent hacker who enters a site out of curiosity without permission, prowls the site, and does no damage but identifies flaws in the security of a company's computer network (*San Francisco Chronicle* 2002b). The extent to which these people are prosecuted appears to depend on what they do with the private and corporate information they gain.

Hoaxes promulgated on the Web may reach millions of people in a very short time and, even though false, may take on a life of their own. They are unethical because they can burden the information system, can decrease the speed of transmission of other information, and are time-consuming to fix. Virus promulgation is another means of willful computer harassment. It is deceitful, damaging, and also a burden in terms of lost data, time, and revenue. Losses from the Melissa virus, which infected millions of computers, were estimated to be in the billions of dollars. The perpetrator was sentenced to 20 months in prison.

Plagiarism

Plagiarism is another ethical issue arising with the use of the Internet. The reader is referred to Chapter 2, "Ethics in Scientific Research," where this subject is discussed at length. It is mentioned in this chapter to reiterate its importance to scientists and engineers, and as a reminder about how easy it is for students and writers to access the Internet to find information that can be cut and pasted into their own work. Large sections or blocks of material can be taken and claimed, inadvertently or not, as one's own, without proper citation. Writers will need to be increasingly aware of how they research, store, and cite information. Computer software programs are now available and becoming increasingly useful to detect passages plagiarized from other work.

Journalism on the Internet

Elsewhere in this text, we cite cases of erroneous or misleading information propagated by journalists (e.g., reporting errors on cold fusion experiments, the Kyoto Protocol negotiations, and the effect of GMO crop toxins on Monarch butterfly larvae). Have the new technologies led to an improvement in the accuracy of journalism? The ability to access information has enhanced the rapid dissemination of news, facts, and views. Along with this greater access to information comes the problem of verifying the truth. The reader is faced with the burden to weed out fact from fiction.

Theft, Deception, and Fraud on the Internet

A professional software engineer might be asked to create methods of spying or even stealing data from competitors using the Internet. Because of the anonymity of cyberspace, the perpetrators can escape detection and avoid being held accountable for their activities. At risk are corporate secrets as well as personal information of unsuspecting clients. The computer engineer or corporate manager asked to act on these requests is faced with the ethical dilemma of duty and loyalty to the corporation vs. duty to the professional code of ethics and personal virtue. Guidance in these situations is gained by considering the Four A's strategy for resolving dilemmas. One of the alternatives is to consider the duty to whistleblow (see Chapter 4, "Business Ethics").

Censorship in Internet Communications

Governments around the world have struggled with whether and how to censor indecent, illegal, offensive, or politically sensitive information placed on the Internet. The principal ethical problem is the conflict between freedom of expression or individual autonomy and utilitarian requirements of non-malfeasance. To what extent does this become censorship of freedom of speech in general? Only a few governments have banned or restricted access to unsuitable materials by legislation. Many government regulations regarding censoring have been challenged in the courts, and overall there are no uniform global regulations, even for censoring pornography. We list some policies that had evolved as of 2005 (Macan-Markar 2005, OpenNet Initiative 2004–2005):

- Government policy to encourage Internet industry self-regulation and end-user voluntary use of filtering/blocking technologies (used by the United States, United Kingdom, Canada, New Zealand).

- Criminal law penalties (fines or jail terms) applicable to providers who make content "unsuitable for minors" available online (used by some Australian states and attempted in the United States, although not legislated presently).

- Government mandated blocking of access to content deemed unsuitable for adults (Australian Commonwealth, China, Saudi Arabia, Singapore, United Arab Emirates, and Vietnam).

- Government prohibition of public access to the Internet (such as China, Cuba, Iran, United Arab Emirates).

Attempts to regulate or censor Internet traffic in the United States have been generally ineffective. For example, three have been ruled unconstitutional (the Communications Decency Amendment, the Child On Line Protection Act, and the Children's Internet Protection Act). The 2002 law Dot Kids Implementation and Efficiency Act provides safe

Internet sites using the www.kids.us address managed by NeuStar. This Web site allows children under the age of 12 to explore certain Internet links but controls others that are inappropriate or forbidden. www.kids.us provides parents with a mechanism whereby they can fulfill their duty to provide their children with a safe tool for exploring educational and entertainment material that will not be indecent or inappropriate for those under age 12. In 2004, ABC was the first broadcast network to offer a Web site for children (www. ABCKids.kids.us).

The Federal Communications Commission (FCC) is forbidden from making any regulation that would interfere with freedom of speech (www.fcc.gov/cgb). The Communications Act prohibits the FCC from censoring broadcast material. However, the FCC has encouraged the wireless industry to educate parents about the content available on mobile phones and other wireless devices and the ways they can limit access to that content. The wireless industry is in the process of developing a ratings system to classify content as either appropriate for users of all ages or recommended only to users at least 18 years old. Yet, in 2006, there were few mechanisms for blocking this information using shields, or for rating it, as is done for movies, television, telephones, and home computers. It is hoped that the pressure from the public and the FCC will lead those in the broadcast and wireless industries to be cognizant of their responsibilities to protect children from inappropriate content.

Unfortunately, attempts at censoring a child's access to the Internet also have many flaws, the principal one of which is that children have access to many computers. This and other problems with it are ably summarized by Ian Betteridge (2005):

> The lack of chat rooms and instant messaging services mean that sites in this domain space will be unattractive to exactly the audience that it is trying to draw. Children, even more than adults, love the chatting and social aspects of using the Internet, so any service that doesn't provide these is unlikely to be of interest to them.
>
> The plan, despite its good intentions, is typical of the kind of half-baked measure intended to protect children that, in fact, does nothing of the sort. Unless you manage to prevent children accessing every other part of the Internet, it will not work. You might be able to limit Web access on a single machine, but kids will always find another computer to use, unfettered.
>
> Technology is also changing so rapidly that measures designed by governments to prevent children from meeting strangers on the Internet are unlikely to work. Consider how phones will change over the next few years. As they get more like computers, instant messaging and chat rooms will come to them as well. Short of monitoring every conversation, there will be little that society can do technically to prevent misuse of the Internet.
>
> Rather than control every technology, the answer lies in education. The best way to prevent children from falling into the hands of paedophiles is to teach them what is and is not acceptable behaviour from adults, to help them understand that the world can be a dangerous place and to show them how to explore it without exposing themselves to dire risk.

Have we, as a society, become so addicted to the notion of innocent children not being exposed to any risk that we will fail to arm them with the knowledge they need to survive?

With the exception of individual autonomy, there seems to be no consistent underlying moral philosophy that has guided the regulation of the information to which the public is exposed through the Internet. A major reason for this general lack of global regulation seems to be the dominance of the principle of freedom of expression over the real or perceived social harms inflicted by access to indecent information, fraud, and predators. The dominance of freedom of expression, protected by the First Amendment of the U.S. Constitution, is shared by New Zealand, Canada, the United Kingdom, and other European countries. This right has resulted in the lack of censorship, with perhaps the major harm being to children. The duty of parents to protect children is challenged by the right to freedom of expression. The case in point is the U.S. Communication Decency Act enacted in 1996 but declared unconstitutional by the U.S. Supreme Court in 1997. Australia has enacted some censorship since 2000. China has a censorship system that includes ordering Internet service providers to screen private e-mail for political content. The compilation of censorship laws around the world, published by Electronic Frontiers Australia (2002), reflects the conflict between the objection to pornography and racial hate messages vs. the right of freedom of expression. Even countries such as Germany that wish to enforce strict control of Internet trafficking of unsuitable or offensive information cannot do so because it is impossible to control incoming traffic from outside the country or to prevent access to servers outside the country.

Spam

Spam is unsolicited and unwanted junk e-mail, usually in the form of advertisements. Spam is an increasingly prevalent and annoying form of anonymous communication on the Internet, often sent to many individuals en masse by spammers using fake e-mail addresses to escape detection. It can clog and may even slow down e-mail systems. In 2003, spam was estimated to account for over half of electronic mail, and its volume continues to increase; it is often fraudulent, deceitful, and/or pornographic; the sheer volume makes it time-consuming to delete and obscures wanted e-mail. This form of junk mail is often not identified by spam-blocking systems because perpetrators have learned to work around the program by using false header information, misspelling words, or inserting symbols to escape recognition of pornographic or unwanted terms.

Congress passed the Controlling the Assault of Non-Solicited Pornography and Marketing Act of 2003, 117 STAT. 2699 PUBLIC LAW 108–187—DEC. 16, 2003 (the "CANSPAM Act of 2003").

Availability of Information Technologies

In the United States, computer systems are found in 80% of homes of those with annual incomes of $75,000 or more. Concerns about a so-called "Digital Divide" have arisen because computers are found in fewer than 40% of homes with incomes between $25,000 and

$30,000, and in fewer than 10 to 15% of homes with annual incomes less than $15,000 (Brown 2003). This is of concern because computers have been characterized as the greatest innovation in the history of mankind. They have changed the way we live and the world in which we live. They provide access to such a wealth of information that those without, the "have nots," are at risk for dropping even further behind socially, educationally, and in the work place. This is also true of organizations and countries that have not kept up with computer technology.

CHAPTER SUMMARY

The ethical theories of autonomy (privacy), non-malfeasance, justice (equity of access), and utilitarianism are pertinent to Internet communications and other technologies that communicate ideas, data, pictures, and music. The consequences of current acts of vandalism (e.g., hacking, virus promulgation), plagiarism, stealing, and invasion of piracy are to some extent considered a part of the technology not likely to be eliminated. These problems are to some extent tolerated, perhaps because of the ubiquitous advantages and ease of access to individuals and the products of individual creations using the Internet.

Other technologies that threaten autonomy are those that communicate video information (cell phones), eavesdrop on wireless communications, and provide surveillance without the knowledge of those being observed. As in Internet communications, it is not easy to control or eliminate the invasion of privacy that these impose. In many cases, this loss of autonomy is accepted as the cost of benefits to the public, such as video surveillance against terrorism and crime, for example.

The professional engineer and the scientist involved in the design and installation of computer technologies have the usual ethical responsibilities to ensure the safety of the product. But these responsibilities go beyond assurance of performance and meeting specifications. The added responsibilities include the development of countermeasures to protect the product from unethical activities, or at least an ongoing anticipation, recognition, and issuance of software updates that will counter the negative consequences of its improper use.

A major ethical problem that has evolved with the Internet is the conflict between the right of freedom of expression and the need to censor some of the freely accessible information that is indecent, emotionally sensitive, politically inappropriate, dishonest, or misleading. This is a global problem for political entities and for parents. Some governments censor communications that are believed not in the interest of the government. The majority of governments attempt to censor inappropriate communications such as racially sensitive Web sites, but most of these efforts are challenged by the right to freedom of expression. Parents have a duty to protect their children from suggestive and misleading information. The only current solutions are to encourage parents to educate their children about what information is acceptable and what is not, and to use methods to prevent exposure to this information if they choose to do so.

REFERENCES

Baase, S. 2003a. *A Gift of Fire: Social, Legal and Ethical Issues for Computers and the Internet.* Upper Saddle River, NJ: Prentice Hall Publishing.

Baase, S. 2003b, pp. 342–352.

Baase, S. 2003c, p. 346.

Betteridge, I. 2005. "Safety catch" (editorial), www.macuser.co.uk, accessed June 25, 2005.

Brown, J. 2003. "Crossing the Digital Divide." In: *Computers, Ethics and Society.* New York: Oxford University Press.

Committee on Intellectual Property Rights and the Emerging Information Infrastructure, et al. 2000a. *The Digital Dilemma: Intellectual Property in the Information Age.*, Washington, DC: The National Academies Press, p. 28–49.

Committee on Intellectual Property Rights and the Emerging Information Infrastructure, et al. 2000b, p. 57.

Committee on Intellectual and Property Rights and the Emerging Information Infrastructure, et al. 2000c, pp. 60–61.

Electronic Frontiers Australia. 2002. "Internet censorship: Law and policy around the world." 25 pp. www.efa.org.au/Issues/Censor/cens3.html, accessed July 1, 2005.

Forester, T. and P. Morrison. 1995. *Computer Ethics.* Second Edition, Cambridge, MA: MIT Press, p. 7.

Johnson, D. G. (ed.). 2001. *Computer Ethics.* Third Edition. Upper Saddle River, NJ: Prentice-Hall, Inc., pp. 125–126.

Macan-Markar, M. 2001. "45 countries smothering Internet freedom." Reporters without Borders, www.comondreams.org/headlines01/0209-02.htm, accessed June 24, 2005.

OpenNet Initiative. 2004–2005. "Country studies." www.opennetinitiative.net/modules. php?op=modload&name=Archive&file=index&req=viewarticle&artid=1, accessed June 24, 2005.

Preston, J. 2004 (September 30). "Patriot Act subpoena of Internet data struck down." *San Francisco Chronicle,* p. A3.

Reiman, J. H. 1995. "Driving to the panopticon: A philosophical exploration of the risks to privacy posed by the highway technology of the future." *Computer and High Technology Law Journal* **2**:27–44.

San Francisco Chronicle. 2002a (February 28). "Hacker pleads guilty to software piracy," p. B2.

San Francisco Chronicle. 2002b (February 28). "S.F. hacker cracks *New York Times'* internal network," p. B1.

Schevitz, T. 2004 (April 4). "Colleges leaking confidential data." *San Francisco Chronicle*, pp. A1 and A9.

ADDITIONAL READING

Casmir, F. (ed.). 1997. *Ethics in Intercultural Communication*. Mahwah, NJ: Lawrence Erlbaum Associates (LEA), Inc., 296 pp.

Langford, D. 2000. *Internet Ethics*. London: Palgrave MacMillan, Ltd., 272 pp.

Neher, W. and P. Sandin. 2006. *Communication Ethics*. Boston, MA: Allyn and Bacon Publishers, 368 pp.

Vacca, J. R. 2003. *Identity Theft*. Upper Saddle River, NJ: Prentice-Hall, 475 pp.

Zoellick, B. 2001. *CyberRegs: A Business Guide to Web Property, Privacy, and Patents*. Boston: Addison-Wesley Professional, 336 pp.

APPENDIX

Appendix 3.1. Committee for the Protection of Human Subjects: Security of Research Subjects' Personally Identifiable Data Held by Researchers (Excerpts from http://chps.berkeley.edu)

4. Definitions

A *research data set* constitutes a body of data elements collected in the course of research with living human beings.

Personal identifiers within a data set are any data elements that singly or in combination can uniquely identify an individual, such as a social security number, name, address, demographic information (e.g., combining gender, race, job and location), student identification numbers or other identifiers (e.g., hospital patient numbers).

A *de-identified data set* refers to original data that subsequently has been stripped of all elements (including but not limited to personal identifiers) that might enable a reasonably informed and determined person to deduce the identity of the subject. For research that requires that data elements later be linked to an individual's identity, the original data set may be partitioned into two data sets: a *de-identified data set* and an *identity-only data set*. The latter should contain any and all personal identity information absolutely necessary for future conduct of the research. For purposes of later merging the identity information with other research data, a researcher-assigned identity code (typically a randomly generated number) that is associated with and unique to each specific individual may be included in both data sets, and later be used to link identity data elements back to the de-identified data set. This identity code should not offer any clue as to the identity of an individual.

Secure location refers to a place (room, file cabinet, etc.) for storing a removable medium, computer or equipment wherein resides data sets with personal identifiers to which only the principal (or lead) investigator has access through lock and key (either physical or electronic keys are acceptable). Access may be provided to other parties with a legitimate need, consistent with the policies below and as disclosed in the research protocol.

Secure data encryption refers to the algorithmic transformation of a data set to an unrecognizable form from which the original data set or any part thereof can be recovered only with knowledge of a secret decryption key.

5. Specific Policies   We recognize that not all research data sets can reasonably be de-identified (for example, an audio recorded interview in which a subject identifies him or herself). In this case, the original research data set must be considered an identified data set and treated accordingly. Identified and identity-only data sets should always be stored in: a) a secure location, or b) secure data-encrypted form.

5.1 Collect the minimum identity data needed and describe in the research protocol exactly what personally identifiable data elements will be collected, and whether the data set will be de-identified, split into a de-identified data set and an identity-only data set, or neither.

5.2 De-identify data as soon as possible after collection and/or separate identifiable elements (create identity key, destroy raw data).

5.3 Limit access to identified or identity-only data set and store it in a secure (locked) location separate from data, or store it in encrypted form, or both. Encrypted form is the only acceptable storage for data stored in a computer or removable medium which is not permanently located in a secure location (e.g., laptop computer or a removable disk which is to be carried in a briefcase) or for transmission across the network (for example as an email attachment).

5.4 The investigator shall develop and disclose to CPHS a plan in writing as to what individuals will have legitimate access to an identified or identity-only data set, either through access to secure location key or to decryption key. This plan must include provision for recovery of a lost decryption key, to insure that a data set cannot be permanently lost.

5.5 When an identified or identity-only data set is stored in personal or university-owned or -maintained computer, investigators are strongly encouraged to ensure that this computer be professionally administered and managed. If this is not possible, investigators should disclose such, and provide CPHS with a plan for how the sensitive data will otherwise be secured.

The opportunity for human error should be reduced through: a) limiting the number of people (both users and administrators) with access to the data and ensuring their expertise and trustworthiness; and/or b) using automatic (embedded) security measures (such as storing data on non-volatile medium only in secure data-encrypted form) that are professionally installed and administered. If this computer is connected to the campus network or to the public Internet, the professional administrator of the computer shall ensure that it complies with all minimum standards for network and data security listed below.

7. Summary of the Acceptable Security Measures for Maintaining Personally Identifiable Information for Research Purposes

The level of security necessary is relative to the risk posed to the subject should personally identifiable data be inadvertently released or released as a result of malfeasance. In an effort to ensure best practice it is always more desirable to have a higher level of security than to risk operating at a minimal standard. CPHS has the authority to decide if the security plan to protect subjects' confidentiality or anonymity is acceptable. For data that retains identifiers, the protocol must describe adequate administrative, physical and technical safeguards. Investigators are encouraged to consult with appropriate information technology and security experts such as their system administrators to develop appropriate data security plans when working with personally identifiable data.

PROBLEM SET

3.1. This problem requires some legwork or Internet work.

A. Team A members: Obtain from Safeway, Costco, or a similar establishment a form to apply for a "club" card or discount card. Summarize any statements about how information about the applicant will be used, in less than 40 words. Also, find a manager and ask her or him a question about how your information is protected. State the question you asked and the answer the manager gave you (keep your answer to less than 60 words).

B. Team B members: For each of five Web sites, find the privacy statement. If there is no statement, e-mail the site to receive privacy policy information. You may have to try multiple sites and may still not elicit the information from five of these; nevertheless, your assignment is to list the sites you tried and the essence of the policy or the no-policy.

3.2. Contact two or three nearby universities to determine the policy each has regarding the release of names, addresses, and telephone numbers of students, either individually or as a list. Start by finding out if there is an online student directory, and if so, how access is granted. State the policy of each and then give your argument for a uniform policy, including what that policy should be and why. You may use the Internet, personal visits, or telephone to get this information. You may represent yourself as part of a research project or as a student who is answering a question on an assignment or anonymously.

3.3. Suppose a graduate student in the School of Public Health starts a research project to investigate the behavior of a virus by actually sending out an e-mail virus to the student body of University of California at Berkeley. The virus will enter the hard drive and reside in a secret location for three weeks. The recipient computer will receive a message that the computer has been infected and that every 10th, on average, e-mail that the recipient sends will propagate the virus until the recipient asks the sender for help in disinfecting the virus. The experiment is designed to determine how fast and widespread the infection will be before it begins to decline.

First consider what would happen if the original infection went to 10 individuals and each of the 10 send 20 e-mails each day, but on the second day the first recipients ask for help to disinfect the virus, on the third day the 20 recipients from the first generation of infected agents send 20 requests for help to the first 10 senders. These first 10 take one day to forward information on how to get help by putting the 20 recipients in contact with the originator of the virus. However, during the third day, the 20 recipients will have sent the virus to 40 new recipients. On the fifth day, these 40 recipients will ask the 20 senders for help, but will already have sent e-mail viruses to 80 recipients. The originator's plan is to sit back and respond to the requests for help with delays of one and three days. The researcher expects to carry out this experiment for one month. He will receive a tremendous input of e-mails, from which he can collect statistics on this type of viral infection and possibly publish a paper reporting the results of this experiment.

Make an estimate of the number of infected computers that would exist after 12 days. The originator can send a virus therapy message at a rate of 300 per day, due to the fact that he has not automated the response and needs 20 seconds to send the remedy. Give an analysis of the ethical behavior of the experimenter in 100 words or less.

3.4. Develop and explore the types of questions engineers should be asking about how to engineer better machines, specifically about how to include procedures in anticipation of ethical consequences. Select two from the following list for a 60-word essay each on the instructions a manager should give to the design engineers.

- Taxation algorithm for states to assess for the use of the Internet for purchasing products and services.
- A wireless technology that will allow a quadriplegic subject to control a computer cursor using the tongue.
- A camera system for viewing side and rear end blindspots and in vehicles.
- A movement detector that will give wireless information on human activity in a home.

4

Business Ethics

INTRODUCTION

This chapter tackles the complexities of business ethics. The goal of enhancing profits for top executives and shareholders without appropriate consideration of all the stakeholders is a common cause of ethical problems throughout history. In today's markets, business decisions need to take into account the interests of a variety of stakeholders both inside and outside the corporation. The prevalence of unethical practices today, such as bribery, self-enrichment, misappropriation of resources, false advertisement, misleading statements, insider trading, and disregard for employees' well-being, challenges the corporation leadership and the public stakeholders to create effective remedies for the future. We model the history of ethics in business in Figure 4.1 from the coupling of business practices with laws and regulations initiated by the Hammurabi Code of 1700 B.C. to the present day, when regulations increasingly govern business practices. In addition, business policies and practices now are looking beyond laws and regulations and are emphasizing the importance of moral behavior. This depiction emphasizes that laws and sanctions are not the principal solutions to current unethical practices.

For example, there are many corporate executives whose superior leadership, personal values, and integrity have created a culture of acting virtuously while improving profitability (George 2003a). Some have learned to avoid the root causes of corporate ethical dilemmas by better anticipating and acting on their responsibilities to their employees, the public, and the environment (Schwartz and Gibb 1999). Another solution has been to provide educational processes that instill virtue, moral thinking, acceptable practices, and ethical responsibility in top-level corporate executives, management, and other professionals (Callahan 2004).

However, these solutions are not easily implemented because they rely not only on guidance regarding morally correct actions but also on real experiences in which the decision maker is challenged to make tough decisions. Education must extend beyond merely teaching what practices are right and virtuous. Bill George (2003b) writes of the value of experiencing defining moments through "... a crucible that tests you to your limits ... you learn who you really are and what you want to become." Ideally, one learns from one's mistakes, becomes prepared for future challenges, and acquires the strength to act according to

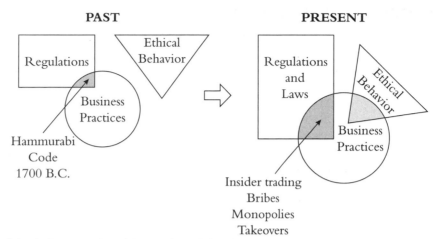

Figure 4.1 The incorporation of laws and regulations in the past and the contemporary respect for moral behavior in business practices.

a set of high values. Molding character and acquiring values take place over time through ongoing mentorship, but mentorship is often lacking in the business world. Frequently, it takes repeated exposure to the challenges of everyday business problems over time to experience and learn the rewards of virtuous behavior. For others, sudden insight occurs after personally experiencing the full brunt of malfeasance, but many professionals may never have experienced these character-building moments. Further, they are difficult to create for use as teaching tools (although some facilitated encounter programs are designed to do so). Moreover, the value of virtuous behavior in the face of dilemmas cannot be instilled by merely educating someone to the benefits. This chapter presents strategies to better equip the professional to make morally correct business decisions when faced with ethical dilemmas.

The plan of this chapter is to broadly outline the historical development of business ethics; to present a modification of the Four A's strategy for use in business situations, pointing out the importance of the stakeholders and the importance of prioritizing their needs in corporate decision making; to present business strategies and cases that demonstrate conflicts between moral theories; and to provide case examples that demonstrate the consequences of poor business practices. The responsibilities of the professional employee to the corporation, to other employees, and to the public are other topics covered by this chapter. As we shall emphasize, the important tools for approaching business ethical dilemmas are:

- Respect for the multiple stakeholders
- Adoption of a credo to guide corporate behavior
- Use of the Four A's strategy

HISTORICAL BACKGROUND

Reviewing the development of business ethics from a historical perspective will help to clarify present business practices and the cases to be discussed. Expectations for fairness in business transactions have existed since ancient times. Hammurabi's Code of Laws bears the name of the king of Babylon who believed he had been chosen by the gods to establish a set of laws for his people that would be just and provide them with better lives. While most of the code was written to accommodate situations of known violations of honesty and faithfulness, many items addressed who would take responsibility for failures of business performance and commitments. This code consisted of 282 specific rules and regulations that pertained to business contracts and to levels of punishments for a wide range of behaviors, including unfaithfulness and stealing. These laws, considered the first laws that pertain to business contracts, were written on stone sometime during Hammurabi's reign that lasted over 50 years, ending in 1780 B.C. It is believed to be one of the first sets of laws to help ordinary people with differing backgrounds live in large groups and to be treated fairly. Some of these laws are explicit guidelines later seen in the Old Testament, such as an eye for an eye and a tooth for a tooth. Others set specific guidelines for land cultivation rights (Laws 49–52), rights for those wronged in the conduct of their daily lives (Laws 102–106), and laws regarding building and construction (Laws 229–232). A few examples are the following (http://eawc.evansville.edu/anthology/hammurabi.htm, accessed February 12, 2005):

> 48. If any one owe a debt for a loan, and a storm prostrates the grain, or the harvest fail, or the grain does not grow for lack of water; in that year he need not give his creditor any grain, he washes his debt-tablet in water and pays no rent for this year.

> 229. If a builder builds a house for someone and does not construct it properly, and the house which he built falls in and kills its owner, then that builder should be put to death.

> 232. If it ruins goods, he shall make compensation for ruined, and in as much as he did not construct properly this house, which he built and it fell, he shall re-erect the house from his own means.

> 265. If a herdsman, to whose care cattle or sheep have been entrusted, be guilty of fraud and make false returns of the natural increase, or sell them for money, then shall he be convicted and pay the owner ten times the loss.

The Greek philosopher Aristotle wrote of the importance of virtue in business dealings, as did the Roman statesman Cicero in the time before Christ. Later, the Romans used justice to limit the amount of land that could be amassed by the wealthy, thereby limiting an aristocratic monopoly on farming and giving others an opportunity to acquire land and to compete. Tiberius (42 B.C.–37 A.D.), the second Roman Emperor, took these laws further by limiting the amount of public land that could be rented to one person; he repossessed all lands that exceeded this, assigned these lands to the poor for a small annual rent, and appointed a land commission to oversee these rights. His grandnephew Gaius Caligula, who succeeded him in 37 A.D., also tried to make improvements for the people (www.yale.edu/lawweb/avalon/medieval/athe1.htm, accessed February 12, 2005). Gaius' code of laws was written to let all know what was expected of them. Laws pertaining to the conduct of business were outlined in detail, and he was among the first to establish just rules about the control of monopolies.

Now, over 2,000 years later, the issues of justice and fairness in corporate practices seem not to have been resolved. Despite efforts to create a more just and reasonable approach to business by Hammurabi and others since then, there are still those today who are attempting to create monopolies to drive competitors out of business, to institute the next hostile takeover, or to do whatever it takes to make a profit.

The prevailing attitude today regarding ethics in business has been summarized as follows (Carr 1988a):

> We live in what is probably the most competitive of the world's civilized societies. Our customs encourage a high degree of aggression in the individual's striving for success. Business is our main area of competition, and it has been ritualized into a game of strategy. The basic rules of the game have been set by the government, which attempts to detect and punish business frauds. But as long as a company does not transgress the rules of the game set by law, it has the legal right to shape its strategy without reference to anything but its profits. If it takes a long-term view of its profits, it will preserve amicable relations, so far as possible, with those with whom it deals. A wise [practitioner] will not seek advantage to the point [of generating] dangerous hostility among employees, competitors, customers, government, or the public at large. But decisions in this area are, in the final test, decisions of strategy, not of ethics. . . .

The modern-day control of monopolies is an attempt to ensure justice by regulations, analogous to attempts by the Romans to give everyone the opportunity to thrive by the fair allocation of land. However, the establishment of guidelines or regulations that would give the employees a fair share of the company profits has occurred only rarely in countries where capitalism thrives. The control over the distribution of company profits is in the hands of the chief executive officer and the board of directors. In the United States, free enterprise is encouraged, and no laws restrict the accrual of personal wealth as long as it is lawfully obtained. However, if that accrual is harmful to the stakeholders, the result has been alienation of the corporate leaders from the stockholders and employees, and reactions to the injustices have included public hostility, lawsuits, and new government regulations.

An effective and comprehensive business school text (De George 1999) points out that a new morality in business ethics emerged in the second half of the twentieth century. Instead of focusing on a shareholder-driven mandate to make, if not maximize, a profit, the new business mandate, spurred by public outcry and legislative changes, has additional objectives. These include focusing on such concepts as labor safety, equal opportunity, environmental protection, protection of child laborers, workmen's compensation, truth in advertising, consumer protection, fair wages, and pension reform while still making a profit. As a result, employees, consumers, the public, and the environment have become entities of major importance in the strategies of many successful corporations.

However, not all agree that the pursuit of "social responsibility" in corporate policies is wise. Some argue that there are dangers in the well-intentioned corporate support of community endeavors. To paraphrase Levitt (1958a), the principal danger is a slippery slope that leads to the corporation becoming more and more responsible for the welfare of the community, accompanied by a "prodigious currying of political and public favor." He argues this will lead to an insidious centralization of power, with the corporation exerting more and more control over noneconomic aspects of society while its major motive is still material.

One might make a strong counterargument that initiatives by business to create a better environment for employees and the communities in which they exist will lead to long-term profitability. There is no doubt that recruitment and retention of valuable employees is in good part based on the goal of making a profit. Nor should Levitt's fears lessen the thoughtful respect a corporation should have for its stakeholders. The prima facie duty of a business is to increase profits while staying within the limits of the law. This duty, along with the duty to share those profits with the stockholders, can be in conflict with a perceived duty to distribute profits to the community as part of social responsibility.

A solution for many problems has been for corporate leaders to adopt policies that make social responsibility a high priority in the corporation (often expressed in a credo), and to take into consideration the many stakeholders who may be affected by their decisions, discussed below. In fact, two studies reported by Kelly (2004) have shown a statistical correlation between the establishment of ethical and social responsibilities by corporate leadership and increased profits.

DESIGN STRATEGY FOR CORPORATE ETHICAL DILEMMAS

The cases presented in this chapter illustrate the variety and seriousness of ethical problems that have arisen in recent years. Could they have been avoided? Once again, we will use the Four A's design strategy, customized for corporate decision making, to identify morally correct solutions.

The design strategy presented in Chapter 1, "Ethical Principles, Reasoning, and Decision Making," requires customization for use in corporate ethical decision making for three reasons.

- First, the responsible decision maker evaluating a business ethical dilemma represents the corporation, an entity beyond the individual decision maker. This person may be a manager, a member of the board of directors, or any other employee. This person's decision is usually guided by corporate goals rather than personal ideals or reasons, but it should also be driven by fair and honest values. Today's leaders and decision makers need to anticipate issues and establish a climate in which ethical actions are the main priority.

- Second, there are at least 13 stakeholders who might be affected by and should be considered in any major corporate ethical decision. The relative importance of, and therefore the prioritization of, the stakeholders may change, depending on the situation or issues confronting the decision maker.

- Third, the company credo becomes an important element of guidance that should not be minimized. If one does not exist, time should be set aside to develop one as it sets important goals for the company. It is equally important that the credo become a document that is valued by management and all employees. It should be discussed often and followed. Everyone working at the company should pledge to uphold it.

The schema below incorporates these additions into an overall design strategy for assessing corporate ethical issues:

Table 4.1   BASIC STRATEGY OF THE FOUR A's (underlined sections indicate modifications for business ethics)

Approach a dilemma by a systematic consideration of the Four A's (i.e., Acquire facts, Alternatives, Assessment, and Action):

- **Acquire facts**
  - Define uncertainties, clarify ambiguities
  - Get the facts
  - Seek advice from others, including some of the stakeholders

- **Alternatives**
  - List alternate solutions
  - Develop alternate plans in parallel

- **Assessment**
  - Assess possible solutions according to the moral theories of virtue, justice, duty, rights, and utilitarianism
  - Assess the situation according to the company's credo
  - Identify and prioritize stakeholders affected by the decision

    Customers

    Responsible decision makers

    Stockholders

    Board of directors

    Employees

    Environment

    Employee families

    Local community

    Global community

    Suppliers

    Distributors

    Competitors

    Government
  - Assess the consequences to each of the stakeholders
  - Perform risk analysis when appropriate

- **Action**
  - Decide on a plan or plans for action
  - Keep alternate action plans under consideration should they be needed
  - Adjust and adapt, recognizing that the process is a dynamic one, that an initial solution may require revision, and that the ranking of the stakeholders might change
  - Keep an open mind to new options

The application of this design strategy and its components will be discussed in various cases as the chapter proceeds.

CORPORATE PRACTICES CHALLENGE ETHICAL THEORIES

This section looks at how moral theories come into conflict in the corporate world, and reviews the special situations of business ethics in contrast to social or individual ethics. The duty of a poker player gives us a perspective on the issue of conflicting ethical theories in the game of winning. Though in some sense the game of poker is an unlikely analogy for corporate business competition, the "rules of conduct" of poker do present important analogies to nonvirtuous behaviors that corporate managers believe they must use to survive and maintain their duty to the stakeholders:

> Poker's own brand of ethics is different from the ethical ideals of civilized human relationships [virtue]. The game calls for distrust of the other . . . [fellow]. It ignores the claim of friendship. Cunning, deception, and concealment of one's strength and intentions, not kindness and openheartedness, are vital in poker. No one thinks any the worse of poker on that account. And no one would think the worse of the game of business because its standards of right and wrong differ from the prevailing traditions of morality in our society (Carr 1988b).

Business has become extremely competitive, especially in the United States. It is also the case that in the pursuit of success, business has become a game of tactics. Business tactics are limited by the law, but when driven by profit, a company's approach may well be in conflict with the Golden Rule and most moral philosophies. Bluffing, cover-ups, ruthlessness, hostile takeovers, and whatever else is done by others in the game of business determine the rules (Carr 1988c).

Ethical dilemmas can be shrouded in a nonvirtuous corporate culture. Engineers, scientists, as well as other professionals (such as accountants or lawyers) can be rendered helpless and lose substantial autonomy in situations that deviate from good practices of justice and virtue. Examples are engineers employed in weapons research, computer scientists directed to intercept proprietary data from another company, and scientists working for companies that use inferior components in the manufacture of their products.

Corporate goals of making a profit can challenge concepts of beneficence, the Golden Rule, and utilitarianism (i.e., doing the best for the majority). Ideally, corporate decision making takes into consideration the best interests of the "most," as well as what is best for local and global communities and the environment. However, the corporate goal of making greater and greater profits frequently conflicts with justice and fairness. The way in which corporate strategies are prioritized and merged into the decision-making process may be vital for the survival of the corporation, although the tactics used could destroy a corporation regardless of the acceptable ethics of the strategy. This can occur because corporations operate in a fast-moving competitive field without clear codes and with few laws to guide decisions. In this environment, corporate tactics might be expedient solutions that can lead to a cascade of consequences. A decision based on quarterly performance and expediency may be the gateway to a slippery slope of more serious consequences, with major departures from acceptable behavior, including unethical, morally wrong, or even illegal activities. Corporate debacles (such as those of Enron and WorldCom outlined in Cases 4.9 and 4.10) have shown how poor decision making can destroy the company and the careers and even the lives of those involved, as well as devastate corporate retirement funds and stockholder investments. Other issues that feed into ethical dilemmas include performance demands by stockholders, cover-ups by employees who have made mistakes, unfair practices by management, and a "group think" mentality whereby viewpoints are molded to conform with what the members believe to be the consensus of the group, leading to decisions that would not otherwise be made and that may be unwise.

There are often special circumstances that prevent the professional from taking preemptive, or even reactive, actions to mitigate ethical problems in corporate life. Corporate employees face difficult dilemmas when endeavoring to effect justice and virtue when encountering questions of product quality, false claims, negligence, safety, or loyalty to the corporation itself. As paraphrased from Ladd (1982), engineers or scientists may feel powerless to influence decisions outside of their control. When ethical problems arise in a corporation, the individual is expected to be accountable to the organization. However, the corporation is not an individual, particularly in the sense that there is no reciprocation of friendship or loyalty, and its priority is to act only in its own self-interest, which is to make a profit. Loyalty to the corporation is expected, and this expectation is often delineated as a part of the employee's contract of employment. When corporate circumstances challenge individual morals, there are no easy answers. The professional must heed his or her own sense about what is right and act accordingly.

Differences in Corporate Tactics

There exists today a wide difference in management philosophies, even among the most famous and successful leaders. Four examples below illustrate some major differences and conflicting philosophies of corporate leadership.

The first builds on the famous Chrysler Corporation bankruptcy situation of the 1980s and the rise to public awareness of Lee Iacocca. He is said to have saved the corporation, which then continued as a successful competitor in the automobile industry. The motto or credo of Iacocca can be inferred from the following quote attributed to him (Schinzinger and Martin 2000): ". . . our most important customers are not the people who buy our cars—our most important customers are the people who buy our shares." This statement sets a clear priority toward the stockholders, one that is in conflict with other priorities that may be more important to the engineer and scientist. That philosophy can be seen to underlie the famous Pinto automobile gas tank design case that occurred after Iacocca moved to Ford Motor Company.

Case 4.1 Ford Pinto Gas Tanks

During crash tests, Ford engineers discovered that the Pinto's fuel system ruptured in nearly all rear-end crash test collisions. Safety was not a major concern to Ford at the time of the development of the Pinto. Lee Iacocca's specifications to be strictly adhered to were that *"The Pinto was not to weigh an ounce over 2,000 pounds and not to cost a cent over $2,000"* (www.fordpinto.com/blowup.htm., accessed February 12, 2005).

There were three engineering design flaws that presented safety problems. The engineers had recognized that a rear-end collision would compress the gas tank onto a hexagonal bolt on the differential and a tank rupture could result. Another design problem was that the frame of the car was not stiff enough to prevent a rear-end crash from jamming the doors, possibly trapping the occupants. In addition, a fire under the car would enter the automobile due to an inadequate barrier between the car underside and the interior.

The engineers performed a risk analysis (see Chapter 12, "Ethics of Emerging Technologies") and computed the cost of preventing the bolt-tank rupture problem would be $137 million. An arbitrary value was assigned to calculate the cost of the number of deaths and burns predicted. The cost of redesign was more than two times the estimated cost of 180 deaths, 180 serious burns, and 2,100 cars burned. Their estimate of the risk was $50 million. Thus, the cost of safety outweighed their estimate of the value of life and suffering. Ford Motor Company executives made the decision not to make modifications (www.fordpinto.com/blowup.htm., accessed February 12, 2005).

The automobile was involved in rear-end collisions that ruptured the gas tank and resulted in fatalities from explosive fires. The engineering flaw could have been corrected if Ford's priority had been safety instead of profit. The company lost millions in settlements and legal fees over a decision that was costly in terms of client faith and trust as well.

The tactics of Larry Ellison, the CEO of Oracle, are examples of a management philosophy that puts company gains above all else, ignoring the pain that might be inflicted on stakeholders. We will return to Oracle later in the chapter, but point out here that Ellison's tactics are said to be akin to "it's not enough that we win; everyone else must lose" (Said 2003). Contrast the policies of Iacocca and Ellison with the credo that guided Johnson and Johnson's response to the Tylenol crisis of 1982.

Case 4.2 Tylenol Cyanide Crisis

Cyanide was criminally introduced into Tylenol capsules manufactured by the giant health care corporation Johnson & Johnson, causing seven deaths. It was determined that the cyanide contamination occurred after the medication had reached stores in the Chicago area. Among the 31 million bottles of extra-strength Tylenol capsules on drugstore shelves around the country, only the Chicago area stores were affected. The entire country became aware of the problem, and Tylenol sales plummeted. Opinion polls indicated that most Tylenol users probably would never buy the product again, and marketing experts predicted that the product would never recover. On short notice, Johnson & Johnson decided to immediately recall *all* Tylenol containers from around the country at a cost of $100 million. A few weeks later, a new product was introduced packaged in triple-sealed, tamperproof packages. These actions showed the company's commitment to safety and to their customers, and consumers responded so positively that within a year the product had regained 90% of its previous market share (www.jnj.com/our_company/history/history_section_3.htm, accessed Feb. 8, 2005; also www.motherservice.org/VitalDifffiles/CHAP3.htm#nr06, accessed Feb. 8, 2005).

Johnson & Johnson's president, David R. Clare, noted: "Crisis planning did not see us through this tragedy nearly as much as the sound business management philosophy that is embodied in our credo. It was the credo that prompted the decisions that enabled us to make the right early decisions that eventually led to the comeback phase" (www.motherservice.org/VitalDifffiles/CHAP3.htm#nr06, accessed Feb 8, 2005). J&J Chairman James Burke indicated that while it had taken two years to write, modify, and acquire employee signatures on the credo, the time taken was well worthwhile. He confirmed it served the company not only in crises but also in decisions made by the Board and other day-to-day decisions as well (Burke 1990). The full J&J credo appears in the next section as an example of a work ethic that has benefited J&J's management, its employees, and the public.

Another illustration of the importance and success of a management strategy that gives high priority to the customer among the stakeholders is known as the "House of Quality" philosophy. This term refers to the philosophy first used by the Japanese in the early 1970s at the Kobe, Japan, shipyard of the Mitsubishi Corporation, a philosophy later adopted by the Toyota Motor Corporation. The philosophy and term "House of Quality" were brought to the United States principally through a 1988 article in the *Harvard Business Review*: ". . . the foundation of the house of quality is the belief that products should be designed to reflect customers' desires and tastes—so marketing people, design engineers, and manufacturing staff must work closely together from the time a product is first conceived" (Hauser and Clausing 1988). This strategy remains a successful approach in today's business world.

Credo vs. Vision Statement

An approach to improving the ethical behavior of executives and employees of commercial organizations, including financial institutions, is to take very seriously the process of developing a credo. Almost all corporations have a vision statement indicating what they stand for and what their policies are with respect to fairness, honesty, and service to the public. However, what is often missing is a statement that encompasses what each employee truly believes about the goals (mission), the values, and the ethics of the corporation, a statement that each employee signs and stands by. This is known as a credo. It should include a statement about vowing to do one's job ethically and honorably, to refuse to do wrong if asked, and to seek advice and protection from others if asked to act unethically. The statement should also give the company's philosophy on the prioritization of the stakeholders.

Johnson & Johnson, the corporation that had a serious situation involving cyanide poisoning of Tylenol, followed this credo and emerged from the crisis in an ethically responsible manner (www.jnj.com/our_company/our_credo/index.htm, accessed October 21, 2003):

Our Credo

We believe our first responsibility is to the doctors, nurses and patients, to mothers and fathers and all others who use our products and services. In meeting their needs everything we do must be of high quality. We must constantly strive to reduce our costs in order to maintain reasonable prices. Customers' orders must be serviced promptly and accurately. Our suppliers and distributors must have an opportunity to make a fair profit.

We are responsible to our employees, the men and women who work with us throughout the world. Everyone must be considered as an individual. We must respect their dignity and recognize their merit. They must have a sense of security in their jobs. Compensation must be fair and adequate, and working conditions clean, orderly and safe. We must be mindful of ways to help our employees fulfill their family responsibilities. Employees must feel free to make suggestions and complaints. There must be equal opportunity for employment, development and advancement for those qualified. We must provide competent management, and their actions must be just and ethical.

We are responsible to the communities in which we live and work and to the world community as well. We must be good citizens—support good works and charities and bear our fair share of taxes. We must encourage civic improvements and better health and education. We must maintain in good order the property we are privileged to use, protecting the environment and natural resources.

Our final responsibility is to our stockholders. Business must make a sound profit. We must experiment with new ideas. Research must be carried on, innovative programs developed and mistakes paid for. New equipment must be purchased, new facilities provided and new products launched. Reserves must be created to provide for adverse times. When we operate according to these principles, the stockholders should realize a fair return. (Copyright © 2003. J&J. With permission, all rights reserved.)

Guidelines and Pitfalls of Corporate Strategies

Up to this point, we have outlined some differences in philosophies and some conflicts between ethical theories in the conduct of businesses. As there exists little in the way of guidance for corporations other than laws such as those controlling monopolies and fraud, and internally derived guidelines such as the corporate credo or policy statements, what then might be a codification of strategies for success in corporations? There are some guidebooks based on the leadership qualities of historical figures from the past that some CEOs have used to underpin their strategies. Examples are *Leadership Secrets of Attila the Hun* by Wess Roberts (1989), *The Five Temptations of a CEO* by Patrick Lencioni (1998), and *Sun Tzu: The Art of War for Managers* by G. A. Michaelson (2001). The latter takes the 50 strategies of war, as described by Sun Tzu in *The Art of War* published over 2,000 years ago, and adapts them to strategic rules for commercial corporations based on Sun Tzu's philosophy that one can avoid fighting if one plans the right strategy before the battle. Analysis of the 50 rules and their interpretation for business managers reveals not one that suggests a departure from ethical behavior if we start from the premise that the war itself is a just war. We list some these 50 strategic rules in Appendix 4.1 of this chapter for the curious reader but remark here that whereas the ethical behavior inherent in the strategies might be found defensible by most moral philosophers, the tactics employed in today's business world to execute these strategic rules frequently are in violation of ethical principles, and some are even amoral. *The Art of War* makes a distinction between strategic planning and the tactics employed to carry out a strategic objective, and this separation will become important when considering corporate ethical dilemmas, alternate solutions, and the decisions made by the individual scientist and engineer employed by a corporation.

Ethical dilemmas in the business world are often the result of fear, greed, peer pressure, and/or the need to make a profit. They may also be the result of poorly thought out decisions or those made for the sake of expediency. Initially, these decisions may not even appear to be unethical but rather seem a reasonable step without negative consequences. An example would be a line manager failing to record a higher than normal one-time temperature reading in the manufacturing process for an antibiotic; while seemingly not harmful, the elevated temperature may adversely affect the product so that it is not effective for the patient. The consequences of not following an approved manufacturing protocol can be extremely serious, not only for management and for shareholders but also for customers who may be harmed. As we shall see in the examples that follow later in this chapter, the existence of the corporation itself may be harmed as a result.

CONFLICTS IN ETHICAL THEORIES

The corporate goal of making a profit can make it difficult for management to think through a problem involving rights, autonomy, beneficence, utilitarianism, and, in some cases, virtue and justice. The difficulty arises because the goal of profit making often invalidates or conflicts with ethical concepts. The corporation, acting on its own behalf, has a primary purpose of making money. Ideally, the corporate decision maker makes a personal effort to ensure that an action serves the best interests for all concerned. However, the pressure of corporate goals can present obstacles that make decisions difficult. There is a difference between taking into consideration what is "in the best interests for all" and what is best for the shareholders and corporate board of directors. An example of this conflict is the finding of a possible defect in a product or a possibility that the product design might cause harm to the customers; here, the decision for a product recall would be in the best interests of the customers, but a corporate decision not to proceed with a $20 million product recall will save money and please the stockholders. If it is decided that a recall is not needed, and this decision is wrong, the consequences could be disastrous. The situation was different in the case of the Ford Pinto gas tank problem. In that case, a comparison was made between the cost of lives that were expected to be lost and the cost of revising the product. Lives were lost and the corporation lost due to the faulty reasoning.

In the Pinto case, a recall to fix the gas tank problem posed no additional risk to the owner. However, consider another instance in which a premature or even ill-founded recall could pose harm to the consumer over and beyond the proximate cause of the recall itself. In this example, a heart valve manufacturing defect is discovered that affects 0.5% of valves. Surgery to replace the valve carries, let's say, a 3% mortality and considerable morbidity in each instance. In a situation such as this, the remedy carries more risk to the entire population than the underlying problem itself. Thus, in this situation it becomes essential to refine as much as possible the population truly at risk so as to avoid unnecessary surgery for those whose valves are not at risk. This is a delicate balancing act with exceedingly serious consequences. Issuing a blanket recall on all valves may alarm, injure, or even kill some patients. Taking too long to figure out which batches are at heightened risk may result in valve failures that are catastrophic for some patients. Ignoring the issue by assuming it is merely a case of random failures is not an option.

In the early days and even months after a problem is suspected, it is not easy to sort out whether the problem is with the valve itself or with some other factor. There are instances where what looks like a product failure turns out to be user error instead (e.g., a surgeon holding the heart valve disk in a surgical forceps against proper procedures, thereby introducing cracks in the disk that later result in breakage and catastrophic failure). Thus, there are times when a measured reaction is preferred. While this may appear to others to be obfuscation and denial, it is really prudence (Citron 2005).

A similar example came to public light when Guidant Corporation announced that "We have determined that a magnetic switch in these devices (implantable defibrillators) may become stuck in the closed position, which in some cases inhibits the device's ability to treat ventricular or atrial tachyarrhythmias and can accelerate battery depletion. Four occurrences have been confirmed out of approximately 46,000 devices" (press release, June 17, 2005). This announcement and subsequent action by the FDA led to a recall. One month later, another 28,000 devices (different models) were also implicated. Patients had to decide with their physicians if their devices should be surgically replaced. Thus, dealing with the discovery of a defect in a product installed in patients is complicated by the fact that a simple recall will put patients at risk. The risk is that the surgical replacement procedure involves general anesthesia that can cause heart, lung, and other complications postoperatively in a patient with cardiac problems. Not all patients have a critical need for a defibrillator, but in some medical groups, the placement of defibrillators has become almost routine for all patients after a heart attack. The decision about whether and when to replace a possibly faulty defibrillator is one that requires careful analysis and the promulgation of detailed information to inform the patient and the responsible physician. The manufacturer has a duty to help in this process. Another example involved the recall of hip replacement protheses. In 2000, Sulzer Orthopedics, Inc., voluntarily recalled approximately 17,500 of its acetabular hip replacement joints. Because hip replacement surgery has a relatively high incidence of morbidity associated with it, especially in elderly patients, the consequences of the recall were significant. Here, the eventual loosening of the prosthesis was a more likely probability than that of the cardiac defibrillator failure. We will learn in Chapter 12, "Ethics of Emerging Technologies," how to approach the risk analyses that will differ for these two situations.

As far as possible, all risks and ramifications of a corporate decision should be thought through, and specific risk assessment analyses done whenever appropriate. There is danger in making decisions in the moment for the sake of expediency. Decisions that are not carefully thought through may put the corporation onto a slippery slope from which it may be difficult to escape. Application of the Four A's will structure the reasoning and help identify the best decision.

Beneficence to the Community vs. Duty to the Stockholders

Up to this point, the concept of duty to the stakeholders has been emphasized as the theme behind the trend toward social responsibility. There is another aspect to the discussion of whether social responsibility is good for society as a whole. Also, one could question whether it is good for the corporation. When the corporation uses profits to give to charity or to support other social programs such as school renovations or the local fire department, corporate executives might be contributing funds without stockholder representation, thereby shirking their duty to the stockholders. On one hand, the use of corporate funds for social programs should be done with the approval of those who might be affected by so doing (both stakeholders and stockholders). On the other hand, shareholders of companies that have foundations as part of their corporate philanthropic efforts are well aware of the mission and governance surrounding the foundation via annual reports and Security Exchange Commission filings. From time to time, a shareholder may even bring the matter of the foundation or philanthropic spending to a vote via the proxy statement, thereby giving shareholders the right to weigh in. Some shareholders actually value corporate philanthropy, since it is a sign of a corporate culture that places values other than profits on its agenda in a measurable way (Citron 2005).

Utilitarianism vs. Justice

In much of this book, the dilemmas that have been discussed have pitted utilitarianism against autonomy (e.g., the duty to do what is in the best interest of the most versus what best benefits the individual). For the purposes of this discussion about business ethics, we equate autonomy with justice. The fundamental concept of utilitarianism is the maximization of happiness and minimization of pain. The even distribution of justice throughout society may not meet this maximization of happiness or even goodness, as exemplified by Case 4.3.

Case 4.3 Distribution of Bonuses

A corporation has made profits beyond its expectations, and the board of directors appropriates large sums to each of the division managers with instructions to distribute the funds as Christmas bonuses. The method of distribution is left to the division managers to determine. The employees of one division have salaries that vary by a factor of three. Some of the employees were hired at low wages relative to their peers. The choices are to 1) divide the bonus allocation equally among all employees, 2) divide it as a percentage increase, 3) divide it in accordance with the value of the employee's contributions to the business, and 4) divide it evenly but add a supplement for the top 10 performers. The moral issues are justice or equity and utilitarianism.

The Four A's strategy is helpful in selecting the just method of distribution. Starting with acquisition of **advice**, the manager will learn which of the precedents had good consequences and bad consequences within the current culture of the company. The alternatives that include the choices listed above are not very helpful for the manager acting alone. An **alternative** in the decision making is to enlist the help of the other division managers to come up with a method of distribution that is consistent across the groups, or to convene a bonus committee to help in the decision making. The **assessment** of the tactic of using a committee will lead to the conclusion that such a committee will give the employees some comfort that the distribution was not based on favoritism from the manager. The tactic should be written down, and care should be taken to avoid discrimination. The **action** should make it clear that the bonuses were given based on the outstanding profits of the year and that a defined procedure was followed in making the determination.

What should guide the distribution of bonuses in a company whose employees have worked equally hard? Is it better to make a majority a little better off or to make a few very happy? In an analysis of justice and utilitarianism by J. C. C. Smart (2001), utilitarianism is seen as "generalized benevolence . . . and justice [is seen] as a means to the utilitarian end . . . it does not matter in what way happiness is distributed among different persons, provided that the total amount of happiness is maximized." Smart's view is that a utilitarian, given the option of one action that would give one unit of pain (e.g., a negative happiness) to one employee and nine units of happiness to another employee, or another action giving two units of happiness to them both, would choose the former even though someone would be worse off. This choice would be in conflict with John Rawls's utilitarian theory of justice that seeks to minimize pain and considers an action right only if it benefits the least advantaged person (Rawls 1999).

How, then, do these examples and the implied conflicts in moral and ethical theories relate to business ethics? In Case 4.1 about the Ford Pinto gas tank problems, the strategy was to benefit the stockholders, leaving the customers at risk. The tactics to accomplish some benefit brings into question the issue of justice. When confronted with the question "Who should benefit when not all can benefit?" one must choose the tactic that is most fair. Does one give to those who would benefit society more than others or does one draw lots?

In Case 4.3, the distribution of bonuses can be resolved by the proper choice of tactics. The tactics are the procedures for judging who should be rewarded. This frequently is decided by a peer group or judges selected by a "fair" process.

Over the past several decades, the focus of many corporations has shifted away from profit making for stockholders to now include consideration of all stakeholders in the corporation. Benefiting all the stakeholders is a utilitarian-based management goal, but the just distribution of benefits or happiness to all the stakeholders is impractical or perhaps impossible in many situations. The managers of a corporation should endeavor to follow a moral minimum. That is, they should follow their duty not to cause harm to any aspect of the corporation.

Another type of problem arises when the arguments of the utilitarian come into conflict with justice and autonomy. An example is the use of genetically modified seeds to increase food availability in parts of the planet where starvation is prevalent. There are ethical issues being sorted out about the acceptability, planting, growing, sale, and exportation of genetically modified plants (GMPs). They are reportedly easier to grow, require less water and fertilizer, are more resistant to diseases, have a longer growing season, and are more nutritious. The U.S. government has championed the development of these plants and their exportation in order to assist starving people in Third World countries. However, there are major concerns about their use. Critics look to the government subsidization of large corporations to produce and provide these crops as examples of unjust decision making. They argue that the testing for safety is incomplete and that small farmers who cannot afford GMPs will be left behind, while those who can afford them will increase corporate profits (Martin 2003; see also Chapter 6, "Ethics of Genetically Modified Organisms").

Duty vs. Justice and Virtue

Concerns for human dignity and justice are raised as moral issues when poor people in a foreign country are used to make clothing for commercial enterprises in a sweatshop atmosphere. Usually, they are paid a pittance of a salary and receive no benefits. This allows the company to maximize profits at the expense of those toiling to eke out an existence. These violations of justice in the United States have been dealt with in the past by laws that regulate fair wages and child labor. However, U.S. laws cannot be extended to other countries, and U.S. corporations contracting for services from a foreign corporation are subject to that country's policies and regulations. If the wage paid by a foreign company is considered fair by its standards, then how can one argue that there is an injustice? Can one argue that hiring relatively inexpensive laborers outside the United States is unjust to U.S. citizens because it denies them the right to employment? Would a restriction on U.S. corporations to hire only U.S. citizens be an infringement on autonomy? The government decided that justice dictates that a U.S. citizen should have priority over a non-U.S. citizen for a job for which both are equally qualified. Foreign labor certification programs permit U.S. employers to hire foreign workers on a temporary or permanent basis to fill jobs essential to the U.S. economy. Certification may be obtained in cases where it can be demonstrated that there are insufficient qualified U.S. workers available and willing to perform the work at wages that meet or exceed the prevailing wage paid for that occupation in the area of intended employment. Foreign labor certification programs are designed to assure that the admission of foreign workers into the United States on a permanent or temporary basis will not adversely affect the job opportunities, wages, and working conditions of U.S. workers. However, the current practice of contracting for services using corporations in other countries gets around this law. In countries such as India, the service corporations hire large workforces to perform the same job as is done by U.S. citizens involved in occupations such as medical and x-ray transcribers, sales personnel for catalog-based merchandising, and consultants for U.S. computer companies. The foreign business pays its employees less than the prevailing wage for the same work in the United States. Is this a just practice?

A conflict between duty and virtue arises for managers of a corporation when their duties to stakeholders might result in actions that are not virtuous, at least according to the concepts of virtue germane to the culture of one or the other of the corporations involved. Two examples illustrate the conflict.

Case 4.4 Working in a Foreign Country

A manager of a U.S. corporation with offshore subsidiaries is negotiating to use a harbor in country A for exporting materials to country B. The worth of this contract to the U.S. company is projected to be $1,000,000,000 over the next five years. The port authority agent in country A informs the manager that he will complete the contract (which is also in competition with another corporation in country A) if the manager of the U.S. corporation transfers $50,000 to the port authority agent's account. The U.S. manager, who lives and works in country A, has a duty to maintain his job in good standing by getting the contract. He must also follow U.S. law. He also has a duty to the corporate board and a duty to the stake-holders. Yet if he were to proceed with the money transfer, he would be performing a less than virtuous act and, indeed, an unlawful act by U.S. standards. Still, such transfers are the norm for country A and the competitor corporation in country A.

Case 4.5 Ramifications of a Hostile Takeover Bid for a Company

In June 2003, Oracle CEO Larry Ellison made a $5.1 billion hostile takeover bid for PeopleSoft. The move was originally made in order to counter PeopleSoft's $1.9 billion buy-out of J.D. Reynolds. The Oracle buyout would block PeopleSoft's attempt to move them-selves into the number two position after the large German independent software company SAP. The Oracle takeover bid also interrupted PeopleSoft's business because clients were reluctant to make deals or open new accounts until the takeover plans solidified because Ellison indicated his disinterest in supporting PeopleSoft's technology. Ellison had also been quoted as believing "it's not enough that we win; everyone else must lose" (Said 2003). Ellison did indeed win in December 2004, after an 18-month fight that involved five offers and a $10.3 billion takeover, and after prevailing in a suit filed by the U.S. Justice Department claiming the transaction would reduce competition (Guglielmo 2004). The CEO of PeopleSoft resigned, and 5,000 workers at PeopleSoft lost their jobs during the merger with Oracle.

This case exemplifies how the negative consequences may be unimportant to the poli-cies of at least some companies. If it were not for the antitrust laws, monopolies would be more common, driving up prices of goods and services, and even decreasing motivation to be innovative because of the lack of competitive pressure. The Oracle case represents another example of how, in a commercial enterprise, perceived duty to establish one's superiority and duty to make a profit conflict with virtue and justice.

Non-Malfeasance vs. Utilitarianism

Cases 4.6 and 4.7 present two corporate actions. One was negligent in reporting problems with procedures, and the other blatantly ignored issues of product safety that could harm individuals. The two moral theories of non-malfeasance and utilitarianism are in conflict in the first case because the benefits of the device and its continuing medical trials could be far more important than the procedural incidents. In the second case, maximizing profits was seen as more important than the potential dangers that might arise from the faulty equipment.

Case 4.6 EndoVascular Technologies

EndoVascular Technologies of Menlo Park, California, was a division of Guidant Corporation of Indianapolis. The company had developed a medical device that offered a less invasive way to surgically correct an abdominal aortic aneurysm (ballooning of the main artery in the body that can suddenly burst with catastrophic results). The $10,000 device was supposed to allow correction of these risky bulges in the aorta with minimal hospitalization and a shortened recuperation time. However, the company failed to notify the FDA of serious problems that occurred during or after its use. Within two years of marketing the product, there were malfunctions in 2,626 of the 7,000 devices sold, 57 patients had to undergo emergency reoperations involving invasive procedures, and 12 people died.

FDA rules require a company to report any death, injury, or malfunction of a device within 30 days. The company's vice president admitted to wrongdoing, unethical procedures, and failure to follow the legal requirements of reporting. As a result, the company was charged with nine counts of introducing a "misbranded medical device with the intent to defraud or mislead" and one count of lying to the FDA about the extent of the problems (Finz 2003). Company employees could go to jail, and the case continues to be investigated by the FBI and FDA. Suits filed by injured patients could total awards in the millions.

The problems were discovered during an FDA inspection of the company's headquarters in Menlo Park, California, in July 2000. Their findings are described in the Department of Justice statement below (www.usdoj.gov/usao/can/press/html/2003_06_12_endovascular.html, accessed February 12, 2005):

> During the inspection, the FDA official asked specifically for all complaints the company had received about one type of malfunction [of the Ancure Device]. The company intentionally misled the inspector by giving him a list of 55 complaints when, in fact, the company knew that there had been hundreds of complaints about this particular malfunction.

> Under federal law, a company is required to report to the FDA any incident in which its medical device may have caused or contributed to a death or serious injury or the medical device experienced a malfunction that would be likely to cause or contribute to a death or serious injury if the malfunction were to recur. The reports to the FDA are called Medical Device Reports (MDRs). In this case, the company was aware of every malfunction because, as a condition of FDA approval, it had a sales representative present in the operating room during each surgery and the reports of those failures were repeatedly tabulated and distributed to company officials. In failing to file thousands of Medical Device Reports, the defendant concealed the true extent of problems with the Ancure Device from patients, doctors and the public.

Anonymous whistleblowers (see section on whistleblowers and their responsibilities later in this chapter) had alerted the government of the problems. The FDA's subsequent investigation found that the company was out of compliance with FDA regulations. Guidant closed its division in Palo Alto, California, and the device was no longer sold after October 2003.

This case exemplifies the extent to which malfeasance can occur and the extent of the consequences for the stakeholders. It is hard to believe that a company would allow such misconduct to occur. The company paid more than $92 million in fines. A number of lawsuits have been filed against the company, not only by patients injured as a result of the device but also on behalf of stockholders who bought stock between June 23, 1999, and June 12, 2003. The allegations assert that Guidant recklessly misled investors by failing to disclose known problems with the Ancure Device (Finz and Walsh 2003). The irony is that it was not the intra-aortic device itself that malfunctioned after insertion but rather the delivery system by which the device was threaded into place. This was a problem that probably could have been remedied, and a valuable product could have been saved.

Case 4.7 Uwatec and a Faulty Dive Computer

In 1995 a design engineer for Uwatec notified company owners that its dive watch computer software was flawed. The problem was that the device underestimated nitrogen levels after closely spaced dives, so a diver relying on the computer was at risk for exceeding safe dive times and for developing the bends due to unsafe nitrogen levels. Even though high nitrogen levels could result in serious disability and even death, the owners decided to continue to sell the product. In 1996, two company managers became aware of the design flaw. They put together and made multiple copies of a recall notice, but the owners again decided not to acknowledge the problem and fired the managers. When the bill for the copies came to the person who had replaced the fired workers, a famed and experienced diver himself, he was told by one of the owners that the notice was the work of disgruntled employees who had been fired, and that there was no problem.

A year later, the company was sold. The new buyer was assured there were no defects in the dive watch, but found out otherwise when one of the fired managers sued the company for terminating him after reporting a defect. The previous owners continued to insist during the trial that there was no defect, but evidence was presented that proved otherwise. The jury voted that the manager had been terminated wrongfully, but still the product was not withdrawn from the market.

It was not until 2001–2002 that an increasing number of divers reported injuries and sued. One of them was an expert diver who had made thousands of dives as part of his work. Both he and his dive partner suffered severe cases of the bends with serious long-term debilitating consequences. Their illnesses occurred after a series of dives when the dive computer indicated blood nitrogen levels were sufficiently low to safely make an airplane flight. One of the lawyers had a copy of the memo pointing out the flaw in the dive computer in 1996. The product was finally recalled in February 2002, the day the suit was filed.

Thus, for seven years a product was marketed that put lives at risk. Clearly, the mandate to make a profit won over the duty to protect the customer and the company's duty for social responsibility. In fairness, it was not until 2001–2002 that the majority of problems with the dive watch were reported. Although the seriousness of the problems should have been predicted in 1996 when the flaw was discovered, the recourses for the design engineer and other company officials might have been limited early on because of uncertainties about the significance of the flaw.

Virtue vs. Duty

The three cases presented below show how corporate duty to stakeholders can lead to misrepresentations in advertisements or false claims of company profits. These nonvirtuous acts resulted in harm to millions of people.

Case 4.8 Misleading Product Advertising by Corporations

Regulators are increasingly prosecuting violations of truth in advertising. Recently, the FDA ordered Gilead Sciences, Inc., and Bristol-Myers Squibb to modify claims exaggerating the benefits of their HIV and cholesterol-lowering drugs, respectively, in what it termed ". . . false marketing messages" and creating ". . . a public health hazard when people are misled by false claims," according to FDA Commissioner Mark McClellan (Tansey 2003). In 2004, the FDA's drug marketing division sent warning letters to a dozen major pharmaceutical companies such as Amgen and GlaxoSmithKline, evidence that these problems were continuing.

Thus, these ongoing business practices challenge virtue by exaggerating advertising or extolling yet to be proved promises. Can a suit be brought against a corporation for lying about the performance of a product? The morality of telling a lie is a clear example of a violation of virtue and possibly non-malfeasance. For example, pharmaceutical corporations have been blamed for prematurely publishing study results that look encouraging before confirming the study results in sufficient numbers of patients. Other problems of exaggeration are in advertisements that inflate a product's effectiveness, make claims that are unproven, or fail to disclose serious adverse events occurring after the product is marketed. An example of unsupported claims in advertisement and the FDA's response is found in Appendix 11.1 of Chapter 11.

Corporations can and have been prosecuted for lying about a product. So, why is there a continuing problem with falsely exaggerating printed, televised, and oral claims for product performance? Is this because the employee, manager, or professional believes loyalty to the corporation is a duty, and this implies a duty to do what is necessary within the bounds of the law to make a profit? The employee has three mechanisms to deal with this situation: first, seek guidance from laws or professional codes that explicitly prohibit exaggerated claims; second, look to the company's credo as a guide (e.g., whistleblowing); and third, act in a virtuous manner to prevent promulgation of exaggerated information. The consequences of false claims about company profits are exemplified by the next two cases.

Case 4.9 Enron

Enron was a fast-growing energy company that by the year 2000 was the seventh largest company in the United States. Their focus was on markets for energy, water, broadband communications, and anything that could be turned into a product. It was an exciting place to work with stimulating projects and the freedom to pursue ideas. But the spring of 2001 brought the end of the corporate bubble. With a fall in the company's profits, Sherron Watkins had changed jobs within the company to work with Andrew Fastow, the company's CFO. As she looked for assets to sell off and ways to streamline, she found instead ". . . fuzzy off-the-books arrangements that seemed to be backed by nothing more than now deflated Enron stock. No one she asked could . . . explain what was really going on" (Morse and Bower 2003). Her findings of improper representations that implicated not only Enron executives but also the auditing company were forwarded early on to the Enron chairman, Kenneth Lay. The record of this e-mail shows that on the one hand, she was alerting management of major discrepancies, a virtuous act. On the other hand, perhaps because of duty or loyalty to Enron and her job, she was willing to become an accomplice, though in the end, she became an admired whistleblower. What was her dilemma during the discovery of corporate wrongdoing in her own company? She was seeking solutions to the dilemmas involving the virtue of exposing a wrongdoing vs. the duty to protect the stakeholders, including her own employment. A quote from her e-mail to Kenneth Lay is evidence of this:

> 1. The probability of discovery is low enough and the estimated damage too great; therefore, we MUST find a way to quietly and quickly reverse, unwind, write down these positions/transactions. 2. The probability of discovery is too great, the estimated damages to the company too great; therefore, we must quantify, develop damage containment plans and disclose. I firmly believe that the probability of discovery significantly

increased with Skilling's shocking departure. Too many people are looking for a smoking gun (www.itmweb.com/f012002.htm, accessed January 12, 2006).

In a subsequent meeting she had with Lay, he promised to look into her concerns, but instead, he asked Enron's law firm to make the evaluation; the lawyers concluded that ". . . the transactions Watkins questioned appeared proper, though they might look bad" (www.forbes.com/2002/02/14/0214watkins.html accessed, accessed February 12, 2005). The collapse of the company was clear in mid-October when the company announced a $618 million third-quarter loss and a $1.2 billion write-off, the very thing that had concerned Watkins. Noteworthy is the fact that the CEO, Kenneth Lay, took 1.5 million in stock options himself just before the third-quarter loss was announced. The Enron collapse ruined the company, employee retirement plans were decimated, and the sharp fall in the price of the company's stock left investors with huge losses. On December 2, 2001, Enron filed for Chapter 11.

The allegations of fraud cast aspersions on other aspects of the business dealings of both Enron and the auditing company. In the end, many others were affected. These included (Gordon 2003):

- Other accounting firms that promoted fraudulent maneuvering of funds and signed off on falsified ledgers.
- Banks that made bad loans and, investment banks that provided financial backing for the huge mergers and other deals that were never on firm ground.
- Brokerage firms that misled investors by promoting certain stocks to win business.
- Individual investment banks suspected of pressuring analysts to ". . . maintain rosy stock ratings."

These investigations led to the indictment of former Enron CFO Andrew Fastow for masterminding complex financial ventures while at Enron, alleging that he and others aimed to defraud Enron and its shareholders with unethical accounting practices that made the company look far more profitable than it was. Further, a 109-count federal indictment with other charges was filed against Fastow's wife, Lea, and seven other former Enron executives. Lea Fastow pleaded guilty to helping her husband hide income from his many financial schemes, and was sentenced to a year in prison in May 2004 (www.usatoday.com/money/industries/energy/2004-05-06-lea-fastow-plea_x.htm, accessed February 12, 2005). After numerous delays, the trial of Lay and Skilling is set to begin on January 30, 2006, as this book goes to press. Lay is charged with seven counts of fraud and conspiracy, Skilling with 35 counts of fraud, conspiracy, insider trading, and lying to auditors. Both have pleaded not guilty (Hays 2006).

The Justice Department and the Securities and Exchange Commission subsequently investigated the role of the accounting firm of Arthur Andersen. In November 2002, the firm received the maximum punishment possible under current law (a fine of $500,000) for having obstructed justice by destroying massive numbers of documents related to the Enron investigation. This decision was ultimately overturned by the United States Supreme Court in May of 2005 (*Arthur Andersen LLP v. U.S.,* case no. 04-0368), but the result was that the firm, once one of the biggest and most respected accounting firms in the United States and abroad, closed its auditing business and sold its assets via mergers and other deals. Two of the nation's largest banks, JP Morgan Chase and Citigroup, paid almost $300 million for manipulating Enron's financial statements and misleading investors as to Enron's true financial status, a violation of the federal securities laws (McClam 2003).

Case 4.10 WorldCom

WorldCom began as a small long-distance communications company in 1983, but by 2001, it had become the largest Internet carrier in the world. Its operations provided Internet services to over 100 countries on six continents, carrying over half of the Internet's total traffic, including about 70 percent of e-mails sent within the United States, and 50 percent of all e-mails in the world. It also served some of the largest agencies of the federal government, including the Defense Department, the State Department, and the General Services Administration. The price of stock in the company grew to the point that employees were dreaming of early retirement. However, by 2001, earnings were adversely affected because other competitors had entered the market.

Vice President Cynthia Cooper became suspicious of financial problems within WorldCom when "... A worried executive in the wireless division told her in March 2002 that corporate accounting had taken $400 million from his reserve account and used it to boost WorldCom's income" (Ripley 2003a). She knew that the accounting firm of Arthur Andersen, the same accounting firm involved with Enron, was also responsible for overseeing WorldCom's audits. It did not take long to find the problem: the company had categorized billions of dollars as capital expenditures in 2001, "... a trick that allowed WorldCom to turn a $662 million loss into a $2.4 billion profit in 2001" (Ripley 2003b). Cooper became a whistleblower (discussed further in this chapter's section on whistleblowing). As a result, the CFO was fired and the company publicized that the reporting of the company's profits had been overly inflated. Initially, the overinflated figure was thought to be $3.8 billion, but later was reported to be $11 billion. According to many estimates, the collapse wiped out about $180 billion in shareholder value. WorldCom sought protection under Chapter 11, filing the biggest bankruptcy in U.S. history.

The negative consequences to WorldCom's stakeholders were enormous, affecting 60,000 employees and more than 20 million customers. WorldCom's MCI phone service handled about 70 million calls every weekend alone. Tens of thousands of businesses depended on its networking services. In 2003, a federal judge and the SEC fined WorldCom $750 million for its $11 billion accounting fraud. The company filed for bankruptcy and agreed to ongoing monitoring by the Securities Exchange Commission. In reorganizing and attempting to distance themselves from the problems of the past, the new leadership elected to retain only the company's MCI business. In January 2005, 10 of WorldCom's former directors agreed to pay $18 million of their own money as settlement for a class action lawsuit filed by stockholders who lost huge sums as a result of the collapse of the company. This amount is far less than the actual amounts lost by the claimants, but the agreement is unprecedented because in the past, collapsing companies have used insurance monies to settle class action claims. WorldCom executives were investigated and charged for their roles as perpetrators in the fraud. The CEO, 60-year-old Bernard Ebbers, agreed to surrender an estimated $40 million in assets to settle with investors and was sentenced in July 2005 to 25 years in prison, and five others were to be sentenced in 2005 (McClam 2005). In February 2005, two telecommunication companies attempted a buyout of MCI.

CORPORATE RESPONSIBILITIES

This next section describes the duties a corporation has to employees, the public, and its business partners. In the United States, these duties are generally well understood and considered standard practices for the most part. However, the globalization of business markets has meant working with companies with business cultures and practices that might differ widely from those found in the United States.

Responsibilities to Employees

Corporate management has a responsibility to the employees for fair, ethical, and nondiscriminatory hiring policies, fair pay and benefits, and safe working conditions. Additional moral responsibilities involve establishment of a culture that encourages respect for employees, and makes unacceptable the harassment of employees, deception, racial inequities, and age or gender discrimination. Salary and promotion practices must be fair and nondiscriminatory. Ideally, a corporation is also expected to support an employee in developing skills and in acquiring further education whenever appropriate.

Ethnic, Religious, and Sexual Harassment Issues in the Workplace

Title VII of the Civil Rights Act of 1964 was the first to prohibit discrimination against employees on the basis of race, color, religion, sex, or national origin. The commission set up to enforce these mandates, the Equal Employment Opportunity Commission (EEOC), has specific definitions to define harassment of various types. Ethnic and religious discrimination still exist in the United States and elsewhere. These issues should be dealt with in the workplace as soon as they become apparent.

Sexual harassment is another type of harassment that is unacceptable in the workplace. The moral basis is malfeasance. It has only recently been more specifically defined and is important to understand since its definition is broader than most realize. The EEOC has defined sexual harassment as follows:

Unwelcome sexual advances, requests for sexual favors and other verbal or physical conduct of a sexual nature constitute sexual harassment when

1. submission to such conduct is made either explicitly or implicitly a term or condition of an individual's employment or academic advancement,

2. submission to, or rejection of such conduct by an individual is used as the basis for employment decisions or academic decisions affecting such individual, or

3. such conduct has the purpose or effect of unreasonably interfering with an individual's work or academic performance or creating an intimidating, hostile or offensive working or academic environment (Bravo and Cassedy 2001a).

Most federal courts, including the U.S. Supreme Court, have accepted this definition.

There are four categories of illegal sexual harassment:

1. Quid pro quo
2. Sexual favoritism
3. Hostile environment
4. Harassment by nonemployees

A quid pro quo situation occurs when an employee must meet a demand for a sexual favor in return for something else such as a bonus, raise, promotion, project assignment, or even to remain employed. Sexual favoritism occurs when promotions, bonuses, or other rewards are given to those who submit to sexual demands. A hostile work environment is one that allows the existence of unwelcome sexual conduct in the form of sexually explicit jokes, pinups, vulgarity, inappropriate touching, *or any other condition* that makes the workplace offensive for an employee; this type of behavior is, by definition, whatever is hostile to whoever is on the receiving end. And lastly, sexual harassment might be coming from outside the workplace, as from customers, contractors, or suppliers over whom the employer should or could have some control (Bravo and Cassedy 2001b). Each company should have a plan of action in place in the event harassment occurs and should take immediate action as soon as such a situation is identified.

Sexual harassment does not depend on intent; it depends on the impact of the behavior. Joe may not intend to offend or harass anyone, but the impact of his behavior is what counts. While one joke will probably not result in a sexual harassment claim, if this type of behavior is pervasive, chances are that someone will find it offensive and unwelcome.

For the academic situation of communications in the classroom, in discussion sessions, campus invited lecturers, debates, academic cultural events, campus newspapers, as well as the speech of students and teachers, the First Amendment right of free speech applies. Title IX is intended to protect students from sex discrimination, not to regulate the content of speech. In cases of alleged harassment, the protections of the First Amendment must be considered. The following case illustrates the exception.

Case 4.11 Sexual Harassment in Schools vs. First Amendment Free Speech (from Revised Guidance for Title IX [*Federal Register* 2001])

In a college-level creative-writing class, a professor's required reading list includes excerpts from literary classics that contain descriptions of explicit sexual conduct, including scenes that depict women in submissive and demeaning roles. The professor also gives writing assignments to students, which are read in class. Some of the student essays contain sexually derogatory themes about women. Several female students complain to the dean of students that the materials and related classroom discussion have created a sexually hostile environment for women in the class. What must the school do in response?

The school is not required to do anything. There is no violation of Title IX as freedom of speech protections are particularly applicable to schools in general. In addition, First Amendment rights apply to the speech of students and teachers. Offensiveness of a

particular expression as perceived by some students, standing alone, is not a legally sufficient basis to establish a sexually hostile environment under Title IX. In order to establish a
violation of Title IX, the harassment must be sufficiently serious to deny or limit a student's
ability to participate in or benefit from the education program. However, since the situation
has made some students uncomfortable, their concerns should be addressed in an appropriate forum.

Responsibilities to the Public

A corporation is responsible for protecting the public and the environment from any
adverse occurrence that might result from its activities. The corporate mission should
include responsible risk analysis to ensure recognition of potential problems and their consequences. It is a general moral and legal responsibility to notify the public of all possible
risks, and appropriate measures should be in place to protect workers and the public should
such problems occur. The company is also responsible for providing ongoing monitoring
to identify new risks that might arise due to changes in manufacturing of established products or in the manufacture of new ones.

The corporation also has a duty to the community to hold meetings with residents to
learn of their concerns, answer questions, and address their concerns up front. It is important for management to find solutions to community concerns that can include parking, the
adequacy of roads to and from the site that must support the volume of daily commuting
employees, the maintenance of views, height restrictions for buildings, use of acceptable
building materials and colors, and minimizing the impact of lighting, noise, traffic, toxic
waste, and hazards. Examples are early warning systems and countermeasures to mitigate
possible accidents such as inadvertent release of toxic substances (e.g., gases, radiation),
power failures, and fires.

One can argue that the community residents are stakeholders with whom the corporation has a duty to negotiate. Frequently, these negotiations result in the establishment of
community benefit projects at the expense of the corporations. These benefits range from
road construction and parking lot expansions to educational programs and community
beautification programs. These acts, while seen by some as acts of charity, are usually
motivated by the long-term goal of improving profits of the corporation. In parallel with
investing in research on future products, investment in improvements of the local community through creation of a better working environment and attractive employment in a competitive personnel market is associated with long-term success. Giving to global charities
is not in the same category, and this sharing in profits should be controlled by the stockholders. Some feel that this is too restrictive. In the wake of the tsunami of 2004, many
companies chose to give financial aid via organizations such as the Red Cross and Red
Crescent. Such giving arguably was consistent with the philanthropic patterns of the past,
but some saw these efforts as a quest for publicity that was a good advertisement for the
company. One can only hope that these good works are the result of trying to do the right
thing rather than to gain publicity or acknowledgment. Of course, both motives can be in
place.

Responsibilities in Business with a Foreign Country

In many countries, corporate responsibilities have been recognized by explicit rules or laws. However, as will become evident, there are many foreign countries in which business practices differ widely from those in the United States. An example is the differing definition of what is just treatment and remuneration for employees. Those working in such environments outside the United States are finding the need to learn about and adjust to these differences, because it is unlikely that corporations in other countries will be open to changing them.

Doing business with other corporations in the United States or some international corporations requires that certain standards will be met with regard to the truthful and just exchange of information, realistic time lines, and respect for confidential information. However, the rules of business in some countries may differ significantly from those of one's own country. One major difference is how various cultures accept or reject the use of bribes. Bribes can occur in various forms, including gifts to negotiators, first class hotels and travel arrangements, and secret bank deposits to accounts of negotiators. Bribes are illegal in the United States but are a way of doing business elsewhere. A lack of understanding about import and export regulations, union controls, licensing requirements, or local cultural and religious customs can severely impact negotiations. Adequate fact finding about business customs and expectations is imperative.

All U.S. companies doing business in a foreign land must abide by U.S. laws. There is a Foreign Corrupt Practices Act that must be signed by all employees of a company doing business outside the United States. If a practice is not legal in the United States (such as bribing officials or paying exorbitant or corrupt customs fees), it is not legal in a foreign country either. There are stiff fines and possible jail terms for company agents violating these laws.

Responsibilities to the Ecosystem

A question a corporation must address is whether its manufacturing processes or the product itself has an undesirable effect on the ecosystem. Other questions are whether there are plans for the use of recyclable materials. Are the labeling and packaging biodegradable or of a quality such that the distributors, customers, and ecosystem will not be harmed? Is there a plan to mitigate or minimize any negative effects on the ecosystem? The balance should lie in favor of the ecosystem. Corporations are responsible for sustaining it and taking no more from it than can be given back, as well as repairing or restoring any damage so that future generations have the same opportunities and resources that we have today.

Other corporate ethical issues having to do with the principles of ecosystem sustainability are outlined by Dorf (2001), who makes the following recommendations to promote preservation of the stability of an ecosystem and to maintain the biodiversity within it.

1. The necessary conditions for processes in an ecosystem must be maintained (for example, DDT was banned because it had lethal consequences for fish and birds).

2. Operation must be within the carrying capacity of the ecosystem (for example, automobile gridlock occurs when the carrying capacity of roads is exceeded).

3. Harvesting rates should not exceed the regeneration rate (a violation would be when the harvest of trees exceeds the regeneration rate).

4. Waste emissions should not exceed the assimilative capacity (an example is the long disintegration factor for plastic six-pack rings).

5. The rate of exploitation of nonrenewable resources should be equal to or less than the rate of development of renewable substitutes (an example is the substitution of solar or nuclear power for fossil fuel sources).

Chapter 5, "Environmental Ethics," discusses how corporations as well as individuals are stewards of the environment. Product development and disposal should reflect this responsibility. To succeed will involve the commitment of employees at all levels of a corporation.

THE ETHICS OF THE PROFESSIONAL EMPLOYEE

Thus far, we have focused on the corporate entity as an ethical agent. Next, we focus on the responsibilities of the engineer and scientist with respect to the corporation. Many of the ethical or moral decisions and conflicts encountered by a professional employee stem from the implied or contractual agreement between the employee and the corporation. This agreement is a declaration or assumption that the employee will be loyal to the corporation. The implications of this loyalty are very broad and affect the behavior of an employee over a lifetime, even when no longer employed by a corporation.

There are other important topics that relate to the moral theories of duty and the rights of others. For the most part, these relate to the concept of respecting the autonomy of the corporation and other employees of the corporation. Along these lines, there are three main topics that are important for the engineer and scientist to fully understand. These include confidentiality, whistleblowing, and conflicts of interest.

Confidentiality

It is expected that corporate employees will protect a company's trade secrets, marketing plans, and other sensitive information from competitors. It is also expected that an employee who leaves the corporation will not reveal trade secrets of the previous employment to the new employer or anyone else. In this section, we discuss trade secrets from the ethical standpoint, that is, what are the obligations of the scientist and engineer to secrecy during and after employment by a corporation?

The justification for confidentiality is based on four premises (Bok 1989a):

1. Individuals have a right to secrets.

2. Individuals have a right to share secrets.

3. A pledge of silence creates an obligation beyond the respect due to an individual.

4. A pledge of silence by professionals (i.e., lawyers, doctors, priests) assigns weight beyond ordinary loyalty to keep secrets. Professional confidentiality gives an individual's secret an expected protection within society.

The classes and protections of the intellectual property of a process or a product are:

1. Secrets
2. Copyright
3. Patents
4. Trademarks

The latter three allow exclusivity without resort to secrecy, and indeed, a patent requires the holder to publish openly the science and engineering of a method. What are the ethical issues underlying trade secrets? One of the major arguments for trade secrets is personal autonomy (Bok 1989a). The individual has a right to have control over one's own thoughts, ideas, inventions, and plans. Consider either a recipe for a chemical polymer or a cookie or plans for a machine that an individual has created or views as an important personal accomplishment. Anyone attempting to discover the secret is attempting to steal property, an action that is unethical. An attempt to discover a trade secret is a violation of the autonomy of the individual, and by extension, a violation of the autonomy of the corporate body. The definition of a trade secret given in Bok (1989b) is:

> Any formula, pattern, device, or compilation of information which is used in one's business and which gives him an opportunity to obtain an advantage over competitors who do not know or use it.

Trade secrets may be vital to protect a manufacturing process from would-be competitors. Two examples include secrets about how to manufacture Chinese porcelain and how to manufacture Coca-Cola. Porcelain manufacture resisted discovery for hundreds of years, and the process for Coca-Cola production remains a secret to this day, years after its invention.

In the history of business practices, one of the first examples of intellectual property theft occurred when the process of manufacturing porcelain was revealed (Porter 1834). The method was a tightly held secret by the Chinese, and this secret benefited the manufacturers by making them an exclusive source of valued teacups and other utensils. A Jesuit priest learned something vital about the method and transferred the information to his colleague in France, thereby stealing the secret process from the Chinese. There is no ethical problem in trying to manufacture porcelain by reverse engineering methods or by trying to duplicate it by trial and error, but stealing the process is unethical.

A reason for the use of patents is that the invention could be discovered by reverse engineering; that is, disassembly of the working invention to discover how to build a competitive product. Frequently, reverse engineering leads to the discovery of subtle details of how a machine is put together, allowing the development of a similar product that is sufficiently different from its predecessor so as not to infringe on the patent. An innovation discovered by a competitor and developed as a minor component of the competitor's product can be considered stealing art or a patent infringement. The patent protection depends on how the part or the application is defined in the claims of the original patent.

The reason that a patent application needs to disclose the engineering details as well as give evidence of the applications of the invention is that these details demonstrate the uniqueness of the invention. It also discloses to the public the methodology so that when the patent expires, the innovation can be used by others. Patents make one person's intellectual property available for use by others through the licensing process that protects the inventor and assignee. Patents expire after 20 years. For pharmaceuticals that require years of human safety and efficacy testing, the 20 years is extended by the time between the patent date and FDA approval, which can be many years.

An employee has a duty to protect the secrets of his or her corporation. The information that must be kept confidential for the protection of a corporation and its clients includes the corporation's strategic plan, experimental procedures, and data associated with the development of new products or processes. These processes might include chemical synthesis steps, specific ingredient ratios, computer codes, clinical trial plans for new drugs or devices, drugs in the pipeline or being considered by FDA for approval, or genetic modification methods. Usually a corporation protects its interests by careful indoctrination of employees in addition to contractual arrangements with each employee. However, the temptation to reveal secrets of the company is ever present, as is the intent of those outside the corporation to steal secrets. It would seem straightforward that the professional would have no problem exercising loyalty to the company and/or exercising good judgment with respect to protecting information that would give competitors some advantage. However, there are circumstances that challenge the professional's ethics and lead to ethical dilemmas. A case history will serve to illustrate a situation that occurs at technical and scientific conferences throughout the world.

Case 4.12 Employee's Protection of Privileged Information

Engineer Rubissow, an employee of a pharmaceutical corporation, has found a way to noninvasively image brain inflammation. This is a key breakthrough in studying Alzheimer's disease because it provides a way to follow the progression of the disease and the effect of medications used to treat it. He is invited to an international conference to discuss his work and replies that he is constrained by confidentiality from giving the details of the chemical synthesis of the agent but has data demonstrating its use in patients that he would be happy to present. After his presentation before an audience of 3,000 engineers and scientists, a member of the audience asks for details about the agent used to reduce inflammation. He replies that the information is proprietary, and the audience voices its displeasure. The chairman calls for other questions, and several scientists approach the floor microphones. The first questioner states that the paper is possibly a major breakthrough, but without some information on the drug used to show before and after changes in patients, it is hard to accept that the method was in fact measuring changes in inflammation. The next questioner blurts out, "But the drug could have caused a decrease in blood flow, and what you show is not a change in inflammation." The chair of the meeting interrupts and asks Engineer Rubissow ". . . can you just give us a clue about the mechanism of the drug and its formulation generically?" Engineer Rubissow then, with great embarrassment and frustration, states, "I am not a chemist; all I know is that we put a methyl group on the neuroreceptor ligand we tried to use in stroke years ago, but it never worked; now it seems to work."

At a press conference afterwards, the scientific reporters pressed the question further to get the name of the previously failed drug, and published an article quoting Rubissow as saying that his biotech company had found a cure for Alzheimer's disease by a simple chemical modification of a compound, naming the compound. Clearly, Rubissow entered the gateway of expediency, and the slippery slope led to a loss of self-respect and the loss of his employment. Could Rubissow have anticipated this chain of events? Should he have given the paper in the first place before the agent could be announced? Could he have handled the questions differently? Was the chairperson acting professionally by asking for additional details? Did the chairperson also err innocently by trying to mollify the agitated audience? Should Rubissow have lost his job? Rubissow's situation shows how a public forum or a private discussion may put the professional in an uncomfortable position of having to honor confidentiality or please his peers or host. The proper action should have been clear.

The theory of Aristotelian virtue is a starting point for the analysis of this case. According to the dictates of virtue, the individual is expected to maintain loyalty to others. The importance of this loyalty to the protection of secrets is vital for the company if it is to capitalize on ideas and innovations. These assets would be lost if disclosed to competitive corporations. Despite the fact that the concept of loyalty to a nonperson entity has been challenged philosophically, the virtue of loyalty can be the basis for proper behavior of the professional when faced with the temptation to reveal company secrets. Truthful answers to probes about company secrets can be framed around: "I am sorry, but I believe this information is still under wraps by the company, and even if I could disclose more information, I am not sure how up to date I am and would not want to mislead you."

Under the category of confidentiality, we include the importance of avoiding any communication that would reveal information about a corporation or its employees that is damaging, privileged, secret, or otherwise not available to the public. The term *insider* is applied to anyone who has this information and communicates it. If it leads someone to act as a result, such as to buy or sell the stock of the company, it is known as insider trading, which is illegal. Although generally attributed to senior management, the term can also be applied to many others, as listed in Table 4.2 (Citron 2005). The end result is that the public stakeholders that own or might buy stock are at an unfair disadvantage.

Table 4.2   Those Who Might Be Involved in Insider Trading

- Members of the board of directors
- Company managers
- Auditors
- Consultants
- Secretaries
- Friends or partners of company executives
- Janitors
- Relatives of a supervisor who learns of an impending public announcement
- Dentist who hears a story during nitrous oxide anesthesia
- Supplier of electronic parts who learns of their application

The concept of loyalty to the corporation or protection of trade secrets should be separated from loyalty to the corporation as a person, no matter what are the circumstances. Clearly, the idea of loyalty to a corporation cannot take priority over the need to reveal wrongdoing when the consequences of the wrongdoing might lead to malfeasance.

Confidentiality When Changing Jobs

When a professional changes jobs from one corporation or industry to another, the professional is obligated to keep in confidence the trade secrets of a former employer. These might include manufacturing processes, company strategies, company prospective clients, and other secrets that would otherwise be considered confidential information during the employment of the professional. Even such information as layouts of the plant, aspects of quality control, customer lists, and marketing strategies are all properties of a business that may come under protection as trade secrets, even though no originality or invention is involved (Baram 1968). There is a moral duty to respect the relationship of trust between employers and a former employee unless consent is given by the employer to the former employee to divulge certain secrets. This duty raises many difficulties for engineers who transfer to another corporation in the same line of business. Probably more violations of confidentiality occur in this situation than in any other, and there is little control or inducement to do the right thing except a reliance on the virtue of the professional.

Corporations have put safeguards into place. One is the contractual agreement signed by the new employee upon hiring. Honoring this contract can be important because any violation could be grounds for prosecution. The likelihood of a prosecution and the success in court are probably small if the secret is just a "little" trade secret. However, if the violation of a trade secret is of major consequence to the corporation owning the secret, the punishments can be severe (Baram 1968). The second safeguard used by some corporations is to require professionals leaving the company to sign an agreement that they will not transfer to a corporation in the same industry or product line. Alternatively, some companies require an employee to agree not to take a position with a competitor for a specified period of time. However, these types of contracts can be contested on grounds that they restrict an employee's freedom to select a livelihood. Enforceability is also state-specific because employment laws vary by state. Nevertheless, employee noncompete agreements have been enforced successfully on many occasions. The employee who is sued for violation of his or her employee agreement must also pay legal costs for the defense. Unless the court finds the lawsuit frivolous and assigns attorney fees to the plaintiff, these costs can be substantial.

These and other means of guarding trade secrets are probably of limited effectiveness. Knowledge of trade secrets is likely to be an inherent part of the professional's activities. So, what is a solution to the prevalent dilemma between one's duty to a previous employer and the duty to do one's professional job at a new corporation? How does the employee, whether engineer, scientist, receptionist, manager, or budget specialist, avoid malfeasance to the previous employer and at the same time deliver the maximum benefit to the new employer? It seems much can be done by instilling in the individual a deep respect for what is right, and perhaps even more importantly, by avoiding situations that might cause moral conflict for the new employee. A case illustrates how a dilemma might be handled.

Case 4.13 Duty to Yourself vs. Duty to Your New Employer

On arrival at a biotech company producing genetically modified plants (GMPs), Engineer MacMillan is assigned to the DNA sequencing laboratory, a new direction of research for her. Her previous work was to develop methods to make the seeds of GMPs nonviable to help keep genetically modified plants separate from non-GMPs. MacMillan sees an announcement about an upcoming symposium on seed infertility. This topic is very interesting to her, and she thinks she might make a major show of her talents by attending and adding ideas. Yet, she senses a danger that she will be tempted to reveal secrets from her former company. She elects not to attend, thereby avoiding the gateway that could lead to a loss of self-respect, professional disgrace, or put her in the position of being marginalized to the DNA sequencing lab for the rest of her employment. The lesson of this example is to avoid situations where the duty to keep company secrets might be in jeopardy.

Loyalty vs. Duty to Report Misdeeds

Confidentiality requirements can come in conflict with the professional's duty to report illegal or dangerous activities. There is no obligation to maintain confidentiality when industrial or private agencies have been involved in the covering up of crimes. For an engineer, the authority for not following contractual rules of confidentiality is the ethical code of various engineering societies (e.g., Amer. Soc. Civil Eng. [ASCE] Canon 1; Natl. Soc. Prof. Eng. [NPSE] Sec. III; Amer. Soc. Mech. Eng. [ASME] Canon 1). The responsibility of the professional who discovers the occurrence of misdeeds within the corporation is to take some action, but the extent and type of action vary, depending on the nature of the problem. The section below on whistleblowing provides guidance on how to proceed.

Whistleblower Responsibility of Corporate Employees

The term *whistleblower* is applied to an individual who observes wrongdoing and reports the situation to persons who can take remedial actions. Guidelines for whistleblowing were presented under scientific misconduct (see Chapter 2, "Ethics in Scientific Research"), and these guidelines are applicable as well to professional activities within corporations. Generally, this reporting is done within the corporation. To go outside the company or to the press may endanger the integrity of the whistleblower and be considered disloyal. It may affect the future trust of mangement toward the whistleblower, as well as the willingness of other employees to work with the whistleblower in the future. Although admired by some, the whistleblower may be identified as a troublemaker or disliked by colleagues. This was the experience of each of the three whistleblowers who disclosed major corruption at Enron, WorldCom, and the FBI. Whistleblowers may also lose promotion or other opportunities even though the allegations are proven to be correct.

There are other reasons a whistleblower should not go public with a case:

- Placing a complex, unresolved issue into the public arena can produce unpredictable results that are not necessarily in the best interests of the whistleblower, the accused, or the enterprise.

- Such publicity may compromise the integrity of an ongoing inquiry or investigation.
- An attempt to circumvent the institutional process might bias those charged with reviewing the allegation.
- Respect for privacy and confidentiality of both the whistleblower and the accused would be compromised.

A charge of slander can result if oral accusations are made, or of libel if written accusations are made; in either situation, the case can be sidetracked to the courts with gag rules imposed. Thus, whistleblowing is never to be undertaken lightly or for the wrong reasons, and not until the certainty of facts can be documented as fully as possible. Because such accusations are serious, the dimension of the wrongdoing should be sufficiently important to justify putting one's reputation or career on the line, since the ramifications of accusing another of wrongdoing are likely to be large. For the same reasons, the whistleblower should have firsthand information about the wrongdoing and should have the facts before proceeding. Other options besides whistleblowing should be considered and exhausted before proceeding.

Management, on the other hand, should make clear the ethical standards to be upheld within their institutions and set an example by living up to these standards themselves and by showing that wrongdoing will not be tolerated. A mission statement or credo can serve to underscore the high expectations for employees, and an open door policy to be apprised of concerns should serve to underscore management's commitment to excellence. Policies and procedures should be in place to handle the allegations as quickly and fairly as possible, which may mean having an independent committee or panel of experts review the charges.

Whenever whistleblowing occurs, it is a generally difficult and career-threatening occurrence for the perpetrator. Sometimes there are also unexpectedly negative consequences for the whistleblower as well. It is also an exceedingly difficult task for those having to investigate the allegations, an emotional time for coworkers of the accused who, even though tangentially involved, might also come to have their reputations tarnished and their work interrupted.

Though the public awareness of the need for whistleblowing and the benefits of whistleblowing in mitigating egregious practices in the financial industry are now well recognized, it is generally accepted that the professional life and general happiness of the whistleblower are negatively affected in spite of the billions of dollars and potentially many lives that the whistleblower has saved. Part of the deterrent to whistleblowing is retaliation by employers. Currently, many states as well as the federal government have laws to protect whistleblowers. The first law of significance was the Civil Service Reform Act of 1978, which protected federal employees from reprisals for disclosures of information on violations or any government regulations, including misuse of authority. Additional sources of relief for whistleblowers include:

1. Government Accountability Project (GAP) (see www.whistleblower.org) is an independent organization that assists government employees and government contractors' employees who need guidance and protection in the whistleblowing process. Founded in 1977, GAP is a nonprofit, public interest organization, and law firm that receives funding from foundations, individuals, and legal fees.

2. Professional societies have a focus on ethical problems and can assist the individual by protecting the whistleblower from negative consequences. They might also assist the whistleblower in finding a new job (see www.ewh.ieee.org).

Obligations to Report Safety Violations or Threats to the Public

How should a professional proceed in the event a misdeed or act by another individual or another corporation could cause a public danger? Generally, the professional who has observed potential malfeasance inside another corporation has a responsibility to make this situation known. Although a person outside the corporation where the misdeed is occurring is not in a strong position to intercede, that professional might be the only one to alert the corporation to the problem. Guidance regarding the expected behavior is also provided by the explicit statements in the code of ethics of professional societies, such as that given below from the American Society for Mechanical Engineers (ASME):

> If engineers have knowledge of or reason to believe that another person or firm may be in violation of any of the provisions of these Canons, they shall present such information to the proper authority in writing and shall cooperate with the proper authority in furnishing such further information or assistance as may be required (Section 1d, ASME Ethical Code).

The relevant canon is that which states that the engineer shall hold paramount the safety, health, and welfare of the public.

Conflicts of Interest

A conflict of interest can derail a project, compromise professional reputations, and disrupt the lives of an entire team of researchers and students. Over the past several decades, it has become more and more important to identify how these conflicts arise and to manage them because of the increase in multidisciplinary projects involving multiple companies and industry-academic partnerships. Aspects of conflict of interest have already been discussed in Chapter 2 relative to academic researchers. Here we present the issues from the perspective of the professional employee.

When an employee engages in professional or personal business activities outside the responsibilities to a company, there is a potential conflict of interest. The reason is that outside activities might result in compromised or less than optimal service to an employer or client. Even the appearance of a conflict of interest must be avoided as this can also bring harsh consequences. Obvious examples of conflict of interest are the employment of a professional as a consultant to a competitive company or the establishment of a new company as a moonlight activity using the tools and clients of the employee's parent company.

The best protection and optimal guide for avoiding conflict of interest allegations against the professional is for the employee to declare all outside interests to the employer. The methods of making that declaration are very convenient in most settings, and even if it appears the outside project or employment might be disapproved, the professional must make the declaration and provide a compelling reason why that activity is consistent with the expectations of the employer.

FEDERAL REACTION TO CORPORATE FRAUD

Corporate fraud and abuse have been so frequent that President G. W. Bush has formed a committee to pursue the problem. The function is as follows (www.whitehouse.gov/infocus/corporateresponsibility, accessed 16 June 2003):

- Exposing and punishing acts of corruption
- Holding corporate officers and directors accountable
- Protecting small investors, pension holders, and workers
- Moving corporate accounting out of the shadows
- Developing a stronger, more independent corporate audit system
- Providing better information to investors.

However, it should be noted that this committee does not have on its agenda the discovery and remediation of the root causes of corruption, which should be an important part of its task. In the cases of corporate fraud presented in this chapter, whistleblowers stepped up to set the record straight. Employees need to be prepared to do the same if they perceive wrongdoing, taking careful note of the need to be reasonably certain of the validity of their concerns. Once the process is initiated, it cannot be turned back.

CHAPTER SUMMARY

The duty of the members of a corporation is to make a profit while avoiding harm by respecting rights of individuals and by practicing virtue in consideration of the many stakeholders. Having an ethical approach to business is as important as it is to have an ethical approach to science or any other field of endeavor. Ethical dilemmas appear in the business world as a result of fear, greed, peer pressure, and the need to make a profit. The penalties for unethical behavior can be extremely serious, not only for management and shareholders but also for customers and competitors. The failure to acknowledge a defect within a product or organization not only can affect the customer who is harmed but also can be fatal to the existence of the corporation itself. Recent debacles are examples of conflicts on multiple levels that ultimately wreaked havoc not only within the corporations and on its unsuspecting employees but also on stockholders and other stakeholders with far-reaching consequences.

The goal of this chapter, and indeed, the goal of this text, is to show that ethical wrongdoing can be avoided by education. Not all decision makers will have had instilled in them through training and experience a strong set of personal values to guide them in decision making. To a large extent, virtue and values are learned, and now that the business world has become globalized, learning what other corporations and cultures value is also a necessity. The Four A's design strategy is helpful when dealing with a technical or business negotiation when the values of the negotiators differ.

It is clear that in the consideration of any ethical dilemma, there must first be the intent to make the best decision and to stand by one's ethical principles. One must be able to see the warning signs of a decision that could go in the wrong direction, and to use reflection as well as consultants if necessary in spotting the potential for wrongdoing. The multitude of dilemmas is such that the experience of a wise manager or employee is invaluable in many situations. The application of the Four A's design strategy is also valuable to discipline even the most successful decision maker to make an organized consideration of all of the conflicting aspects of a dilemma, allowing better choices. Otherwise, an instantaneous or expedient decision in the absence of adequate consultation and reflection might begin the step onto the slippery slope of wrongdoing. Had time been taken to draw on the experience of others and to think through the consequences, things might have turned out differently. Indeed, there is currently a policy of employing an ethicist consultant in corporations, particularly those involved in biotechnology, but this consultant should not be considered a substitute for management's making a full assessment of the ethical problems.

An effective mechanism to prevent or discourage the occurrence of unethical situations is the establishment of a company credo that considers the importance of the multitude of stakeholders in actions of the corporation. Managers and employees should agree to the tenets of the credo. It is also important to establish a corporate culture that encourages open communication and establishes avenues by which employees can report issues of safety, harassment, or unethical acts should they arise. These issues should also be addressed in the employee handbook and in required educational programs that review these issues at intervals.

Because scientists and engineers today may well be faced with an ethical dilemma in the course of their careers, we synopsize the customization of the design strategy for making ethical decisions in business dilemmas:

- Acquire facts and advice from peers and other corporations.
- Alternative solutions need to include consideration of the credo.
- Assessment must consider the consequences of solutions for all stakeholders.
- Action plans must be adapted to the dynamics of the situation, particularly with regard to the many stakeholders.

Of further importance is the need for corporations to adopt policies of social responsibility, and these policies need to be with representation from the stockholders. The successes of the companies with the best reputations are attributed to policies that in addition to serving the interests of the shareholders also give some priority to the major stakeholders, who are the customers, the employees, the public, and the environment.

REFERENCES

Baram, M. 1968. "Trade secrets: What price loyalty?" *Harvard Business Review* **46**(6):66–74.

Bok, S. 1989a. *Secrets: On the Ethics of Concealment and Revelation*. New York: Vintage Books Edition, a Division of Random House, Inc., pp. 119–122.

Bok, S. 1989b, p. 136.

Bravo, E. and E. Cassedy. 2001a. "Sexual Harassment in the Workplace. In Chapter 6: "The Modern Workplace: Transition to Equality and Diversity." In: *Business Ethics: Readings and Cases in Corporate Morality.* W. M. Hoffman, R. E. Frederick, and M. S. Schwartz (eds.). Fourth Edition. New York: McGraw-Hill Press, p. 329.

Bravo, E. and E. Cassedy. 2001b, pp. 329–331.

Burke, J. 1990. Personal communication.

Callahan, D. 2004. *The Cheating Culture: Why More Americans Are Doing Wrong to Get Ahead.* Orlando, FL: A Harvest Book, Harcourt, Inc..

Carr, A. Z. 1988a. "Is Business Bluffing Ethical?" In: *Ethical Issues in Professional Life.* J. C. Callahan (ed.). New York: Oxford University Press, p. 71.

Carr, A. Z. 1988b, p. 70.

Carr, A. Z. 1988c, pp. 70–72.

Citron, P. 2005. Personal communication.

De George, R. T. 1999. *Business Ethics.* Fifth Edition. Upper Saddle River, NJ: Prentice-Hall, Inc..

Dorf, R. C. 2001. *Technology, Humans and Society: Toward a Sustainable World.* San Diego: Academic Press, p. 11.

Federal Register. 2001 (January 19). "Revised Sexual Harassment Guidance: Guidance of Students by School Employees, Other Students, or Third Parties." Title IX. Office of Civil Rights. *Federal Register* 66(13). www.ed.gov/offices/OCR/archives/shguide/index.html, accessed March 1, 2005.

Finz, S. 2003. "Guilty plea in medical fraud—12 patients die. Bay area branch of Guidant fined $92.4 million in criminal and civil penalties over malfunctions." *San Francisco Chronicle,* p. A1.

Finz, S. and D. Walsh. 2003 (June 17). "Menlo Park medical firm to close over fraud." *San Francisco Chronicle,* pp. A1 and A5.

Fleddermann, C. 1999a. *Engineering Ethics.* Upper Sadle River, NJ: Prentice-Hall, p. 89.

Fleddermann, C. 1999b, pp. 88–90.

George, W. W. 2003a. *Authentic Leadership.* Hoboken, NJ: Jossey-Bass/John Wiley & Sons.

George, W. W. 2003b, pp. 27–29.

Gordon, M. 2003 (May 14). "Corporate fraud investigations turn focus to bankers." *San Francisco Chronicle,* p. B3.

Guglielmo, C. 2004 (December 30). "Oracle takes control of PeopleSoft." *San Francisco Chronicle,* p. C1.

Hauser, J. R. and D. Clausing. 1988 (May–June). "The house of quality." *Harvard Business Review* **66**:63–73.

Hays, K. 2006 (Jan. 6). "Enron execs seek new trial venue." *San Francisco Chronicle,* p. C5.

Kelly, M. 2004 (Winter). "Holy grail found: Absolute, definitive proof that responsible companies perform better financially." *Business Ethics Magazine,* p. 4.

Ladd, J. 1982 (June). "Collective and individual moral responsibility in engineering: Some questions." *IEEE Technology and Society Magazine* **1**(2):3–10, p. 6.

Lencioni, P. 1998. *The Five Temptations of a CEO.* Hoboken, NJ: Jossey-Bass/John Wiley & Sons.

Levitt, T. 1958a. "The dangers of social responsibility." *Harvard Business Review* **36**:41–50.

Levitt, T. 1958b, p. 44.

McClam, E. 2003 (July 29). "Banks settle Enron charges with SEC." *San Francisco Chronicle*, p. B5.

McClam, E. 2005 (July 14). "Former WorldCom CEO gets 25 years." *San Francisco Chronicle*, pp. E1 and E6.

Martin, G. 2003 (June 24). "Agriculture secretary pushes new crops." *San Francisco Chronicle*, p. A3.

Michaelson, G. A. 2001. *Sun Tzu: The Art of War for Managers.* Avon, MA: Adams Media Corporation.

Morse, J. and A. Bower. 2003 (December 30, 2002–January 6, 2003). "The party crasher." *Time Magazine*, p. 55.

Porter, G. R. 1834. *A Treatise on the Origin, Progressive Improvement, and Present State of Manufacture of Porcelain and Glass.* Philadelphia: Carey, Lea & Blanchard, pp. 20–26.

Rawls, J. 1999. *A Theory of Justice.* Second Edition. Boston: Harvard University Press, pp. 131–133.

Ripley, A. 2003a (December 30, 2002–January 6, 2003). "The night detective." *Time Magazine*, p. 46.

Ripley, A. 2003b, p. 47.

Roberts, W. 1990. *Leadership Secrets of Attila the Hun.* New York: Warner Books Publishers.

Said, C. 2003 (June 10). "Ellison hones his 'Art of War' tactics. Swashbuckling CEO of Oracle draws early blood in his ruthless attempt to take over PeopleSoft." *San Francisco Chronicle,* p. 1A.

Schinzinger, R. and M. W. Martin. 2000. *Introduction to Engineering Ethics.* New York: The McGraw-Hill Companies, Inc., p. 174.

Schwartz, P. and B. Gibb. 1999. *When Good Companies Do Bad Things: Responsibility and Risk in an Age of Globalization*. New York: John Wiley & Sons.

Smart, J. J. C. 2001. "Distributive Justice and Utilitarianism." In: *Business Ethics: Readings and Cases in Corporate Morality*. W. M. Hoffman, R. E. Frederick, M. S. Schwartz (eds.). Fourth Edition. New York: McGraw-Hill Publishers, Inc., pp. 66–67.

Tansey, B. 2003 (August 9). "FDA slaps drugmakers for misleading claims: HIV and cholesterol treatments under fire." *San Francisco Chronicle*, p. B1.

ADDITIONAL READING

Badaracco, J. L. 2003. *Harvard Business Review on Corporate Ethics*. Boston: Harvard Business Review Paperback Series, Harvard Business School Publishing Corporation, 208 pp.

Beauchamp, T. L. and Bowie, N. E. (eds). 1988. *Ethical Theory and Business*. Third Edition. Upper Saddle River, NJ: Prentice-Hall, Inc., 596 pp.

Callahan, J. (ed.). 1988. *Ethical Issues in Professional Life*. New York: Oxford University Press, Inc., 470 pp.

Dorf, R. C. 2001. *Technology, Humans and Society: Toward a Sustainable World*. San Diego: Academic Press, 500 pp.

Harris, C. E., M. S. Pritchard, and M. J. Rabins. 2000. *Engineering Ethics: Concepts and Cases*. Second Edition. Belmont, CA: Wadsworth Publishing Co., 300 pp.

Herkert, J. (ed.). 2000. *Social, Ethical and Policy Implications of Engineering*. New York: IEEE Press, 339 pp.

Hoffman W. M., R. E. Frederick, and M. S. Schwartz (eds.). 2001. *Business Ethics: Readings and Cases in Corporate Morality*. Fourth Edition. New York: McGraw-Hill Companies, Inc., 638 pp.

Jennings, M. M. 2001. *Business Ethics: Case Studies and Selected Readings*. Fourth Edition. Mason, OH: South-Western College Publications, 416 pp.

Johnson, D. G. (ed.). 1991. *Ethical Issues in Engineering*. Upper Saddle River, NJ: Prentice-Hall, Inc., 392 pp.

Johnson, R. A. 2003. *Whistleblowing: When It Works and Why*. Boulder, CO: Lynne Reinner Publishers, Inc., 171 pp.

Michaelson, G. A. 2001. *The Art of War for Managers*. Avon, MA: Adams Media Corporation, 202 pp.

Rampton S. and J. Stauger. 2001. *Trust Us, We're Experts! How Industry Manipulates Science and Gambles with Your Future*. New York: Jeremy P. Tarcher/Putnum (Penguin Putnam, Inc.), 360 pp.

Schlossberger, E. 1993. *The Ethical Engineer*. Philadelphia: Temple University Press, 284 pp.

Schwartz, P. and B. Gibb. 1999. *When Good Companies Do Bad Things: Responsibility and Risk in an Age of Globalization*. Hoboken, NJ: John Wiley & Sons, Inc. 194 pp.

Velasquez, M. G. 2001. *Business Ethics*. Fifth Edition. Upper Saddle River, NJ: Prentice-Hall, Inc., 524 pp.

Whitbeck, C. 1998. *Ethics in Engineering Practice and Research*. Cambridge, UK: Cambridge University Press, 330 pp.

Web Sites Helpful for Business Ethics Issues

www.business-ethics.com/current_issue/index.html

www.business-ethics.org/index.asp

Center for Ethics and Business at Loyola Marymount University in Los Angeles, found at www.ethicsandbusiness.org/whoweare.htm

APPENDIX

Appendix 4.1 Some Strategic Rules from Sun Tzu (Michaelson 2001)

Thoroughly Assess Conditions

Compare Attributes

Look for Strategic Turns

Build a Sound Organization Structure

Apply Extraordinary Force

Coordinate Momentum and Timing

Marshal Adequate Resources

Make Time Your Ally

Everyone Must Profit from Victories

Know Your Craft

Plan Surprise

Gain Relative Superiority

Seek Knowledge

Be Flexible

Win Without Fighting

Strength Against Weakness–Always

Beware of "High-Level Dumb" (those at the scene should prevail, those at the scene should be experienced)

Obey Fundamental Principles

Maneuver to Gain the Advantage

Achieve the Critical Mass

Deceive Your Competitor

Develop Effective Internal Communications

Gain the Mental Advantage

Be Invincible

Attain Strategic Superiority

Use Information to Focus Resources

Consider Tactical Options

Prepare Adequate Defenses

Avoid the Faults of Leadership

PROBLEM SET

4.1   Your duty as a manager of a biotech company in the marketing division is to make a profit this year. One way to do this is not to pay the invoices for advertisements until the next fiscal year, three weeks away. What are the ethical considerations you might think about in making a decision on how to proceed?

4.2   Create a credo for a biotechnology company that produces wireless biomonitoring equipment for hospital and home care. The credo should contain elements that bear on the goal of the company, respect for employees, the methods for disclosing wrongdoing to management and to the public, and the quality of the product. In short, the credo should reflect the corporation's responsibilities to all of the stakeholders.

4.3   A U.S. company had problems carrying on its business in the congested port of Klong Toey, the principal port of Bangkok. Not only was the harbor severely congested with traffic, but the short land approach to the port itself was so congested that it took hours to cover the distance and longer still to move from the entry of the port to the loading area because of the numerous "customs fees" that had to be paid along the way. As a result, shipping schedules could not be maintained, there were delays in the delivery of goods to clients, and earnings were affected as well.

The company was able to identify a Thai port abandoned after the Vietnam War that was sufficiently close and of a size that it could be used as a substitute port to avoid the severe congestion of Klong Toey. The use of this other port would invigorate the area, and its use would be a "win/win" situation for both parties. Yet there were lengthy and unexplainable delays in finalizing negotiations with the Thai Transportation Agency and other officials involved. One day, a call came to the U.S. negotiation delegation that a donation to the coffers of the Transportation Minister in the amount of $250,000 would close the deal. This was not done because it is illegal by U.S. law, although another competing company did make the contribution and won the contract. Discuss alternate solutions the U.S. company could have considered.

4.4   A dilemma regarding employee promotions challenges ethical theories of duty and justice. An employee with severe asthma cannot work on the remodeled floor of your office building because of reactions to the carpet glue or material. The only position open in another department appropriate for this individual's skills would entail a promotion, but this individual has less seniority than another employee in the same department. How should the manager responsible for both employees approach the problem? What ethical principle(s) is (are) involved?

4.5   The National Academies have as their missions the service of national needs. They are supported by gifts, government funding for specific studies, and funding by private institutions seeking to evaluate social and educational issues (examples of which might be: what are the current curricular issues in undergraduate bioengineering? or, will adult stem cells suffice for tissue regeneration?). The state of California wishes to conduct an unbiased and pristine review of several proposals competing for $40 million in funding to enhance the state's health and education status. The state offers to pay the National Research Council (NRC) $0.5 million to conduct the review. The NRC is a nonprofit institution whose specific mission is to conduct studies in the national interest. The state of California is interested in improving the wealth of the state, and the competition is an exclusive to state of California businesses and institutions.

A. Should members of the National Research Council, who conduct reviews and workshops in the interest of the nation, be involved in reviewing proposals?

B. If you were the manager of the NRC, how would you deal with this request from a single state, having recognized it would establish a precedent for use of NRC and members of the Academies to perform reviews?

4.6   A professor, in trying to encourage a failing student by being friendly, calls on her in class for trivial questions, and on occasion makes friendly remarks such as "You did really well today" or "You handled my question well." The student continues to fail, and after she receives a D for the course, she makes a formal complaint of sexual harassment. What are the steps the professor and the university might take in this case?

4.7   Whistleblowing: A vice president of a 50-employee publicly traded engineering company learns that the board is considering selling the company to a large corporation with inducements of stock options and stipends for the board of directors. Each of the five board members would receive $50,000/year for his or her participation. A stipulation in the negotiations requires secrecy of the details until approval by the Security and Exchange Commission officials. The credo of the company is very specific on issues of fairness and disclosure to employees and the public. The vice president objects, informing the president that to go forward would be a violation of the company's credo. The president tells the vice president that disclosure of the pending deal to employees or the public would drastically perturb stock prices. What should the vice president do? (During the conversation, the president boasts that he has just bought 100,000 shares in his own company.) (80 words or less.)

4.8 Conflict of interest: A chief scientific officer of a biotechnology company making a vaccine for Alzheimer's disease learns that in the initial safety study of the vaccine, 2 of 10 patients have died. The presumption is that the vaccine killed them. He has 20,000 shares of the corporation stock now valued at $60 each. He knows that the stock value will drop as soon as there is a public announcement of the failure of the vaccine. What should he do? If he holds onto the stock, it will go down. If he sells it, it will be sold to the public who will lose when it goes down. How should he proceed? (Use 20 words or less to describe moral as well as legal responsibilities.)

4.9 Credo vs. vision statement: What are at least three major elements of a credo? Discuss the components that should be in a credo according to the text, and the components you might add. (60 words in bullet format.)

4.10 A professor consults with company A. Company A pays him a stipend in addition to giving the university a grant of $200,000 per year. The grant funds pay the summer students' stipends associated with research the company has approved, allow the purchase of equipment, and provide travel funds to support research goals that may benefit company A. What are the potential conflicts of interest and how should they be managed? (Answer in 20 words or less.)

4.11 Johnson Controls manufactures batteries and employs women and men in an industry where lead toxicity is a high probability. Because lead poisoning might result in deformed fetuses, the company decided to restrict the battery-processing jobs to men. In 1977, the company realized it would be in violation of the Civil Rights Act of 1964 by denying women the right to perform the required job based on gender. As part of its health-monitoring program for women and men, the company discovered high levels of lead in the blood of eight pregnant women. In 1982, the company decided to reinstitute its policy, but this time it allowed women to work if they could demonstrate their inability to have children by medical evidence. Two women sued the company on the grounds that it was up to them to decide on the risks, that gender discrimination had denied them job opportunities, and that demonstration of the medical evidence of sterility was a violation of their privacy. The district Court and the Court of Appeals ruled in favor of the company, but the U.S. Supreme Court ruled against the company.

Blue Team – Present the case for the women employees.

Gold Team – Present the case for the company.

4.12 Below is a belief statement written by a student in response to an assignment to create a credo. Evaluate what elements might be missing and what elements are essential to any company you might imagine, large or small:

Our first responsibility is to consumers who use our products and services. We strive for the highest quality in our products and services while constantly trying to reduce our costs and the cost to the consumer. Customers' orders must be fulfilled promptly and accurately. We will evaluate our products in a rigorous and well-recorded manner and will provide full and fair disclosure of product ingredients. Consumers will be promptly informed of any changes in product ingredients and other services via public news releases.

We maintain a responsibility to our employees for providing a clean, orderly and safe work environment and a sense of security in their jobs. Employees of our company have a duty to respect each individual's dignity, and the company will recognize an employee's merit by providing compensation that is just and fair. Each qualified individual must have equal opportunity for employment, development, and advancement. Employees have the right to refuse to engage in activities that he/she believes is unethical and unacceptable without repercussion. Employees shall respect the confidentiality rules of the company and have a duty to report any unethical conduct. Use of company resources for personal affairs should be kept to a minimum.

We pledge to provide competent management whose actions are just and ethical. New ideas and individual accomplishments will be recognized. Technical publications shall honor the highest standards of ethical conduct and should be approved by management before they are sent for publication. Management and employees should have open communication lines, and in cases of allegations of misconduct, management will investigate claim discreetly. We guarantee that communications with management will remain confidential, and there will be no retaliation for coming forward. Grievances will be addressed in a prompt and fair manner by management.

We have a responsibility to our stockholders, and they will be informed of advancements and mistakes in our research in a timely and honest manner. We have a responsibility to our community and will encourage civic improvements, education, and protection of the environment.

_____   _____
   Signature   Date

4.13 The eight categories in the table below represent situations wherein an ethical problem might arise. We have listed a score for whether a particular moral theory might or might not be in conflict with a given category. The problem is to mark with an X or O those empty boxes under "Business." As the duty to make a profit and to benefit all the stakeholders is a given, we entered O as a correct answer.

For the box denoting autonomy, your answer must specify whose autonomy is being respected or not respected. For this box, explain your answer in less than 40 words.

Application of Ethical Theories and Principles to Business Ethics

	Business	Profession	War	Sports	Poker Game	Society	Medical	Person
Virtue		0	X	X	X	X	0	0
Justice		0	0	X	X	0	0	0
Duty (Kant)	0	0	0	0	0	X	X	0
Utilitarianism		0	0	0	0	X	0	X
Rights		0	0	0	0	X	0	0
Autonomy		0	X	X	0	X	0	0
Beneficence		0	X	0	X	0	0	0
Non-malfeasance		0	X	0	X	0	0	0

0 = applicable to activity.
X = at odds or not applicable to the activity.

5

Environmental Ethics

Some 20 centuries ago [there was] the story of two men who were out on the water in a rowboat. Suddenly, one of them started to saw under his feet. He maintained that it was his right to do whatever he wished with the place that belonged to him. The other answered that they were in the rowboat together—the hole that he was making would sink both of them (Vayikra Rabbah 4:6).

INTRODUCTION

In earlier chapters, virtue, justice, rights, duty, and utilitarianism have been applied to situations affecting individuals or communities of individuals. This chapter extends the ethical community from individuals, corporations, professional societies, and governments to include animals, plants, and the environment. Though concerns about local environmental degradation existed in ancient cultures of the world, by the last half of the twentieth century, these concerns extended to the preservation of the global environment. Table 5.1 shows a broad overview of humanity's relationship with the environment. Prior to the tenth century, efforts were consumed with survival in a relatively hostile environment. However, during the tenth to twentieth centuries, the major activities were in expanding control of the environment, until it became clear that resources were running out and must be sustained for the future.

Today's world can no longer ignore the fact that the Earth's human population is using up natural resources, which will lead to their depletion for future generations. The pollution of air, land, and water threatens to damage the planet in ways that cannot be reversed by generations of our successors.

Sustainability has become the watchword of this millennium, with nations around the world being encouraged by the U.N.'s Agenda 21 to conserve or restore environmental resources, and to use only what is necessary. Scientists and engineers will need to make decisions about what actions to take in the face of the ethical dilemmas to balance human needs and sustainability. Their input will also be important for technical solutions, such as finding methodologies for efficient energy production, water conservation, food production and distribution, waste management, and pollution control.

The principal factor depleting resources and at the same time contaminating the environment is the increasing requirement for and use of energy. Not only are there more people requiring energy, but also each individual is consuming more energy per year due to

Table 5.1   Evolution from Taming to Sustaining the Environment

The Environment before the Tenth Century	Mastering the Environment Tenth to Twentieth Centuries	Sustaining the Environment for the Future
Dense forests	Clear	Decontaminate
Floods	Dam	Conserve/recycle
Arid deserts	Irrigate	Restrict
Insects/pestilence	Expand	Control emissions
Dangerous animals	Build	Control water use
Ice	Pave	Search for alternatives
Heat	Transport	
Cold	Manufacture	
Geographic isolation	Mechanize	
	Excavate	
	Provide energy	

advances in technology that generally improve the quality of life. Examples of these technologies include refrigeration units, air conditioners, automobiles, computers, and cell phones. The downside is that despite efforts to make these technologies more energy efficient, they contribute to further pollution. Sixty percent of the carbon dioxide (CO_2) produced in the United States is from burning fossil fuels for transportation and manufacturing. Fluorocarbons, toxic waste, and sludge are contaminating the land, rivers, and wetlands.

In addition to using up resources, the need to balance sustainability with increasing energy demands has become a global problem. Thus, this chapter includes an examination of current and possible future methods of meeting and sustaining the energy needs of Earth. To understand better the thinking of different cultures, we review the perspectives, ideologies, and moral thinking relative to the use and neglect of the environment by human beings. We have also summarized the international efforts to date (2006) to create solutions and to regulate our use of the environment. Even in these efforts, we will find serious ethical conflicts between utilitarianism and individual rights. No matter what the philosophy of the individual engineer, manager, or scientist, there is an essential responsibility to consider the environmental aspects of a project with respect to the rights of others in the short term and the intergenerational consequences in the long term.

Some ethical questions to consider while studying this chapter are:

- Who decides what is good or bad for the ecosystem?
- How can one put value on the loss of a forest or an endangered species?
- Are instruments such as the Earth Summits of 1992 and 2002 and the Kyoto Protocol sufficient to reach the goal of sustainability?
- When is resource decline justifiable for the goal of civilization development?

- Is the degradation of the environment simply the result of the conflict between individual rights and utilitarianism?

- Why is society able to manage and preserve some common resources, yet unable to manage and preserve others?

We start with two examples of environmental disasters. One illustrates the consequences of societal destruction of the environment on Easter Island; the other shows one of the consequences of modern technology.

Case 5.1 The Collapse of Easter Island

This is an example of an environmental disaster that occurred in the microcosm of tiny Easter Island in the South Pacific. The sequence of events and their consequences for its inhabitants caused this civilization to collapse, bringing it to near extinction in the relatively short time from A.D. 900–1600. This story brings an important message for today's world, as chronicled by Jared Diamond in his book *Collapse* (2005).

Easter Island, tiny and isolated, is located 1,200 miles from its nearest Polynesian island neighbor to the west and 2,300 miles from Chile to the east. Though its subtropical location is similar to Florida in the northern hemisphere, Easter is cooler and windier because it is far from a continental land mass and surrounded by colder water. Before the arrival of the first settlers, the island was entirely forested with a variety of large trees, many birds, and an abundance of porpoises and other fish that could be accessed by large sea-going canoes. The population at its peak is estimated to have numbered around 15,000. Today, there are no trees, few birds, only a small number of Easter Island descendants, and many hand-carved stone statues that appear to have had a spiritual or protective presence for the clan that erected them. The average statue was 10 feet tall and had a mass of about 10 tons (the largest statue successfully moved into place was 32 feet in height with a mass of about 75 tons, although others still in the rock quarry were 70 feet long and weighed 270 tons). The stone platforms on which the statues were erected weighed between 300 and 9,000 tons each. Competition between rival chiefs on the island was probably the reason later statues were larger still, and also included the placement of a cylinder of red rock, which weighed up to 12 tons, on top of the statues' flat heads.

Their preoccupation with constructing, moving, and setting up the platforms and statues required enormous food resources for the 50 to 500 people involved (stone carvers, transport crew, erecting crew), as did the celebratory feasting for the clan. These activities are estimated to have added 25% to the food requirements for the population. Aside from food, their most needed resource was trees, which were used as rails to move the statues, as large canoes for safe transport to their distant fishing waters, for cooking, and for fires to warm them in their generally cool, windy climate. Massive amounts of wood were also used for cremations. By about 1600 A.D., there had been complete deforestation of the island.

The consequences included severe soil erosion and decreases in yields from crops, problems that could not be resolved. The lack of trees meant they had no wood to burn for warmth. Their cold winters, isolated geographical isolation, and lack of resources made it impossible to foster trade for needed supplies. With no large canoes, they could not get to their fishing waters or to other islands. The land could not support them, nearby fish were scarce, and birds and land animals (except for rats) had dwindled to extinction. They had nowhere to turn or go. Starvation set in, the chiefs and priests were overthrown, and the people swiftly resorted to cannibalism, civil war, purposeful toppling of many of the statues, and anarchy. Epidemics brought by the few European visitors to the island killed off many, and others were abducted as slaves. By 1872, only 111 islanders were left on the island, which has since been annexed by Chile.

Diamond compares their plight, and the plights of other failed civilizations of the past, with those of the Earth today. We, too, are destroying the forests, polluting the environment, using up more and more resources, and destroying many species that depend on these elements. We, too, have "nowhere to go," and are in the midst of political, social, and religious upheavals that are of growing concern. With our more advanced tools and machinery, we have a far greater capability for destruction of our planet than did the Easter Islanders. As with Easter Island, our planet and its environment are becoming far more fragile as a result of the increasing pressures of encroaching civilization. Now that we, too, are outstripping our resources so quickly that the next generations will not have access to them, we must be courageous about adopting a more sustainable way of living. Otherwise, we may fall to a similar fate.

Case 5.2 How Recent Technology Is Contributing Environmental Harm

A more recent example of harm that can come to the environment involves discarded computers and other electronic devices from the United States being sent to China for disassembly to scavenge some of the metals used to make them. The workers are exposed to noxious gasses and other toxic materials but are wearing no protective gear. The waste materials are thrown into rural rivers and land areas, littering the landscape, poisoning the air and drinking water with mercury, lead, and dioxins, and creating toxic sludge. Great harm is resulting, not only to the environment but also to the health of over 100,000 workers eking out an existence from these activities in the poorest sections of the country. Yet there is no public policy in place to stop or clean up this problem that is sure to affect future generations (Fackler 2002).

This case is but one example of the numerous ethical dilemmas facing our society on the environmental front. It is not only a matter of how each individual contributes to the problems, but also how each individual can become responsible for protecting the environment. Most important is how we all, individually, locally, and globally, can make decisions that support a healthy planet.

These first two cases have underscored the environmental problems we face. We next turn to review the greatest factor contributing to environmental pollution today, the current energy-producing technologies. We also provide a review of emerging alternative technologies that have the potential to meet energy demands and diminish environmental degradation.

CURRENT ENERGY PRODUCTION AND USE

Most of the production of energy today comes from burning fossil fuels (i.e., coal, oil, and gas). These fuels power steam turbines for electricity production and engines for transportation. The current energy-producing technologies have four negative consequences with ethical ramifications that include the potential for changing life on this planet as we know it:

- Depletion of the nonrenewable (nonsustainable) sources of energy.
- Damage to public health due to emissions and atmospheric contamination.
- Degradation of the environment.
- Deleterious changes in climate due to increasing levels of CO_2 and other emissions.

Ethical dilemmas have resulted from these situations:

- The conflict between the autonomy of the individual or a particular sector of society vs. the utilitarian needs of another individual or sector of society.
- The unequal distribution of resources and unequal distribution of responsibilities for stewardship of the environment.
- The possible long-term intergenerational risks to people and the planet.

Two of the most pressing problems for the planet are related to the combustion of fossil fuels. First, the use of these fuels is depleting nonrenewal resources of the Earth. Second, the emissions from burning fossil fuels have contaminated the atmosphere with CO_2 to the extent that they have the potential for causing global changes in the Earth's climate. The rising levels of CO_2 contamination of the atmosphere are due primarily to the burning of fossil fuels for the production of electricity and for transportation. CO_2, oxides of nitrogen and sulfur, as well as halocarbons released from refrigeration processes, are greenhouse gases that act to trap heat and prevent its escape into outer space. Rising temperatures, melting ice caps, droughts, and other extreme climate changes are believed to be the result of the accumulation of these gases in the atmosphere. The increases in these emissions are recognized as threats to the planet that are in need of global regulation. However, not all countries agree that the methods for regulating these emissions are just, citing the unequal distribution of responsibility for reducing emissions that would put some governments at an economic disadvantage relative to other countries (see the discussion of the Kyoto Protocol later in this chapter).

Predicted Energy Needs for the Future

The importance of developing alternative energy sources is seen from estimates that in the next 200 years, there will be a global increase in the demand for energy by a factor of eight. This is calculated from an estimate of population growth by a factor of two, and an average increase in the per capita energy use in the world equal to the per capita use in the United States today, an increase of five times the current use. The issue is not that fossil fuel reserves will be depleted in 200 years but that their extraction will be more difficult and possibly prohibitively expensive. This is because the readily accessible oil and gas resources will have been mined, and those remaining will be more difficult to recover. Figure 5.1 shows the relative predicted use of various sources of energy in the future.

It will take perseverance and conviction to move to a greater use of the new energy sources because fossil fuels are currently an inexpensive source of energy, and gearing up the alternative sources will be an expensive undertaking. However, the costs for their installation are rapidly decreasing and engineering developments will improve efficiency. Not only are the alternative sources becoming economically viable but attitudes toward these sources are becoming more positive (see below). The United States and other countries are

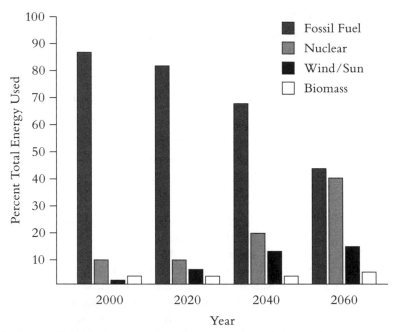

Figure 5.1 Current (World Resources Institute 2003) and one model of predicted future consumption of energy sources by percent. These numbers assume the acceptance of nuclear reactors as a source of power and an increase in the use of wind and solar sources from 0.65% in 2001 by 10% per year to 2020, and by 5% per year to 2040, remaining constant thereafter. The prediction is dependent on the assumption solar and wind energy will decrease in cost by a factor of four in the next 20 years. Note that even though in the future there is a breakthrough in new methods for clean, inexpensive energy production, we will still require fossil fuels for almost 50% of our energy needs.

growing wary of dependence on foreign sources of oil. Further, there is a growing desire for cleaner sources of energy and financial incentives are being provided by governments to ease the transition from fossil fuel-based electricity to solar and wind electricity generators, alternative technologies that may gain favor in the future.

The annual energy use by the United States is 30% of the total energy used in the world (World Resources Institute 2003). The United States consumes five times the per capita average energy use in the world, an electrical power consumption of 1.3 kilowatts per person (see Appendix 5.1 for definitions of dimensions and terms). Of all the energy consumed in the United States, 87% comes from nonrenewable fossil fuel resources (oil 40%, coal 24%, and gas 23%). Sixty percent of the oil used by the United States is imported. Fossil fuels in general are the predominant source of electricity in most developed countries of the world. Nuclear energy currently provides 9% of the U.S. energy consumed per year but as much as 41% of the energy consumed in France. The negative consequences of burning fossil fuels for transportation and production of electricity are well known. There is also substantial information regarding the risks of fission reactors that generate nuclear energy. However, little is known about the consequences of other technologies that may be increasingly used in the future for energy production, such as solar panels, fuel cells, wind, ocean waves and currents, and the burning of biomass. However, these technologies provide only about 4% of the total energy used in the United States. Figure 5.2 shows six sources of electrical power.

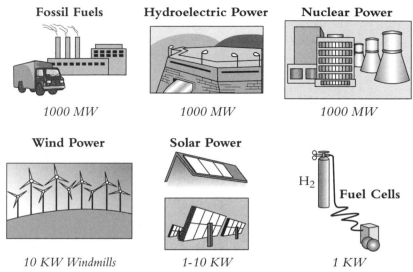

Figure 5.2 Of the six sources of electricity, the most common is steam turbines powered by oil, gas, or coal. Electricity is the major commodity produced by the methods outlined here. Fossil fuels are also the major source of energy for transportation.

Transportation: the Greatest Consumer of Fossil Fuels

Carbon dioxide emissions in the United States increased by 40% in the five years from 1990 to 1995. Transportation, including air travel, is the technology that is the fastest growing contributor of CO_2. Transportation-based emissions account for about one-third of the total carbon dioxide emissions of the United States, an amount about equal to those from industry. Each automobile contributes 5 *tons* of carbon dioxide to the atmosphere each year, assuming that the distance driven is 12,000 miles. The 200 million gas-fueled automobiles contribute about 10% of the carbon dioxide produced by the United States. In addition to carbon dioxide, other emissions have also had a detrimental effect on the planet, as evidenced by acid rain and pollution of the land, rivers, and lakes. The number of passenger cars per capita in the United States is six times that of the average for the rest of the world. Relative to other developed countries (e.g., Germany, Switzerland, France, Italy, Austria, and Great Britain), the United States has about 25% more automobiles per capita.

Despite its great benefit to us, the automobile also produces smog, a form of environmental pollution that is a major contributor to health problems. This is an ethical issue because while an individual driver might accept the risk of an accident in exchange for the benefit the automobile provides, environmental pollution puts others at risk from emissions. The increase in air pollution is directly related to a marked increase in asthma and other lung diseases over the past decade, diseases that contribute to over 40,000 deaths per year (Kay 2000). The negative consequences of transportation-based emissions on health and mortality in the United States result in an estimated annual uncompensated health cost of $500 billion, which does not adequately reflect the value of lost or diminished human life (www.pewclimate.org/policy_center/policy_reports_and_analysis/brief_energy_policy/picture.cfm, accessed April 29, 2005). In addition, these emissions might cause genetic mutations (Somers et al. 2004, Samet et al. 2004).

The future prospects for wind, solar, and nuclear energy to replace fossil-fueled energy production have the potential, in principle, to provide all the electrical needs of the United States. But they cannot satisfy the energy requirements for all forms of transportation. Ship and aircraft transportation will not be possible without using a burnable fuel. Ship engines require oil or coal, but nuclear propulsion could be substituted. Aircraft jet engines require petroleum fuels for propulsion, and propeller-driven electric aircraft are not feasible because the mass of the batteries would be prohibitive. Development of transportation based on burning hydrogen and methane produced from electricity generated by nuclear reactors, solar panels, or wind turbines will be an engineering challenge.

Energy Use by Communications Technologies

The communication technologies impact the environment in a number of ways:

- Telephone poles are aesthetically unappealing.
- Electromagnetic fields created by electricity passing in wires overhead have given rise to health concerns, though currently unsubstantiated.
- Energy requirements for new technologies such as cell phones are significant, and industry continues to manufacture energy-dependent products that require the use of more and more nonrenewable resources.
- Many new products negatively affect the environment by noise pollution.
- Disposal of communication devices pollutes the environment.
- Diversion of attention during cell phone communication while driving puts the public at risk for accidents.

Cell phone technology has other consequences not often considered. Cell phones require energy, and this consumption of energy is growing. In New York, there are about six million cell phone users. At 5 watts power average between standby and transmitting, the average energy consumed is 0.1 kilowatt-hour per day. Thus, six million persons will consume 600,000 kilowatt-hours per day of energy for cell phone communications. The electrical energy comes from electrical generators, the same amount as the energy used by 100,000 homes if the use is 6 kilowatt-hours per day. Relative to the total usage of electrical power, cell phone usage by New Yorkers is only 0.4%. Nevertheless, the global use does add up. For example, the number of cell phone users projected for China in 2007 is 500 million. As the current Asian use of power is only 0.22 kilowatt per person, cell phone usage could be 2.3% of the consumed electrical power. Some might argue that this energy consumption is trivial relative to other uses, but the cumulative amount of energy used in the future may make a significant impact on gas emissions and pollutants from fossil fuels that are burned to provide energy to charge them.

Other significant but hidden power consumption comes from appliances, charging devices, and communication and entertainment technologies that remain in standby or readiness mode for quick turn-on. For example, television sets have transformers and filament heaters that allow for an instantaneous picture without having to wait for the device to "warm up." As long as the TV set is connected to the power outlet, it consumes on the average 15 watts, which can add up to a total energy use of 0.4 kilowatt-hours per day. When one adds to this the power consumed by the multitude of devices connected to electrical power sources around the home (computers, printers, scanners, chargers, and so on), the power used can easily add up to around 200 watts. A study of German household appliances showed that standby power (off-mode power consumption) was 17% of the total power consumed by appliances in the house (IEA 2000). This consumption can be minimized by improvements in engineering, the use of sleep circuitry to turn off all nonessential charging and entertainment devices after a predetermined period of nonuse, and consumer conservation behavior.

Alternative Energy Technologies

Approaches to circumvent the problems relating to the use of fossil fuels range from developing alternative sources of electricity to finding methods for limiting pollutants from burning these fuels. One method of limiting the emission of CO_2 is to separate this gas from effluents and bury or sequester it deep into the Earth's crust (Andersen and Newell 2004). Engineering designs for CO_2 capture systems are currently aimed at power plants and natural gas purification plants. However, one might consider development of a CO_2 extractor as part of the exhaust system of automobiles, trucks, and buses. It is estimated that the cost of deploying CO_2 capture systems will increase by 20% the cost of a kilowatt-hour to the consumer (e.g., from \$0.10 to \$0.12) (Socolow 2005).

There are six main alternative sources of energy production: fuel cells, nuclear fission reactors, nuclear fusion reactors, wind turbines, hydroelectric generators, and solar photon conversion systems. These sources and their ethical implications are discussed below.

Fuel Cells

Fuel cells operate by converting the chemical energy of a fuel, such as hydrogen, and an oxidant, such as oxygen, to electricity. This alternative depends on a supply of hydrogen. However, hydrogen is most commonly obtained by separation from natural gas, which itself is a cheaper, more efficient fuel. There are other sources of hydrogen. One is the classical process involving electrolysis of water, with the electricity for electrolysis supplied from photovoltaic methods. Another method involves the use of freshwater algae, which produce hydrogen, if starved, from oxygen and sulfur (Greenbaum 1982, Zhang and Melis 2002). Risk analyses will be required for both of these sources of hydrogen because their manufacturing and distribution technologies involve the danger of fire and explosions associated with gaseous or liquid hydrogen.

Nuclear Reactors

Nuclear reactions heat water to produce high-pressure steam to power generators similar to those used in hydroelectric production of electricity. This technology has a great potential for providing a cheap and practically inexhaustible source of electricity. The production of energy from nuclear fission presents two important issues for the health and safety of the public and the environment: the safety of reactor itself and the possibility of radiation exposure from reactor fission products. Each of these is discussed below.

The likelihood of a reactor accident that would cause fatalities or result in cancer-causing radiation release into the environment has received extensive evaluation using probabilistic risk analysis (see Chapter 12, "Ethics of Emerging Technologies"). A panel of scientists and engineers in conjunction with the Nuclear Regulatory Commission used this form of risk analysis to set guidelines regarding the probability of a human fatality from an accident involving a nuclear reactor compared to the probability of death from all other accidents. The engineering guidelines for safety are to set the risk of death from a reactor core accident at 1,000 times below that from other accidents. Specifically, the NRC specified:

- The probability that a person living near a nuclear power plant will die within a few weeks or months after a nuclear accident from the radiation released in the accident must be less than 0.1% of the total probability that a person will be killed in any accident.

- The probability of death from cancer for any member of the public following an accident must be less than 0.1% of the total probability that a person will die of cancer from all causes.

About 500 persons per million die from all accidents each year in the United States. Thus, the first safety goal provides that the probability per year that the person living next to a nuclear plant will die soon after a nuclear accident from the radiation released in the accident must be 1,000 times less, that is, less than one in two million. The accident safety goal therefore is 5×10^{-7}. An analysis of cancer risks gives the safety goal of 2×10^{-6}. The most extensive study of reactor safety ever conducted, NUREG-1150, was completed by the NRC in 1990. It was determined that the likelihood per year of an accident large enough to cause at least one early fatality to the public is in the range of one in one million to one in one billion per year. Thus, it was concluded that fission reactors are safe alternatives to burning fossil fuels for energy.

The second issue is the potential contamination of the environment from the fission products of nuclear reactors. This is the major focus of ethical debate. Originally, the idea in the 1950s–1960s was that the uranium mined from the earth and used in fuel rods to power nuclear reactors could be partly recovered from the spent fuel rods, a process known as reprocessing. Although reprocessing results in nine times less radioactive waste requiring storage compared with nonreprocessing, it also recovers radioactive uranium and extracts radioactive plutonium with a half-life of 24,000 years that could be used for nuclear weapons. The argument is that if the United States reprocessed spent nuclear fuel, then other countries would do so as well, leading to a large supply of plutonium in the world with the widespread potential for nuclear weapons proliferation. The U.S. decision was not to reprocess it. Thus, since the mid-1970s, most of the spent but highly radioactive waste in the United States is temporarily being stored at the nation's 110 nuclear power plants and at the weapons facilities until a more permanent facility can be built. One 1,000-megawatt nuclear facility generates 2 cubic meters of waste per year. One can project the requirement for storage of waste from 1,000 reactors needed to supply the current U.S. electricity use as the volume of a four-story building 40 feet on each side.

Possible storage sites for this material under consideration by the United States and other countries are listed below:

- Store in granite or other rock caverns deep in the ground.
- Store in containers and drop into deep ocean trenches.
- Store in deep salt beds free of water that become self-sealing due to the heating from radioactive decay.

- Launch into space.
- Transmute it in accelerators or other reactors.
- Transport to the surface of the Antarctic ice sheet and allow the warm (from radiation) containers to melt through the ice until they reach the ice-rock interface.
- Store on Native American reservations.

Each of these options presents risks and consequences of unknown probability. For example, dispersion into space has the risk of a failed launch; deep-sea burial requires the engineering of containment systems with a half-life of 24,000 years, systems that are not yet available. Land-based burial of fission products is being planned by all 15 countries with active reactors. For example, in the early 2000s, Germany decided to phase out all nuclear plants over a period of about 10 years; storage is to be in a salt deposit facility at Gorleben. France depends on nuclear plants for 41% of its total power consumption, but a phase-out is also planned, with storage options in granite, clay schist, or salt deposits. However, in Japan, nuclear power plants are necessary to sustain the needs of 100 million people on a small island. There, the plan is to bury fission by-products in granite and dense volcanic ash deposits.

In the United States, because of the interest in becoming independent of foreign oil, there have been government discussions to increase the number of nuclear plants. However, there has been marked debate on the subject because of public concerns about the safety of nuclear plants, and because it has proven difficult to find a place to store nuclear wastes from the nuclear reactors currently in use due to widespread objections by those living nearby. The Nuclear Waste Policy Act of 1982 specified that the U.S. Department of Energy (DOE) would be responsible for retrieving, transporting, and disposing of the used fuel in geologically sound underground facilities by 1988. However, this schedule was delayed because it has been difficult to find acceptable storage sites that people could agree upon. The most recent plan to store the waste in the Yucca Mountain facility in Nevada was met with such debate and public outrage there that the planned completion date of 1998 was not met. Critics say the site is unsafe, pointing to potential volcanic activity, earthquakes, water infiltration, underground flooding, nuclear chain reactions, and fossil fuel and mineral deposits that might encourage future human intrusion. The ethical discussions involve utilitarianism vs. the rights of the inhabitants of Nevada to choose not to be exposed to possible health risks from ionizing radiation. Efforts to find a place for storage that would have insignificant risks of environmental contamination have been extensive, as has been the debate regarding this storage.

Delays in the project have meant that the nuclear waste facility was not ready as expected in the early 2000s, and currently (2006) it is still unclear whether it will open. Interim storage plans have been proposed at multiple sites, including a Native American reservation in Utah. These delays and the inability to properly store nuclear wastes are likely to interfere with any plans to move ahead with nuclear power (Wald 2005):

> Rep. David Hobson, R-Ohio, chairman of the House Energy and Water Development Committee, said: "If we want to build a new generation of nuclear reactors in this country, we need to demonstrate to Wall Street that the federal government will live up to its responsibilities under the Nuclear Waste Policy Act to take title to commercial spent nuclear fuel."

In fact, the problem of storage of wastes, whether from unprocessed fuel rods or the waste material from reprocessing plants, has curtailed an already wary public's interest in nuclear reactors as a method of supplying electricity. Two prior nuclear plant accidents at Chernobyl and Three Mile Island and subsequent fears about terrorist attacks have also fueled public concerns about the safety of nuclear reactors. The explosion at the Chernobyl nuclear power plant in the northern Ukraine April 26, 1986, sent 50 tons of radioactive dust into an area of nearly 140,000 square miles, covering parts of the Ukraine, Belarus, and Russia. The radioactivity was 200 times the amount in the atom bombs dropped on Hiroshima and Nagasaki combined. It is estimated that as many as 4.9 million people were exposed to significant doses of radiation. The 1979 reactor accident at Three Mile Island in the United States did not result in radiation exposure or injury to the public but raised fears that it could have. Thus, even if the storage problem is resolved, there is a continuing public perception of risk from this technology.

The nuclear power industry is an example of a technology developed without thoroughly thinking through the problems and possible consequences from waste disposal. It shows the lack of foresight and analysis of the trade-offs between the known risks of storage problems vs. the presumption that not allowing processing will be an effective deterrent to nuclear weapons terrorists. Other technologies also have toxic wastes with long-term consequences for the environment. These wastes include heavy metals with infinite half-lives that are dispersed in the effluent of manufacturing processes and are present in many electronic products. Metals such as chromium, zinc, beryllium, arsenic, lead, and mercury are biological hazards, yet until recently, the United States disposed of them in landfills. California has begun a mandatory program requiring the recycling of computers and television sets funded by the consumer, and two other states have followed suit, although funding is coming from manufacturers; 24 other states are planning similar programs, and Congress is assessing the need for a national program to mitigate pollution from discarded electronic appliances (Epstein 2005).

Nuclear Fusion as a Source of Energy

The sun and some stars produce light and heat through nuclear fusion. This occurs when two atoms of hydrogen or hydrogen isotopes (deuterium and tritium) fuse together to form energy released as heat and light. This energy is equivalent to $E = mc^2$, where m is the difference between the mass of the two hydrogen nuclei and the mass of the by-products, helium and a neutron. The energy is released as kinetic energy of the products. A neutron comes out of the reaction with most of the energy (14.1 MeV), and an alpha particle (helium nucleus) with the rest (3.5 MeV), but it has proven difficult to capture the energy in the neutron. Because deuterium is naturally available (one out of every 6,500 hydrogen atoms in water is deuterium), and tritium can be generated by the neutrons interacting with a lithium-doped encasement around the reaction, this mechanism for energy production has almost inexhaustible fuel. A temperature of 100 million degrees is required to achieve an ignition in which the energy of the helium nuclei will sustain ongoing fusion for the process of inertial confinement and laser-driven fusion. A method of confinement is needed so the nuclei are held together long enough to achieve fusion.

Currently, fusion has been achieved but not a sustained reaction. Success resulted from the use of magnetic confinement, and improvements in this method should allow success with sustaining the burn. Research in this area has been underway for 40 years, and some successes have spurred a multinational effort to build an experimental reactor, which should be operational in 2007. There is also a U.S. effort underway at the National Ignition Facility in Livermore, California, which expects to reach ignition in 2010. Large fusion reactors (producing thousands of megawatts) are expected to have significant economies of scale such that the cost of electricity will be competitive. The minimum competitive size of a fusion plant is one producing from 1,000 to 1,500 megawatts of electric power. As the per capita use of electrical power in the United States is 1.3 kilowatt per person, a 1,300-megawatt facility could supply enough electrical power for a city of 1 million persons.

What are the ethical problems? Nuclear fusion is a technology of the future whose risks seem minimal at present, but a risk analysis similar to that used for fission plants may show unacceptable negative consequences not yet appreciated. Although fusion is a nuclear process, the products of fusion are helium and neutrons. Helium is not dangerous, and the neutrons that could be problematic are contained in the shielded confinement space. Fusion plants have significant advantages over fission reactors, coal- or oil-burning power plants, and even hydroelectric dams. Fusion systems, while necessarily comparable in size to a large fission reactor, will not have the major problems that have prevented the acceptance of fission reactors by international societies. There is minimal waste material, but what waste is produced cannot be reprocessed or recycled into new fusion plants. The major waste is contained in the activated walls of the confinement cave, which receives neutrons generated by the fusion reaction. This material does not have the long-lived radioactive products of a fission reactor, and the material need not be recycled except during refurbishing of the containment facility, which is required only at 10-year intervals. There are likely to be very low levels of radioactive tritium in the environment because of some leaks, but the shorter half-life of the low levels of released tritium will amount to less background radiation than one normally receives from thorium and radium in the natural environment. Still, the possibility of local radiation leaks, as well as the estimates of the likelihood of a maximum credible accident (the worse accident which can reasonably be expected to occur), needs to be included as part of a thorough risk analysis.

Will successful fusion plants replace other alternate energy sources? Probably not, particularly for small towns and single-family dwellings that can be supplied by wind, photo-voltaic devices, or fuel cells. However, a major advantage of fusion plants could be to supply hydrogen for fuel cells by hydrolysis. This hydrogen could be a source of energy for transportation and fuel cells for other electricity needs. The most likely use of fusion would be to supply electrical energy for industrial needs, which require energy in amounts larger than are likely to be generated in sufficient amounts by solar or wind power. Storage of electricity from fusion reactors would be the same as storage with steam- and hydro-powered turbines.

Fusion reactors are more likely candidates to provide energy in areas of dense populations. This is so because they are capable of providing large amounts of energy that would be wasted in less-dense population areas. Fusion reactors will be expensive to operate, but these costs will be covered by multiple users. One of the ethical problems associated with electrical distribution systems, including those expected from fusion reactors, involves the controversy over the deleterious effects of the electromagnetic fields associated with high-current transmission lines. These possible risks need further assessment, although after years of investigation there are no convincing data to show that such risks are real.

One problem is how to store electrical energy during low demand periods. For this, some have suggested using a superconducting magnetic storage ring. However, these types of storage rings could have environmental consequences with some risks that are difficult to predict and may be unacceptable. These risks include magnetic field effects on migrating birds as well as regional plants and animals. For example, will high local magnetic fields change or reset the navigational system of migrating birds? Will small currents generated by animals moving in a magnetic field gradient change their behavior, growth, and development? Because this approach requires underground high magnetic fields, the risks to the environment need further assessment.

Wind Production of Energy

Wind power for the production of electricity was popular on farms around the world before distributed electrical power was available. The industry all but died prior to the oil crisis of the 1970s, but now there is an increasing demand for small (20 watts to 1,000 watts) to large turbines (greater than 30 megawatts). Wind energy is the world's fastest growing energy technology. In 2005, the total U.S. wind-generated power was 6,300 megawatts, which is equivalent to only six conventional power plants. The Wind and Hydropower Technologies Program of the U.S. government works with industry to develop the advanced technology needed to convert more of the wind into electricity, and currently, the industry is subsidized at 1.7 cents/kilowatt-hour. However, the impact of wind power on the power requirements of the United States is unlikely to exceed 6% of the total requirements for electricity in 2020 because the wind speed and open space availability are not uniform throughout populated areas, and the supply of electricity is dependent on the reliability of robust wind speeds. This statement does not imply that wind power cannot play a major role in supplying electricity for local and regional use, as the role of wind power for improving the quality of life of individuals around the world is significant. However, here, too, there is a need for further improvements in technology for storing power in the absence of wind.

To better understand the extent to which wind power could play a role in supplying energy, one can calculate how much electrical power can be derived by a small windmill placed anywhere the wind has a speed of about 10 miles per hour, or 4.5 meters per second. The fundamental concept is that wind of velocity V, turning fan blades covering an area A, will produce power in accord with:

$$\text{Power} = 1/2 \, \rho \, A \, V^3 \qquad\qquad 5.1$$

where ρ is the air density, and A is the area covered by the blades of radius R. Using a wind speed in meters per second and a radius of the blades in meters, the conversion is simply:

$$\text{Power} = 1.92 \, R^2 \, V^3 \qquad\qquad 5.2$$

Thus, for a wind velocity of 10 miles per hour (4.5 m/s) and a blade diameter of 3 meters ($R = 1.5 \, m$), the theoretical power would be 394 watts. (Note: a watt has units of ML^2T^{-3}, see Appendix 5.1). The efficiency is less than 20%, so this system would supply only a single appliance such as a radio, television, or light, while at the same time charging a battery for use when there is no wind. A household system of about 10 kW would require a blade diameter of 7 meters and a wind speed of 12 m/s. Wind power plants with blades 26 meters in diameter can supply 250 kilowatts of power, or about 50% of the theoretical amount calculated from equation 5.1.

Despite these shortcomings, the potential of wind as a source of electrical power is seen by some cities as an attractive supplement to coal-fired steam turbine electrical production. For example, London expects to supply 10% of the city's electrical power needs by installation of offshore wind-driven turbines with a power capacity of 1,000 megawatts. Environmental costs include the destruction of significant numbers of birds, the loss of significant open space, and the insertion into the landscape of large objects that are not aesthetically pleasing.

Hydroelectric Production and Storage of Energy

Hydroelectric power, generated by moving water sources, provides 12% of the electricity consumed in the United States currently. This source is unlikely to increase, and in fact, as other sources improve, it is predicted that there will be a relative decline in the contribution from hydroelectric production. However, its efficiency can be increased by finding ways to store its energy during times of low demand. Better efficiency through energy storage methods, such as superconducting magnetic storage, could increase the efficiency of hydroelectric power use. However, as noted previously, the use of superconducting storage rings will have environmental consequences with risks that are difficult to predict and may be unacceptable.

The ethical aspects of hydroelectric plants currently in use have not changed for years. Installation of dams for purposes of serving electricity to populations remote from the dam is based on utilitarianism. Autonomy is challenged for those whose lands are compromised, those whose irrigation waters are jeopardized, and those who are at risk of a dam failure through natural or unnatural causes (engineering flaws, terrorism).

Solar Source of Electricity and Fuel

The power (energy per time) available from the sun during daylight hours is a major direct source of electricity and fuel. The sun's photons can be converted to electric current or chemical substances using special materials or natural biological substrates. While most solar energy conversion is currently for the production of electricity, there is the potential for conversion of the sun's photons to produce hydrogen and methane gases for use in transportation through photoelectro-chemical processes. We discuss here the conversion of photons from the sun to electricity by a simple device known as a photovoltaic (PV) cell.

A typical silicon PV cell consists of a thin wafer of boron-doped silicon known as the P-type layer. On the side facing the sun is an ultra-thin layer of phosphorus-doped silicon (N-type). When sunlight strikes the surface of this wafer (usually called a cell), the electric field provides momentum and direction to electrons, which are stimulated by the energy of the incoming photons (a few electron volts). When wires from the P-type and N-type layers are connected to an electrical load, such as an incandescent light or even a television set, the result is a flow of electrical current. Regardless of size, the PV cell produces 0.5–0.6 volt under no load conditions (i.e., no attached device to consume energy). Commercial solar panels, each comprising 30 or more solar cells, come in areas ranging from a few square centimeters to square meters. A panel with the area of an 8.5 by 11 inch sheet of paper (0.06 m^2) can deliver nearly 5 watts at peak incident sunlight. A panel with an average power of 120 watts is about 1 m^2 in size and costs about $500. At an installed cost of $6 per watt, the capital cost is $6,000 per kilowatt. As will be discussed under cost benefit analysis in Chapter 12, the cost to buy and install solar photovoltaic energy for a system operating for 20 years is $0.20 per kilowatt-hour, which is two times greater than the average cost of electricity in the United States. However, once the solar unit is installed, it is practically maintenance free. Wind power electricity depends on the average speed of the wind, which makes wind power less reliable than solar power. Wind turbines also require more maintenance than solar panels due to their moving parts. In 2005, the cost to install a single wind turbine with a sufficiently large (3 meter) blade to power the needs of a two-person family was about two times higher than the cost for solar power installation.

Commercial solar panels operate with only about 15% efficiency, but in theory could have 38% efficiency. Only 1/6 of the available photons from the sun are absorbed by conventional solar panels because the photovoltaic devices are sensitive to only part of the energy spectrum of incoming photons (e.g., near infrared photons are not absorbed by contemporary solar panels). It is possible to use the photons from warm objects radiated during the day and at night by photovoltaic devices sensitive to infrared energy photons. An effort has been underway to achieve greater efficiency by capturing photons in dyes that allow use of near infrared photons. Thus, in spite of the current high costs for PV cell electricity of about $0.20 per kilowatt-hour (i.e., five times the cost of oil and gas), there is promise for major reductions in cost. This can be accomplished either through improved PV cells or through the use of very large panels made of materials that can make use of the entire sun's spectrum, including photons from warm sources at night.

There are two billion people in the world without electricity, principally due to a lack of production and distribution networks in developing countries. Solar photovoltaic systems or wind generators could generate energy for immediate use at the installation site, but most of those without electricity cannot afford the initial installation costs. In 2004, worldwide solar cell installations produced 1/14 the energy generated worldwide by wind-produced energy. The rate of increase in installations is approximately 30% per year for both technologies.

MORAL, SPIRITUAL, AND IDEOLOGICAL PERSPECTIVES ON THE ENVIRONMENT

Having considered more environmentally friendly alternative ways of providing energy for transportation and other needs, we now turn to review the alternative ways the environment is viewed by others, focusing on how moral, spiritual, and ideological approaches differ. These differences are important to appreciate because one or more of these viewpoints might be in conflict in the conduct of public, engineering, or scientific projects located in a particular community. We briefly review the views of Greek, Utilitarian, and other philosophers, followed by the spiritual approaches of Buddhism, Hinduism, Judaism, Christianity, and Islam. In addition, there are ideological approaches to the environment such as anthropocentrism, biocentrism, ecocentrism, deep ecology, ecofeminism, and modern pantheism. Table 5.2 summarizes these various approaches.

Table 5.2   Moral, Spiritual, and Ideological Approaches to the Environment

Moral	Spiritual	Ideological
Virtue (Aristotle)	Confucianism	Anthropocentrism
Utilitarianism	Buddhism	Biocentrism
Rights	Hinduism	Ecocentrism
Duty (Kant)	Daoism	Deep ecology
Justice	Judaism	Ecofeminism
	Islam	Pantheism
	Christianity	

Using vs. Saving the Environment

There are a number of ways of looking at the environment. Consider the tree hugger vs. the logger; those wanting to dam rivers vs. those wanting to hold on to the property that the dam will flood; the hog farmer trying to make a living vs. those affected by the water and air pollution this industry produces; the animal preservationists vs. those wanting to build a road or an airport through the habitat of an endangered species. There is this constant pull in different directions between those who advocate more expansion to accommodate our growing population and those who want to preserve the planet unencumbered as much as possible. We need to be careful that, in the many situations that seem in direct conflict between using the environment and preserving it, we have a clear understanding of the facts regarding the consequences of the contested use. For example, many believe that rain forests should not be cut down because they utilize carbon dioxide and contribute needed oxygen to the atmosphere through photosynthesis. However, the tropical rain forests do not produce a net increase in oxygen because they may consume as much oxygen as they produce through "cold burning," a part of natural decay (Whitmore 1980). Thus, the main reason to preserve the rain forests is to preserve the habitat of birds and other animals, as well as the aesthetic value of the rain forests as a wonder of nature.

The Dimensions of the Moral Community

The concept of a moral community usually refers to a community of human beings, and it used to be that moral theories focused only on human-to-human interactions. However, as we shall see, many segments of society now extend the dimensions of the moral community to include all elements of the environment. Even if one does not accept the inclusion of earth, water, and sky as major elements of the moral community, the environment has become more and more a subject of strife between human beings. This is so because in the last century, pollution, contamination, depletion of resources, and the expanding footprint of human activity are destroying ecosystems, threatening animal species, and negatively impacting numerous aspects of the environment that are treasured by many. As a result of the strife arising over the effects of these negative consequences on individuals, communities, nations, and now the planet, we are being forced to examine these actions anew. The environment is no longer a distant concept but an object requiring care and remediation in a sense similar to the care and rehabilitation an injured person deserves.

For example, there has been an acceptance of extending one's responsibilities beyond the home to the public by not smoking around others and not burning a fire on one's property or even in a fireplace. This extension of responsibility not to harm others is only part of the concept of an extended moral community. The exploitation of the environment by the increasing population of the Earth is jeopardizing the rights of individuals. Thus, there is a case for extending the human-to-human moral ethic to include not only human beings but also all sentient and nonsentient beings, as well as the environment (such as forests and rivers) of the planet. In addition to the recognition that local actions have an effect on others outside the local community, there has been a change in what is considered valuable in the environment. In the 1700s, we valued the whale as a source of whale oil for lanterns; now most people of the world value the whale as an awe-inspiring wonder of nature to be preserved. One could say species such as the whale, eagles, owls, and all other admired parts of nature have become part of our moral responsibility, thus part of the moral community.

People have different moral standards and feel strongly about their choices (for example, to eat meat or be vegetarian, to perform medical science research on other species or to ban all animal experimentation). It is important to respect varying moral and cultural viewpoints whenever the scientist or engineer is involved in a project affecting others, especially their environment. In the following section, we review some of the major moral, spiritual, and ideological approaches to the environment.

Moral Theories and the Environment

Utilitarian ethics can be applied to environmental dilemmas in which the good of the majority is seen to outweigh the harm to a minority. Cutting down forests has the utilitarian motive of providing fuel and building materials for the current generation, and food for the current and future generations. Deforestation has consequences, which include a threat to animal species that make the forest their home, a loss to those who respect and admire the forest, a loss to recreational hunting, and most importantly, a degradation of the environment. A utilitarian considering the balance between the good of sustaining the forest for its long-term reward of providing fuel and food and cutting the forest down in order to create farming land will rely on relative utilitarian values. These values vary with individuals and with the circumstances. The need to create land that could be farmed and the abundance of forested land in the eastern United States led to a deforestation of 50% of the tree cover. The farmland cleared two centuries ago still serves society for the generation of food. But there are differing opinions about the ethics of clearing forests in the Amazon valley, where land might be cleared for commercial interests quite apart from the regional needs of the indigenous peoples, depending on whether one values the lumber or values the rain forest.

Kant would say that an action of one human being forced upon another is unethical, and any behavior that is cruel to humans must not be tolerated. Kantian philosophy would conclude that if a species with no interest to humans becomes extinct, it is not a bad thing. However, if there is a human attachment to the species, then by harming that species, one is harming another and the act is morally wrong. In addition, harming or treating another species badly says something bad about human nature. Kant also speaks of an object, a thing, or a condition that has a price as having relative worth; something that has intrinsic worth, a worth beyond any price, has dignity and immediate respect (Kant 1785). If we are cruel to animals, how much further is it to behave with cruelty to humans? This echoes Aristotle's concerns.

Religious or Spiritual Approaches to the Environment

Among the many spiritual and religious approaches to the environment, we discuss as examples the perspectives of Buddhism, Hinduism, Judaism, Islam, and Christianity.

Buddhism says that all creatures and all life deserve reverence and respect. Buddhism rejects the concept that humans should be valued over all else, or that they should have dominion over other species, as described in Genesis. The approach to the environment is based on Buddhist principles of nonviolence, tolerance, respect, and compassion for other living things. Buddhists apply these principles to encourage better stewardship of the planet, and to encourage a commitment to the environmental work that responsible stewardship demands. Buddhists consider it important to be clear about our interactions with the natural world and to support and repair it. Previously, we had to survive (see Table 5.1). Now that we know we can survive, we can be more respectful and less demanding toward nature. Surely, consideration of a dam in China will need to be approached with an appreciation and respect for these values, just as a project in India will be enabled by a deep respect for all elements of the environment.

Hindus believe that humans, animals, and nature are integral parts of one "organic whole." The universe is considered a divine creation; thus all elements, including animals, trees, plants, mountains, forests, rivers, and all living things, are worshipped for the noble qualities they possess. Cows, for example, are so highly revered that killing them is banned, and those that no longer produce milk are retired, not slaughtered. Hinduism teaches that people should use the world unselfishly in order to maintain the natural balance and to repay God for the gifts he has given. The concept of nonviolence and respect for life prevents a Hindu from causing harm to any creature. In India, Hindus take an active part in checking government projects, such as building of large-scale dams that could cause damage to the environment. In the Assisi Declarations, issued by a gathering of world religious leaders in 1986, the Hindu statement on the environment was as follows (Singh 1986):

> The human role is not separate from nature. All objects in the universe, beings and non-beings, are pervaded by the same spiritual power.

> The human race, though at the top of the evolutionary pyramid at present, is not seen as something apart from earth and its many forms. People did not spring fully formed to dominate lesser life, but evolved out of these forms and are integrally linked with them. Nature is sacred and the divine is expressed through all its forms. Reverence for life is an essential principle, as is ahimsa (non-violence).

There are many similarities between Hinduism and Buddhism. The traditions in China and India are ancient, and the teachings have not changed; however, these teachings have not been strong enough to resist the pressures of industrialization, as evidenced by some of the worst pollution in the world is in the Delhi and the Ganges Rivers.

Judaism has a two-tiered, or dualistic, approach to the environment. As an example, two viewpoints are that humanity is both a part of nature and apart from it. That both viewpoints are espoused within Judaism is like ". . . having two truths in one's pocket" (Fink 1998). These opposing viewpoints are also seen in interpretations regarding the significance of God creating humanity last, on the sixth day, after light, water, land, plants, and animals. Some believe that God created the environment before creating Adam and Eve so that nature was ready to provide them with sustenance and enjoyment, an act that indicated the superiority of humanity over the environment. Others contrast this position with another Jewish viewpoint that the order of creation serves to remind human beings of their inferior position relative to nature and animals. Modern Jewish scholars have presented the concept of humanity having a stewardship over the garden that is the environment, and that this assignment came from its owner, God (Fink 1998). These concepts differ from those of Hinduism and Buddhism, where the relationship is more that of an integrated whole, with human beings and the environment as partners. Judaism holds that human beings are not just a part of nature but that they should enjoy and use it as well as protect it. In short, the Jewish tradition is complex and contains both "green" and "non-green" elements. The Jewish stance carries two messages: that it is not appropriate to overemphasize protection of the environment on the one hand, and on the other, that it is not appropriate to exploit the environment to its detriment. The analysis by White establishes a perspective of Judaism quite different from that of the conservationist: "Even more so, we are told, the entire earth awaits your finishing touch. Your labor is welcome, and its results are pleasing to me, says the Lord. For this reason, Judaism is prouder of man's skyscrapers than of God's swamps, and prouder of man's factories than of God's forests" (White 1967).

The Christian perspective has similarities to the Jewish perspective in that it is more of an anthropocentric (human-centered) philosophy, as can be concluded from the Bible:

> So God created man in his own image, in the image of God created he him; male and female created he them.

> And God blessed them, and God said unto them, Be fruitful, and multiply, and replenish the earth, and subdue it; and have dominion over the fish of the sea and over the fowl of the air, and over every living thing that moveth upon the earth.

Some claim this passage gives man dominance over the earth and all things on it. However, with six billion people on the planet, this interpretation is outdated. Dominion can be interpreted as protective stewardship rather than dominance and exploitation. Another reference to dominance comes in this passage from the Bible:

> And the fear of you and dread of you shall be upon every beast of the earth and upon every fowl of the air, upon all that moveth upon the earth, and upon all the fishes of the sea; into your hand are they delivered.

The theologian Thomas Aquinas incorporated these tenets of Christianity with the ancient Greek concepts expressed by Aristotle. He saw no possibility of sinning against nonhuman animals or against the natural world, only against God, our neighbor, and ourselves (Singer 2000).

The tradition of converting wilderness areas to other uses for humans clearly now has another position forced by the threat of human activity to engulf these wilderness areas. But even though it has become clear that these areas need to be protected, the old anthropocentric mindset is proving difficult to change, as is the concept of protecting the future.

Islam, some say, was developed because there were loopholes in Christianity. Allah created the world, and we must all protect it. Allah would say taking the life of an insect could equate to taking the life of a human. There is a built-in respect for the environment, and harm to the environment is considered harm to Islam. Islam looks at Allah with God as Master, whereas the approach of Christianity is that it is we who are the stewards of God's creations. There are countries in which Islam is a part of the government, with government and religion tied together in a way that is entirely different from England, Germany, the United States, and other countries.

Thus, various religious traditions and ideologies have shaped our views of environmental values. Some of these traditions have made it difficult to reconcile exploitation of the environment for the use of human beings with respect for the environment as an entity of intrinsic value.

Ideological Approaches to the Environment

The descriptions of the ideologies that follow may be similar to some religious or spiritual concepts of the environment but are free of religious ties that might otherwise cause some not to espouse them.

Anthropocentrism

This theory states that the world centers on humans, that only humans have worth, and that they have the right to benefit from the environment as much as possible. Whereas some believe that animals have the right to exist in the environment and to be protected, anthropocentrism does not give value to nature beyond its use to humanity. Humans are the source of all worth and value, and nature is valuable only as a means to their ends. Examples of religions with anthropocentric views include Christianity and Judaism. As we will see later, this view does not conflict with contemporary moves toward regulation of the use of the environment in order to prevent irreversible degradation, because regulation ensures continued use of the environment. This view, though not explicitly disrespectful of environmental values, does place them (including animals, earth, air, and water) on a lower plane than human-centered values.

Biocentrism

While anthropocentrism favors a worldview centering solely on humans and sees value only in human beings, biocentrism sees everything in nature as having value. Biocentrism rejects the idea that humans are superior to animals and should have dominion over them. The environment should be shared between humans and animals (http://library.thinkquest. org/26026/Philosophy/biocentrism.html?tqskip1=1&tqtime=0420, accessed April 23, 2003). Examples of those with biocentric views include Buddhists and Native Americans. Some have equated biocentrism with deep ecology.

Ecocentrism

This concept views the environment and all living things in it as the center of the world. Those who share this view, such as the Native Americans, believe that bears, deer, and trees, for example, must be respected. Buddhists believe in reincarnation and hold to the ecocentric view of the environment, believing that all life should be protected and that karma can be affected if all life is not respected.

Deep Ecology

Arne Naess, one of Norway's best-known philosophers, developed deep ecology around 1970. The motivation was the recognition that our current use of the environment needs to be redirected to a theme of conservation and respect for nature. He originally defined it using the word "ecosophy" and thought of it as an evolving philosophy involving thinking and acting with ecological wisdom and harmony (www.resurgence.org/resurgence/185/harding185.htm, accessed January 11, 2006):

> For Arne Naess, ecological science concerned with facts and logic alone, cannot answer ethical questions about how we should live. For this, we need deep ecological wisdom. Deep ecology seeks to develop this by focusing on deep experience, deep questioning and deep commitment. These constitute an interconnected system. Each gives rise to and supports the other, whilst the entire system is, what Naess would call, an ecosophy: an evolving but consistent philosophy of being, thinking and acting in the world that embodies ecological wisdom and harmony.

A principal tenet of the deep ecology movement is that both human and nonhuman lives have intrinsic value. The inherent value of nonhuman lives is different from their instrumental value seen in the service they provide to human beings and is independent of their usefulness for human purposes. Human life can use nature for vital needs; however, it is ethically wrong to take too much or to waste nature. The movement espouses a philosophy of appreciating life without striving for a higher standard of living. Included in the philosophy is that the flourishing of human life requires a substantial decrease in the human population, a decrease that is necessary if human and nonhuman life is to flourish. The followers of deep ecology also encourage a decrease in energy dependence, policies to ensure efficient utilization of resources, and a culture of giving back to the environment. In many respects, deep ecology is similar to Buddhism. One of the dissimilarities is that deep ecology advocates population control and encourages political actions.

Ecofeminist Approach to the Environment

Ecofeminism describes a movement to create an interconnected community where all beings of the organic world have intrinsic value and are part of the same living organism, the Earth. The movement was given its name by the French feminist Francois d'Eaubonne in 1974. Although in some contexts it is described as a women's movement to protect the environment, it is really an effort to end all forms of oppression, whether it be dominance of women by men, ethnic discrimination, or excessive exploitation of nature and the Earth. Ecofeminists advocate a philosophy whereby all will benefit by working together. The reason that the term includes *feminism* is that the movement uses gender as a starting point for working to end the oppression. Some espousing ecofeminism feel that acts against nature and the environment are the result of the dominance of males in decision-making roles. Ecofeminists seek a gentler, more feminine approach to the environment.

Modern Pantheist Approach

Pantheism broadly holds that the world is either identical with God or in some way a self-expression of his nature: "God is everything and everything is God" (Owen 1971). Although pantheism had roots in the early societies of Egypt, India, and Greece (e.g., stoicism was a form of pantheism in the Greco-Roman world), it is more widely thought of today as a modern philosophical movement. While some have termed it a religious movement, it has no church. The pantheist's environmental ethic has important similarities to Confucianism, Buddhism, Hinduism, and Daoism in that it extends our moral community to other living and nonliving things. "The land ethic simply enlarges the boundaries of the community to include soils, waters, plants, and animals, or collectively, the land. . . . A thing is right when it tends to preserve the integrity, stability, and beauty of the biotic community. It is wrong when it tends otherwise" (Leopold 1949).

Pantheists see great value and beauty in nature, and hold that one must preserve, protect, and respect the natural order of things, not only for their own but also for others' well-being. This sympathetic approach to nature and the environment gained favor in the United States in the 1820s at the time the land had become 50% deforested on the east coast, which some saw as an environmental disaster. Pantheism was the philosophy of some of the nineteenth century poets, such as Wordsworth and Whitman. It was also fostered in the twentieth century by John Muir, Robinson Jeffers, D. H. Lawrence, and the poet Gary Snyder who identified with, extolled, and exclaimed people's close association with nature.

A Conceptual Structure for Environmental Ethics Based on Relative Value

One can establish a conceptual framework for discussion of environmental dilemmas by considering the environment, including animals and plants, from three additional perspectives (Sagoff 1991a):

1. as an instrument to benefit human beings and at the same time to be preserved for its usefulness, or its ability to meet a need;

2. as an aesthetic object universally admired for its beauty; and

3. as a moral object deserving of reverence, affection and respect.

These additional perspectives are useful tools with which to approach ethical situations without having to choose a spiritual or ideological concept to guide judgments or actions. The importance of perceiving the environment as an instrument of value to benefit human beings should be put in balance with the perspectives of respect and aesthetic value. The balance and relative priorities of these three perspectives will vary depending on the situation. For example, deforestation, while seeming to ruin a beautiful entity, nevertheless has an important result that is instrumental in advancing the quality of life for many. The guiding concept of relative value will help in the discussion of the utility of exploiting the environment, including animals, vs. the value of preserving the environment. The following case is drawn from the book *Charlotte's Web* (White 1952a).

Case 5.3 Pork Chops, a Champion Pig, and a Spider's Love

Wilbur was a pig owned by a farmer named Zuckerman who initially valued Wilbur because of his market value in pork chops and bacon. But the spider, Charlotte, was fond of Wilbur and started a campaign to save him. Charlotte spun webs over the pigsty in the design of words such as "terrific," "radiant," and "humble," which she copied from newspaper advertisements brought by a helpful rat. Farmer Zuckerman, faced with the dilemma of preserving Wilbur, a champion pig at the state fair, or sending Wilbur to the butcher for his instrumental value, was moved by the spider's signs and decided to save Wilbur. At the end of the story, Wilbur asks Charlotte why she did this for him as he never had done anything for her. She replied, "You have been my friend. That in itself is a tremendous thing. I wove my webs for you because I liked you. After all, what's a life, anyway? We're born, we live a little while, we die. A spider's life can't help being something of a mess, what with all this trapping and eating flies. By helping you, perhaps I was trying to lift up my life a little. Heaven knows, anyone's life can stand a little of that" (White 1952b).

This parable also illustrates three perspectives of appreciating nature and the environment noted at the beginning of this section. The pig has value as a useful instrument (pork chops and bacon), as an aesthetic good in and of itself (a champion, a perfect specimen), and as an object of love and respect (Sagoff 1991a). The analogy to the environment can be made by considering a river descending through the mountains into a valley. The river contains potential energy of the mass of water flowing under the influence of gravity, and this energy can be converted to electricity to serve the local community by building a dam. This is its instrumental value. However, the natural beauty of the river as well as the valley will be destroyed if the river is dammed to create a reservoir, a loss of its aesthetic value. Destruction of the river and valley will result in the loss of land and homes for the native people who love living there, a loss of its value as a place that is loved in and of itself.

When a decision concerning use of the environment for the benefit of human beings pits one object, such as the value of timber, against the value of another, such as the preservation of the white owl, we need some guidance for decision making. The strategy of assigning value is difficult in a situation when one is trying to decide moral trade-offs. The current quantitative value of timber vs. the owl would rule in favor of the value of logging. However, others' aesthetic and moral values would rule that it is far more valuable to preserve a species. The value of the forest as an instrument for building and fuel is easily quantified. But to quantify the value of the loss of a forest or a species is far more difficult. How does one argue that the existence of the forest has more value than the advancement of society from its use? We can put a value on a forest being harvested, but it is more difficult to put a value on having preserved it.

We now have the technology to do great things or do great harm. We are getting closer to becoming more responsible for our planet, but we are not there yet. It will need to be a global responsibility, and progress will most likely require concerted efforts at the local and individual level. In addition to these global commitments to preserve the environment and improve the quality of life for those in need, it is also important to restrain our avid consumerism and desire for a standard of living that takes more and more resources to sustain. We may come to "... measure wealth not in terms of what we can consume but in terms of what we can do without—what we treasure for its own sake" (Sagoff 1991b).

The Global Environment

There is no "right" global approach to the environment. In fact, there are many highly charged disagreements about how to accomplish sustainability, and the disagreements about solutions must be respected. Yet we have a duty to implement some solutions in order to prevent harm to the next generations. Some would disagree with the idea that we need to find a way to change things. Complex ecosystems are more resilient than predicted (for example, the recovery from the Exxon Valdez oil spill disaster in Prince William Sound was more complete than predicted [Sagoff 1991c]). On one hand, it may be acceptable to continue to exploit, pollute, and contaminate the environment. On the other hand, maybe what we are doing is unethical and we will indeed destroy everything viable on the Earth. In a million years, there would be no sign that we had been here. Perhaps another species would come to inhabit the Earth, and it has been this way for many eons. Why should we care? Some say that impending disasters have been predicted for hundreds of years, and many times, they have not materialized, or if they have, humanity has found solutions. However, most have come to a realization that now we no longer have the luxury to take from the environment without consequences.

Up to a point, we have been able to do a certain amount of environmental damage and still repair it or absorb the damage. But there is a point at which oil will run out and cannot be replaced, and there is a point at which the excess CO_2 released into the environment by manufacturing and transportation can no longer be sequestered by the oceans. There may be breakthroughs in technologies for alternative energy production, manufacturing, and transportation that will provide some relief from the reliance on fossil fuels as well as technologies that will sequester the greenhouse gasses. Until technological advances provide this relief, the movement toward sustainability and responsible stewardship of the planet is gradually gaining attention at international, national, and local levels, as will be covered in a later section.

RESPONSIBILITY FOR TECHNOLOGIES WE CREATE

This is a critical time for civilization and for the planet. A team working with Harvard biologist Edward Wilson calculated that as of 1999, humanity was making demands of our environment that overshot by 20% the earth's capacity to sustain them (Wackernagel et al. 2002). This figure was also documented in the Living Planet Reports issued in 2002 (Loh 2002) and 2004 (Loh and Wackernagel 2004). This figure of excess use is projected to grow even larger in the next 50 years. Because of growing populations, dwindling resources, the rapid growth of technology, and the capacity for mass destruction, there is now a moral responsibility to shape new directions of caring for the environment.

We can make just about anything on a macro- and micro-scale. But the question is no longer *can* we make something, but *should* we make something. If the decision is to make something, there is increasingly a need for creating it in an environmentally responsible manner, with the least waste, the least pollution, and with an eye to recycling materials. The ethics of these decisions, and how they will affect the future and impact the environment, must now be among the first considerations about a project.

The responsibility to consider the environmental consequences of our actions can be thought about in a number of different ways. There is the classical concern about how we should be interacting with the environment. There is also a spiritual way of looking at the environment. There is what have we been taught from childhood, our heartfelt feelings about our interaction with the world. There is no specific answer or set of answers, but there is the need to understand our responsibilities about what we do as scientists and engineers. Tree huggers who oppose land developers who would sacrifice a forest are examples of how differently we think about the environment, but our pluralistic society requires that we share ideas and develop an acceptance of a variety of opinions and reach out to others to find acceptable answers.

The Tragedy of the Commons

From Aristotle to modern times, it has been recognized that when the public has access to an environmental resource (designated as a common pool resource or CPR), that resource is generally less well cared for than if it were owned privately. It is also recognized that it is not unusual for people given access to a common resource to use it indiscriminately. For example, it is the tendency for cattlemen given access to a public grazing land to want to increase the size of their herds. This is not done with malicious intent, but the problem is that each animal added to the herd degrades a common resource to some extent. Since the resources in this public grazing land are limited, they will ultimately be exceeded. Thus, the behavior of individuals to increase their personal pleasure or wealth can lead to the degradation of public lands or "commons."

This is known as the "tragedy of the commons," aptly brought to the attention of the world by ecologist G. Hardin in 1968 (DeYoung 1999). This concept of the overuse of federal or public lands applies to a great many environmental-use problems (e.g., overfishing, deforestation, overuse of water resources for irrigation, contamination of our land and oceans, to name a few). Ways must be found to avert the these tragedies. If not, disasters will occur such as those described in Rachel Carson's *Silent Spring* (1962), Jonathan Schell's *The Fate of the Earth* (1982), and Jared Diamond's *Collapse* (2005).

Where We Go from Here

Some of the environmental decisions facing us today include whether to build dams that may harness hydroelectricity but displace people and animals, whether to log forests, drill for oil in pristine areas, or permit new manufacturing that will pollute the environment. How should we best manage global water resources? These are not easy questions. A recent (2003) report from the United Nations found that even after 25 years of international conferences on proper water management to ensure the survival of the planet, there have been "hardly any" solutions. The report points out that:

> . . . 20% of people in the world lack good drinking water and 40% lack sanitation; global per capita water supplies decreased by one-third between 1970 and 1990 and are likely to drop by a third over the next 20 years unless countries overcome their inertia at the leadership level (*Science* 2003).

Some examples of specific things that can be done immediately include the following (Dorf 2001):

1. The necessary conditions for processes in an ecosystem must be maintained (for example, DDT was banned because it had lethal consequences for fish and birds).
2. Operations must be within the carrying capacity of the ecosystem (for example, automobile gridlock occurs when the carrying capacity of roads is exceeded).
3. Harvesting rates should not exceed the regeneration rate (a violation would be when the harvest of trees exceeds the regeneration rate).
4. Waste emissions should not exceed the assimilative capacity (an example is the long disintegration factor for plastic six-pack rings).
5. The rate of exploitation of nonrenewable resources should be equal to or less than the rate of development of renewable substitutes (an example is the substitution of solar or nuclear power for fossil fuel sources).

The list of environmental issues is lengthy and increasingly complex, primarily due to population growth and dwindling resources. There are no "right" answers, and each issue will require careful thought, patience, and an open mind to new solutions. It is important to decide one's own approach to environmental ethics, which may be one of those described above, or a combination of several or many of them. The important thing is to be sensitive to the vulnerability of the environment and to make the environment part of our moral community. We can no longer deny how fragile it is, and must make decisions about issues that will make the most of scarce resources and at the same time benefit the future generations inhabiting the planet.

Societal Efforts to Prevent Environment Degradation and Promote Sustainability

Global efforts, ranging from Greenpeace to the Kyoto Protocol, have been underway to find ways to ensure sustainability and a cleaner environment while supporting economic growth. A list of the major environmental conferences involving governments around the world is found in Table 5.3. Additional global efforts are underway to control the introduction into the environment of genetically modified organisms (GMOs), discussed in Chapter 6, "Ethics of Genetically Modified Organisms."

Table 5.3   International Conferences on Sustaining the Environment

Year/Site	Conference	Outcome
1985, Vienna	Convention on the Protection of the Ozone Layer	Set up a legal frame-work for action in Montreal
1987, Montreal	Protection of the Ozone Layer	Montreal Protocol Protection of ozone layer
1992, Rio de Janeiro	U.N. Conf. on Environment and Development	
	Convention on Climate Change	Reduction in CO_2
	Convention on Biological Diversity (led to the Cartagena Protocol*)	Conservation and sustainable use of biological diversity
1997, Kyoto	U.N. Framework Convention on Climate Change	Kyoto Protocol Reduction in "greenhouse" gases
2000, New York	United Nations Millennium Summit	Preservation of the environment
2002, Johannesburg	U.N. World Summit on Sustainable Development	Agenda 21 Sustainable development
2003–2005	U.N. World Environment Conference of Mayors	World mayors vow local support for environmental accords
1998–2005	G8, annual meeting of heads of eight major countries	Addresses a wide range of international economic, political, and social issues

*Cartagena Protocol is also discussed in Chapter 6, "Ethics of Genetically Modified Organisms."

SUMMARY OF INTERNATIONAL CONFERENCES TO PROTECT AND SUSTAIN THE ENVIRONMENT

The Vienna Convention on the Protection of the Ozone Layer, Vienna, Austria, 1985, and the Montreal Protocol on Substances That Deplete the Ozone Layer, Montreal, Canada, 1987

In 1985, scientific concerns about damage to the ozone layer prompted governments to adopt the Vienna Convention on the Protection of the Ozone Layer, which established an international legal framework for action. Two years later, in 1987 in Montreal, international negotiators met again to adopt the *Montreal Protocol on Substances That Deplete the Ozone Layer,* legally binding commitments that require industrialized countries to reduce their consumption of chemicals harming the ozone layer (www.undp.org/seed/eap/montreal/montreal.htm, accessed January 11, 2006).

The agreement was signed initially by 25 nations, and later by 168 nations, including the United States. This protocol set limits on the production of chlorofluorocarbons (CFCs), halons, and related substances that release chlorine or bromine to the ozone layer of the atmosphere. Because of increasing scientific knowledge about the effects of CFCs and halons on the ozone layer, the original protocol has been amended several times. At meetings in London (1990), Copenhagen (1992), Vienna (1995), Montreal (1997), and Beijing (1999), amendments were adopted that were designed to speed up the phasing out of ozone-depleting substances. Not all parties to the main protocol are parties to these amendments. The production and consumption of halons was phased out by January 1, 1994, and of CFCs, carbon tetrachloride, methyl chloroform, and hydrobromofluorocarbons by January 1, 1996, subject to an exception for agreed essential users. Methyl bromide was phased out in 2005, and hydrochlorofluorocarbons are to be phased out by 2020. (Phase-out dates are later for developing countries.)

Under the protocol, the ozone-depleting potential, or ODP, of any substance is measured with respect to an equal mass of CCl_3F, or CFC-11, which is assigned a value of 1.0. Hydrochlorofluorocarbons, which are being used as transitional replacements (until 2020) for CFCs in refrigeration, have ODPs that are generally less than 0.5. Hydrofluorocarbons, which are also replacing CFCs as refrigerants, have ODPs of zero.

Currently, this protocol is still in place and active. A report on its status was published on the 10-year anniversary of the Montreal Protocol (www.unido.org/doc/18265, accessed June 7, 2005). Efforts are continuing to remove pollutants that cause ozone depletion, to make refrigeration more environmentally compatible, to control air pollution and ground water contamination, and to replace the use of toxic substances with more natural ones.

U.N. Conference on Environment and Development, Rio de Janeiro, 1992

In 1992, the United Nations Conference on Environment and Development (otherwise known as the Earth Summit) was held in Rio de Janeiro, Brazil. Two agreements resulted. One was a binding agreement of the world's industrial nations to reduce CO_2 emissions, the Convention on Climate Change. After it became clear that the conditions of this convention were not being fully met, it was replaced by the Kyoto Protocol of 1997 (see below).

The second outcome of the Rio meeting was a global agreement on the conservation and sustainable use of biological diversity, which led to the Cartagena Protocol. This protocol remains active in 2006, and its activities cover ". . . all ecosystems, species, and genetic resources. It links traditional conservation efforts to the economic goal of using biological resources sustainably. It sets principles for the fair and equitable sharing of the benefits arising from the use of genetic resources, notably those destined for commercial use. It also covers the rapidly expanding field of biotechnology, addressing technology development and transfer, benefit-sharing and Biosafety" (www.biodiv.org/doc/publications/guide.asp, pp. 2–3, accessed July 9, 2005). Thus, the Cartagena Protocol is diverse, and its provisions are proving especially helpful in conservation and resolution of issues relating to genetically modified plants and animals, including safety, transfer across national borders, and equitable availability (see also Chapter 6).

U.N. Framework Convention on Climate Change, Kyoto, Japan, 1997

The Kyoto Protocol, written and adopted in Kyoto in 1997, came into effect in February 2005. It has as its objective the reduction of greenhouse gas emissions such as CO_2 to prescribed levels below the 1990 measured emissions by 2008–2012. Reduction of hydrocarbon emissions is to be based on 1995 levels. The quotas set for the so-called developed countries can be met by either lowering their emissions or by buying credits from those countries that have emissions below their quotas. Those countries that exceed their quotas will be fined. The protocol lists developed nations and developing nations as separate entities with separate responsibilities.

This global agreement is a substantial step toward world cooperation in lowering greenhouse gas emissions; however, the Kyoto Protocol has what some believe are fatal flaws. With respect to the possibility of reaching the 2008–2012 quotas, some countries have *doubled* their emissions since 1990. For example, Canada must decrease its actual emissions by 25 to 30% to reach the goal of 5% or less of its 1990 emissions. The provisions of the Kyoto Protocol would require the United States to return to the greenhouse emission quantity of 1990 by 2010. This goal, while predicted by some to be impossible to reach without rationing combustible fuels, was not what prevented ratification of President Clinton's agreement by the U.S. Congress. The Clinton agreement was based on the contingency that China, India, and Brazil would be included as accountable developing countries. The U.S. Congress agreed that the Clinton conditions must be met. The principal problem in the debate about the United States and the Kyoto Protocol is that the journalists have not fairly represented the root cause of the U.S. stance on the requirements of the agreement. Australia is the only other developed nation not to join, saying that until major polluters such as the United States and China are taking part, it would be harmful for them to do so. Issues of justice regarding the per capita emission ranking in some countries, and the need to increase industrial wealth and, therefore, emissions for other countries, were part of the protocol development and are a continuing concern to participating and nonparticipating countries.

Another problem with the Kyoto Protocol agreements is that the predictions of possible lucrative credits made by countries in 2000 or 2002 might not hold up. An example is what might happen in New Zealand. In 1997, New Zealand predicted it would have fewer emissions than the target for 2008–2012, so it could sell credits and earn as much as $500 million. However, in 2005 it was realized that the numbers of automobiles and industries that produce greenhouse gases are increasing beyond expectations, and that by 2008, it might have a deficit in credits costing as much as $500 million.

An overlooked problem in the assessments of the CO_2 emissions is that countries with major aviation activities did not predict the increase in greenhouse gas contributions from commercial air traffic. In Great Britain, for example, projected trends for "unconstrained" aviation growth will result in 2030 emissions that are equal to 60% of the United Kingdom's 2050 CO_2 quota in the Kyoto contract (Upham 2003).

United Nations Millennium Summit, New York, New York, 2000

This was the largest gathering of representatives of the nations of the world in the year 2000. A number of resolutions directly or indirectly had to do with preserving the environment, not the least of which was to integrate the principles of sustainable development (defined as meeting today's needs without compromising the ability of future generations to meet their own needs) into national policies throughout the world. This summit also resulted in a plan to reverse the loss of environmental resources. Other major resolutions were to reduce by half the proportion of people without sustainable access to safe drinking water, and to achieve significant improvement in the lives of at least 100 million slum dwellers by 2020.

On environmental protection, the declaration states that no efforts should be spared to counter the threat of the planet being irredeemably spoiled by human activities. The participants of the summit resolved to adopt a new ethic of conservation and stewardship, making efforts to ensure the implementation of the Kyoto Protocol. The document encourages better management, conservation, and sustainable use of forests and water resources. It also presses for the full implementation of conventions on biological diversity and desertification (the degradation of formerly productive land due to human pressure that stresses the environment over its tolerance level) (http://pubs.usgs.gov/gip/deserts/desertification/, accessed June 15, 2005).

United Nations World Summit on Sustainable Development, Johannesburg, South Africa, 2002

In August of 2002, the 4th United Nations World Summit on Sustainable Development opened in Johannesburg, South Africa. As explained below, the Rio and Johannesburg conferences led to Agenda 21 (referring to the twenty-first century).

Agenda 21 is summarized by the U.N. Division of Sustainable Development as follows (www.un.org/esa/sustdev/documents/agenda21/index.htm, accessed May 18, 2005):

Agenda 21 is a comprehensive plan of action to be taken globally, nationally and locally by organizations of the United Nations System, Governments, and Major Groups in every area in which humans impact on the environment.

Agenda 21, the Rio Declaration on Environment and Development, and the Statement of Principles for the Sustainable Management of Forests were adopted by more than 178 Governments at the United Nations Conference on Environment and Development (UNCED) held in Rio de Janeiro, Brazil, 3 to 14 June 1992.

The Commission on Sustainable Development (CSD) was created in December 1992 to ensure effective follow-up of UNCED, to monitor and report on implementation of the agreements at the local, national, regional and international levels.

The full implementation of Agenda 21, the Programme for Further Implementation of Agenda 21 and the Commitments to the Rio principles, were strongly reaffirmed at the World Summit on Sustainable Development (WSSD) held in Johannesburg, South Africa from 26 August to 4 September 2002.

President Bill Clinton dutifully formed by executive order the President's Council on Sustainable Development to comply with these initiatives. Agenda 21 is a dictum of social behavior with good intentions of protecting the environment and creating sustainability, but some fear that it threatens the very core of human rights and democracy because it will result in tight controls of what constitutes sustainability and what is harmful to the environment (Lamb 2005):

> The concept of sustainable development is constructed on the foundational belief that government must manage the affairs of its citizens in order to balance the three-legged stool of sustainable development: (1) environmental protection, (2) economic development, and (3) social equity. . . . The fatal flaw in the concept of sustainable development is the absence of the principles of freedom.

Thus, plans to promote sustainability have found harsh critics. Examples of the U.N. plan being enacted at the local level to restrict outward expansion include governmental "locking up" of undeveloped lands by declaring them parks, and city planning boards putting restrictions in place to prevent outward expansion past current city limits. Some are concerned that eventually, populations will be restricted to urban areas where their movements and life choices will be regulated in virtually all aspects. But who is to decide what activities will sustain the environment and what will not? Sustainable development is, in fact, the process by which societies are being reorganized around the central principle of protecting the environment. It is a process that originated in the international community, and is now sweeping across America, encompassing states (e.g., Minnesota) and small towns (e.g., Santa Cruz, California). Time will tell the extent to which these efforts are so restrictive to those who must live with them that they become intolerable.

U.N. World Environment Conference of Mayors, San Francisco, California, June 2005

Another approach to increase the success of the U.N. mission to improve the environment has been to bring together the mayors of the world to engage them in the fight for environmental reforms. Their leadership at the local level has a more likely chance of resulting in changes such as conserving energy and reducing pollutants. This conference, an annual event, was held for the third time in 2005 and was attended by 68 mayors from around the world. Over the seven days of the meeting, seven areas were identified for change, including energy conservation, waste reduction, urban design, urban nature, transportation, environmental health, and water, and 21 accords were agreed upon, some of which are listed below (Martin 2005):

- To add renewable energy sources
- To reduce greenhouse gas emissions 25% by 2030
- To achieve zero waste to landfills by 2040
- To implement user-friendly composting programs
- To mandate a "green building" rating system
- To support organically grown food production and consumption
- To create more environmentally beneficial jobs in poor neighborhoods
- To ensure the presence of a city park or some green space within one half km of every city resident
- To greatly increase trees in cities
- To increase affordable public transit

- To reduce and eliminate harmful chemicals within cities
- To provide more access to potable water

Fifty of the 68 participants signed the accords. Those not signing cited the need for ratification by their city councils or other reasons usually not related to a disinclination to support the goals. All saw that the accords could have a direct impact on improving life in their cities. It is believed that this local connection to environmental needs will make a difference in implementation of the accords in a way that the global meetings have not.

The G8 Annual Conferences

Since 1975, the heads of state or government representatives of the major industrial democracies began meeting to discuss major economic, political, and social issues facing the international community as a whole. The list of those participating has gradually expanded, and now includes France, Germany, Japan, the United Kingdom, the United States, Italy, Russia, and Canada. The first G8 Summit was held in 1998, and the group has met annually since then. A variety of issues have been major agenda items, but by 2005 the major agenda item was global warming. The July 2005 G8 meeting in Scotland was an attempt by the United Kingdom to lead a consensus between the United States, China, India, and the United Kingdom regarding agreements to lower greenhouse gas emissions. However, behind-the-scenes efforts by the United States pressured negotiators to weaken key sections of the proposal; thus, the U.S. government continues to resist adopting stricter restrictions on greenhouse gas emissions, citing uncertain scientific information as justification to do so (Eilperin 2005).

CHAPTER SUMMARY

The fragility of our environment and the depletion of its resources have come under increasing scrutiny the last several decades. Society has had to face and is beginning to accept that current demands on the resources of the Earth are 20% greater than the Earth's capacity to sustain them. We are starting to come to grips with the fact that continued pollution of our land, rivers, oceans, and air can have long-term effects that will be difficult, and in some cases impossible, to reverse. We are already seeing the accumulation of neuro- and other toxins from pollutants in our food sources, which are especially detrimental to children, who are at risk for accumulating dangerously high levels of these substances. This is so because their intake of food is about the same as adults, but toxins are more concentrated in children because of their smaller body size. The cumulative effect of these environmental problems has resulted in a growing determination to mount individual, local, and international efforts to improve the health of the environment and to sustain its resources for future generations.

This chapter emphasizes the importance of alternative energy sources because the pollution resulting from use of fossil fuels for energy is the main cause of contamination of the atmosphere. There are technological developments for wind and solar energy production that will sustain the Earth's resources and at the same time will become available to underdeveloped countries. However, until these alternatives become practical, the international community is taking action to develop a global approach to lessen pollution, conserve environmental resources, and decrease harmful emissions, among other things. These efforts began in 1985 with the agreement to reduce halocarbon emissions. The Kyoto Protocol of 1997 is, in 2006, the key instrument for encouraging nations to reduce greenhouse gas emissions such as CO_2. These international efforts have raised ethical questions of utilitarianism versus individual rights, and the rights of individuals, societies, and countries. The implementation of the Kyoto Protocol raises ethical questions of justice and equity. On the one hand, those countries that produce the most emissions need to reduce emissions because of harmful effects on others in the common environment. On the other hand, imposition of severe restrictions on the highest emission producers must be done justly.

The United States (which produces the greatest amount of greenhouse gases) and Australia (which has one of the highest per capita CO_2 emissions) are the only two nations not to have ratified the protocol. Both countries said it would not be in their economic interests to do so. China, the world's second largest producer of greenhouse gases, approved the Kyoto treaty, but along with India has no obligation to cut carbon dioxide emissions until 2012. This is an unfair exemption in the eyes of the United States, which cites this as one of the reasons it walked away from the agreement in 2001. The full protocol went into effect in February of 2005. The outcome, if achieved, will have great benefits in decreasing greenhouse emissions and for sustainable development.

Agenda 21 evolved as a result of international conferences held under the aegis of the United Nations in Rio and Johannesburg. This agenda has as its prime focus achieving sustainability. Concerns are being voiced as to who will decide those activities that will sustain the environment and those that do not. An extension of these concerns is that these decisions have the potential to result in reorganization of our present way of life and may result in governmental regulations that are overly restrictive of life choices. While it is important that there is increasing recognition that we are depleting the resources of the planet, a balance needs to be struck to ensure that Agenda 21 does not unduly impinge on the currently accepted freedoms of citizens of many countries.

The principal method of maintaining a sustainable supply of energy, food, and water is through the establishment of a global commons. New technologies exist to increase the availability of food and water, which are discussed next in Chapter 6, "Ethics of Genetically Modified Organisms." The world's nations have taken up the challenge of environmental sustainability. However, there needs to be some vigilance regarding the extent to which government control of communities might restrict freedom—a challenge to the moral theory of rights ethics. Up until about the tenth century, people used the environment in whatever way was necessary to survive; now to survive, the communities of the world need to do whatever is necessary to promote a global stewardship of the environment.

REFERENCES

Andersen S. and R. Newell. 2004. "Prospects for carbon capture and storage technologies." *Annual Review of Environment and Resources* **29**:109–142.

Carson, R. 1962. *The Silent Spring.* 1994 Edition. New York: Clarion Books/Houghton Mifflin Company.

DeYoung, R. K. 1999. "Tragedy of the commons." In: *Encyclopedia of Environmental Science*. D. E. Alexander and R. W. Fairbridge (eds.). Boston, MA: Kluwer Academic Publishers, pp. 601–602.

DeYoung, R. K. and S. Kaplan. 1988. "On averting the tragedy of the commons." *Environmental Management* **12**:273–283.

Diamond, J. 2005. *Collapse: How Societies Choose to Fail or Succeed*. New York: Penguin Group Division of Viking Press, pp. 95–119.

Dorf, R. C. 2001. *Technology, Humans, and Society: Toward a Sustainable World*. San Diego, CA: Academic Press, p. 11.

Eilperin, J. 2005 (June 17). "G-8 bows to U.S. on climate change. Action plan weaker after Bush team exerts pressure." *San Francisco Chronicle*, p. A14.

Epstein, E. E. 2005 (June 6). "E-waste: Congress examines disposal options for electronics." *San Francisco Chronicle*, pp. A1 and A7.

Fackler, M. 2002 (February 28). "Chinese villages poisoned by American high-tech trash." *San Francisco Chronicle*, accessed via www.sfgate.com/cgi-bin/article.cgi?f=/news/archive/2002/02/28/international1604EST0735.DTL, May 6, 2005.

Feeny, D., F. Bikret, B. J. McCay, and J. M. Acheson. 1990. "The tragedy of the commons—22 years later." *Human Ecology* **18**:1–19.

Fink, D. B. (1998, Fall–Winter issue). "Judaism and ecology: A theology of creation." Originally published in *Earth Ethics* **10**(1). In: *Forum on Religion and Ecology* (2004). Harvard University Center for the Environment,

Greenbaum, E. 1982. "Photosynthetic hydrogen and oxygen production: kinetic studies." *Science* **196**:879–880.

Hardin, G. 1968. "The tragedy of the commons." *Science* **162**:1243–1248.

IEA. 2000 (Jan.). "Reducing standby power: Opportunities & challenges. IEA 2nd International Workshop on Standby-Power." Accessed via www.iea.org/standby/Brussels.htm, June 11, 2005.

Kant I. [1785]. *Foundations of the Metaphysics of Morals and What is Enlightenment?* Second edition (revised, 1997), translated by L. W. Beck. Upper Saddle River, NJ: Prentice-Hall, Inc., p. 52. Leads to this aspect of Kant's thinking on the environment were provided by M. Sagoff in his article "Zuckerman's dilemma: A plea for environmental ethics," referenced below.

Kay, J. H. 2000 (June 15). "Cars are key to global warming." *Progressive Populist*. In: *Energy Alternatives: Opposing Viewpoints*. H. Cothran (ed.). San Diego, CA: Greenhaven Press, p. 168.

Lamb, H. No date. "Sustainable Development Is Not Sustainable." Accessed via www.freedom21santacruz.net, May 15, 2005.

Leopold, A.1949. "The land ethic." In: *A Sand County Almanac and Sketches Here and There*. New York: Oxford University Press, pp. 224–225.

Levine, J. S. (ed.). 1991. *Global Biomass Burning: Atmosphere, Climatic, and Biospheric Implications*. Cambridge, MA: MIT Press.

Lloyd, W. F. [1833]. "Two lectures on the checks to population." In: *Managing the Commons*. G. Hardin and J. Baden (eds.). San Francisco, CA: W. H. Freeman & Co. (1977), pp. 8–15.

Loh, J. (ed.). 2002 (July). "The Living Planet Report 2002." World Wildlife Fund. Accessed via www.panda.org/livingplanet/LPR02/index.cfm, June 9, 2005.

Loh, J., and M. Wackernagel, et al. (eds). 2004. "The Living Planet Report 2004." World Wildlife Fund. Accessed via www.panda.org/downloads/general/LPR2004.pdf, June 9, 2005.

Martin, G. 2005 (June 6). "U.N. conference: Accords promote taking responsibility locally." *San Francisco Chronicle*, pp. A1 and A7.

Ostrom, E. 1990. *Governing the Commons: The Evolution of Institutions for Collective Action*. New York: Cambridge University Press.

Ostrom, E. 1992. "The rudiments of a theory of the origins, survival, and performance of common-property institutions." In: *Making the Commons Work: Theory, Practice and Policy*. D. W. Bromley (ed.). San Francisco, CA: ICS Press, pp. 293–318.

Owen, H. P. 1971. *Concepts of Deity*. London: Macmillan Press, p. 74.

Sagoff, M. 1991a (September–October). "Zuckerman's dilemma: A plea for environmental ethics." *Hastings Center Report*, p. 32.

Sagoff, M. 1991b, p. 34.

Sagoff, M. 1991c, p. 35.

Samet, J. M., D. M. DeMarini, and H. V. Malling. 2004 (May 14). "Do airborne particles induce heritable mutations?" *Science* **304**:971–972.

Schell, J. [1982]. *The Fate of the Earth*. Palo Alto, CA: Stanford University Press (September 1, 2000), 173 pp.

Science. 2003 (March 14). "Global water dilemma." **299**(5613):1657.

Singer, P. 2000. *Writings on an Ethical Life*. New York: Ecco Press/HarperCollins, p. 89.

Singh, K. 1986 (September 29). "The Hindu declaration on Nature." In: *The Assisi Declaration. Messages on Man and Nature from Buddhism, Christianity, Hinduism, Islam and Judaism*. Basilica di Francesco, Assisi, Italy. WWF 25th Anniversary, pp. 15–18.

Socolow, R. H. 2005 (July). "Can we bury global warming?" *Scientific American* **293**(1):49–55.

Somers, C. M., B. E. McCarry, F. Malek, and J. S. Quinn. 2004 (May 14). "Reduction of air pollution lowers risk of heritable mutations in mice." *Science* **304**:1008.

Upham, P. 2003 (December). "Climate Change and the UK Aviation White Paper." Tyndall Briefing Note No. 10, accessed via www.tyndall.ac.uk/publications/briefing_notes/note10.shtml, June 20, 2005.

Wald, M. L. 2005 (June 5). "Alternative to deep burial of nuclear waste is considered. Plan calls for storing spent fuel in casks on Indian reservation." *San Francisco Chronicle*, p. A6.

Wackernagel, M., N. B. Schulz, D. Deumling, et al. 2002 (July 9). "Tracking the ecological overshoot of the human economy." *PNAS* **99**(14):9266.

White, E. B. and G. Williams (illustrator). 1952. *Charlotte's Web* (reissue edition). New York: HarperCollins Publishers, p. 162.

White, L., Jr. 1967 (March 10). "The historical roots of our ecologic crisis." *Science* **155**:1207.

Whitmore, T. C. 1980. "The conservation of tropical rain forests." In: *Conservation Biology: An Evolutionary Perspective*. M. Soule and B. Wilcox (eds.). Sunderland, MA: Sinauer Press, p. 313.

World Resources Institute, 2003. Accessed June 22, 2005 via www.wri.org. (Note: Calculations from tables available at www.wri.org show the United States uses 94 megawatt-hours per person per year of total energy [transportation, heating and cooling, and other energy needs]. This is five times greater than the per capita average energy use in the world. The electricity consumed in the United States is 12 megawatt-hours per year per person, which is 13 times that used per capita in China in 2004.)

Zhang, L. and A. Melis. 2002. "Probing green algal hydrogen production." *Philos Trans R Soc Lond B Biol Sci* **357**(1426): 1499–1507.

ADDITIONAL READING

Andelson, R. V. (ed). 1991. *Commons without Tragedy: Protecting the Environment from Overpopulation—a New Approach*. London: Shepheard-Walwyn Publishers, 198 pp.

Ausubel, J. H. and H. D. Langford. 1997. *Technological Trajectories and the Human Environment*. Washington, DC: The National Academies Press, 214 pp.

Light, A. and H. Rolston, III (eds.). 2003. *Environmental Ethics: An Anthology*. Oxford, UK: Blackwell Publishers Ltd., 554 pp.

Newton, D. E. 1995. *The Ozone Dilemma*. Santa Barbara, CA: ABC-Clio, Inc., 195 pp.

Swanson, T. (ed). 1996. *The Economics of Environmental Degradation: Tragedy for the Commons?* Cheltenham, UK: Edward Elgar Publishing Ltd., 192 pp.

APPENDIX

Appendix 5.1 Physical Dimensions and Units

Energy is the amount of work performed. It is the power times the time the power is applied.

$$Energy = Power \times Time$$

$$Power = Energy / Time$$

The dimension used in this text for power is the watt, and the dimension used for energy is watt-time. For example, the amount of energy consumed from burning a 100-watt light bulb for 24 hours is 2,400 watt-hours, or 2.4 kilowatt-hours. The internationally used dimension for energy is the joule, but watt-hours, calories, and the British thermal unit (B.T.U.) are terms also used for energy. Power in international units is joule per second, but watt or horsepower are other terms frequently used for power. When calorie is used to describe the energy available in foods, it is actually the kilocalorie, also known as the large calorie. A kilocalorie is the amount of heat (energy) needed to raise a kilogram (about 2 pounds) of water 1 degree Celsius. Conversions between these dimensions are shown in Table A.5.1.

Table A.5.1   Units of Energy and Power

Energy (dimensions are ML^2T^{-3})

Calorie (kilocalorie) = 4,168 joules

watt-hour = 3,600 joules = 0.86 kilocalories (nutrition calories) = 3.4 B.T.U.

kilowatt-hour = 1,000 watt-hour = 860 Calories = 3,400 B.T.U.

megawatt-hour = 1,000,000 watt-hour

joule = 1.0 watt-sec = 1/4,168 kilocalories

kilojoule = 0.95 B.T.U.

MeV (million electron volt) = 1.6 x 10^{-13} joules

Daily food energy of 2,640 calories = 3 kilowatt-hours

Bicycle exercise at 100 watts = 86 calories for one hour. Because the efficiency of the body for delivery of 100 watts is only 20%, the body will burn 5 × 86 = 430 Calories for one hour of exercise.

Power (dimensions are ML^2T^{-3})

watt = joule per second (J/s) = 1/746 horsepower

kilowatt = 1,000 watts = 1.3 horsepower (the work per time a horse can apply to drag 330 pounds 100 feet in one minute).

megawatt = 1,000,000 watts

PROBLEM SET

5.1 The United States has consistently rejected the Kyoto Protocol, an international effort to preserve and benefit the environment. What are the issues upon which this decision has been made? Is this the right decision? (The 1997 Kyoto Protocol can be found on the Internet.)

Team A - Take the position of the U.S. government and present the arguments against the Kyoto Protocol.

Team B - Present the arguments in support of the Kyoto Protocol.

5.2 Your firm is proposing a project to build a new dam. You are the engineer on the project. Outline how you will approach the report describing the environmental impact of this project.

5.3 You are aware of soil contamination with toxic substances on a prime piece of land under consideration for extensive residential development. As a project manager, how would you approach the following dilemma: The property's location is ideal relative to transportation, schools, malls, and other conveniences. However, the extent of the contamination and preliminary estimates of the cost for cleanup are four times greater than the value of the land if you use the most strict cleanup guidelines. If you use other guidelines, the cleanup cost will be only 1.5 times greater. Outline in less than 200 words your approach to this problem.

5.4 You are a member of a multiuniversity, multination team to assess and make recommendations about global warming. You have evidence that certain nations are contributing more to greenhouse gasses than are others. How would your team approach these nations? Consider the different metrics of greenhouse gas per capita and total gaseous emissions per year.

5.5 The text refers to Agenda 21 as a possibly dangerous movement. Although intended as a method to protect the environment, it might conflict with current protections of human freedom. In less than 100 words, give your views on this subject after reviewing the Internet for current views on the subject. Consider whether the model of planned communities practiced by societies that are organized as a "commons" can be a workable model of Agenda 21.

5.6 What would be an economic cost per kilowatt for installation of solar panels if the household requires 2 kilowatts for 6 hours each day? The solar panels will last 20 years. The alternative is to buy electricity at $0.04 per kilowatt-hour. Assume the interest on the initial solar installation investment could be compounded at 4% per year. (Hint: You can set this up as a problem to solve for the installation cost based on equaling the cost per kilowatt-hour of $0.04, or you can iterate by guessing at the installation cost of $ 2.00 per watt.)

5.7 Transportation is responsible for energy use as well as for environmental contamination. Air transportation produces an important component of "earth-warming" gases. Research the relative importance of this form of transportation as a contributor to the annual CO_2 and halocarbon emissions. (Hint: England has information on the importance of air transportation emissions on its budget for gas emission reduction to conform to the Kyoto Protocol.)

5.8 Hard physical activity over the course of a day requires about 4,000 calories of energy. This energy comes from food. Calculate the average power to sustain an individual assuming that the energy is averaged over 24 hours. How does this compare with the average power per capita in the United States for all activities supplied by fossil fuels? (Hint: See Appendix 5.1.)

6

Ethics of Genetically Modified Organisms

INTRODUCTION

Genetically modified organisms (GMOs) are created using genetic engineering to manipulate the gene structure to permanently encode a desired trait. Historically, the development of desired traits in plants, animals, and even human beings has been discussed and practiced since the time of Plato. Simple plant selection, crossbreeding to eliminate or incorporate a desired trait, and the creation of hybrids by cross-pollination to provide desired characteristics are centuries-old natural means of improving a plant species. The breeding of domestic animals and the human species through selective mating, in vitro fertilization, and in the past by castration are other forms of genetic engineering.

After Watson and Crick determined the sequence of nucleotides that makes up the genetic code of DNA in 1953, scientists quickly realized that the introduction of certain genes into a plant or animal could result in a broad range of beneficial effects. Advances from this technology have occurred in three main areas. One has been to develop genetically modified plants that require less water and fewer pesticides, have seeds with a shorter germination time, have better crop yields, and produce foods that have extended shelf lives, better taste, and improved nutrition. Some can even supplement needed vitamins and provide immunity through antigens that stimulate formation of antibodies against diseases (Galun and Galun 2001). Another advance has been the development of genetically modified animals engineered to mature rapidly, to be larger and leaner, and to produce drugs or rare human proteins vital for people who lack them. The other significant advance has been the development of genetically modified animals that have characteristics that mimic human genetic diseases. The genetic changes are produced by "knocking out" or "knocking in" the genes associated with a particular malady. The availability of these animal models has allowed research into disease etiology in ways previously unavailable.

The two main reasons that the genetic modification of plants is seen as an essential technology for the future of Earth is to increase the food supply by methods that promote better sustainability of the land, and to conserve water by producing GM plants that utilize it more efficiently. The world population in 2005 was 6.1 billion, a figure that is expected to grow to 9.2 billion by 2050 (see Appendix 6.1). Currently, over 0.8 billion people in the world do not have adequate food, as measured by the number of people with less than 2000 calories per day of available food (Food and Agriculture Organization 1992, 1996). The number of malnourished people could more than triple by 2050 or even sooner if there is not a major improvement in food production and distribution. Already the current annual rate of cereal yield (the major measure of basic food availability) in developing countries is less than the rate of population growth (U.S. Census Bureau 2005). Pasture and crops take up 37% of the Earth's land area. Increasing the amount of land under cultivation will impose undesirable environmental consequences, including the further depletion of available water. About 70% of available water in the world is being used for agriculture (see Appendix 6.1), and many U.S. farmers experience water shortages. In many countries, river waters no longer reach the sea because they are diverted to irrigation, and 50% of the world's wetlands have disappeared as a result (Somerville and Briscoe 2001). Thus, there are important reasons to find better ways to produce food, conserve water, and better sustain soil resources. A major goal of the so-called Green Revolution of the 1960s was to increase crop yields to feed the populations of developing countries (Conway 1998), and major changes did occur in the 1970s through improved irrigation, fertilization, and pesticide use. Further advances in the breeding of rice and wheat plants have continued the Green Revolution of today. Major advances in crop productivity have lessened the differential between world food needs and food production. A more recent revolution is underway to advance the availability of better plants through gene transfer.

Despite the apparent benefits of GMOs, their development, testing, and use have given rise to a number of safety and ethical questions, not only in the United States but also in the global community (Mephram 2001). The multitude of issues and concerns are highlighted by the following questions:

- What, exactly, are the risks to the environment? To habitats? To wild populations of animals? For example, will faster maturing, larger fish cause native fish to disappear? Will the heartier plants and animals eliminate natural ones and cause a loss of species diversity?

- What are the implications for natural food crops whose genome might also be modified by genes from nearby (and not so nearby) genetically modified plants? Will transgenes enter the genome of other species and even weeds? Will there be a loss of identity of the stable food source or a loss of the valued food source altogether?

- The separation and labeling of GM foods to distinguish them from natural foods has been inadequate and unreliable. Thus, those who wish to avoid GM foods have not been able to do so.

- What are the long-term effects of GMOs on humans who eat them? Have these been sufficiently explored? The U.S. Food and Drug Administration decided early on that genetically modified food was "essentially equivalent" to non-genetically-modified food and was "not inherently dangerous." Was this prudent? Are we sure there are no risks to humans? There is now evidence that enhancement of soybeans by transfer of the Brazil nut gene for methionine makes these soybeans allergenic (Nordlee et al. 1996). Are there other unknown risks?

- What are the short-term and long-term risks of gene transfer from species to species? Might GMOs cause unpredicted harmful generational and intergenerational effects to plants, animals, and human beings? Could society be susceptible to infections from bacteria that have received resistance to antibiotics through transfer of the antibiotic marker genes placed in foods?

- The current climate about their use is to impose regulations that restrict their use. Is this premature? Will this lessen their benefits to society?

- How are the issues of risks and benefits being addressed by national and international forums that are now offering opinions about GMOs?

GMOs are often at the center of debates about environmental ethics, the ethics of animal experimentation, the ethics of enhancement, and the ethics of emerging technologies. However, many of the issues bulleted above have not been discussed in peer-reviewed literature. GMOs have the potential to improve food supplies for starving populations and have ignited commercial developments of extraordinary dimensions. Unfortunately, GMOs have also ignited fears of their unforeseen consequences that are similar to those associated with the introduction of fluoride in drinking water or vaccinations to prevent diseases. Many of the concerns are not realistic. Others involve a conflict between utilitarianism and the Kantian duty to respect human autonomy.

It is important to put GMO ethics in balance with the ethics of natural breeding. To many experts, the public fears of GM foods are unnecessary because many of the processes do not involve the transfer of foreign genetic material to stimulate the improvements; rather, the changes result from changing the plant's environment in the laboratory and inducing the genetic changes much in the same way nature has done it for thousands of years. Further, when a GM plant is made by transferring desirable genes from one plant into another, the number of genes transferred is exceedingly small compared to the plant's own genes. Others compare GM crops to the introduction of taxonomically distant species; thus they argue for stringent regulation.

The debate on this issue has led to the recognition that conventional plant modification procedures carry with them the same risks and potential consequences as do GMOs (National Research Council 2002a). To understand better the basis of the concerns about GMOs, we begin by first presenting some of the technologies used to make the genetic modifications of plants and animals, then review the arguments for and against GMOs and the likelihood of serious consequences from them, and then discuss some of the ethical problems.

PROCESSES INVOLVED IN GENETIC ENGINEERING OF PLANTS AND ANIMALS

The genetic engineering of plants, animals, and organisms can be accomplished by several new technologies. The focus here is on plant and simple organism modification technologies, although some of the basic techniques used for plant modifications can also be used for animals. The genetic modification of animals through gene transfer using vectors, cloning, and "knock-out" or "knock-in" techniques for the study of genetic diseases is provided here, in detail in Chapter 8, "Ethics of Human and Animal Experimentation," and in Chapter 11, "Ethics of Enhancement Technologies".

Genetic Modification of Plants without the Introduction of Foreign DNA

Stimulation of Mutations in Plants by Chemicals or Ionizing Radiation

The discovery that some chemicals and ionizing radiation could induce mutations in bacteria and yeasts led to the use of these stimuli to induce mutations in plants and animals. These methods produce offspring that by random changes in their genome might have increased value. This method makes use of the plant's innate DNA without the introduction of foreign DNA. Those exhibiting traits that are more desirable are selected and multiplied. As the changes are entirely random, these methods require the screening of populations of tens of thousands of candidates. Depending on how rapidly a plant or animal matures, finding the improved characteristics and determining the stability of the genetic modification of a particular plant species could take years.

Stimulation of Transposons to Induce Changes in the Plant Genome by Stress

Until about 50 years ago, scientists thought that all the genetic material of an organism contributed to its form, function, and well-being. Then it was discovered that there are marked similarities in the DNA between animals such as mice, *Drosophila* (fruit fly), and human beings. It was also found that in human beings, the number of genes were far fewer than previously believed (i.e., 30–40 thousand rather than previous estimates of hundreds of thousands). The rest of the DNA was labeled as "junk DNA" until it was shown that much of this DNA consists of a class of mobile pieces or fragments of DNA that actively relocate randomly from one chromosomal location to another. These fragments have come to be called transposable elements, transposons, or hopping genes. They have been found in vast numbers in virtually every organism studied, and their random movements are believed by many to have major roles in the natural changes that occur in the evolution, structure, and function of plants and other organisms. Transposable elements make up an estimated 30% of the plant's genome and 40% of the human genome.

This method of inducing mutations takes advantage of the mobile capabilities of the otherwise inactive fragments of the plant's own DNA, the transposons and retrotransposons. When physically stressed, plants induce the production of enzymes called transposases that free the plant's own mobile genetic material (transposons) in the genome. As a result, these mobile elements can move to new locations, thereby changing the gene expression of the plant without the introduction of outside or foreign DNA. Also when stressed, the plant's retrotransposons are stimulated to move, but they do so by replicative transposition without the need for transposases. A copy of the transposon is made at the original site, using reverse transcriptase. The copy moves to another site while the original DNA sequence is preserved. The insertion of transposons at another location in a chromosome is independent of complementary base sequences. This random movement of fragments of an organism's own DNA to other parts of the genome may activate genes that result in better or more desired traits that enhance (or in some cases detract from) the plant's value. Thus, both transposons and retrotransposons cause genetic reshuffling of a chromosomes genome.

In Situ Microsurgery to Stimulate Transposon-Induced Changes in Plants and Animals

Another approach also utilizes transposons to induce changes in an organism's genome, this time by using a form of in situ microsurgery on the plant's own genes. This technique was developed to mimic the findings in nature that even minor changes in a plant or animal's own genes can result in major changes in phenotype, including appearance, robustness, and yield. These changes can be achieved by utilizing gentle micromanipulations of the plant's own DNA to stimulate changes in the plant's gene expression.

Thus, transposons increase the potential to change the plant's genome or gene expression by a natural response to stress. The movement of transposable elements can be effected by artificially stressing the plant using temperature changes, drought conditions, and light-dark cycling, or by gentle microsurgical manipulations of the plant's own DNA. Both processes induce changes in the genome of a plant by mimicking what a plant would do under a condition of stress in its natural environment. Because the changes are random, large numbers of plants must be screened to identify those with desired characteristics. Both methods induce genetic changes that are permanent and passed down in subsequent generations of the species. Since the process involves random changes in the genome, thousands of plants resulting from such stimulation and their progeny must be screened to identify desired changes. Large-scale screening can be readily implemented if the desired phenotype is easily recognized (color, size) (www.uga.edu/columns/000911/front3.html, accessed February 28, 2005).

Examples of possible modifications include:

- Transposon inserts in the middle of a coding sequence, thus destroying that gene's function or modifying it by the coding region of the transposon that inserts into and possibly interrupts the original gene, leading to a new gene product.

- Transposon inserts in a regulatory portion of the genome, thus increasing or decreasing the gene expression. In this way, plants can be genetically modified to produce a protein that was not expressed in that plant before.

- Transposon carries a code for a gene that, when inserted downstream from a regulatory promoter, leads to expression of the dormant gene. This is an infrequent phenomenon.

More information on the biology of hopping genes can be found in Campbell and Reese (2002).

Since this technology involves finding ways to switch on the plant's own innate genetic capabilities without inserting any foreign DNA, commercial producers hope that the U.S. Department of Agriculture (USDA) will be willing to consider foods resulting from this technology as "organic," a labeling of increasing importance in the marketplace. However, as of 2006, no GM foods can be labeled "organic." There remain concerns that these plants may have some of the risks of GM plants produced by gene transfer. The random tweaking of plant genomes could result in unwanted mutations that might also be considered a danger to other crops or species (e.g., generation of a robust species that will make other species extinct). However, these fears seem for the most part to be unwarranted (Stewart 2004, Parekh 2004).

Other methods of plant modification such as cross-pollination and hybridization result in genetic changes and species changes that can introduce consequences not unlike those thought to be a problem for transfer of genes into plants as discussed below. A major focus of the debate is whether the safety and environmental hazards are any different between genetically modified crops and crops obtained by conventional breeding methods.

Genetic Modification of Plants and Animals by Transfer of Foreign DNA

Three of the common methods for transferring new recombinant DNA sequences to produce a desired characteristic in a plant or animal are:

- Microinjection into eggs or embryos (animals and sometimes plants)
- Bacterium-plasmid transfer (plants and microbes)
- Shotgun or biolistic method (plants and bacteria)

These methods are diagrammed in Figures 6.1, 6.2, and 6.3, respectively. These methods allow the direct insertion of a functioning gene from another source to produce the protein needed for a desired trait.

Genetic Modification by Microinjection of DNA into Cells

Microinjection involves the injection of the desired DNA sequence into an egg or embryo by a needle or other method (e.g., microparticle gun or electroporation). This method has been used successfully in mammals since 1980 (Houdebine 2005) and in fish since the 1984 (Maclean and Talwar 1984). The goal is to equip the animal (mammal or fish) with a known sequence of DNA that will produce the protein related to the desired trait. A functioning gene needs a region that is called the promoter that allows the commencement of the messenger RNA synthesis. A terminator region is also needed to stop the RNA synthesis. Once formed, this RNA moves to the cytoplasm, where ribosomes make the specific protein that gives the desired trait. Thus, the functioning gene consists of more than the specific sequence that produces a specific protein. Figure 6.1 illustrates the production of a transgenic mouse using a pipette to inject the desired gene into a fertilized egg. The insertion into the chromosome is random, and the desired trait might or might not be produced.

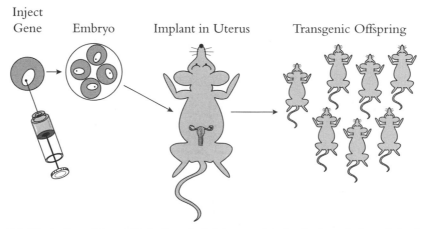

Figure 6.1 Illustration of the method of producing transgenic animals through direct injection of the desired gene into the nucleus of the fertilized egg.

The insertion of the gene is often facilitated by a carrier such as a plasmid (see below), or a virus *vector* (usually a virus that can penetrate the cell wall and enter the cell nucleus) into which the desired gene has been placed. After the fertilized egg grows into a small embryo, it is inserted into the uterus of the animal. The offspring are screened to determine whether they manifest the expected new trait. This offspring is known as a knock-in animal. The techniques of genetic modification as well as cloning of animals and human beings are discussed further in Chapter 9, "Ethics of Assisted Reproductive Technologies."

Genetic Modification by Gene Transfer Using Plasmids and Bacteria

The most commonly used method for plant genetic engineering relies on special capabilities of a bacterium that can penetrate the cell wall and nuclear membrane to transfer the desired gene to the genome or chromosomes of the plant. The history and molecular biology of this method are found in Galun and Galun (2001a). The most commonly used bacterium is *Agrobacterium tumefaciens.* It can penetrate plant cell walls, and part of its DNA can enter

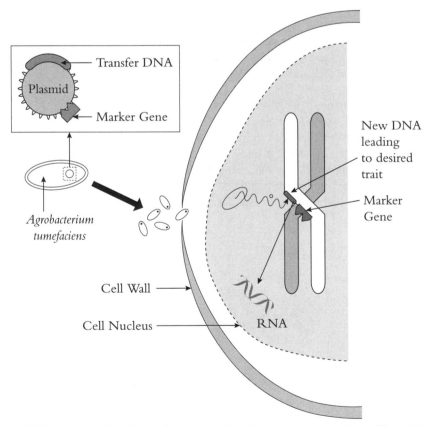

Figure 6.2 Commonly used technique for the insertion of a genetic trait into plants. The vehicle is the bacterium *Agrobacterium tumefaciens*, which has the ability to penetrate cell walls. The cargo or gene to be inserted is carried by a plasmid that moves from the bacterium to the cell nucleus. Assisted by proteins in the plasmid, the transfer DNA is inserted randomly into the plant's chromosomes, along with the marker gene. Plasmids carrying the DNA can be injected directly into the nucleus as well.

the nucleus of the cell. This bacterium carries the gene in a plasmid, a plant virus with no protein coat. This plasmid contains a DNA sequence with a specific region known as the transfer DNA (T-DNA) region. The plasmid is joined with DNA sequences of a desired gene along with a promoter sequence and terminator sequence needed to functionalize the gene once it is inserted into the plant genome. Those sequences are incorporated into the T-DNA region (see Figure 6.2). In addition, a marker gene such as a fluorescent protein or an antibiotic resistance protein is also included. The marker allows identification of the presence of the gene in the plant cells. The gene of interest in the T-DNA region will be transferred to the plant's genome, assisted by proteins in the plasmid. Those plants that have taken up the marker gene, and therefore have also taken up also the associated gene of interest, are identified and selected for further multiplication.

It should be noted that the use of *Agrobacterium tumefaciens* is patented, making it difficult for others with commercial interests to use this tool. Thus, other bacteria with similar attributes were developed in 2005. These are readily available as open source materials to commercial and public sector endeavors without the limitation of patents on that technology (Boothaerts et al. 2005).

The bacteria-mediated approach to genetically modified foods produces changes faster than traditional breeding, but when the time it takes for crossbreeding and the regulatory steps for GM approval are taken into consideration, the two processes take about the same time to reach the consumer. It does allow greater precision in selecting the desired characteristics than traditional breeding. The analog in human medicine is preimplantation embryo selection (see Chapter 9) and gene therapy (see Chapter 11).

Biolistic Impaction Method of Gene Transfer into Plants

Another method of gene insertion is the use of tiny gold or titanium particles as projectiles that have the desired gene precipitated on the surface. These projectiles are propelled into

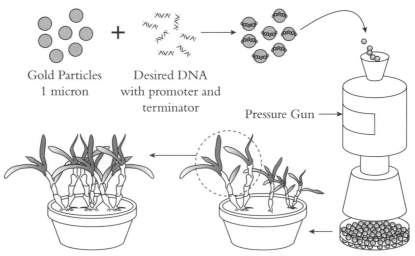

Gold Particles
1 micron

Desired DNA
with promoter and
terminator

Pressure Gun →

Figure 6.3 The biolistic method of gene transfer uses physical force provided by helium gas or gunpowder to propel small gold particles coated with the desired DNA sequence into the plant cell nuclei.

plant cells at high speeds utilizing a "gun" that uses helium under pressure to provide the penetrating energy (see Figure 6.3). The cells that take up the new DNA are identified by a marker gene and are cultured to grow into new plants. This method has proven very effective in plant engineering.

GROWTH IN THE USE OF GMO PRODUCTS

There has been a steady increase in the number of countries using GMOs. There has also been an increase in the land area used to grow genetically modified crops, with an increase of 20% between 2003 and 2004 (James 2004, see also www.isaaa.org and www.ornl.gov/sci/ techresources/Human_Genome/elsi/gmfood.shtml). A synopsis of GM crop activities is:

- Genetically modified plants are grown commercially or are under evaluation in over 17 countries on six continents.
- In the year 2004, there were nearly 200 million acres planted with transgenic crops of soybeans, corn, cotton, and canola in both industrialized and developing nations.
- Countries that grew 99% of the global transgenic crops were the United States (68%), Argentina (23%), Canada (7%), and China (1%).
- In the United States, 79% of the cotton crop acreage is now genetically modified, as is 87% of soybean and 52% of corn crops.
- In recent years in the United States, about 80% of all soy, 40% of corn, and most canola have been genetically engineered, either with a bacterial gene to help the plant fight insect pests or with a gene that makes it resistant to a weed killer.
- Other crop modifications include squash, sugar beets, potatoes, papaya, and summer squash.
- On the horizon are bananas that produce human vaccines against infectious diseases such as hepatitis B (Galun and Galun 2001b) and fruit and nut trees with increased productivity.

Genetically modified products (e.g., corn, canola, cottonseed) are being used increasingly in processed foods in the United States. One of the earliest examples of the use of GM organisms was as a spray containing genetically modified bacteria. Developed in 1986, these GM bacterial organisms when sprayed on strawberry crops protected them from frost damage; however, these were never used commercially.

Genetically engineered or transgenic animals have been developed in increasing numbers that allow new ways of studying human and other diseases in large numbers of identical animals. Millions of vertebrate GM animals are used each year for a variety of pharmaceutical and medical studies, both in the United States and abroad. Other kinds of genetic modifications of higher mammals, including farm animals, have resulted in cows and other animals that grow more rapidly, have higher milk production, and have leaner meat. GM fish have been developed that grow and mature more rapidly than their wild counterparts. In 2005, there were no GM animals on the market. Sheep and other animals have been modified to produce rare human proteins in their milk in order to treat rare human diseases in patients who lack them. Treatment currently relies on proteins harvested from human plasma, but the supply is limited. The proposal is to increase the availability of these proteins so that more patients needing them can be treated; however, these have not been approved or commercialized as of 2006. The use of GM technology is expected to expand exponentially in the next 10 years as researchers find new ways to adapt genomic resources to the needs of the planet.

BENEFITS AND RISKS OF GMOS

Proponents of this technology argue that bioengineered food could provide the most important step forward toward relieving world hunger in our lifetime, but opponents fear there may be unknown consequences. The usual way to approach this problem would be to assess whether GMOs have sufficient benefits to outweigh the risks. However, in this situation, it is difficult to balance the known benefits of GMOs when the risks and the probability of their occurrence are currently not known. In fact, many have stressed that caution is needed *because* there are unknown effects that cannot currently be predicted. Thus, it is important for the global community to conduct ongoing evaluations and debates about GMOs so that the risks and their ethical implications can be identified and discussed.

Benefits

Current estimates are that 0.8 million people are malnourished (see Appendix 6.1). There is enthusiastic support for the idea that bioengineered food could help alleviate the critical shortage of food for those in need. Countries with difficult climates and poor growing conditions could dramatically benefit from genetically modified foods that grow faster, are drought resistant, tolerate aluminum in the soil, are resistant to pests, and have greater yield and more nutritional value (e.g., Slabbert et al. 2004, Gressel et al. 2004). Populations in general can benefit from these and other qualities of genetically modified plants (GMPs). As we have noted, GMPs also provide increased food security and crops that taste better, mature more rapidly, are grown using more environmentally friendly growing techniques, and have a longer shelf life. Plants that are resistant to pests require fewer pesticides, thereby reducing the human health effects of these often-toxic substances. Pesticides are of particular concern in young children who eat three times as much on a weight basis compared with adults but are more at risk for problems because the pesticides are distributed in a smaller volume of tissue, resulting in concentrations as much as three times higher in children than in adults (Wilson et al. 2004).

GM products have also revolutionized the approach to other common foods, an example of which is genetically engineered chymosin, a substance that has replaced the use of rennet in 80% of cheese making in the United States. This represents a major transformation in the way cheese is made, as previously the rennet had to be obtained from the ground up stomachs of young calves or other animals. The lives of thousands of animals have been spared as a result. Other GM technologies are being used to create plants and organisms modified to produce vitamins, medications, or other unique products. Overall then, these products offer substantial advantages to crop security, to farmers, to the economy, to Third World countries suffering starvation, and to sustaining the environment.

The potential benefits of GMOs are summarized in Table 6.1.

Table 6.1   Potential Benefits of GMOs

- GMOs potentially will grow faster, thus allowing more efficient utilization of land through multiple plantings in some climates. (National Research Council 2002a).

- GMOs taste better, resist drought and pests, have higher yields, have longer shelf lives, and provide greater profitability.

- GMOs have the potential to provide new foods that are more nutritious or produced specifically to meet the needs of individuals with allergies or restrictive digestive capabilities.

- Beneficial products can be developed and made available sooner than by waiting for changes by evolution or more traditional means if field testing and regulatory requirements allow rapid commercialization.

- GMOs have the potential to expand crop availability in harsh environments (e.g., salty soil), providing more food to feed the 2.3 billion malnourished people expected by 2050.

- GMOs can contribute environmental benefits, including smaller irrigation requirements and fewer pesticides.

- GMOs provide new ways of producing vitamins, proteins, and pharmaceuticals, including vaccines needed to treat diseases.

- Many of the same attributes of GM plants apply to genetically modified animals: they mature faster, are of larger size, and are more profitable in the marketplace. They can be genetically modified to produce more milk, rare proteins needed to treat some diseases, or animal models of a genetic disease for study. Insulin was made available through genetically modified bacteria in 1982.

Risks of Genetically Modified Organisms

The leading risks and concerns have to do with the safety of GMOs. The public and some scientists worry that the ingestion of a foreign protein in a plant or organism could interfere with natural metabolic processes, or result in untoward effects on the immune system (such as allergic reactions). It is possible that allergic reactions can develop in individuals exposed to plants that have been engineered to produce specific proteins to which the individuals are allergic. However, only scant reports of this problem have appeared in the literature, one of which was a severe dermatitis of the hands from handling the plants (Galun and Galun 2001b). This reaction can also occur in conventional breeding due to increased psoralens (compounds found in limes, lemons, figs, and celery).

A popular concern among the public is that the antibiotic resistance marker, inserted during gene transfer to identify plants that have taken up the intended genetic modification, could be transferred from the plant tissues to pathologic bacteria. If bacteria with the potential to infect become resistant to antibiotics, treatment of animals and human beings by antibiotics will be compromised. However, the resistance is not that to the antibiotics in current use. Furthermore, it is extremely rare for this to happen if at all. Other fears of health problems with GMOs are related to the introduction of genes and their promoters into the food chain; however, there is no substantiated evidence that this has been problematic (Stewart 2004).

Another concern is that the use of genetically modified insects to neutralize or counteract an invading pest and avoid the use of pesticides could be problematic. Even the introduction of non-GM organisms to contain some threat to the environment has had serious consequences in the past. For example, kudzu, a climbing and spreading ground cover from the pea family, was introduced into the United States from Japan in 1876. From 1935 to the mid-1950s, farmers in the South were encouraged to plant kudzu to reduce soil erosion, and the Civilian Conservation Corps planted it widely for many years. This plant became a major pest, taking over acres of land, bringing down trees, and spreading voraciously into public and private lands. It was called "the vine that ate the south," and by 1953 was removed from the list of permissible cover plants. These problems could also occur with GM organisms, with the added risk of introducing a virus, bacterium, or insect that is protective of some crops but could be an environmental or health problem for animals and humans. Again, these appear to be unlikely.

However, there is reason to be concerned that GMOs may have major ecological impacts. There is the possibility that certain vigorous, non-native GM plants and other organisms (insects to counter destructive pests, for example) may overwhelm, overrun, or even cause extinction of indigenous species (www.ornl.giv/sci/tecgresiyrces/Human_ Genome/elsi/gmfood.shtml, accessed March 2, 2004). It is important to keep in mind that these kinds of unintended consequences have resulted in the past with modified plants developed by conventional breeding processes. For example, a gene used to create corn hybrids was found several years later to make the corn susceptible to a toxin from the fungus *Bipolaris maydis*, the disease organism causing a serious blight called Southern corn leaf blight (Dewey et al. 1986). Natural hybridization between the introduced New World cord grass and the native *S. maritime* in Britain gave rise to *S. angica*, an invasive species that drastically altered the coastal ecosystems of the British Isles (Thompson 1991). Potatoes bred for insect resistance were found to be toxic because of high levels of glycoalkaloids and had to be withdrawn from commercial production (Hellenas et al. 1995).

There is also concern that natural forces already present in the environment could contribute to movement of GM genes into other species of both plants and animals. As we shall see below, the consequences of gene transfer are not known presently and will take time to ascertain.

Concerns about Physical Escape of GMOs into the Wild

The genes of all plants, whether genetically modified or not, disperse into the environment by three methods.

Dispersion of seed (by wind or accidentally).

Pollination (by pollen carried by wind, birds and insects).

Horizontal gene transfer (see below).

The seeds and pollen from genetically modified plants are now known to disperse into the environment, as documented by finding GM marker genes in crops and foods that have not been genetically modified. The consequences of uncontrolled dispersion into the environment of GM plants, and also of GM organisms such as salmon, beetles, bacteria, and yeasts, may have unintended worldwide consequences not yet recognized. However, it is not known whether bacteria modified by horizontal gene transfer can transfer genes into vertebrates.

The containment of genetically modified plant foods to prevent their dispersion into the wild has not been successful to date. This has made it difficult to label accurately whether products do or do not contain GMOs, discussed later in this chapter. While the causes of these difficulties are not fully understood, they are likely due to factors such as horizontal gene transfer (discussed below), cross-pollination, wind and animal dispersion, spillage of seeds, and dominance over other species if the attributes of the GM plants give them an advantage over the less robust indigenous species. The National Research Council of the National Academies of Sciences recently reviewed this problem. Their report confirmed that containment techniques have not been fully effective, and recommended further efforts to find methods to prevent genes from GM plants from spreading to other species, perhaps by making the seeds sterile (National Research Council 2004a).

Concerns about the Biology of Horizontal Gene Transfer

The basis for concerns regarding the containment of GMOs is the fact that the genetic modification can be transferred to other species in the environment through the process of horizontal or lateral gene transfer (Lorencz and Wackernagel 1994). It had long been believed that mating or conjugation occurs only between members of the same species. But it is now known that a special small strand of DNA known as a plasmid, usually found in bacteria, can be transferred at mating (conjugation) of bacteria. In addition, transposons can be transferred through mating wherein the fertility plasmid can drag along genomic material from one bacterium to another. Among these jumping genes are special kinds of genes that support the efficient transfer of antibiotic resistance genes. Pathways for genetic exchange include the passage of DNA from living somatic cells of a plant to bacteria. This transfer is accepted as part of infrequent events in evolution, but it is not a commonly occurring event that can be demonstrated by scientific observations. The passage of genetic material can be associated with any pathway by which bacteria travel in a viable form, such as by means of inoculation of soils, movement in ground water and surface water flows, on soil particles that are wind-blown, or in the gut of blood-feeding insects and other parasites.

If genetic material is subsequently transferred horizontally from bacteria to other organisms such as plants or fungi, then the pathways for genetic movement will include the normal movements of those organisms and the normal dispersal mechanisms of organisms such as pollen, spores, and vegetative reproductive structures (e.g., budding in yeast). Concern over the spread of genetically modified plant pollen has arisen because natural plants can be cross-pollinated (bred) by the pollen of modified plants. Pollen can be dispersed over large areas by wind, animals, and insects; thus, modified genes could be carried some distance away from the source. These pathways for the movement of genetic material could be constrained by a gene that would render seeds ineffective, but this mode has not been widely used. Thus, any release of a GMO outside of a contained laboratory entails a potential release of modified genetic material into the wider environment. The potential consequences from this transfer include:

- Antibiotic resistance due to marker genes spreading to pathogenic bacteria.
- Disease-associated genes spreading and recombining to create new viruses and bacteria that cause diseases.
- Transgenic DNA insertion into human cells, triggering cancer. There is no evidence for this.

Although there have been reports of GM DNA in some mammals (e.g., cows' milk), there are many studies where this was not found. No evidence exists that horizontal gene transfer has occurred into vertebrate cells. Evaluation of tissues from chickens fed GM corn did not show lateral gene transfer (Aeschbacher et al. 2005). However, there is ample evidence that horizontal gene transfer occurs in bacteria. Transfer of antibiotic resistance between bacteria is now well established (Wolska 2003). Genetic exchange between bacteria and single cell organisms has also been observed (Mukai and Endoh 2004, Zhu 2004).

The discussion above reviews possible mechanisms that could lead to risks, but the evidence for widespread horizontal gene transfer even among same animal species is sparse. A balanced review of the possibilities for gene transfer among GMOs and between GMOs and weeds is found in Stewart (2004), as well as in National Research Council Reports (2002, 2004). There are concerns that GM crops have the potential to threaten native species of crops by gene transfer. The possible gene transfer from GM corn imported from the United States into ancient and diverse native varieties of corn in Mexico has attracted attention from the opponents to GM crops. Despite a moratorium in Mexico making it illegal to grow modified corn since 1998, in 2001, 7.6% of corn in 188 corn-growing communities in Oaxaca showed evidence of genetic modification. The fear expressed by some is that if this single gene transfer is allowed to continue, there could be a loss of the unique diversity associated with certain species of ancient Mexican corn. However, scientists have found that genes from modern hybrid corn have been moving into other species for years without resulting in extinction, and further counter that a single gene added to 730,000 other genes would not cause a loss of diversity (Lemaux 2005). How this modified corn came to Oaxaca is not clear, though it is possibly from illegal planting. It is a potential problem that will be followed carefully. In the meantime, the North American Free Trade Agreement (NAFTA) has resulted in the importation from the United States of 70% of Mexico's fresh edible corn, of which 50% is genetically modified (Leary 2004). This genetically-modified corn is meant to be used for food or animal feed only.

One of the more notable controversies in the scientific literature also involved Mexican corn and U.S. GMOs. A report in *Nature* (Quist and Chapela 2001) described GM-induced changes after a certain gene from GM corn produced in the United States was found in native corn in Mexico. This was reportedly due to introgression (i.e., the incorporation of a gene from one organism complex into another as a result of hybridization). However, the genes did not introgress, based on the genetic definition of this term (Lemaux 2005). It involved the cauliflower mosaic virus promoter for U.S. GM corn that appeared in native corn of the Mexican state of Oaxaca. This report was challenged by two letters in *Nature* as well as by the editor of *Nature* on the grounds that some of the methods were not appropriate and the results could be artifacts (Metz and Futterer 2002, Kaplinsky et al. 2002). The Quist and Chapela conclusion that was cited as most objectionable was that promoter sequences and transgenes are genomically unstable. This conclusion, if it were true, would be counter to previously accepted theories of plant biology: that is, the Quist and Chapela paper claimed that transgenes introduced through biotechnology behave differently than natural gene sequences. The validity of the Quist and Chapela paper is in question, but it brought to the fore a greater concern, which is how U.S. GM corn appeared in Oaxaca in the first place as such importation for cultivation is illegal (see above). It also fueled the issue of what are the consequences of gene flow of a particular transgene.

Concerns about Biodiversity and Perturbations in Biologic Evolution

Biodiversity is considered essential to the survival of many plant and animal species. The genetic manipulation of plants and organisms has the potential to influence dramatically the biodiversity of plants, which if widely used and invasive of other species, could result in the creation of monocultures and decreasing species richness. If an area's food source depended on only a single species of corn or wheat, it is possible that a viral, bacterial, or fungal scourge could eliminate an entire food type, making extinct a specific food that a human population has come to rely upon.

Another concern about genetic modification is that a characteristic such as increased resistance to pesticides might be transferred to weeds, making their eradication more difficult. Still another concern is that cross-pollination of weeds might result in a change detrimental to the foraging material for animals (e.g., weeds that are less tall in the winter and therefore hidden by snow, or less resistant to drought), thereby risking the eradication of species dependent on that material. On the other hand, cross-pollination could lead to grasses that are more robust and a boon to the indigenous animal population. Because the effects of genetic modifications are not known and may be completely opposite than predicted, it is difficult to do a proper analysis that will allow a decision of whether to proceed with it or not.

Those who oppose GM crops raise the issue that tampering with the natural inheritance of a plant or organism may change its inherent value. Individuals and environmental groups who oppose GM foods argue that a species deserves protection and preservation. However, this argument cannot single out GM crops as the only cause for the problem because, as we observed earlier, species modifications are also the result of conventional breeding (Ammann 2005):

> . . . the development and introduction of GM crop varieties does not represent any greater risk to crop genetic diversity than the breeding programs associated with conventional agriculture. After all, the quantity of its product is the result of thousands of genes and the genetic background is almost always more important for the questions dealt with in this review than a single transgene.

Concerns about Marker Genes and Other Modifications

As we have disscussed, genetically engineered plants are often given a "marker" gene that confers resistance to a specific antibiotic but otherwise has no known effects. The purpose is to simplify the laboratory identification of a product that has been genetically engineered, and allow an assessment of the stability of the new gene over time and in differing environments prior to commercialization. A consequence of these markers might be that the ingestion of sufficiently large amounts will result in the accumulation of physiologically significant concentrations of antibiotics, leading to allergic reactions or antibiotic resistance (Wuthrich 1999), although this seems highly unlikely. Antibiotic marker genes are not being used in new GM crop development programs as companies have found other ways to identify plants that have taken up the desired genetic modification.

Concerns about Genetically Modified Animals

The development of transgenic animals has raised ethical concerns regarding their general treatment, but in general, these animals are highly valued and protected. Nevertheless, there are concerns about how genetic modification procedures might affect animal well-being, which have been categorized as follows (Orlans 2000):

1. Animals might suffer *directly* as a consequence of the manipulation.

2. Animals might suffer *indirectly* as a consequence of the modification.

3. Animals might suffer from ingestion of or treatment with genetically modified products.

4. Genetic transfer results in a change in the nature of animals in substantial ways not for the benefit of the animals themselves, but for ours.

5. Animal suffering might not be minimized unless standards of care and rules for experimentation are followed.

These and other concerns underlie principles to govern the humane use of farm animals promulgated by the United Kingdom Ministry of Agriculture, Fisheries and Food in 1995. The principles are to avoid "harms," and if some harm is inflicted, it must be justifiable and steps taken to minimize it:

1. Harms of a certain degree and kind ought under no circumstances to be inflicted on an animal.

2. Any harm to an animal, even if not impermissible, nonetheless requires justification and must be overweighed by the good that is realistically sought in so treating it.

3. Any harm which is justified by the second principle ought, however, to be minimized as far as is reasonably possible.

These rules are the direct result of the dictum that imposition of suffering on animals must be avoided in both animal farming (Mephram and Crilly 1999) and animal experimentation for medical science, as will be discussed in Chapter 8.

Concerns about Who Will Benefit

Countries with the resources to pursue GM technologies are those with a high proportion of scientists per capita, or that have control over patents in biotechnology for the production of GM food. For example, the United States and Japan have 70 scientists per 10,000 people, while Africa has less than 1 scientist per 10,000 people. Countries having no GM technologies of their own may not be able to develop them or afford to be licensed to produce them. This access to and ownership of world food production has come to be dominated by a few companies that control the market, forcing farmers and countries out of the market. Therefore, an important ethical consideration in developing this technology will be to ensure that those countries needing food the most are not put at a disadvantage by and benefit from this technology. The first patents for genetically engineered microbes, plants, and transgenic animals were in 1980, 1985, and 1988, respectively. A 2004 case regarding whether patents on genetically modified genes and cells should be afforded the same protection as other entities protected by patents was heard on appeal by the Supreme Court of Canada (*Monsanto of Canada v. Schmeiser* 2004 SCC 34, CanLII). The court ruled in favor of patent protection. This case could have an important effect on the availability and scope of GMO patents. In some developing countries, there is not a legal requirement to respect an outside patent for a crop. Thus, if the methodology or the seeds for growing these products could be accessed, growing them would be allowed without penalty. However, crops grown in these countries will not be freely exportable because of limitations from international trade agreements.

The genetic engineering of plants has tremendous potential for revolutionizing nutrition. Two examples of products that would benefit Third World countries are improved nutritional content of golden rice, which contains carotenoids to stave off blindness, and rice with higher amounts of iron (King 2002). Many will benefit from genetically modified (transgenic) animals that are models of human genetic diseases because they allow the study of various aspects of disease in ways never before possible. An example is the study of the complex nature of atherosclerosis enabled by rodents that have modifications of genes responsible for the transport of lipids.

Concerns about GMO Labeling

In countries where GM foods and crops have been introduced, there are increasing demands for their accurate and clear labeling, and insistence that they remain separate from non-genetically-engineered food forms. Under the U.S. FDA's biotechnology food policy, GM foods must be labeled if the source of the gene is from one of the common allergy-causing foods (e.g., cow's milk, eggs, fish and shellfish, tree nuts, wheat, and especially peanuts and soybeans) unless they have been proven not to be allergenic through additional safety testing (Metcalfe et al. 1996). However, these restrictions have proven difficult to follow. Analyses have shown that GM crops have made their way into the food chain despite efforts to keep them separate. An example is the use of StarLink corn in 1998 in human food products when it had been licensed only for use as animal feed. The discovery was made by a citizen who sampled and analyzed corn-based products such as tacos made by Kraft foods in 2000. The StarLink corn was engineered to have a small protein that was an insecticide. As that protein met some of the criteria of the USFDA for a possible allergen, it was not allowed to be marketed for human consumption. The impact was significant because this StarLink corn protein had contaminated a large percentage of corn produced in some areas of the United States (e.g., 50% of Iowa corn).

Foods labeled "substantially equivalent" might also include oils or plant products from the GM plant that are equivalent to food or ingredients that have been present in food for a long time, and are substantially equivalent to these foods in nutritional characteristics. Foods or food products labeled "no deliberate" genetically modified material may still contain genetically modified material. The Food and Drug Administration's labeling policy is the same for foods produced using biotechnology and conventional foods. The purpose is to ensure the consumer is given information about a nutritional value of the food as well as information on any possible allergens in the food. The process by which the foods reach the consumer is not controlled by the USFDA. The Department of Agriculture, however, does oversee how crops are grown and will not allow organic farmers to grow GM crops. Thus, in principle, GM crops are denied the coveted "organic" label.

Many consumers insist that their autonomy to avoid ingesting GMOs has been denied because of errant labeling. This is especially problematic because many consumers are reluctant to eat them due to perceived safety concerns. The issue has not been fully resolved, but accurate labeling would go a long way to allay concerns for those who want to avoid them.

Cumulative Effect of Concerns about GMOs

While it is possible although highly unlikely that GMOs pose a risk to human health and welfare, the previously listed possibilities led to widespread concern about the safety of these products in Europe. The entry into the market in Europe and England came at about the same time that Europe was immersed in fears of mad cow disease. GM products did make it to the marketplace in Great Britain, but rising concerns about safety now prevail there as well. In 1998, the EU banned the entrance of GM foods into those markets, even though some EU countries have been involved in field-testing of some GM crops to evaluate the safety and value of these products. The EU has more recently decided to lift this ban and to replace it by "traceability and labeling" rules that demand detailed records of GMOs "from farm to fork," documenting the origins and movements of GMOs, and also insisting on specific labeling to identify GMO foods. These new restrictions took effect in late 2004. They appear to be even more contentious for U.S. and other GM exporters than was the EU ban. Thus, concerns about GMOs have not been allayed as of 2006. The major concerns about GMOs are summarized in Table 6.2.

Table 6.2   Summary of Concerns about GMOs

- Concerns about safety (allergy to antibiotics used as markers, antibody resistance, compromised nutrition, compromised immune responses, and other currently unknown safety issues)
- Escape of GMOs through dispersion of seeds or through horizontal gene transfer into other plants and organisms.
- Loss of biodiversity if robust GMOs overwhelm natural plants, leading to monocultures that might destabilize the food supply if struck by a natural disaster, or lead to a loss of species richness.
- Loss of soil nutrients or soil fertility, either due to overuse because GM plants grow rapidly and can be planted more often or because of GM modifications that may cause a loss of properties that return fertility to the soil (e.g., loss of nitrogen-fixing bacteria).
- Possible toxic effects on humans and other species.
- Alteration of the inherent value of a natural plant or organism.
- Development of unwanted expression in other organisms or species.
- Loss of effectiveness of pest and weed herbicide resistance due to mutations or spread of pesticide resistance to weeds and other nuisance plants that would require even stronger pesticides.
- Genetically modified arthropods or other animals used to contain pests in one environment may result in unforeseen damage to other crops or species.
- Failure to label all products containing GMOs as such.
- Concerns of GM growers that that the "organic" label cannot be used on GMOs. This is a potential problem for marketing because this is a coveted label that attracts consumers.
- Loss of access to and ownership of world food production, whether Third World countries would truly benefit or would be locked out because of patents or because they cannot afford either the technology or the price of the seeds.

ONGOING EFFORTS TO EVALUATE AND PROVIDE INFORMATION ON GMOS

Local and international societies are acting to collect facts, encourage research, and educate the public with respect to conventional and unconventional methods of modifying plants that lead to foods. A major organization is the International Service for the Acquisition of Agri-Biotech Applications (ISAAA), whose mission is to contribute to the alleviation of poverty and hunger. The focus is to increase crop productivity and income generation, particularly for resource-poor farmers, and to bring about a safer environment and more sustainable agricultural development. Another prominent organization is the professional organization of scientists and technical personnel known as the Institute of Food Science and Technology (IFST). This group has its own code of ethics, and its operations are independent of government, industry, and special interest groups. It has made major contributions in the ongoing evaluation of the safety and environmental aspects of GMOs.

In addition, a Washington advocacy group called the Union of Concerned Scientists recently commissioned laboratory tests of natural U.S. crop seeds to see if strands of engineered DNA could be detected. The results showed that more than two-thirds of the 36 conventional corn, soy, and canola seed batches tested contained traces of DNA from genetically engineered crop varieties. While the actual amount identified was relatively small and within the amounts deemed safe by regulators for consumption, the concern is that federal rules and farm practices must be tightened if the United States is to guarantee that specific portions of its food supply are free of gene-altered elements. The inability to do so could disrupt the export of U.S. foods, seeds, and food extracts as well as harm the domestic and rapidly growing market for organic food. Further, because many crop varieties are also being engineered for the production of pharmaceutical and industrial products in their leaves and stems, there may also be health risks. The report calls for the Department of Agriculture to conduct a thorough assessment of the extent of genetic contamination of the U.S. seed industry (Weiss 2004).

Comprehensive studies of the environmental effects and safety of GMOs and a discussion of approaches to assessing unintended health effects were conducted by the Institute of Medicine and the National Research Council with extensive evaluative reports (National Research Council 2002c, National Research Council 2004b). These studies generated recommendations that focus on the need for federally guided studies of each product before commercialization, the need for standardization in analytical methods, and the need for evaluative studies, including population studies, to determine if there any unintended health effects from new foods or animal products. The facts revealed in the National Research Council's 2002 report argue the importance of applying the same standards of evaluation and regulation for both genetically engineered crops and crops from conventional breeding.

There are a number of major university-based information and educational resources devoted to the science of agriculture and related biotechnology. The University of California, Berkeley administers a major resource for educational material and scientific references relevant to breeding plants and issues surrounding GM plants (http://UCBiotech.org). The University of Maryland AgNIC gateway (http://agnic.umd.edu) is an easy to use resource for researchers, information specialists, educators, and members of the public seeking information on domestic animals, plants, and food processing, including genetic engineering. An Internet site administered by the European Initiative for Biotechnology Education (EIBE) seeks to promote skills, enhance understanding, and facilitate public debate throughout Europe (www.eibe.info).

U. S. Government Oversight and Regulations

In the United States, the major agencies currently involved in the evaluation and regulation of GMOs include the Food and Drug Administration (FDA), the U.S. Department of Agriculture (USDA) and its Animal and Plant Health Inspection Service (APHIS), as well as the Environmental Protection Agency (EPA) (http://vm.cfsan.fda.gov/~lrd/biocon.html, accessed March 8, 2004). These different agencies are involved because each has specific oversight and regulatory responsibilities for the dissemination of materials into the environment, the production of foods, and their safety. Their roles are reviewed below.

FDA's Biotechnology Policy

The early official policy of the FDA regarding GMOs appeared in the *Federal Register* of May 29, 1992 (57 FR 22984). This policy describes the agency's application of the Federal Food, Drug, and Cosmetic Act with respect to human foods and animal feeds derived from new plant varieties, and provides guidance to industry on regulatory issues.

Acceptability is established by determining whether a novel food is "substantially equivalent" to an analogous conventional food with respect to composition, nutritional value, toxins, allergen content, and the effects of its consumption on vulnerable groups such as infants and the elderly.

Foods are assigned to three categories:

- Products that are shown to be substantially equivalent to existing foods or food components.

- Products that are substantially equivalent to existing foods or food components except for defined differences.

- Products that have not been shown to be substantially equivalent to existing foods or food components.

Testing must include monitoring of any modifications to ensure safety over time, and if there are any differences found between the GM product and the conventional food, FDA requires additional extensive testing comparing the two. This definition of substantial equivalence is used in many countries, including the United States and Canada (www.ifst.org/hottop10.htm, accessed March 12, 2004).

The FDA approves the marketing of GMO foods through a process called consultation. Two or three years before proceeding with plans to commercialize a GM food, the developer meets with the agency to discuss the planned development of the product and any relevant safety, nutritional, or other regulatory issues, and then proceeds with the agreed-upon development plan. After a written summary of this information is submitted, the FDA evaluates the submission. Once any questions have been resolved, the FDA notifies the company by letter whether approval has been granted.

The FDA continues to reevaluate and update its guidance for industry for new plant varieties intended for food use. The latest statement on this guidance is dated November 19, 2004 (www.fda.gov/bbs/topics/ANSWERS/2004/ANS01327.html, accessed February 28, 2005):

> While FDA has not found and does not believe that new plant varieties under development for food and feed use generally pose any safety or regulatory concerns, this guidance is consistent with FDA's policy of encouraging communication early in the development process for a new plant variety. Such communication helps to ensure that any potential food safety issues regarding a new protein in such a new plant variety are resolved prior to any possible inadvertent introduction into the food supply of material from that plant variety. . . . An alternate approach may be used as long as it satisfies the requirements of applicable statutes and regulations.

USDA/APHIS Regulation of Bioengineered Plants

Most bioengineered plants are considered "regulated articles" under regulations of the Animal and Plant Health Inspection Service (APHIS) of the U.S. Department of Agriculture (USDA). APHIS regulates the interstate movement, importation, field testing, and environmental release of genetically engineered plants and microorganisms through permitting and notification procedures. USDA's APHIS has regulated agriculture biotechnology since 1987, ensuring the safe field testing of more than 10,000 genetically engineered (GE) organisms and overseeing the deregulation of more than 60 GE products. Over the past several years, the Bush Administration has taken steps to strengthen USDA's biotechnology regulations through the creation of the biotechnology regulatory services program, enhancements to its permitting system for plant-made pharmaceuticals and industrials, and the development of a compliance and enforcement unit to ensure adherence to the agency's regulations. APHIS is responsible for the safety and containment of GMOs during movement and field testing. Its regulations continue to be updated because there is continuing expansion of GMO development into new products of unknown risks and limited experience to guide regulatory oversight. For example, APHIS's regulations recently required permits rather than notification for the introduction of plants, such as those engineered to produce industrials. APHIS regulations are found as federal law 7 CFR 340 (see www.aphis.usda.gov).

Environmental Protection Agency Functions

The safe use of pesticides is regulated by the Environmental Protection Agency (EPA). Thus, a bioengineered food that is the subject of a consultation with FDA may contain an *introduced pesticidal substance* that is subject to review by EPA. FDA has identified each bioengineered food that contains an introduced pesticidal substance with an asterisk (*) on its Web site. For additional information about EPA's regulation of bioengineered foods that contain a pesticidal substance, see www.epa.gov/pesticides.

Involvement of the U.S. Congress

Congress is also responding to the need for further scrutiny of the safety and labeling of GM foods, and there are bills pending that may result in Congressional actions to tighten regulations on genetically engineered foods. In July of 2003, Congressman Dennis J. Kucinich (D-OH) introduced six bills to provide a comprehensive regulatory framework for all genetically engineered plants, animals, bacteria, and other organisms. The main thrust of these bills was proper labeling, but from our information as of 2006, none has been acted upon.

The United States vs. the European Union Stance on GMOs

The United States insists these foods are essentially equivalent to non-GM foods, that they have no harmful effects, and that they are safe and acceptable for public marketing. Further, the United States views the EU ban on the importation of GMOs, put into effect in 1998, as without scientific basis. It unfairly restricts trade and is a violation of World Trade Organization (WTO) rules to which the EU agreed. Additional U.S. concerns are that the EU is standing in the way of the availability in the European Union of more nutritious foods, more vaccines, higher livestock reserves, and higher-yield crops made with fewer pesticides and fertilizers. The EU stance is also affecting decisions in other needy Third World countries that decided not to accept U.S. GMOs because their trade with EU countries might be affected as a result.

The EU concerns about GMOs are mainly about safety. Their adoption of a moratorium on the importation of GMOs in 1998, and the subsequent adoption of the precautionary principle (PP) in 2000, came after episodes of mad cow disease (bovine spongioform encephalopathy), foot and mouth disease, and dioxin contamination threatened their overall food supply. The PP requires that before food can be produced or sold, it must be proven safe beyond a doubt with no unanswered questions, a stance that could delay approval for the import of GMOs for years.

Following the U.S. filing of litigation that the EU ban on GM foods was a violation of WTO rules, the EU substituted instead "traceability and labeling" rules. These rules demand detailed records of GMOs "from farm to fork," documenting the origins and movements of GMOs; they also require specific labeling to identify GM foods. The United States sees this action also as a barrier to fair trade, and fears that not only will the documentation requirements raise the cost of GM food by up to 50%, but the stigma of the GMO label will reduce demand for the product and profitability as well.

Reacting to inadequate labeling of GMOs, in April 2004 the European Union nations established the European Network of Genetically Modified Organisms Laboratories. This network sets up and monitors methods of labeling GMOs. The ethical basis is that consumers have the right to choose between products that contain or do not contain GMOs.

International Regulatory Approach to GMOs: U.N. Cartagena Protocol on Biosafety

At a meeting in Montreal in January 2000, 130 countries signed the Biosafety Protocol to the United Nations (U.N.) Convention on Biodiversity that was developed as a result of the U.N. Conference in Rio de Janeiro in 1998. It later became known as the Cartagena Protocol on Biosafety (see Appendix 6.2) and remains under the auspices of the U.N. The entire Cartagena Protocol can be seen at www.biodiv.org/biosafety/articles. It became law in September 2003 upon ratification by more than 50 countries. Ratification procedures have continued since that time, and as of December 2005, 127 of the 195 member countries of the U.N. have ratified the protocol.

Specifically pertaining to GMOs, the protocol provides the following:

- It defines GMOs as different from conventional organisms.
- It provides regulations regarding the movement of GMOs across international boundaries.
- It includes a provision to prevent contamination of GM seed products.
- It provides international rules on biosafety and trade of GMOs.
- It specifies ongoing risk assessment regulations.
- It reaffirms the precautionary principle, allowing member countries to decide whether or not to allow importation of GM products if it is deemed there are insufficient safety data.
- It establishes a procedure for ensuring that countries are provided with the information necessary to make informed decisions before agreeing to the import of such organisms into their territory.

Although the four countries growing 99% of the GMO crops in the world (the United States. 68%, Argentina 23%, Canada 7%, and China 1%) signed the Cartagena Protocol, it was not ratified by the United States, Argentina, Canada, Australia, or New Zealand. China ratified it in 2005. Those not ratifying the protocol cite export-import conditions that might limit trade of important commodities, and for some, there are definitions in the protocol that are contested. For example, the statement that GMOs differ from conventional organisms conflicts with the U.S. position that GMOs are not different from the plants and animals from which they are derived.

Since the United States is the main exporter of GMOs, pressure on countries to accept these products may intensify. Signatories of the Cartagena Protocol have not been able to control the import and general dispersion of GMO corn seed, as evidenced by Spanish feed farmers who demand and use GMO corn for animal feed. As recently as June 20, 2005, at a meeting of member countries, no agreement could be reached about the transportation of GMOs across national boundaries. Another meeting is scheduled in 2006.

ETHICAL ISSUES

The ethical issues that GMO technologies have raised include those related to the environment, animal and human experimentation, human health, and the ethics of balancing major benefits against risks of unknown consequences. These include **rights** ethics, or the autonomy of the individual to choose whether or not to consume GMOs; **rights** of farmers whose seeds might be controlled by others; **rights** of nations whose foods are monopolized by other nations; the **duty** to protect animal and plant species from the consequences of gene transfer; the **duty** to prevent harmful consequences of bringing modified species into the environment; and **utilitarianism**, which seeks to provide benefits to citizens with minimum negative consequences. As with many other ethical problems, the moral theories of utilitarianism and autonomy or individual rights come into conflict. When considering a potential risk of GMOs, the following principles should be considered when arriving at an action:

- Utilitarian principle of general welfare of the citizens.

- Duty to promote the safety and health of citizens.

- Duty to maintain the rights of the people, including the right to freedom of choice for consumers.

- Justice, requiring the burdens and benefits of policies and practices to be fairly shared among those affected, including farmers and food producers.

- Respect for cultural ethics, including ideologies of animal humane treatment and use of animal-derived products.

- Duty to maintain a sustainable environment.

Critics who are concerned GMOs may be unsafe advocate the precautionary principle. However, prudence dictates caution against allowing the fear of possible risks to take precedence over the multitude of previously noted benefits. In fact, there may be no fundamental differences between most of the new GM technologies and all the other ways used by humans to modify plants and their environment over time. However, many continue to focus on the lack of thorough testing to ensure that these crops are safe, and are concerned that the main reason biotechnology is pushing for these crops is to increase corporate profits. A continuing concern is that small farmers will be left behind because they cannot afford the costs for GMOs, widening the gap between rich and poor (Martin 2003).

There is a major schism between those who look to GM foods to increase the food supply and those who resist their use. The concerns of the opponents are that GMOs may negatively impact biologic diversity, human health, and natural evolution, among other things. Other scientists claim these consequences are unlikely. It is not possible to accurately assess these future risks, and fears have unnecessarily evoked the precautionary principle. Engineers and scientists will be directly involved in assessing these issues. Even though we have listed the pros and cons of GMOs in Tables 6.1 and 6.2, these lists do not capture the full extent of complexity of the issues. Two cases below illustrate some current issues.

Case 6.1 Risks of GM Salmon

One of the most controversial concerns is that GM salmon have been engineered to have a 40-fold increase in growth hormone, resulting in fish that are three times the size of wild salmon of the same age. This controversy is frequently confused with the controversy over the dangers of farmed salmon wherein the farmed salmon are not GMO but do present an ethical dilemma in and of themselves due to accumulated toxins, which are dangerous (Hites et al. 2004). Clearly, both issues are coupled because the genetically modified salmon are also farmed salmon, and this presents a second dilemma regarding the possible escape of the GM salmon into the environment and the resulting consequences.

With respect to the consequences of fish farming, there is strong evidence that compared with wild fish, farmed fish contain high concentrations of toxic substances such as PCBs (polychlorinated biphenyls) and mercury. The ingestion of growth hormone GM salmon should not be detrimental as this should be digested and made nonfunctional in the human digestive tract. However, it has been shown that among the children of women who consume contaminated salmon, there was a statistically significant effect on cognitive performance. It cannot be concluded at this time that there is a direct association with PCB as a fish-borne neurotoxin, because animal studies show that the effects from PCBs alone require much higher tissue concentrations than the concentration in tissues of salmon. Further study is needed to decide if there are other toxins which either alone or synergistically with PCBs cause neurodegeneration (Seegal 1999, Winnecke et al. 2002).

With the above information in mind, how would one approach the dilemma of producing beneficial foodstuffs for the world when there are substantial reasons to believe the known and unknown risks could be dangerous? We return to our Four A's design strategy to systematically approach this complex problem.

Acquire Facts: First, we gather more facts, including the risks of the contaminants themselves, such as:

- The origin of the contaminants in farmed salmon.
- The natural occurrence of contaminants in wild salmon.
- The methods of measuring the contaminants.
- The methods to minimize the contamination.
- The amount of contaminated fish that would need to be eaten to be toxic.

In parallel, we evaluate facts concerning the risks of genetically modified salmon such as:

- The consequences of the ingestion of hormones and proteins not naturally found at the same concentrations in wild salmon.
- The differences in spawning habits, migration, mating, aggression, dominance, life span, and occurrence of disease between farm and wild salmon.
- The presence or absence of mutations in offspring.
- How important a role will salmon availability play in relieving malnutrition and starvation?
- Is it still a source of omega-3 fatty acids?
- Is it necessary to expose the population to unknown risks of GM salmon?

A third category of inquiry, in parallel with the evaluation of the facts, is the investigation of the consequences of the inadvertent release of GM salmon to the environment, such as:

- The consequences to other species through a horizontal gene transfer.
- What is the potential for the dominance of GM salmon over wild salmon? What is the potential for the extinction of wild salmon as a result?
- The results of breeding of GM salmon (e.g., can wild salmon compete for mates? Is there an increase in malformed, weaker, or more aggressive salmon?).
- The potential jeopardy to the giant salmon because of size-related limitations in their capabilities to return to spawning grounds.
- The consequences of risking the unknown.

Alternatives: Next, we engage in identifying possible proposals for alternative solutions, which might include one or more of the following:

- Identify and remove toxins in the food supply of GM and farmed salmon.
- Undertake experimental demonstrations of the results of breeding GM salmon with wild salmon.

- Consider creating sterile females so if there is a release to the wild, the consequences would be minimized.
- Evaluate the limitations, if any, in the ability of the GM fish to swim or navigate through rapids and up ladders to spawning grounds.
- Allow commercialization of GM salmon but require strict labeling as a GM food.
- Pass legislation to control the use of farmed and GM salmon for a specified time after which legislation would impose more or fewer restrictions depending on the experience during the trial period; this alternative requires a method to detect the occurrence of unintended consequences.
- Evaluate whether GM salmon present too much of a risk to those eating it.
- Assume the risks of the unintended consequences are minimal and allow the ingestion of the salmon as "essentially equivalent" and no different from non-GM salmon.

Assessment: We must also evaluate the consequences of the alternative action approaches listed above if they were to be put into action, such as:

- Is it safe or safe enough?
- How risky is it? What do we know from other experiments? What sort of risk assessment might be helpful?
- Is the risk due to the rDNA technology used in GM salmon? What are the risks of rDNA-based approaches?
- Evaluate the consequences of the sterilization of female GM salmon in a simulated wild environment.
- Will sterilization of female GM salmon result in other problems such as levels of chemicals that might be problematic for human growth and development?

Action: It is important also to recognize that the development of actions and further investigation of alternatives in this case are dynamic and uncertain. National approaches to the regulation of GMOs vary widely, and the flow of information and changes in governments make the situation unsettled. The public and others are not well informed and are often guided by emotional responses to limited or faulty information. Public perceptions appear to be guided mainly by journalists, along with consensus forums consisting of mixtures of informed and uninformed experts, legislatures, and lay public. Caution and conservatism are guiding principles. In fact, the real ethical dilemma, once there has been sufficient scientific discovery to understand the risks and remediation for the known problems, might be the conflict between depriving large numbers of people of an important food source vs. allowing them access to GM food that carries with it unlikely risks of the unknown.

Case 6.2 GM Corn and Accusations of Harm to Other Species

There was wide media coverage in 2000 of the possible toxic effects of genetically modified *Bt*-corn on the larvae of butterflies. A major product of GMO technology is the insertion of a gene that makes a toxin to the European corn borer insect. This toxin was first discovered in the bacterium *Bacillus thuringiensis (Bt)*. The gene was spliced into corn, potatoes, and cotton (see the methodology previously described in this chapter for *Agrobacterium tumefacien*). This corn is protected from the borer and does not require chemical spraying with pesticides. However, the news stories described the destruction of monarch butterfly larvae in the vicinity of GM crops, which served to kindle rising popular opinion against the use of GM foods. However, a subsequent collaborative study by United States and Canadian scientists showed there were no significant toxic effects (Sears et al. 2001). Risk analysis studies (Chapter 12, "Ethics of Emerging Technologies") showed that although monarch populations share their habitat with corn ecosystems, the portion of the monarch population that is exposed to toxic levels of *Bt*-corn pollen is negligible and declining. In fact, planting of the particularly toxic 176 hybrid was phased out through 2003 (Sears et al. 2001). Another report indicated that *Bt*-corn produced no significant risk to the monarch butterfly or to other beneficial insects or to natural enemies (Gatehouse et al. 2002). An informative Web site that answers frequently asked questions concerning the origin and impact on the environment of *Bt*-corn can be found at www.ext.colostate.edu/pubs/crops/00707.html.

This is a case in which the media presented such a harsh early indictment of modified corn that its value was tarnished. Well-meaning journalists influenced the public perception of the technology, and the negativity surrounding GM corn probably extended to other GM foods as well. In addition, there is no evidence that journalists promulgated the results of the scientific findings that showed *Bt*-corn was not a threat to the monarch butterfly. However, answering the questions about the butterfly is not the end of the story because of another finding having to do with the *Bt*-toxin.

It had been thought, but not proven, that *Bt*-corn toxin in the soil after harvest was biodegradable and the toxin was not expected to accumulate in the environment. However, *Bt*-toxin has been found in the soil for a number of months after harvest of a GM crop. When Richelieu River watershed sediments in Canada were analyzed, *Bt*-toxin was found in elevated amounts in the *Bt*-corn field watershed over the toxin concentration in the watershed from the non-*Bt* corn fields. A formal report for this study could not be located in the refereed literature. The implication is that there is a gene transfer from the roots of the *Bt*-corn plants, to soil bacteria that then are a continuing source of the *Bt*-toxin or that it is exuded into the soil directly. The claims in the non-refereed literature that *Bt*-toxin is lethal to earthworms have been countered by experimental evidence that earthworms and similar species thrive in soils from *Bt*-corn crops (Saxena and Stotzky 2001). It should be noted that the same toxin isolated from *Bacillus thuringiensis* has been used as an externally applied insecticide for 30 years, and the accumulation described above cannot be immediately ascribed to *Bt*-corn.

In 2005, a Korean team described a newly discovered bio-insecticide that could provide an alternative to *Bt*-corn (Seo et al. 2005). They were able to produce an effective protein-based toxin for the corn borer using *E. coli* as a host, a process frequently used to produce pharmaceuticals. Whether this pesticide can supplant the usefulness of the *Bt*-corn will be determined in field trials. It is not clear that the newly proposed bio-pesticide is any more biodegradable than the *Bt*-toxin.

The point of these two examples (i.e., PCB toxins in fish and toxins from GM corn affecting butterflies) is that premature assignment of causative mechanisms for observed harms results in needless costs and negative public opinion. The association between PCBs and GM fish is not based on the fact that genetic modification has caused an increase in toxin levels, but on the fact that farmed fish accumulate PCBs more than wild fish. The causation here is the presence of PCBs in the feed used for the farmed fish (Carlson and Hites

2005). There is some relationship to fat content and PCB accumulation in wild and farmed fish; thus, one might argue that a genetic modification that includes minimizing lipid content relative to protein content could be preferred, because this in principal would decrease the toxin load of fish products. The second example is one of journalistic amplification of some nonscientific observations.

CHAPTER SUMMARY

The debate around the world regarding GMOs has been two-dimensional. Critics have demonized GM foods as "Frankenfoods," while others see GM foods as a way to better utilize water and increase the food supply of the world without substantially increasing farm acreage at the expense of the environment. Either we accept that many people of the world will starve, or we engineer and distribute foods more efficiently. The only method currently known to increase the productivity and nutritional value is genetic modification by gene transfer. Even with transposon-based approaches that do not use foreign DNA for improving foods, the public debate has still concerned itself with dispersion and lateral gene transfer. Theories of utilitarianism, duty, rights, autonomy, and non-malfeasance are among the ethical moral principles involved.

If we accept the duty to feed the people of the world, then we must first gather the facts and better inform the public so that controversies can be dealt with in a rational manner. Most of the controversies, other than justice to farmers and import-export issues, stem from how thoroughly scientists have evaluated the safety of GMOs and their impact on the environment. However, the public concerns might be out of proportion to reality. Problems such as the transfer of genes into unintended plant species by dissemination of pollen could be detrimental due to growth of robust weeds, but this has not occurred. From the early 1980s to the present time when GM foods constitute approximately 70% of foods consumed by the U.S. public, major problems have not emerged, although the dissemination of dangerous allergens to foods that do not normally carry these allergens is an accepted possibility for which regulations and testing must continue. Non-GM foods have been found that contain evidence of genetic modification, and better labeling is needed so that those who wish to avoid GM food can do so. With reference to transfer of genes to vertebrates and human beings, there is no evidence that this has occurred or will occur, although this concern remains prevalent among the public. Most of all, there is a need for better ways to assess the risk associated with emerging technologies of modern biology.

The opposing views of benefits vs. risks of GM technology are based in good part on the fact that we do not know the probability of a proposed risk, or whether a supposed risk will lead to a negative consequence. Thus, many of the arguments have to do with how to balance the precautionary principle with the importance of the benefits. GMO technology presents the dilemma of known benefits vs. unknown risks, and this is why the subject has sparked major ethical debates.

The ethical questions of this fast-moving field are being addressed by governments, with progressive legislation in the United States, Australia, and the United Kingdom. In the United States, the primary agencies currently regulating biotechnology are the U.S. Department of Agriculture (USDA), the Food and Drug Administration (FDA), and the Environmental Protection Agency (EPA). Most agree that the laws and regulations governing this biotechnology need active review and revision because they are not comprehensive and span too many agencies. The need for modern regulations is a global problem. Risk assessment in the presence of many unknowns is difficult. Appendix 6.2 at the end of this chapter outlines plans for risk assessment as defined by the Cartagena Protocol. The protocol provides one approach to the varying ethical issues around this subject. Also see the section on genetically modified animals in Chapter 8 under animal experimentation, and Chapter 12, which further addresses the ethics of emerging technologies.

In sum, it appears that genetic modification presents no new categories of risk compared to conventional methods of crop improvement, but specific traits introduced by either of the approaches can pose unique risks (National Research Council 2002b). Both methods of modifying plants that are introduced into the environment and processed into foods need a balanced scrutiny. The real ethical dilemma might be the conflict between depriving large numbers of people of an important food source and allowing them access to GM food that carries with it unknown risks.

REFERENCES

Aeschbacher, K., R. Messikommer, L. Meile, and C. Wenk. 2005. "Bt176 corn in poultry nutrition: physiological characteristics and fate of recombinant plant DNA in chickens." *Poultry Sci* **84**:385–394.

Ammann, K. 2005. "Effects of biotechnology on biodiversity: herbicide-tolerant and insect-resistant GM crops." *Trends in Biotechnology* **23**(8):388–394.

Boothaerts, W., H. J. Mitchell, B. Weir, et al. 2005 (February 10). "Gene transfer to plants by diverse species of bacteria." *Nature* **433**(7026):629–633.

Camazine, C. L. and R. A. Morse. 1988. "The Africanized honeybee." *American Science* **76**:465–471.

Campbell, N. A. and J. B. Reece. 2002. *Biology*. Sixth Edition. San Francisco: Pearson Education, Inc., of Benjamin Cummings Press, p. 345.

Carlson, D. L. and R. A. Hites. 2005. "Polychlorinated biphenyls in salmon and salmon feed: global differences and bioaccumulation." *Environ Sci Technol* **39**(19):7389–7395.

Conway, G. 1998. *The Doubly Green Revolution: Food for All in the Twenty-First Century.* Ithaca, NY: Comstock Publishing Associates, a division of Cornell University Press, pp. 44–65.

de Maagd, R. A., A. Bravo and N. Crickmore. 2005 (July). "Bt toxin not guilty by association." *Nat Biotechnol* **23**(7):791.

Dewey, R. E., C. S. Levings and D. H. Timothy. 1986. "Novel recombinations in the maize mitochondrial genome produces a unique transcriptional unit in the Texas male-sterile cytoplasm." *Cell* **44**:439–449.

Food and Agriculture Organization (World Health Organization). 1992. "World Food Supply and Prevalence of Chronic Malnutrition in Developing Regions as Assessed in 1992" (Document EES/MISC/1992); FAO. 1996, Sixth World Food Survey, Rome, Italy. Food and Agriculture Organization.

Galun, E. and E. Galun. 2001a. *The Manufacture of Medical and Health Products by Transgenic Plants*. London: Imperial College Press, UK, pp. 37–81.

Galun, E. and E. Galun. 2001b, p. 249.

Gatehouse, A. M., N. Ferry, and R. J. Raemaekers. 2002. "The case of the monarch butterfly: A verdict is returned." *Trends Genet* **18**(5):249–251.

Gressel, J., A. Hanafi, G. Head, et al. 2004. "Major hertofore intractable biotic constraints to African food security that may be amenable to novel biotechnological solutions." *Crop Protection* **23**:661–689.

Hellenas, K. E., C. Branzell, H. Johnsson, and P. Slanina. 1995. "High levels of glycoalkaloids in the established Swedish potato variety Magnum Bonum." *J Science and Food Agriculture* **23**:520–523.

Hites, R. A., J. A. Foran, D. O. Carpenter, et al. 2004. "Global assessment of organic contaminants in farmed salmon." *Science* **303**:226–229.

Houdebine, L. M. 2005. "Use of transgenic animals to improve human health and animal production." *Reprod Domest Anim* **40**:269–281.

James, C. 2004. "Global Status of Commercialized Biotech/GM Crops: 2004." International Service for the Acquisition of Agri-Biotech Applications (ISAAA). Briefing Note No. 32. Accessed July 22, 2005 via http://ucbiotech.org

Kaplinsky, N., D. Braun, D. Leisch, et al. 2002. "Maize transgene results in Mexico are artifacts." *Nature* **416**:601.

King, J. C. 2002. "Biotechnology: a solution for improving nutrient bioavailability." *Int J Nutr Res* **72**:7–12.

Leary, W. E. 2004 (March 12). Modified genes in native corn worry Mexico. *San Francisco Chronicle*, p. A7.

Lemaux, P. G. 2005. Personal communication.

Lorencz, M. G. and W. Wackernagel. 1994. "Bacterial gene transfer by natural genetic transformation in the environment." *Microbial Reviews* **58**:563–602.

Maclean, N. and S. Talwar. 1984. "Injection of cloned genes with rainbow trout eggs." *J Embryol Exp Morphol* **82**(Suppl):187.

Martin, G. 2003 (June 24). "Agriculture secretary pushes new crops." *San Francisco Chronicle*, p. A3.

Mephram, B. 2001. "Novel Foods." In: *Ethics of New Technologies*. Ruth Chadwick (ed.). San Diego: Academic Press, pp. 299–313.

Mephram, T. B. and R. E. Crilly. 1999. "Bioethical issues in the generation and use of transgenic farm animals." *Alternatives to Laboratory Animals* **27**:1–9.

Metcalf, D. D., J. D. Astwood, R. Townsend, et al. 1996. "Assessment of the allergenic potential of foods derived from genitically engineered crop plants." *Critical Reviews of Food Science and Nutrition* **36**(S):S165–186

Metz, M. and J. Futterer. 2002. "Suspect evidence of transgenic contamination." *Nature* **416**:600–601.

Mukai A. and H. Endoh. 2004. "Presence of a bacterial-like citrate synthase gene in Tetrahymena thermophila: recent lateral gene transfers (LGT) or multiple gene losses subsequent to a single ancient LGT?" *J Mol Evol* **58**:540–549.

National Research Council, 2002a. *The Environmental Effects of Transgenic Plants.* Washington, DC: The National Academies Press, pp. 34–35.

National Research Council, 2002b. *The Environmental Effects of Transgenic Plants.* Washington, DC: The National Academies Press., p.79.

National Research Council. 2004a. *Biological Confinement of Genetically Engineered Organisms.* Washington, DC: The National Academies Press, 254 pp.

National Research Council, 2004b. *Safety of Genetically Engineered Foods: Approaches to Assessing Unintended Health Effects.* Washington, DC: The National Academies Press.

Nordlee, J. A., S. L. Taylor, J. A. Townsend, et al. 1996. "Identification of Brazil-nut allergen in transgenic soybeans." *N Engl J Med* **334**:726–728.

Orlans, B. 2000. "Research on animal, law, legislative, and welfare issues in the use of animals for genetic engineering and xenotransplantation." In: *Encyclopedia of Ethical, Legal, and Policy Issues in Biotechnology.* T. H Murray and M. J. Mehlman (eds.). Hoboken, NJ: John Wiley & Sons, pp. 1020–1030.

Parekh, S. R. (ed.). 2004. *The GMO Handbook: Genetically Modified Animals, Microbes, and Plants in Biotechnology.* Totowa, NJ: Humana Press.

Quist, D. and I. H. Chapela. 2001 (November 29). "Transgenic DNA introgressed into traditional maize landraces in Oaxaca, Mexico." *Nature* **414**(6863):541–543.

Roth, D. and N. L. Craig. 1998. "VDJ recombination: A transposase goes to work." *Cell* **94**:411–414.

Saxena, D. and G. Stotzky. 2001. "Bacillus thuringiensis (Bt) toxin released from root exudates and biomass of Bt corn has no apparent effect on earthworms, nematodes, protozoa, bacteria, and fungi in soil." *Soil Biology and Biochemistry* **33**:1225–1230.

Sears, M. K., R. L. Hellmich, D. E. Stanley-Horn, et al. 2001. "Impact of Bt corn pollen on monarch butterfly populations: A risk assessment." *Proc Natl Acad Sci U S A.* **98**(22):12328–12330.

Seo, J. H., J. S. Yeo, and H. J. Cha. 2005. "Baculoviral polyhedron-Bacillus thuringiensis toxin fusion protein: A protein-based bioinsecticide expressed in Escherichia coli." *Biotechnol Bioeng* (E-pub ahead of print).

Seegal, R. F. 1999. "Are PCBs the major neurotoxicant in Great Lakes salmon?" *Environ Res* **80**:S38–S45.

Shiklomanov, I. A. 1999. *World Water Resources and Their Use: a Joint SHI/UNESCO Product,* http://webworld.unesco.org/water, accessed July 20, 2005.

Slabbert, R., M. Spreeth, and G. H. J. Kruger. 2004. "Drought tolerance, traditional crops and biotechnology: breeding towards sustainable development." *South African J Botany* **70**(1):116–123.

Somerville, C. and J. Briscoe. 2001 (June 22). "Genetic engineering and water" (editorial). S*cience* **292**(5525):2217.

Stewart, Jr., C. N. 2004. *Genetically Modified Planet*. New York: Oxford University Press.

Strauss, S. H. 2003 (April 4). "Genetic technologies: Genomics, genetic engineering, and domestication of crops." S*cience* **300**(5616):61–62.

Thompson, J. D. 1991. "The biology of an invasive plant." *BioScience* **41**:393–401.

U.S. Census Bureau, 2005. Population Division, International Programs Center. www. census.gov/ipc/www/world.html, accessed November 19, 2005.

Weiss, R. 2004 (February 24). "Study finds seeds tainted with engineered DNA strands." *San Francisco Chronicle*, p. A2.

Wilson, N. K., J. C. Chuang, R. Iachan, et al. 2004. "Design and sampling methodology for a large study of preschool chidren's aggretate exposures to persistent organic pollutants in their everyday environments." *J Expo Anal Environ Epidemiol* **14**:260–274.

Winneke, G., J. Walkowiak, and H. Lilienthal. 2002. "PCB-induced neurodevelopmental toxicity in human infants and its potential mediation by endocrine dysfunction." *Toxicology* **181**:161–165.

Wolska, K. I. 2003. "Horizontal DNA transfer between bacteria in the environment." *Acta Microbiol Pol* **52**(3):233–43.

Wuthrich, B. 1999 (April 1). "Food additives and genetically modified food—a risk for allergic patients?" *Schweiz Rundsch Med Prax* **88**(14):609–614, 616–618.

Zhu, X. Y. 2004. "Phylogenetic analysis indicates bacteria-to-apicoplast lateral gene transfer." *Yi Chuan Xue Bao* **31**:1316–1320.

ADDITIONAL READING

Atherton K. T. (ed.). 2001. *Genetically Modified Crops Assessing Safety*. New York: Taylor and Francis/CRC Press, 208 pp.

Conway, G. 1998. *The Doubly Green Revolution: Food for All in the Twenty-First Century.* Ithaca, NY: Comstock Publishing Associates, a division of Cornell University Press, 335 pp.

Fedoroff, N. and N. Brown. 2004. *Mendel in the Kitchen: A Scientist's View of Genetically Modified Food.* Washington, DC: The National Academies Press, 352 pp.

Fox, M. W. 1999. *Beyond Evolution: The Genetically Altered Future of Plants, Animals, the Earth . . . and Humans.* New York: The Lyons Press, 256 pp.

Lambrect, W. 2001. *Dinner at the New Gene Cafe: How Genetic Engineering Is Changing What We Eat, How We Live and the Global Politics of Food.* New York: St. Martin's Press, 383 pp.

Liang, G. H. and D. Z. Skinner. 2004. *Genetically Modified Crops.* Binghamton, NY: Haworth Press, 363 pp.

National Research Council, 2002a. *The Environmental Effects of Transgenic Plants.* Washington, DC: The National Academies Press, 342 pp.

Nottingham, S. 2003. *Eat Your Genes: How Genetically Modified Food Is Entering Our Diet.* Second Edition. London: Zed Books, 256 pp.

APPENDICES

Appendix 6.1 Estimating the Future Increases in World Population

The arguments for introducing GM foods are in the main based on the current expected increase in malnutrition of the world population from the current estimate of 13% to 50% or more, depending on regional population growth. The death rate from starvation is 0.15%, or 9.1 million per year.

There are very large differences in the projections for Earth's human population increases to 2050. The expectation is that there will be an increase from the current population of 6.1 billion to over 9 billion by 2050. Figure A.6.1 summarizes data from the U.S. Census Bureau. The rate of population increase is adjusted for various factors such as current age distribution and estimates of birth and death rates.

The projection of the number of malnourished and starving people in the world is dependent on assumptions regarding the birth rate and death rate in different regions of the world, and on the supply or distribution potentials of foods to these regions. Estimates use an average birth rate in the world of two offspring per woman per lifetime. However, the population of young women of childbearing age in India is much greater than in the United States. In addition, there is a wide variation in death rates, which depend on longevity, nutrition, diseases, and war (e.g., in Zimbabwe, the life expectancy is 38 years, while in Singapore the life expectancy is 80 years). There is also a wide variation between countries in the rate of population increase. For example, the rate of population increase for Mexico is 2.1% per year, for the United States it is 0.6 % per year, but for Sweden it is close to 0%. In some countries, the death rate is decreasing more than in other countries. The impact of these rates on world population growth depends on the base population in each region.

Thus, whereas the model of population growth can be simply stated as the sum of the growth rates of all regions, the complexities in estimating the change in this rate over a period of, say, 50 years are great, as are the unknowns such as epidemics, the development of future remediation (e.g., mosquito control, AIDs therapy), and food availability. These complexities are the reason one finds major differences in projections of the Earth's population.

The population increases also have a major impact on the water resources of the world. An analysis of the past and projected consumption of water in the world is shown in Figure A.6.2.

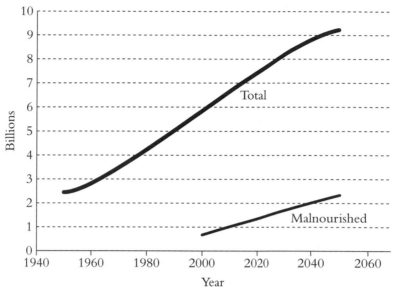

Figure A.6.1 Data from the U.S. Census Bureau are plotted from 1950 to the present and projected thereafter to 2050 (upper curve). The estimates for the number of malnourished people of the world (lower curve) are based on the data of 0.8 million in 2005 and a projection that assumes that crops can be increased by only 30% due to current practices of irrigation. Thus, 2.3 million will be malnourished in 2050.

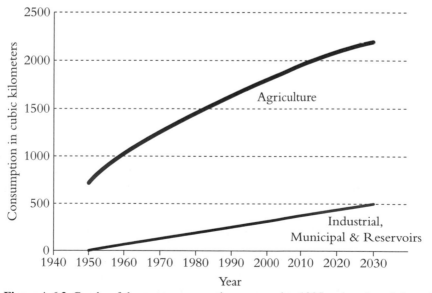

Figure A.6.2 Graphs of the water consumption measured to 1995 and projected thereafter from data collected by Shiklomanov (1999).

Appendix 6.2 Cartagena Protocol on Biosafety: Risk Assessment

Objective

1. The objective of risk assessment, under this Protocol, is to identify and evaluate the potential adverse effects of living modified organisms on the conservation and sustainable use of biological diversity in the likely potential receiving environment, taking also into account risks to human health.

Use of risk assessment

2. Risk assessment is, inter alia, used by competent authorities to make informed decisions regarding living modified organisms.

General principles

3. Risk assessment should be carried out in a scientifically sound and transparent manner and can take into account expert advice of, and guidelines developed by, relevant international organizations.

4. Lack of scientific knowledge or scientific consensus should not necessarily be interpreted as indicating a particular level of risk, an absence of risk, or an acceptable risk.

5. Risks associated with living modified organisms or products thereof, namely, processed materials that are of living modified organism origin, containing detectable novel combinations of replicable genetic material obtained through the use of modern biotechnology, should be considered in the context of the risks posed by the non-modified recipients or parental organisms in the likely potential receiving environment.

6. Risk assessment should be carried out on a case-by-case basis. The required information may vary in nature and level of detail from case to case, depending on the living modified organism concerned, its intended use and the likely potential receiving environment.

Methodology

7. The process of risk assessment may on the one hand give rise to a need for further information about specific subjects, which may be identified and requested during the assessment process, while on the other hand information on other subjects may not be relevant in some instances.

8. To fulfill its objective, risk assessment entails, as appropriate, the following steps:

   (a) An identification of any novel genotypic and phenotypic characteristics associated with the living modified organism that may have adverse effects on biological diversity in the likely potential receiving environment, taking also into account risks to human health;

(b) An evaluation of the likelihood of these adverse effects being realized, taking into account the level and kind of exposure of the likely potential receiving environment to the living modified organism;

(c) An evaluation of the consequences should these adverse effects be realized;

(d) An estimation of the overall risk posed by the living modified organism based on the evaluation of the likelihood and consequences of the identified adverse effects being realized;

(e) A recommendation as to whether or not the risks are acceptable or manageable, including, where necessary, identification of strategies to manage these risks; and

(f) Where there is uncertainty regarding the level of risk, it may be addressed by requesting further information on the specific issues of concern or by implementing appropriate risk management strategies and/or monitoring the living modified organism in the receiving environment.

Points to consider

9. Depending on the case, risk assessment takes into account the relevant technical and scientific details regarding the characteristics of the following subjects:

(a) Recipient organism or parental organisms. The biological characteristics of the recipient organism or parental organisms, including information on taxonomic status, common name, origin, centres of origin and centres of genetic diversity, if known, and a description of the habitat where the organisms may persist or proliferate;

(b) Donor organism or organisms. Taxonomic status and common name, source, and the relevant biological characteristics of the donor organisms;

(c) Vector. Characteristics of the vector, including its identity, if any, and its source or origin, and its host range;

(d) Insert or inserts and/or characteristics of modification. Genetic characteristics of the inserted nucleic acid and the function it specifies, and/or characteristics of the modification introduced;

(e) Living modified organism. Identity of the living modified organism, and the differences between the biological characteristics of the living modified organism and those of the recipient organism or parental organisms;

(f) Detection and identification of the living modified organism. Suggested detection and identification methods and their specificity, sensitivity and reliability;

(g) Information relating to the intended use. Information relating to the intended use of the living modified organism, including new or changed use compared to the recipient organism or parental organisms; and

(h) Receiving environment. Information on the location, geographical, climatic and ecological characteristics, including relevant information on biological diversity and centres of origin of the likely potential receiving environment.

PROBLEM SET

6.1 A specific location of transposable elements of DNA within a chromosome might beneficially alter gene expression and introduce traits like drought tolerance, pest or virus resistance, or increase in nutrient content in agricultural crops such as rice, according to research of Susan Wessler now at Georgia Agricultural University. The importance of rice is recognized by the fact that about two-thirds of the peoples of the world depend on it as a food source. The new technologies have enabled the development of a new variety of rice that is resistant to drought and at the same time has a higher concentration of iron. These technologies range from ethically acceptable breeding or hybridization to modern biology approaches. The assignment is to use the library or Internet to find two of the newer methods of modifying plants and list three plausible yet distinct ethical concerns for each of these methods. Your answers should describe each method in 60 words or less (a diagram is acceptable and may be preferred). After the method, list the three ethical concerns (10 words for each concern) that you believe are most relevant to the particular technique.

Do not waste words on the arguments *for* genetically modified crops as these are covered in: ". . . proponents of GM crops claim advantages to farmers, national economies, the Third-World poor, and the environment. GM crops will, they assert, sustain and improve the productivity, profitability and competitiveness of crop production. Food production will be able thus to keep pace with the growing world population. At the same time, lower volumes of chemical pesticides will be used, with consequent benefits to the environment, food safety and human health. Food quality will improve, and new foods will become available. Non-food products from crops will replace those currently derived from petrochemical raw materials, thus saving scarce resources and reducing pollution" (www.uccf.org.uk/external/acf/acf_tech.htm., accessed April 22, 2004).

6.2 Mad cow disease (bovine spongioform encephalitis, or BSE) does not appear to affect horses or dogs. Your professor decides to investigate the gene involved in protecting these animals so that she can develop a transgenic cow that would not harbor the prion that causes BSE. This breed would potentially be the only safe beef in the world. Investment bankers flock to this opportunity, and your professor starts a company. You are asked to join this company but have some misgivings regarding the enterprise.

List at least six ethical problems and elaborate what moral or regulatory issues are of concern in 20 words each. Start with:

1. Conflict of commitment to university obligations.

6.3 As editor of the *Journal of Environmental Research*, you receive a manuscript that concludes that gene transfer from transgenic aphids has resulted in mutated gnats that are killing cows in New Zealand. The experiments were conducted in an enclosure of 30 meters × 30 meters with climate control and 20 cows. Discuss your dilemma regarding responsibility to the author in terms of confidentiality and proper review of the manuscript vs. consequences to the transgenic pest producers and to the public should the science be flawed (100 words).

6.4 The use of GM corn for food in human beings has raised many ethical concerns, but the use of GM corn as animal feed might present other concerns. Through the Internet, find information regarding GM corn and animal feed. Synthesize in 80 words or less the possible consequences that might be used in an ethical argument against GM corn for animal feed.

6.5 List 10 uses of genetic modification of plants. (Hint: pharmaceutical production of an antibody.)

7

Medical Ethics

BACKGROUND

This chapter explores a new horizon in our study of ethics. Unlike business ethics or scientific ethics, in which scientists and engineers are confronted with problems where justice, non-malfeasance, utilitarianism, and duty to multiple stakeholders are the main considerations, medical ethics presents situations in which a person's physical and mental welfare demand the principal focus of the professional. Nowhere in contemporary human relations are the ethics of duty and autonomy more important than in the practice of medicine. Medical ethics are ever evolving, as new advances in medical treatment have brought new ethical challenges not envisioned even 10 years ago.

The autonomy of the patient or the autonomy of the participant in medical research reigns supreme in decision making. Virtue (i.e., the Golden Mean) becomes important in physician-patient communications. Some of the main ethical questions to be addressed in this chapter are:

- Does the autonomy of the patient limit the capabilities of the medical professional to render care?

- When medical criteria are equal, who should receive a heart transplant, kidney dialysis, or another scarce medical resource that is not available to all who need it? Is it ethical to make a choice based on lottery and random chance when not all can be treated? Or should exceptions be made in certain cases?

- Is physician-assisted suicide ever an option?

- If a patient begs that treatment be stopped for a life-threatening illness or after a devastating accident that will affect the patient for a lifetime, who should decide whether it is ethical to do so?

We shall discuss these and many other issues in this chapter.

GENERAL MEDICAL ETHICS

Codes of Professional Conduct

The code of professional conduct for the medical profession, the *Oath of Hippocrates*, is perhaps the oldest professional society code, and it, along with the *Belmont Report on Human Experimentation* issued in 1979, remain two important guidelines for medical ethics.

The Hippocratic Oath originated nearly 2,400 years ago. Its basic premise is that physicians must, to the best of their ability, use their judgment to act for the benefit of the sick, keep them from harm and injustice, refrain from giving a deadly drug or from causing an abortion, maintain confidentiality, and do what is best for the patient (Markel 2004a). Thus, even 2,400 years ago, the Hippocratic Oath specified that a physician should not cause an abortion or assist a patient to die. The additional admonishment "to do no harm" has been mistakenly attributed to the Hippocratic Oath, but it appears in another of Hippocrates' treatises entitled *Epidemics*: "As to diseases, make a habit of two things—to help, or at least, to do no harm" (Markel 2004b). It can be argued that even avoiding harm requires learning what is harmful, and that some may be harmed in the process of obtaining this information through human experimentation. Today, the oath is usually administered as part of medical school graduation ceremonies and is highly esteemed as a daily source of guidance for physicians.

In the United States, the most comprehensive code of conduct for physicians is the American Medical Association's Code of Medical Ethics. The first standards evolved in 1847. Currently there is a Standards Group with activities divided into three programs: Council on Ethical and Judicial Affairs; Ethics Resource Center (erc@ama-assn.org), where practitioners can find assistance on ethical problems; and the Institute for Ethics (ife@ama-assn.org), which researches emerging issues and develops updates for the Code of Medical Ethics.

Another guide for medical ethics, the *Belmont Report of 1979*, addresses the ethics of human experimentation. Here, autonomy, beneficence, and justice serve as guiding principles (non-malfeasance became a fourth principle later). The report requires that beneficence should be perceived as an obligation: ". . . one should not injure one person regardless of the benefits that might come to others" (National Commission for the Protection of Human Subjects of Biomedical and Behavioral Research 1979). This imperative challenges one to decide when it is justifiable to seek certain benefits despite the risks involved and when the benefits are not sufficient to outweigh the risks. The ethics of human medical research will be discussed in Chapter 8.

Patients' Rights to Information

Prior to the 1960s, the role of the physician was more authoritative, and a patient's involvement in his or her own medical care was relatively minimal. Now a dialogue exists between patient and physician, wherein it is the physician's responsibility to describe treatment options, to outline what adverse events might occur, and to work with the patient to find the best possible solution for a particular situation. Ideally, the patient decides which treatment options to pursue.

Further, informed consent is now required before a treatment regimen begins. This means that the physician must decide what to tell a patient about an illness and provide sufficient information for the patient to make an informed decision regarding treatment. How should one balance informed consent with telling the patient everything? For example, when a patient is diagnosed with what the physician knows will be an extremely painful disease, should the patient be told all the details? Aristotle's theory of virtue, the Golden Mean described in Chapter 1, would have the physician find the appropriate balance between giving every detail of the truth and giving no information except the treatment the physician is recommending. The Golden Mean is achieved by giving the patient sufficient information for decision making without overwhelming the patient, as illustrated by the case described below.

Case 7.1 Disclosure Dilemma

A patient is terminally ill with extensive cancer metastases to vertebrae, ribs, and hip bones. Whereas the doctor has a duty to inform the patient of the nature of her disease and the alternatives for treatment, details regarding the painful consequences of the disease and the fact that there is no hope for survival could harm the patient. In fact, the doctor may misjudge how well the patient will understand the information, and this lack of understanding could lead to denial or despair, with subsequent refusal of what might be optimum treatment for a maximum quality of life. The doctor faces the dilemma of telling the patient all the information she has a right to know vs. the harm this information might bring to the patient's mental state and quality of life. We can begin to structure a solution to this dilemma by first considering the Four A's of our design strategy (Acquire Facts, Alternatives, Assessment, and Action from Chapter 1).

Acquire Facts: There are two aspects to resolving the ambiguities in this situation. First, recall from Chapter 1 the concept of prima facie priority of professional obligation, and second, consider the specifics of the particular disease in this patient. As to the professional obligation, the physician is required to discuss the disease and offer treatment options in order to inform the patient, but some patients require more information than others. Informed consent is required, but veracity is not a requirement in so far as all the details of a disease need be disclosed. The patient must receive at least sufficient information about the proposed treatment options so an informed decision can be made. In each case, we need to collect information regarding the best ways to combat pain, lessen the spread of disease, and provide the best possible quality of life.

Alternatives: Patients have vast opportunities for therapy to lessen pain, ranging from simple anti-inflammatory drugs to spinal cord interruptions. Interventions such as surgery, chemotherapy, and radiation, particularly in the case of bone metastases, can create relief from pain for a year or more as well as extend the patient's life in many cases. Alternative medicine offers techniques such as biofeedback that might also be helpful.

Assessment: Patients vary in how much each needs to know, and in how each responds to the information given. The physician must ascertain what additional information the patient should have, so that treatment options can be selected depending on the course of the disease. The method of assessing the need for information is through direct discussion with the patient, and sometimes with the spouse and family if the patient gives permission for these discussions. Assessment requires more than one conversation with the patient.

Action: The doctor should present the alternatives that are practical and relevant to the patient's condition. A patient who realizes there are many possibilities for dealing with the cancer progression will be diverted from despair. It is in this information sharing that informed consent becomes important so that the patient is aware of the possible consequences of treatment options. For example, the consequence of electing morphine rather than palliative radiation treatment might be drug addiction. Thus, the action in this case is the explanation of options and their consequences, but here one would avoid divulging too much information and speculation on the "what ifs" by concentrating on the current needs for care. Sometimes it is helpful to refer the patient to a patient group for support and further discussion of their mutual problem. The doctor must gauge the patient's ability to absorb the medical details, and the amount of information and timing of the discussions are essential aspects of this communication. Advice from a hospital ethicist, the hospital ethics committee, another physician, or an other appropriate professional should also be sought.

The main ethical considerations in this situation are virtue, autonomy of the patient for confidentiality, and non-malfeasance. The dilemma involves the conflict between the duty to fully inform the patient and the virtue of prudent communication to inform but not unduly frighten the patient. The goal is to achieve the Golden Mean in communication between physician and patient. In addition, this case introduces the reader to the concept of informed consent wherein alternatives and risks of treatment options must be presented to the patient at an understandable level of communication.

Confidentiality

Respect for patient rights to control or keep private their medical information should prevail, unless the patient gives specific permission to the physician to disclose medical information. The reasons for maintaining confidentiality are to maintain privacy and to avoid divulging information that might compromise the patient at work, financially, or socially. We will return to the financial concerns later, but let us first consider why a breach in confidentiality can disbar a lawyer and lead to revocation of a license to practice medicine by the state licensing authority.

The duty to do the right thing from Kantian philosophy extends to respect for the rights of the patient. Thus, the control of information about one's self would be violated if a physician revealed personal information about a patient. The moral imperative is so strong that even married couples who are physicians do not discuss a patient's illness when the identity of the patient might be evident or guessed at.

From the time the Hippocratic Oath was written over 2,000 years ago until the 1980s, there were no established guidelines to protect the autonomy of the patient outside of professional codes and individual discipline. Concerns about a patient's right to be protected from inappropriate disclosure of health information spurred the development of federal guidelines, resulting in the proposal of a Uniform Health-Care Information Act in 1988. This act proposed strong prohibitions to the dissemination of private information and was a step toward laws and regulations that now formally protect patient medical information from unauthorized disclosures (Banach 1988):

- It prohibited the health caregiver from releasing any information pertaining to a patient without the patient's written consent.

- It restricted the power of a subpoena to acquire confidential information, and protected the patient as well as the custodian from being required to release private information summoned on a subpoena.

- It gave the patient access to his or her medical records, something that had never before been allowed.

These rules exemplified logical and ethical considerations, which, after public discussion and revision, evolved in the 1990s to the current federal regulations. Additional safeguards were added when it became apparent in the 1990s that private information, increasingly collected and stored electronically on multiple databases, could no longer be controlled by the treating physician. These data were being provided to or accessed by a wide variety of agencies not involved in the care of the patient, and these violations of individual privacy had widespread consequences, including bias in eligibility for employment, financial aid, and insurance. Thus, the Uniform Health-Care Information Act served as the cornerstone for the Health Information Portability and Accountability Act of 1996 (Public Law 104-191, HIPAA). It was the first comprehensive law outlining federal protection pertaining to the privacy of electronically accessible patient information

Additional regulations have since been developed, including the regulation "Standards for Privacy of Individually Identifiable Health Information" issued by HHS and applicable to entities covered by HIPAA. While HIPAA and this HHS law (*Federal Register* 67: 53181, 2002) were passed in 1996 and 2002, respectively, they both took effect in 2003. Covered entities include (1) health plans, (2) health care clearinghouses, and (3) health care providers who electronically transmit any health information in connection with transactions (e.g., billing, insurance coverage) for which HHS has adopted standards.

The Office for Civil Rights (OCR) is responsible for implementing and enforcing privacy regulation. Among other things, the law requires the establishment of rules for the privacy of personal health information, as well as national standards for electronic health care transactions. Covered entities can be researchers, institutions, organizations, or persons. A research institution that performs research on human subjects is not necessarily a covered entity, but if it is, information about a patient cannot be forwarded without the patient's consent.

Researchers are covered entities if they are also health care providers who electronically transmit health information. The Privacy Rule protects the privacy of such information, while providing ways in which researchers can access and use health information when necessary to conduct research. For example, physicians who conduct clinical studies or administer experimental therapeutics to participants during the course of a study must comply with the Privacy Rule if they meet the HIPAA definition of a covered entity (http://privacyruleandresearch.nih.gov, accessed February 15, 2005). While the privacy rules have generally been helpful in protecting patients, in some instances they have hindered research activities. For example, they prohibit e-mail usage, even between doctor and patient, without explicit permission by the patient. The rules also require medical findings in the course of a research protocol to be communicated only through the patient's physician. These standards were put in place not only to protect patients but also to improve the efficiency and effectiveness of the nation's health care system by encouraging the widespread use of electronic data interchange in health care.

Responsibilities of the Physician Regarding Disclosure of Patient Information

In addition to the federal legislation described above, there are ethical considerations and also state laws regulating and enforcing confidentiality. Since confidentiality is a foundation of proper medical care and is a primary professional concern, the physician may not divulge the condition or treatment details of a patient to anyone without the patient's permission. However, the physician must report a medical problem under the circumstances listed below, with some circumstances requiring a court ruling:

- If the safety of another individual or the public is at risk (for example, a psychiatrist reporting a psychiatric patient who is threatening to kill someone or bomb a public building).
- If the patient is suicidal.
- If the benefit to society outweighs the autonomy of the patient (the reporting of severe acute respiratory syndrome [SARS] was done to minimize the consequences of this highly contagious infectious disease to society).
- If there has been physical abuse and/or rape, even if the victim decides not to report it. The decision to report or not has now been removed from the victim's and physician's hands.
- If the patient is not mentally competent, to overrule the patient's objections to treatment by court order.
- If there is concern about the health status of a minor being denied treatment, in which case a court ruling is usually sought.

Ownership of Medical Information

The doctor's notes regarding a patient have evolved from private notations in a diary and notations by nurses on charts kept at the foot of a patient's bed to formal records of the medical history and physical exam that are now often encoded electronically. The doctor's notes were once considered private notations, and it was generally accepted that these documents were the property of the doctor; it was also understood that the information in the notes must be held in confidence. Contemporary thinking is that all of the medical information regarding a patient belongs to the patient and that no part of the medical record or results of medical tests can be divulged without permission from the patient. Even parents calling to inquire about the status of their hospitalized college student will be denied information unless the student has signed a waiver that this information may be disclosed. Although this contemporary notion seems ethically correct and in keeping with autonomy, there are some instances in which the medical information of one patient might have important consequences for the near term as well as the long-term welfare of other individuals in society. This raises the following questions:

- Who owns a medical history?
- Can a person's medical history be divulged? To whom? Under what circumstances? What if the person's identity is withheld?
- What if the information about a population would advance medical science?

- Who owns an individual's blood cells taken as part of a routine clinical examination? What about the patient's genetic information?

- What about a surgical specimen removed from a patient if a unique cell line with health benefits and marketing potential is discovered?

These questions are only beginning to be addressed. The forums for debate are currently the institutional review boards of universities and medical centers throughout the world (i.e., committees of peers who review human subject research protocols, discussed below). For example, the societal importance of genetic information could become so great that there will be a challenge to the rights of the individual for privacy. The value is in the medical knowledge gained from relating information in public databases, such as variations in gene sequence and their association with specific diseases. This association can lead to major advances in understanding disease mechanisms and treatment. The privacy threat is that genetic variations can also be used to identify individuals, as is commonly done in forensic identification. If someone has access to an individual's genetic data and finds a match in a public data bank, it would allow the detection of present or potential genetic diseases in the individual and in his or her offspring. Such data could be misused in the determination of qualification for employment or health insurance. Currently, there are no guidelines regarding the use of human genetic data (Lin et al. 2004).

This idea that people have certain innate rights that must be respected was advanced by John Locke and became codified through the civil rights movements in the United States. One of the advancements included the right to the protection of an individual's personal and medical information discussed above. There are a number of diseases that represent threats to the public and are required by law to be reported to health authorities. However, it is important to ensure that this precedent is not extended to allow the reporting of other health information such as an individual's genome.

Refusal of Treatment

Suppose that a physician has a clear idea of what the patient's treatment should be, but the patient wants nothing to do with the physician's recommendations. How far can the physician go with persuasion before it becomes coercion? Can the physician involve family members to convince the patient? How about manipulation? What is allowed? In these cases, unless mental incompetence can be shown, there is little the physician can do other than continue to contact the patient and provide easy access to the needed treatment. But usually there is a wide range of possible interactions between the doctor and the patient that would lead to optimizing patient care. The following example illustrates some of the steps doctors may take to optimize patient treatment while behaving ethically.

Case 7.2 Patient Refusal of Treatment

A patient needs coronary bypass surgery, has had one before, and does not want another. The physician knows the patient could be saved by this procedure, yet the patient refuses and even decides not to tell his wife. What can the physician do to convince the patient that he should undergo the surgery? By approaching this problem as a design strategy problem and laying out the alternatives, some provocative concepts emerge. The reasons the proposed treatment has been rejected need to be explored. This is a frequent problem in the practice of medicine. In fact, patients failing to follow medical advice to lose weight, exercise, or stop smoking are also likely to have underlying reasons for their actions. If the patient insists upon physician confidentiality, even a small suggestion to the spouse under any circumstances would be unprofessional and a violation of the prima facie duty to respect confidentiality. Below is a review of how the Four A's design strategy can be used here.

Acquire Facts: Knowledge of the reasons for the denial of further procedures can be a segue to helping the patient to understand the safety and appropriateness of the proposed therapy. The circumstances surrounding the previous experience should be explored. The patient's concerns about financial liabilities and medical consequences of possible surgical errors might be the mental barriers. As in other dilemmas, asking for advice from other physicians can give ideas for approaching the patient. If this is done anonymously, the doctor is not breaking confidentiality.

Alternatives: A repeat of the previous mode of therapy may prove not to be the optimum treatment for this patient. In fact, it might be possible to propose another procedure or another approach to the problem that would be more acceptable to the patient. Consult with other physicians and the patient to discuss alternatives that might be acceptable. An ongoing interaction with the patient will evolve new information, particularly regarding the mental attitude of the patient concerning the proposed solutions to the medical problem. This should not be a game of the physician winning over the patient but a respectful search for consensus between the physician, consultant, and patient regarding the appropriate course of therapy. In the end, the physician's duty is to advise the patient regarding the risks and benefits of alternative procedures, not to tell him what he must do.

Assessment: This might include further discussion with the patient to understand his reasons for refusing to proceed with treatment, referral to another physician for a second opinion, or a consultation with the proposed surgeon to discuss whether a new or modified procedure might help. Almost all diseases and injuries are associated with diagnostic procedures and treatment options, with some procedures more painful but also more effective than others. For example, sigmoidoscopy, a procedure that allows direct visualization of the large bowel using a fiber-optic probe, is more accurate than procedures such as virtual colonoscopy, a procedure to visualize the bowel by painless noninvasive imaging techniques, or the use of a miniature camera that can be swallowed to provide images as it moves through the bowel. However, the consequences of all these efforts could be disastrous if the delay were to lead to death or serious compromise of the patient's health.

Action: Maintain an ongoing interaction with the patient to decide on the treatment and to keep abreast of new information as it evolves. The patient can elect the preferred procedure given the facts, but as we discuss below, medical insurance restrictions might limit the choices. The underlying principle is the autonomy of the patient.

The dilemma described in the following case illustrates the conflict between beneficence and the patient's autonomy. The physician is in the role of a parent who knows what is best for his or her child, but unlike the parent-child relationship, the rights of the patient to be informed and to accept or refuse therapy supersede any assumption of paternalistic authority by the physician.

Case 7.3 Autonomy vs. Professional Standards

Suppose that a tumor in the lung of a woman patient who has previously been treated for breast cancer is discovered by x-ray. The physician requires a biopsy to confirm the histology of the tumor so that a specific treatment plan can be developed. The patient refuses.

Here is a case that challenges the duty of the physician to treat the patient and the duty to provide the correct treatment. What are the reasons for the patient's refusal? Can the physician better describe the need for a correct diagnosis and the consequences without it? Can she be sufficiently reassured to allow the procedure? What *is* sufficient information? How far can a physician go to get help in a refusal situation when confidentiality cannot be breached? Can the physician ethically support the administration of treatment without a specific diagnosis? If the patient refuses the biopsy, it is not a simple matter of going forward with treatment for a suspected tumor. It may become necessary to refer the patient to another physician if he cannot ethically recommend or support the treatment options she has chosen.

What then are the possible strategies by which to resolve this dilemma? The options suggested by the Four A's are to obtain advice from other physicians, discuss alternatives to surgical biopsy for diagnosis, and be certain the patient understands the need for a confirmation of the diagnosis before initiating treatment. The goal is to protect the autonomy of both the patient and the physician.

Ethics of Rendering Medical Aid

Case 7.4 Accident Victim's Autonomy

A citizen or a physician encounters a patient bleeding on the street following a car accident. The patient's autonomy and the right to refuse treatment must be respected. The duty to render aid is overridden by the injured's autonomy unless it is determined that the patient is in an incapacitated state from the accident and unable to properly assess the severity of the injury.

Case 7.5 Rights of an Emancipated Minor

When does a child have the right to make an independent decision regarding his or her health care without parental consent? Sometimes children have declared emancipation from their parents, in which case they are responsible for their own medical decisions. With regard to the physician taking on the role of an advocate for a child, the family has the final word on what treatment option is chosen, which may include the option for no treatment. However, as noted previously, in some cases where a condition is life-threatening, the physician may obtain a court order to proceed with treatment.

Case 7.6 Rights of the Unborn to Medical Care

It is now possible to improve the survival of a fetus with a life-threatening defect by opening the uterus and performing corrective surgery, after which the uterus is closed and the pregnancy continues to term. In addition, it is becoming apparent that in the future it will be possible to deliver gene therapy to correct a genetic defect. What are the rights of the fetus if the mother refuses treatment? Does the fetus have a right to survive? How should the rights of the mother and the rights of the fetus be balanced? A physician faced with this dilemma does have recourse to the hospital ethics committee and other consultants. There is a duty here for the physician to get help. However, the rights of the fetus have not yet been fully explored. One principle that overrides the rights of the fetus is that the mother cannot be put at risk against her will. Where ethical problems could arise is when the mother insists on remedial in utero surgery, but the physician deems the surgery too dangerous to the fetus. The physician has two routes of action in this case. Either convince the mother of the inappropriateness of the surgery, or help her find another surgeon.

Case 7.7 When Parents Refuse Treatment for a Child

A child has suspected meningitis and needs a lumbar puncture to confirm the diagnosis. The parent declines. What are the options? One possibility is to assume that the patient has meningitis and treat anyway. Another is to try to explain further the importance of obtaining the diagnosis and that the procedure is usually not problematic. Another is to get a second opinion. In a similar case at the University of Washington, the Ethics Committee decided on the spot to take away parental rights and allow the physician to do the lumbar puncture. The ethics consultants agreed that the physician should intercede and do the lumbar puncture when there is a reasonable belief that the condition is life-threatening. Therefore, certain circumstances demand intervention. The physician may make a decision if the condition is immediately life-threatening, or may go to court to obtain legal authorization for treatment. The court then decides when parental rights supersede lifesaving medical treatment.

Case 7.8 Forcible Medical Treatment of a Prisoner

Charles Sell was a dentist who had been committed for delusional behavior, including claims that God had told him that for every FBI agent he killed, a soul would be saved. Soon thereafter he was accused of submitting false insurance claims. He was set free on bail but was later jailed and indicted on charges of fraud and attempted murder of the FBI agent who had arrested him and another who was to testify against him. The court judged Sell mentally incompetent. He was hospitalized for a period of four months, during which time his ability to stand trial was evaluated. The staff of the psychiatric facility recommended that Sell take antipsychotic medication, which he refused. Nonetheless, the medical personnel pressed for permission to administer the drugs against his will because he was dangerous; they also hoped he would become competent to stand trial. The lower courts ordered that Sell be given antipsychotic medications, but the Supreme Court reversed this decision. The basis for the reversal was the failure to evaluate whether the fairness of his trial would be compromised by the use of sedatives or other drugs that might interfere with Sell's ability to properly communicate with his attorney, respond to trial events, or show proper emotion. Other issues were raised by this case. Does the autonomy of the prisoner take precedence over the need of the government or society to follow due process? Does the administration of medication against his will interfere with his liberty?

Under conditions of prison safety and courtroom order, the U.S. Supreme Court has ruled that forcible medication is allowed. The dilemmas are that without the process of courtroom prosecution and defense, the prisoner cannot be judged guilty or innocent; without the medication, an orderly procedure cannot occur. Thus, justice to the prisoner is the overriding moral theory, but this means that the autonomy of the prisoner, including his right to refuse treatment as protected by law, is in conflict. Helpful perspectives in understanding these cases come from Annas (2004) and from the decision of the Supreme Court of the United States in an opinion written by Justice Stephen Breyer (*United States v. Sell* 2003). Four conditions are given as a guide to when it is permissible to forcibly medicate prisoners:

1. When important governmental interests are at stake.
2. When the prescribed medications are substantially likely to render the defendant competent to stand trial and unlikely to render the trial unfair (e.g., to cause the defendant to become emotionally indifferent to the alleged morbidity of the crime).
3. When there is no alternative or less intrusive treatment.
4. When administration of the drug is medically appropriate considering the medical condition.

In summary, forced treatment needs to be examined not only with respect to the autonomy of the accused but also with respect to the system of justice and the ethical issues for the health care professional being asked to treat a prisoner who refuses treatment. So how should one proceed? The four principles listed above can be integrated into the strategy for action during fact-finding and the development of alternative approaches.

Case 7.9 Autonomy vs. Duty to Prevent Suicide

The girlfriend of a physician's patient wants to commit suicide. She says she does not want any help. The patient (her boyfriend) comes to his physician to discuss what options there are to intervene. The physician's responsibilities are limited here to giving advice to his patient and to offering to be a resource to the patient's girlfriend, as the girlfriend may have a medical problem such as depression that can be treated. The physician can help the boyfriend clarify ambiguities and acquire more facts to help think through this difficult situation. The physician's duty to his patient and to his girlfriend's autonomy must be foremost, but some positive action may be possible. One alternative is to advise the boyfriend that he could have his girlfriend placed under psychiatric observation for 72 hours or longer, based on the evaluation of her condition. The laws in the United States provide that a person who is gravely disabled as a result of a mental disorder and dangerous to himself or herself, or dangerous to others, may be involuntarily detained for 72 hours of evaluation and treatment. The physician could aid his patient by referring him to an attorney, a judge, or directly to a psychiatric hospital that accepts involuntary commitments.

Case 7.10 Medical Treatment in Dementia

With treatment, many Alzheimer's patients are doing better in the year 2004 than in 1980. However, in 1980 after seeing some of her friends with Alzheimer's, Grandmother puts in writing her decision that should she develop dementia, she does not want any therapy beyond minor treatment (e.g., no ventilator if in a coma, no interference with the natural processes). Grandmother becomes demented but is very happy. At one point she is in an auto accident and is taken to the emergency room, where a doctor does a tracheotomy and places her on a respirator. Her son arrives and indicates her intent not to have this treatment. The physician says this is a temporary state if she is treated with steroids for minor brain swelling, and he cannot ethically discontinue treatment. Should the son let the doctor continue to the next step? A priest says it is a violation of her autonomy to treat and ignore her wishes expressed 24 years ago.

One option is to do nothing, another is to defend her autonomy, and still another is to allow treatment to continue. But when she made her decision in 1980, she probably thought that if she were a demented patient, she would be in a terrible state with no treatment options and a burden to her family. This type of case will be taken up below when the concept of the Advance Health Care Directive is introduced.

When Options for Continuing Care Conflict with a Physician's Ethics

In the event that a set of ethical standards in a society or culture differs from the physician's individual ethics, the physician must respond by informing the patient of all options. An example is a physician who is opposed to abortion and considers it the taking of a human life. If the patient is a pregnant mother with a condition for which one of the options is abortion, the physician must offer it as an option and arrange competent care for her. Another example is the physician's duty to save a life by reasonable and conventional means. This may mean bypassing certain religious or cultural prohibitions, by court order if necessary.

Sometimes there are circumstances when it is best to refer a patient to another physician. Examples include a patient who has become angry, unreasonable, or uncooperative. Professional ethics mandates that the physician is obligated to treat the patient until other care is found; otherwise, it is abandonment. Further, the referral must be done without prejudice. Patients cannot be labeled as management problems because this may affect the standard of care that the patient receives. All patients are entitled to the same level of care. It is the law in some states (e.g., California) that every patient entering an emergency room must be triaged within five minutes of arrival to determine the severity of the symptoms and whether there needs to be immediate treatment. The patient must be triaged, even if the patient is a known malingerer.

Ethics of Medical Mistakes

When does the physician need to tell the patient a mistake has been made? Does every mistake need to be communicated or only those with a legal implication? Ethically, the physician is required to report to the patient any mistake that may impact the health of the patient. A mistake not affecting the health of the patient may be reported at the discretion of the physician. If a physician observes or has knowledge of an error in the treatment of another patient, there is no specific guidance as to the responsibilities of the observer in such a case. However, the honor system enters into the decision, as does professional duty. These cases are usually handled according to the individual circumstances. Licensing boards in each state and sometimes the local medical boards are the appropriate bodies to guide the physician in most circumstances. Depending on the seriousness of the consequences of not interceding when harm is likely, the bystander physician could be admonished for complicity by an ethics board or licensing authority if that physician does not intercede to minimize harm to a patient.

Physician Conflicts of Interest

There are three specific areas where a physician may encounter potential conflicts of interest:

- Conflicts may arise when there is an opportunity for personal gain beyond ordinary reimbursement for the care of a patient. For example, a physician with a financial interest as an owner or stockholder in a laboratory or other facility (blood work, x-ray, MRI, pharmacy) may be biased in the number or frequency of tests or medications ordered from that facility. Such situations are highly discouraged.

- Conflicts may arise in providing care for patients who are undergoing research procedures when the physician has an interest in the outcome other than that of acquiring unbiased scientific data. Though these interests are sometimes financial, they can also be based on ambitions to succeed by being first with a new treatment or hypothesis. Such a conflict of interest may lead to bias in the conduct of the study. Physicians involved in evaluating the effectiveness of a new drug may be tempted to overlook outlying data or to soften an opinion about a cluster of adverse events (e.g., unwanted drug reactions) in favor of the company if they are generously paid or likely to become stockholders in the company as a reward for their ongoing work. A good example of how investigator bias can have serious consequences occurred in 1999 when patient Jesse Gelsinger died while participating in an early gene therapy clinical trial. The subsequent investigation cited two important findings. One was the failure of the responsible physician to take into proper account the adverse reactions occurring prior to and during the experiment; the other was his failure to recognize the conflict of interest that existed between his part ownership in the company sponsoring the trial and his ability to conduct an unbiased study.

  The major committee that reviews all human research protocols is called an institutional review board (IRB) (this aspect of medical ethics will be discussed in Chapter 8 under "Ethics of Human Experimentation"). Conflict of interest assessments are required for all physicians involved in human subject research. Medical societies have instituted procedures that require members to disclose possible conflicts. There is real concern that if there is a financial interest in the outcome of a study, there may be bias in the conduct of the clinical study itself or in the reporting of the data. This type of conflict occurs when a paid consultant of a drug company participates on an advisory panel making recommendations to FDA on the use of drugs the company manufactures.

Physicians conducting studies, submitting manuscripts for journal publication, or making a presentation at a scientific meeting are now required to declare and describe any association with a commercial funding source such as a pharmaceutical or device company so that their work may be assessed with this in mind.

- Conflicts may arise between a physician's goal of providing the best care and the goal of the health care corporation to make the most profit. Examples of these types of conflicts have been reported by physicians involved in certain Health Management Organizations (HMOs) where there may be monetary motivations to see as many patients as possible, restrict treatment options, use only generic drugs, keep to time limits, or undergo peer review of tests ordered. If the physician wants to treat outside these guidelines, permission or approval must be sought from the HMO.

If an HMO restricts treatment options, should the physician limit the information given to a patient about all the options? The answer to this is the American Medical Association's assertion that a patient must be fully informed. Laws dating from the late 1990s in many states protect physicians from termination if he or she disobeys HMO restrictions regarding disclosure of information on alternative treatments. Indeed, it is within the general guidelines of fully informing patients of the options and circumstances of their treatment to disclose the financial incentives between the physician and the HMO. Nevertheless, some physicians must decide whether they can maintain ethical standards while working in an HMO, and if not, to practice elsewhere. Most HMOs are reputable, but a few unethical ones can generate much negative publicity. Some are involved in illegal practices to make a profit, as shown by the Case 7.11 below, which presents the unfortunate occurrence of both corporate and medical malfeasance. Stockholders were harmed, patients were harmed, and two even died, presumably as a consequence of the profit goals of a hospital and the associated corporation. We present this case to make the point that these types of conflicts and unethical actions can and do arise, with what can be devastating consequences.

Case 7.11 Medical Care Corporation Violates Medical and Corporate Ethics

A patient underwent a cardiac stress test at a medical center owned by Tenet Healthcare in Redding, California. Even though he passed the test, the Tenet Healthcare physician recommended cardiac catheterization. The results of that test, he was told, indicated the need for emergency triple bypass surgery. Because the patient had a friend in Las Vegas who could render care after the surgery, he decided to have the surgery done there instead. The heart specialist there could find no evidence of disease.

The subsequent probe into Tenet's operations indicated that the number of cardiac catheterizations done at the center were at least four times the number performed by other centers of similar size in northern California. Also uncovered were overly aggressive laboratory testing procedures, questionable diagnoses, unnecessary operations, multiple medical complications, at least two patient deaths, triple the amount of special Medicare billings compared with the number the hospital made three years earlier, and excessive pressure on physicians to make unrealistic profits from their specialty divisions (Eichenwald 2003). Subsequent investigation by the government found fraud allegations dating back to 1994, and Tenet has paid over half a billion U.S. dollars in complaint settlements.

The Ethics of Futile Care

Futile care is defined as treatment that is unlikely to result in improvement in the condition of the patient or that does more harm than good. Decisions about what represents futile care are among the most difficult decisions a physician, a patient, or the patient's family must make. Moreover, these types of decisions arise more frequently than might be expected. Decisions about these cases must be based on the best possible scientific and medical assessment of the diagnosis and treatment being given, as well as the likelihood that this treatment will, over a reasonable period of time, result in improvement. Futile care brings false hope, burdens family resources, and can consume limited resources that might be used by others.

Futile care also is complicated by issues surrounding passive and active paternalism. An underlying tenet of a physician's care for the patient is beneficence, that is, to ease the suffering and, in so doing, to avoid any harm. This concept may imply that the professional health care giver knows best how to treat and what to disclose to the patient about the consequences of that treatment. However, this perceived responsibility and authority can be in conflict with the patient's autonomy unless it is in the best interests of the patient and consistent with the patient's preferences (Pellegrino and Thomasma 1988, Savulescu 1995). The issues of beneficence vs. autonomy are discussed by Beauchamp and Childress (2001a), who argue that ". . . beneficence provides the primary goal and rationale of medicine and health care, whereas respect for autonomy (along with non-malefeasance and justice) sets moral limits on the professional's actions in pursuit of this goal." The intervention by a physician in the affairs of an individual who cannot reliably care for himself or herself becomes paternalism, just as a parent has justifiable rights to guide, discipline, and care for a child.

The physician assumes a parental role to some extent for the patient who is incompetent and in need of care, but to bring paternalism into the relationship between the physician and the autonomy of the patient is potentially a violation of the rights of the patient. Thus, paternalism can transform the physician-patient relationship into that of a ". . . loving parent with dependent and often ignorant and fearful children" (Beauchamp and Childress 2001b). A connection between the problems of paternalism and futile care becomes clear in the following case.

Case 7.12 Paternalism vs. Violation of Duty to Inform Patient

A patient suffering from severe back pain for many years has been diagnosed with lumbar disc disease. Now at 80 years of age, he has new pains in the back and seeks medical aid. He also has congestive heart failure and has refused aggressive therapy for this condition in the past, which is likely to cause death in the next two years. His physician learns from new tests that his patient also has prostate cancer, but this patient has a tremendous fear of cancer and is likely to panic or become depressed if told he has this dreaded disease. Based on the physician's philosophy of beneficence, he does not reveal to the patient the high probability of cancer. The rationale is that this information would not lead to any action that would reverse the disease and would detract from the quality of the remaining life of this patient.

What are the ethical problems in this case? How would you proceed if you were the physician? If you were on the medical board of inquiry when the facts of this case were revealed, how would you proceed?

This would be seen as a case of physician paternalism. While the physician's actions might seem consistent with the optimum care of this patient, the physician's peers would see it as a violation of the patient's autonomy and, in particular, as a violation of the now accepted obligation to disclose a diagnosis of cancer to the patient. This responsibility is clearly stated in the guidelines of the American Medical Association's Code of Medical Ethics, Section E-8.12 (American Medical Association 1994):

> It is a fundamental ethical requirement that a physician should at all times deal honestly and openly with patients. Patients have a right to know their past and present medical status and to be free of any mistaken beliefs concerning their conditions.

In the physician's defense is the unrestricted discretion traditionally given to the physician with regard to what patients should be told. While some such as Kant and Mill might have argued that temporary paternalism is allowable in some situations, in the final analysis, a strict adherence to veracity is demanded by autonomy and a right to knowledge about conditions for which the patient is at risk. The method by which the information is disclosed is under control of the professional. An approach that involves informing the patient in stages or enlisting help and advice from others might be employed. But a justification for withholding vital health information based on the judgment that this information would upset or depress the patient is not a tenable justification.

Case 7.13 Pursuing Futile Therapies

A patient with severe liver disease due to cancer metastases from colon cancer is now in a coma and the family requests that the physician to do everything possible to bring the patient back to health. The patient's advance health care directive gives no guidance for this situation. However, the physician concludes that the use of liver transplantation or a bioreactor (i.e., an artificial liver that has suspended liver cells in a container through which the patient's blood is perfused to temporarily replace normal liver function) is futile care in view of the presence of extensive disease and the terminal state of the patient. The physician need not communicate the futile care options to the family, and to do so is considered malfeasance or, at the minimum, imprudent. However, if the family is aware of possible extreme experimental protocols or other treatment alternatives and asks about their use, then it is the physician's duty to explain the basis for his or her opinion that their use would be futile.

The ethical concepts here are virtue, non-malfeasance, and duty. Virtue dictates that information dispensed to the family must be based on the Golden Mean. Non-malfeasance here means not to put the patient or the family through physical or financial suffering by attempting futile care. Duty here means that the physician must explain the futility of care, as well as find other appropriate care if he or she is unable or unwilling to follow the family's request for extraordinary measures deemed futile. However, the pressure to guard against malpractice suits can present a dilemma that may enter the decision-making processes of the caregivers.

Use of Medical Marijuana

The use of marijuana as a treatment aid in certain medical conditions is gaining acceptance due to mounting evidence that it is helpful to patients with certain diseases, including AIDS, cancer, glaucoma, and multiple sclerosis. Patients report that it stimulates the appetite, suppresses the nausea and vomiting associated with chemotherapy, and enhances mood. The active ingredient, tetrahydrocannabinol, is known to have anti-emetic and appetite-stimulating properties. Its effectiveness in substantially improving the comfort of cancer patients has been generally recognized. In addition, there is evidence that it can lower intraocular pressure and may help in controlling glaucoma.

However, the 1970 Federal Controlled Substances Act makes it a felony to possess or sell controlled substances, of which marijuana is one. In 1996, California passed Proposition 215, allowing the possession, distribution, and consumption of small amounts of marijuana for medicinal uses. In addition, several states, including Maine and Nevada, have plans to set up state-sanctioned medical marijuana distribution centers, and six other states (Arizona, Alaska, Colorado, Hawaii, Oregon, and Washington) have passed initiatives seeking to amend state law to allow the use of marijuana under medical supervision for specific conditions. Thirty additional states were considering acts favorable to the "compassionate use" of cannabis (Josefson 2001).

But in May 2001, the U.S. Supreme Court unanimously rejected the medical use of marijuana because it is an illegal substance under federal law with no permissible exceptions. The justices ruled against the possibility of the use of medicinal marijuana because it would violate the Controlled Substances Act of 1970. This act classified marijuana as a schedule I drug with addictive potential and no medicinal value, making the possession, sale, distribution, and consumption in violation of federal statutes. States have tried to no avail to separate state and federal law to accommodate compassionate exceptions for the distribution, sale, and consumption for certain medical conditions.

Compassionate use initiatives continue to be issues at the state level, and the Supreme Court decision citing the lack of medical advantages for marijuana is being challenged. Even though California and several other states have passed laws making it legal to prescribe marijuana for medical reasons, the sale and possession of it are still considered violations of federal law. The Justice Department has argued that exceptions to the federal law create a double standard, thereby compounding the difficulty in law enforcement. In 1999, the U.S. Justice Department filed an injunction against California doctors who prescribed medical marijuana. A federal appeals court unanimously ruled that the government could not take action against them (Josefson 2002). Some have argued that the federal government is obligated to take several specific steps to legalize medical marijuana as its use has major benefits in certain patients:

. . . given the intransigent position of the federal government on this issue, state governments are justified in unilaterally legalizing medical marijuana as an act of civil disobedience. . . . I argue that legalization is justified even if one were to grant both that the harms of legalization outweighed its benefits and that utilitarianism is true. This requires a subtle and somewhat extended discussion of utilitarian moral and political theory (Barnes 2000).

There are two issues in debate. First, is it the distribution of marijuana that is illegal, or its possession? Second, is the physician who recommends marijuana to patients committing a crime or merely prescribing a drug? With regard to the first question, does federal law trump states' rights when it comes to public health and safety? With regard to the second, can the federal government rule that a physician cannot give advice about the appropriateness of a remedy? Arguments can be made in favor of medicinal marijuana based on the principles of beneficence, duty, virtue, rights, and autonomy. The arguments against medicinal marijuana are based on utilitarianism and the slippery slope theory. It has proven exceedingly difficult for patients, physicians, and lawmakers to reach a consensus on the subject.

THE ETHICS OF DYING

The ethical dilemmas involved in preserving life can be discussed relative to the situations shown in Figure 7.1:

- Terminally ill but mentally competent individuals.
- Terminally ill but cognitively impaired patients.
- Psychologically competent patients who wish to live or to die.
- Cognitively impaired patients who are demented, in a prolonged coma, or in a persistent vegetative state (PVS) for whom the decision is whether life support is indicated.

Few would argue with the need to initiate life support in an acute situation of severe illness or trauma when the full extent and outcome for the patient cannot be known. This might include the use of a ventilator, a cardiac pacemaker, blood pressure medications, drug-induced coma, intravenous feedings, intensive care, or any other means to keep the patient alive until the patient stabilizes.

It is important to differentiate between the use of lifesaving approaches described in the acute trauma case above, when it appears that the patient will recover, and the use of life support measures in a longer-term situation where recovery may be a matter of months or even years. The problem is that it is often impossible to predict a patient's course and physicians must institute lifesaving measures that ensure survival. The problem becomes complicated, and even excruciatingly difficult, for the family, physicians, and the hospital administration when questions of discontinuation of life support measures arise after it becomes apparent that it is unlikely the patient will recover. Many patients who will never recover normal cognition can be weaned from a ventilator, yet they can live for years in this state with reasonable care. Once ventilation assistance and intravenous feeding are

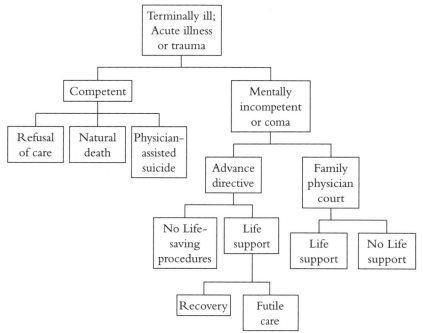

Figure 7.1 Various stages for decisions in dying.

begun, it becomes exceedingly difficult to withdraw them, as seen in the cases of Nancy Cruzan (Case 7.14) and Terri Schiavo (Case 7.15), where removal of a feeding tube would lead to death. The ethical questions involve the patients' rights to elect or consent to medical procedures and the duty of family and physicians to respect the patients' autonomy.

However, when a patient is cognitively impaired, in a persistent vegetative state, or in a coma, the wishes of the patient are frequently unknown. Then it becomes necessary to seek some evidence of the patient's will, particularly a documented expression of what the patient wishes to have happen relative to lifesaving procedures, or a consensus from the family. As we shall see, written expression can be through documents such as the advance health directive or living will. In court proceedings when no such information is available, the decision may rest on prior conversations with friends or relatives. In this section, we also address the issues of purposeful termination of life by a physician's active or passive actions.

Who Decides on Life Support and Death?

Under conditions of incompetence and no advance health care directive, a decision to withhold or withdraw life support involves the moral principles of non-malfeasance and utilitarianism as well as the concepts of futile care and professional duty. On the one hand, the continuation of medical care that is unlikely to improve the patient's status will take a toll on the resources of the family and the health care givers; on the other hand, removing life support equates to killing the patient. A distinction is made between initiating life support and removing this support once initiated. The Cruzan and Schiavo cases described below exemplify how difficult these situations can become. However, there are some guidelines to help determine who has the authority to end life support or determine a patient's death. First, physicians should defer to state law to identify the appropriate surrogate decision maker. One such law from the state of Virginia is very specific, stipulating that there must be agreement between the attending physician and the following, listed in order of priority: 1) judicially appointed guardian, 2) patient-designated decision maker, 3) spouse, 4) adult child or a majority of the adult children reasonably available, 5) parents of the patient, and 6) living relative of the patient (Virginia Natural Death Act 1985).

In the absence of state law specifying either appropriate surrogate decision makers or a process by which to identify them, circumstances will determine who should become the surrogate decision maker. It may be the person with whom the patient is closely associated, such as a close friend or partner. When there is no family, but there are persons who have some relevant knowledge of the patient, such persons should participate in the decision-making process. In all other instances, a physician may wish to utilize an ethics committee or the court to aid in identifying a surrogate decision maker or to facilitate sound decision making.

Further, there is a Uniform Determination of Death Act, drafted and approved in 1980 by the National Conference of Commissioners on Uniform State Laws. This act was recommended for enactment in all states, and was later approved by the American Medical Association and the American Bar Association in 1980 and 1981, respectively. It articulates and expounds the common law definition of death. In order to be declared clinically dead, a patient must exhibit total failure of the cardiorespiratory system or irreversible loss of entire brain function, including the cortex, neocortex, and brain stem.

The reader will recognize the danger of this act in situations where the medical methods for determining brain functionality might be deficient. Further, no given stipulation quantifies the duration of signs of brain death as one of the determining criteria. Because of this, a patient may be pronounced dead erroneously, because some measurements might reveal absence of functioning for an interval due to brain swelling that after a period of treatment may subside and allow brain activity to be detected. What if a pronouncement of death is contested, perhaps because some brain activity is still discernable?

The landmark case below exemplifies how parental wishes were in conflict with the judgments of the patient's health care professionals (*Cruzan v. Director, Missouri Department of Health*, 497 U.S. 261, 110 S.Ct. 2841 [1990]). This example also underscores the importance of making decisions in advance about medical treatment in the event that one becomes incapacitated and unable to communicate these wishes.

Case 7.14 *Cruzan v. Director, Missouri Department of Health*

This case was brought before the court when the parents of a young woman were denied permission to remove her feeding tube after an automobile accident rendered her incompetent in a persistent vegetative state, a condition in which a patient has motor reflexes but no evidence of significant cognitive function. Although she did not require a ventilator, she required a gastric feeding tube for nutrition and hydration. Over a period of several years, the family tried to make some meaningful contact through stimulations such as stroking, talking, and telling her jokes. While some insisted that she exhibited some meaningful responses such as blinking her eyes or smiling at her father's joking, the family saw nothing that seemed either consistent or encouraging that she was improving. Finally, in 1986, three years after the accident, the parents petitioned a state court for the right to remove the feeding tube, an act that would cause her to die.

However, doctors and other intervenors opposed their wishes, seeing her life as valuable even in her present condition. They felt it was morally wrong to remove the feeding tube, not only because this act would result in her death, but also because it would lead to the slippery slope of killing others whose lives were deemed hopeless. In 1988, a state court gave permission to remove the feeding tube, but this decision was reversed by the Missouri Supreme Court (*Cruzan v. Harmon*, 760 S.W.2d 408, 411 [Mo. 1988]), citing the State Living Will statute that embodied a state policy strongly favoring preservation of life, that her parents were not entitled to order withdrawal of hydration and nutrition without clear evidence of her wishes, and that evidence that she would have wanted this was unreliable. Finally in 1990, in *Cruzan v. Director, Missouri Department of Health*, 497 U.S. 261, 110 S.Ct. 2841 (1990), the U.S. Supreme Court addressed whether a doctor, the family, or the governing state of an incompetent patient could determine whether the patient could be taken off life support. In a 5 to 4 vote, the Missouri Supreme Court decision was upheld, noting that it was better to err on the side of life in the absence of any specific evidence of the patient's wishes.

Eventually, new evidence surfaced when two friends, who had worked with her at a school for mentally impaired children, documented her wishes. The family's petition was granted, and in 1990, Nancy Cruzan died. The decision to remove the feeding tube came only after an excruciatingly long and difficult battle for her family, her doctors, her many friends, and others who became emotionally involved in her story. The media kept the case alive, mostly to the detriment of what the family felt was in the patient's best interest. However, Nancy Cruzan's death and the publicity surrounding the case had the positive effect of motivating legislative changes described below that ensured the availability of options for end-of-life decisions.

Advance Health Care Directives

After the Cruzan case, the U.S. Congress passed the Federal Patient Self Determination Act of 1991. This law requires that hospitals, nursing homes, hospices, and other health care providers inform their patients of their rights under state law to make their own medical care decisions, to refuse life-sustaining procedures, and to prepare an advance health care directive. The health care providers must put in place policies for responding to these directives.

Advance health care directives are legal documents that communicate in writing the wishes of individuals regarding their medical care should they become unconscious or cognitively impaired. They are similar to a medical power of attorney that is recognized under state laws as a document that communicates wishes about medical care decisions and are binding in the event the patient becomes unconscious or incompetent. The document also includes the designation of a representative to make decisions for the patient in medical situations that one might not have anticipated or specified. This representative is allowed to make decisions for the individual with knowledge of and in accordance with that individual's values and wishes. There are primarily two reasons for having an advance health care directive: it lets the family know one's decisions for future health care, and it designates someone as the health care advocate to ensure that choices or decisions are honored. An example of an advance health care directive for the state of California appears in Appendix 7.1 at the end of this chapter. Other resources include Web sites such as partnershipforcaring.org and agingwithdignity.org.

A living will is a legally binding document that describes the individual's specific wishes as to which procedures, types of care, treatments, and life support measures may be used in the event of an inability to communicate these wishes in person. Since it is the most flexible, it is the advance directive that most people choose. It is important to note that one can still make one's own health care decisions once an advance directive has been created. The individual is in control of the decision making as long as the he or she is able to do so. Advance health care directives may be particularly useful for decision making in situations of anticipated dementia such as in patients with early Alzheimer's disease (Vollmann 2001). Recall Case 7.10 that depicts a grandmother's situation that predated the advance directive process. Had the son or some advocate been appointed as part of the directive, this person could have acted officially on behalf of the grandmother.

Case 7.15 Case of Terri Schiavo in Florida

Unfortunately, Terri Schiavo did not have an advance health care directive. This is a case of a brain-damaged woman in a persistent vegetative state after a cardiac arrest that occurred in 1990, possibly related to a metabolic imbalance associated with an eating disorder. She was able to breathe on her own but received nutrition through a feeding tube, a tube her husband wanted removed since her condition had not improved in the 15 years that the tube had been in place. He indicated that she had told him she would never want to be kept alive by artificial means. But Terri's parents and other relatives doubted that as a Roman Catholic she would hold this view. They believed she was brain-damaged but showing responsiveness by following items with her eyes and smiling at them. They believed she would improve with proper therapy and fought the removal of the tube.

After years of argument and litigation, a Florida judge ordered the tube removed, but the governor stepped in and ordered the tube reinserted, acting on a newly instated law (Terri's Law), which allowed him to do so. The ethics of his actions, in the absence of a health care directive, raised a number of concerns about who should be making decisions for those who cannot speak for themselves. Of particular concern is whether legislatures or elected officials should be involved, and if so, to what extent. After review of this case by the Florida Supreme Court, the justices found that the law illegally gave authority to the governor to act on Terri's behalf and was a clear violation of the doctrine of the separation of powers between the judicial and executive branches of government (Dahlburg 2004). In January 2005, the U.S. Supreme Court declined without comment to review the Florida Supreme Court decision, halting efforts by the Florida governor to deny the husband's wishes to remove the feeding tube. Ultimately, the tube was removed and Terri died two weeks later, despite efforts to find a way to rescue her by her parents, members of the U.S. Congress, President Bush, and others.

While at the time of her death there were still questions about this case, an autopsy issued three months later confirmed the husband's position that she had irreversible, massive brain damage (Goodnough 2005). What is not in question is how valuable an advance health care directive can be in such circumstances. Had one been in place in this case, the years of legal battles might have been prevented. The advance health care directive is considered an iron-clad, durable legal document that must be respected by the family and those treating the patient. Even so, as seen in Cases 7.16–7.17 below, there are certain circumstances when it can be considered null and void.

Case 7.16 Advance Health Care Directive and Attempted Suicide

A 45-year-old male with a long history of depression but otherwise in good physical health was brought to the emergency room with an acute drug overdose. Because the patient was unresponsive and had labored respirations, he was connected to life support and transferred to the intensive care unit. Later in the day, the patient's brother came to the hospital and presented the attending physician with the patient's advance health care directive. His directive stated that he did not want life support measures under any circumstances. However, the physician is not bound to comply with the patient's health care directive under these circumstances, since the patient was not terminally ill and, with proper care, would recover completely. Furthermore, a patient who attempts suicide can be considered to have reversible mental incompetence, and, therefore, to discontinue life support would not be in his best interest.

Case 7.17 Physician Compliance with an Advance Health Care Directive

A middle-aged patient states in her advance health care directive that under no circumstances does she wish to be put on a ventilator. However, she later develops treatable pneumonia that requires ventilator support for a short time while being treated with antibiotics. The treating physician feels that as a matter of conscience in this situation, treatment cannot be withheld and it is in the patient's best interest to proceed. The patient's sister insists that she will sue both the doctor and the hospital if the wishes of the advance health care directive are not honored. Below, the Four A's design strategy is applied to the dilemma faced by the doctor.

Acquire Facts: The patient likely did not consider when the advance directive was executed that there were treatable short-term conditions that could require the use of a ventilator. Thus, the exact wording on the directive must be studied for clues that the directive is meant to be applicable only in a situation from which there is no hope of recovering. Advice from the hospital ethics committee and the legal office should be sought.

Alternatives: The only possible actions are to continue treatment or to stop treatment. The treating physician has the option of contacting another physician who is willing to assume responsibility for cessation of treatment if the hospital or another authority decides to honor the advance directive. The sister may insist on finding another physician or even another hospital that would accept her sister as a patient and honor the advance directive.

Assessment: This is a dynamic situation requiring ongoing reevaluation. The moral assessment of this dilemma is that the physician has a duty to save lives in spite of the paper declaration banning ventilator assistance. The physician's assessment is that the present medical emergency is not one that was anticipated at the time the patient made the declaration and that the intent was to prevent prolonged ventilator use in a circumstance unlikely to lead to recovery.

Action: Continuation of treatment is the correct action until the hospital advises otherwise, while simultaneously paving the way for ethically sound treatment. Active communication with others who might have had similar problems may provide guidance for the next steps. Introducing the sister to the hospital attorney should be considered. A search for other relatives and close friends may yield evidence regarding the patient's wishes as well as the sister's motives. Initiating contact with the court through proper legal channels will facilitate a court decision if the case reaches that point. While pursuing these actions, the patient's treatment might result in a return to consciousness or to a worsening in her health status. Under the latter circumstance, the ethics committee as well as the court may need to take action and possibly honor the advance directive.

There may be a situation in which alienation has occurred in the relationship between the appointed health care advocate and the patient, as in the case of a bitter divorce. In such an instance, the former spouse might make a decision that would be biased and not in the patient's best interest. The physician can override such a decision.

One proposed amendment to advance health care directives is to limit the time such a document is binding, but usually the decisions specified in the directive have been carefully thought through and are unlikely to change. The assumption is that the individual is sufficiently able to assess accurately the type of treatment desired in a future scenario of incompetence, and that the directive is, therefore, ethically enforceable. A possible drawback of the advance health care directive is that an individual has not accurately assessed possible situations that could occur in the future, or that the future scenario of incompetence is a condition that may be medically reversible. In situations where incompetence is temporary, family members insisting on previous mandates for "incompetence" can make treatment more difficult. In such cases, the mandate to maximize autonomy and minimize harm would be detrimental to the patient if it leads to further complications or death that could have been avoided (Ryan 1996). Respect for autonomy requires that individuals should be permitted to make risky choices about their own lives as long as these do not impinge on others (Luttrell and Sommerville 1996). Recall the example of Case 7.10, in which the improvements in quality of life of Alzheimer's patients were not anticipated by Grandmother 24 years before the accident.

Arguments against Advance Health Care Directives

Sometimes advance directives are formulated so far in advance of incompetence that there is a disconnect between the person who prepared the directive and the person on whom the document is being enacted (Kuczewski 1994a). Over a period of intervening years, there may be strong evidence that the goals and/or philosophy of the individual have changed to the point that the directive no longer represents how the person would want caregivers to respond. In addition, since advance directives cannot anticipate every situation, designated decision makers may lack a sufficient respect for and understanding of what the incapacitated individual would want, raising questions about the usefulness of the advance directive (Kuczewski 1994b). Furthermore, as conditions considered hopeless at one time in the past might become treatable, temporary life support and application of the new treatment might lead to recovery. Because of these objections to their use, arbitrary evaluations not contained in the original act of 1991 may be required.

Sacrifice of One Life to Save Another

Sometimes a person, a family, a physician, or the courts must decide if one person should be sacrificed to save another. This problem may arise during a multiple gestation pregnancy and is often a problem when considering the separation of conjoined twins. For example, if conjoined sisters who share a single heart are separated in the absence of a transplanted heart, the sister without a heart will surely die. Presuming that the continuance of conjoined existence would be fatal to both sisters, can one rationalize the killing of one to save the other?

Some sets of conjoined twins grow to adulthood without separation. However, if it appears that both will die if separation is not attempted, or if there is a demand for separation (as in a recent case of Iranian conjoined twins), then the morality of separation must be investigated before endangering the life of one or both twins. All options must be examined as precisely as possible with regard to the twins' chances for continuing health if not separated, as well as the chance for life and possible disabilities for each individual twin if separated. Both of the joined individuals must consent to separation if each is cognitively capable. Of importance is the requirement that both individuals be fully informed of the consequences of the separation, including the expected quality of life after separation. Two cognate individuals might wish to remain together and die together rather than have a separation with one dying and the other having a psychologically or physically miserable life. Moral arguments for performing a separation operation in which one twin will die are made in the example below.

Case 7.18 Separation of Conjoined Twins

Jodie and Marie were British conjoined twins who shared a single heart and could not be separated so that both would live. Because the heart sustaining the twins was growing weaker, the surgeons felt it would be unethical not to operate and save one, rather than allow both to die. The parents could not bear to sacrifice one child for the other and refused to give permission for the operation. The surgeons, who knew one twin would die, feared being arrested for murder if they proceeded with the surgery without approval from the court. Ultimately, the court decided in favor of the surgery that resulted in the death of one twin, but not without wrestling with the moral issues of the case.

The English courts allowed parental wishes to be overruled. In this case, the high court justices rationalized saving one by an act that would necessarily kill the other was a decision "to choose the lesser of two evils." Though the three justices agreed with the decision to separate the twins, they differed in their reasoning, and interestingly, the analysis of this unprecedented case drew not on ethical principles, explicitly, but rather looked for legal rationales (Annas 2001). For example, the conclusion of the court was: "She would die, not because she was intentionally killed, but because her own body cannot sustain life" (Lord Justice Robert Walker cited in Rachels 2003a).

Two alternate and plausible reasons also support the actions taken in this case. First, one must look to the dictum that one should make the most happy or save as many as one can. To do nothing would cost two lives; to do the separation would cost only one life. James Rachels (2003b) presents the second reason in his argument for an exception to the Sanctity of Life:

> In rare situations, it may be right [to kill]. In particular, if (a) the innocent human has no future because she is going to die soon no matter what; (b) the innocent human has no wish to go on living, perhaps because she is so mentally undeveloped as to have no wishes at all; and (c) killing the innocent human will save the lives of others, who can then go on to have good full lives—in these rare circumstances, the killing of the innocent might be justified.

What are the ethical concepts involved in such cases? Concepts of justice, virtue, and rights seem inapplicable, unless the decision is made to let both live until the course of events takes both lives through natural processes. Kantian duty provides no guidance, as there is a duty to protect each individual. Utilitarianism could be used in the context of doing whatever will give the most happiness or the least pain, but even this principle does not lead to a clear decision when all of the stakeholders are considered. In the case of the conjoined twins, the parents were also stakeholders, and in this particular case, the parents did not want the twins separated despite the advice from the physicians, which confounded the situation further.

Another situation related to the issue of killing one to save another, now famous in mountain climbing lore, occurred when two climbers were at risk of both dying. While tethered together during a treacherous descent in a blinding snowstorm, one of them slipped and fell over a cliff. Although the tether held by other climber broke his fall, he was left dangling above a crevasse of unknown depth. He was unable to gain a foothold and, due to the howling wind, was unable to communicate with the climber holding the tether. The supporting climber was unable to retrieve his partner despite numerous attempts, and the supporting climber was himself in jeopardy of dying as he was slipping toward the edge of the precipice. The supporting climber cut the rope and saved himself, letting the other fall into the crevasse (Simpson 1989). This true story presents to both climbers a major moral dilemma between the duty to survive and malfeasance. Under somewhat different circumstances, the decision to cut the rope might be that of the fallen climber facing the void, in which case he would also be cutting the rope to save the life of the other climber. Thus, the choice of who should die so the other might live, and who has the authority to make that decision, is not easy in specific cases.

Consider the possibility that the dangling climber decides to take his chances with a fall, and does the best he can to maneuver into a favorable position while cutting the rope. Here, the moral thinking is not to sacrifice one's life for another but to do the best one can to save oneself, which in this scenario would save the supporting climber as well. In fact, the dangling climber fell into the crevasse when the rope was cut, but he lived and managed to make his way out to safety.

A more common example of sacrificing one life to save another occurs during pregnancies complicated by medical crises, as illustrated in the case below.

Case 7.19 Fetal Death to Save the Mother

In the seventh month of an otherwise normal pregnancy, the mother develops eclampsia (severe edema and high blood pressure during pregnancy that can lead to coma and death). The usual treatment is bed rest and medications, but if these fail, delivery of the pregnancy usually leads to rapid resolution of the crisis. In this case, the mother's condition is not adequately controlled by drugs, and continuation of the pregnancy for even a few days might result in the death of both mother and fetus. The fetus is of normal size and position, and by ultrasound appears entirely normal but too small to live if delivered at this early stage of gestation. The doctors recommend that the mother abort the fetus, but she refuses. Eventually, the mother becomes comatose, and the hospital ethics committee concurs with the physicians and the family in the decision to deliver the fetus. The mother lived and the fetus died.

The moral problems encountered in this example include the autonomy of the mother, the autonomy of the fetus, and the challenge to the sanctity of life. Rachels (2003b) helps us resolve these issues regarding the sanctity of life: ". . . the innocent human has no future because she is going to die soon no matter what . . . killing the innocent human will save the lives of others, who can then go on to have good, full lives." However, the medical ethics of proceeding with an abortion against the will of the mother requires more analysis. While the mother was capable of deciding her fate and that of the fetus, the doctors and ethics committees felt they must respect her autonomy and allow her to continue the pregnancy. Once a comatose state ensued, the hospital ethics committee elected to deliver the fetus to protect the life of the mother. This action on the part of the ethics committee may have been based on the supposition that the mother would have changed her mind had she been conscious at the time the medical condition had deteriorated to the extent that coma and death for both mother and fetus was inevitable. Others might have decided differently.

Physician-Assisted Suicide

This section addresses the conflict between a patient's desire to die, the social tenets against suicide, and the medical professional's duty to do no harm. This text does not seek to reach definitive conclusions of right or wrong regarding these issues, but to present questions and examples to point out some of the ethical dilemmas. To what extent does autonomy grant the individual the right to determine the time of death? What is the duty of the physician to prevent suicide? Does a physician have a duty to help a patient reach a peaceful end of life? The historical points in Table 7.1 help put physician-assisted suicide in perspective.

Table 7.1   Historical Perspective on Physician-Assisted Suicide

~370 BC	"I will not prescribe a deadly drug to please someone, nor give advice that may cause his death" (Hippocratic Oath).
1516 AD	A patient should have the option to die by opium or by starvation (Thomas Moore in the book *Utopia*).
1650s	Confiscation of estates of suicide victims (Providence Plantation Colonies)
1828	New York State statute outlawed assisted suicide.
1935	Voluntary Euthanasia Society founded in England
1936	King George V with terminal cancer killed by morphine from Lord Dawson.
1997	U.S. Supreme Court decided the Constitution does not provide a right to physician-assisted suicide.
1997	Oregon passed the Death with Dignity Act
1999	Dr. Jack Kervorkian convicted of second-degree murder in Michigan for assisting a suicide.
2001–04	U.S. Attorney General Ashcroft unsuccessful in restricting states' rights to decide on physician-assisted suicide despite repeated appeals.
11/09/05	Ashcroft appeals the case to the U.S. Supreme Court.
02/22/05	U.S. Supreme Court agrees to hear the appeal during its next term.
01/17/06	U.S. Supreme Court upheld Oregon's physician-assisted suicide laws.

In 1997, the U.S. Supreme Court addressed the question of which laws apply to physician-assisted suicide and whether death is a fundamental right. Further, the Court addressed whether laws against physician-assisted suicide unduly burden personal liberty or go too far to remove personal liberty. The principal argument presented in favor of physician-assisted suicide was that if it is permissible to withhold or refuse treatment (for example, chemotherapy), why is it not permissible to help someone die? The Court decided that physician-assisted suicide is not upheld by the Fourteenth Amendment, which prevents all states from depriving any person of life, liberty, or property without due process of law. The difference is in taking action to help the patient to die versus letting the patient die without such assistance.

However, what is illegal may or may not be ethical. If a law states that it is illegal to help a patient die, is a physician acting unethically in helping a terminally ill patient who is in great pain and without hope of cure to die? The fact is that it is happening. Most commonly, a prescription for a cocktail of sedatives is provided, either to the patient or to the family, along with information as to what the lethal dose would be. Another approach is causation of unintentional drug-induced respiratory death by escalating doses of pain medications. Physician-assisted suicide by lethal injection (such as provided by Dr. Jack Kervorkian) is rare.

One of the main arguments against physician-assisted suicide is that it leads to a slippery slope. The slippery slope argument assumes that if one takes the first step in making a decision that may have ethical consequences, it encourages actions or decisions that lead to consequences far worse than intended at the time of the original step. In the case of physician-assisted suicide, the fear is that if one takes the first step to assist a terminally ill patient escape great suffering through a merciful death, might it then be permissible to do the same for patients less ill but suffering nevertheless? Then what? What about a patient in a prolonged coma or a persistent vegetative state? Or a patient with amyotrophic lateral sclerosis or Alzheimer's disease? Where will one draw the line? In addition to the fear that physician-assisted suicide will escalate alarmingly, there are fears that its use will be extended to severely disabled patients, to patients unable to speak for themselves, to patients who might benefit from palliative care, to severely premature neonates whose many months of intensive care and use of limited medical resources can cost up to $1 million (Rosner 2004), or to patients who have not requested it. Others fear that palliative care interventions would cease to evolve or that a worst-case scenario would recur that is similar to Hitler's program of euthanasia for the "mentally incompetent" of the late1930s. At that time, Hitler supplemented the ongoing sterilization program with euthanasia and is reported to have killed 50,000 incurably insane, feeble-minded, and deformed patients in gas chambers (Haller 1963). This preceded the systematic effort to exterminate the Jews of Europe.

Opponents argue that in Oregon where physician-assisted suicide is legal, no evidence of abuse has been observed since it went into effect. A physician may issue a prescription for a lethal amount of drug, but only after a patient presents one written and two oral requests (the latter separated by 15 days).The patient must be sufficiently capable to communicate these decisions and to ingest the medications independently, and must have been informed of all possible alternatives to care. In the first three years, 140 patients received prescriptions for lethal medications, and 91 of these patients died as a result; further, palliative care appears to have improved (Steinbrook 2002), contrary to what opponents predicted. In the rest of the world, the only country in which physician-assisted suicide is legal is the Netherlands, where, after having been practiced for decades, the law permitting it was passed in April 2002.

The view of practicing physicians around the world concerning assisted suicide is one of helping patients die in an optimum social and family environment without fear, without remorse, and without pain. During the dilemmas arising during the last days of a patient's life, medical practice draws on many elements of our design strategy. The presence of advance health care directives has eased the decision-making process during this time. Still, it is estimated that only 10% of the population has taken advantage of this legislation, as reported by the AMA in 2004, although this figure is probably higher in 2006 because of the Terri Schiavo case. Will these directives include requests for physician-assisted death?

Many would argue that there should be no need for legislation or for the arguments around legislation regarding physician-assisted suicide. However, when there seems to be no relief for patients suffering from a disease that requires frequent painful and even humiliating treatments that render them in utter despair, one seeks some way to help the patient end this misery. Sedation is not a solution, nor is psychiatric drug therapy likely to help their day-to-day unbearable experiences. Unfortunately, these types of situations occur commonly in nursing homes, and the ethical dilemma is clear: it is the autonomy of the person who wishes to die vs. the duty of the caregivers to keep the patient alive.

Case 7.20 Unbearable Misery of a Terminally Ill Patient

Mr. A is 82 years old and has been in a nursing home for three years. He has lower-back problems that cause intermittent severe pain, as well as muscle spasms and urination problems including incontinence. He has incurable prostate cancer, for which he has refused chemotherapy. Mr. A has had pneumonia three times in the last six months, and each time he is brought to the hospital for treatment, then returned to the nursing home. Because catheters are used by the nursing home, bladder infections occur that require several days of hospitalization for intravenous treatment. Mr. A is not in constant pain but is so miserable that he pleads with the nurses and his family to help him die. The family physician cannot in clear conscience give morphine in excess of that needed for the back pain, so he calls on a psychiatrist to treat the despair and depression. Suppose that you are asked to give advice to the family.

Acquire Facts: In getting the facts, you ascertain that on the one hand, Mr. A is in an intractable situation of discomfort and despair, and wishes to induce death rather than wait for natural causes; however, his physician is uncomfortable with assisted suicide. You also learn that the nursing home personnel feel it is their duty to treat his symptoms aggressively by rushing him to the hospital for care of his recurrent bouts of pneumonia and urinary tract infections. Had the nursing home personnel had not reacted so aggressively, Mr. A would have died long ago.

You also learn that Mr. A has two daughters and a son who live about 90 miles from the nursing home. One daughter lives on a large farm, which was previously the home of Mr. A. He longs for the comfort of this farm, but the family has not considered this as a possibility. It may be that the family feels his medical problems are too complex for them or even a live-in nurse to manage. But are there resources that could be provided at the farm that would allow care to be given there? Advice from the attending physician reveals that Mr. A is unlikely to live through another bout of pneumonia or another urinary tract infection without hospitalization

Alternatives: If it is Mr. A's intention to die peacefully at home, would this be an option? Mr. A should be asked what he wants other than "to die." How frequently does the family visit? Get input and advice from other caregivers. Four alternatives might be considered. First, one could write orders for the nursing home to provide only supportive care during the next acute illness. Second, the family could take Mr. A home with nursing support (e.g., catheter care) and let him live out his life surrounded by family with whatever medications are indicated for depression and pain. Third, one could add to his medical care by also giving prophylactic medications to prevent the infections from occurring. Fourth, one could consider moving him to another nursing home closer to the family.

Assessment: The dynamics of this situation largely depend on how the family comes together to give Mr. A what he wants most of all: the love of his family and a move to the farm. Alternatively, more frequent family visits might be possible if Mr. A were moved to a more accessible nursing home. By discussions with Mr. A and family members, the effectiveness of alternative care can be assessed.

Action: Gather the family members, the physician, and the nursing home caregivers to discuss the alternatives. Consider that the orders not to treat will be unacceptable to the staff at the current nursing home. Consider moving Mr. A to the family farm, even if this means certain death within a few months. Approach Mr. A about where he would like to be, ascertain what aspects of his situation are most distressing for him, design options that would best help alleviate this distress, work with the family to provide support for their efforts, and review Mr. A's situation with him and his family periodically.

Summary of the Ethics of Physician-Assisted Suicide

At the outset of this section, we posed questions regarding the right of the individual to end his or her suffering and the duties of the physician or others in helping an individual die when living becomes intolerable and death inevitable. Certainly one cannot argue with an individual's right to ask for relief from suffering. So, what is the physician's action when faced with the prima facie duty to help a patient, if the only help appears to be in prescribing drugs that will end the patient's life? Most will agree that in addition to the seriousness of the disease, there are family, social, environmental, financial, and other forces that can contribute to the patient's state of despair. A step-by-step application of the design strategy will often help identify ways to avoid the need for assisted suicide. For example, one might discover that nursing home personnel have been overly aggressive in applying invasive treatments, which, though sustaining life, might be changed to allow a natural death.

An important distinction can be made between an act that will likely lead to the death of an individual and an act that is motivated by the intent to kill an individual. If an Inuit Eskimo who is elderly and a burden to the family group walks off into the Arctic night in order to give others a better chance to survive, this act is not considered immoral or even a suicide. Would one judge the Inuit's family as having assisted suicide by not searching for the elderly member? When an individual dies trying to save another, it is considered heroic and not suicide. Self-killing is suicide only when the intent is to die. Sacrificing one's self for others is not suicide. The relevance of these observations is that sometimes, patients refuse treatment or rescue because they feel virtue requires them to no longer burden the family. Their rights should be respected. These are some of the ethical considerations associated with physician-assisted suicide that have not entered fully into the current debate.

RESOURCE ALLOCATION AND ORGAN TRANSPLANTATION

Chronic organ shortages plague the thousands of patients awaiting transplantation, necessitating allocation of these scarce resources to a limited number of patients who need them. Allocation of these scarce resources is based on a number of complex factors. Thus, allocation protocols have developed that involve ethical concepts of justice and utilitarianism. Each transplant candidate enters a transplant pool of the nationwide transplant network, and as a donor becomes available, each patient is then evaluated based on the individual's unique situation. Due to these multiple factors, it may not be the patient in greatest medical need who receives the organ in short supply. In this section, we review the extent of the limited resources, using organ transplantation as an example, and discuss some of the ethical issues that have arisen as a result.

- Is it ethical to buy or sell organs? What constitutes undue inducement if someone with little or no money is offered a large sum for his or her kidney or some other organ? Is it right and just that organs go to only those who can afford to pay?
- Fetal transplantation, that is, the transplantation of tissues or organs from an aborted fetus, may be able to meet some transplant needs. What might be the consequences of this? What are the ethical issues of this approach?
- What if a dying patient is about to be disconnected from life support, but the person to whom the heart/lung transplant is supposed to go is not immediately available or needs time to stabilize a medical condition pretransplant? Is it acceptable to postpone disconnecting life support until the recipient is available? Who should decide? Who should pay for the cost of this?
- Should organ distribution rules be local, regional, or national?

The reason that organs for transplantation are in exceedingly short supply is that there is an increasing need for organs. For example, over 550,000 new cases of heart failure are reported each year. In the absence of adequate ways to treat this condition, at least one-third of these patients could benefit from a heart transplant. However, only about 2,500 hearts are available for transplant. Table 7.2 indicates the number of patients awaiting transplant as of October 23, 2003, according to data from the Organ Procurement and Transplant Network (OPTN at www.unos.org). The most striking figure is that of the almost 83,000 patients on the transplant waiting list; only about 15,000 will receive the transplant they need.

As of July 12, 2005, there were nearly 89,000 transplant wait list candidates, nearly 80% of whom had not received a transplant. The organization of a national network for overseeing, coordinating, and allocating available organs that become available has been a major step forward in assuring fair organ allocations. The United Network for Organ Sharing (UNOS) formulates policies that strive for equitable organ allocation to patients

Table 7.2   Transplant Waiting List October 2003

All*	82,958
Kidney	56,062
Pancreas	1,451
Kidney/Pancreas	2,398
Liver	17,231
Intestine	168
Heart	3,558
Lung	3,889
Heart/Lung	184

(All* candidates will be less than the sum because some are awaiting multiple organs.)

registered on the national waiting list. This group coordinates donor procurement for over 1,000 donor hospitals and 255 transplant centers that make up the OPTN.

UNOS stores data for all patients on a computerized database, and only when an organ becomes available is a ranked list of eligible recipients generated that takes into account information about the donor and each individual patient's situation. In this way, the individual factors unique to each patient's situation are assessed against all other patients awaiting that type of organ. Factors entering into the decision about who receives an organ are summarized in Table 7.3.

After printing the list of potential recipients, the procurement coordinator contacts the transplant surgeon caring for the top-ranked patient (i.e., patient whose organ characteristics best match the donor organ and whose time on the waiting list, urgency status, and distance from the donor organ adhere to allocation policy) to offer the organ. Depending on various factors, such as the donor's medical history and the current health of the potential recipient, the transplant surgeon determines if the organ is suitable for the patient. If the organ is turned down, the next listed individual's transplant center is contacted, and so on, until the organ is placed (www.optn.org/about/transplantation/matchingProcess.asp, accessed July 12, 2005).

Because the availability of organ donors does not nearly meet the demand, it is important to be aware of various options. These include the use of an animal organ, cadaver organ, artificial organ, an organ from a donor who has recently died, or an organ donated by a living donor, or a bridging or temporary support therapy. Ethical issues surrounding their use complicate some of these alternatives, whereas others are less controversial. Figure 7.2 outlines these options.

Despite the numerous possibilities, a number of patients are still unable or unwilling to opt for alternative procedures and go on to die of their disease. These deaths might be avoided in the future if advances in stem cell research live up to expectations. As will be discussed in Chapter 10, it may be possible in the future for organs to be fashioned from pluripotential embryonic stem cells (cells with the capacity to form any tissue under the proper conditions, or from autologous adult stem cells taken either from the patient or from an unrelated tissue-matched donor) capable of replacing damaged tissue. It may also be curative to inject stem cells into or around the diseased portion of the organ. However, the extent to which stem cells will prove useful is still being decided, so for now, issues relating to the short supply of organs for transplantation need to be addressed.

Table 7.3   Factors Affecting Ranking of Potential Transplant Recipients

- Age
- Blood type (some types are rarer than others)
- Tissue type
- Medical urgency
- Waiting time
- Geographic distance between donor and recipient
- Size of the donor organ and the size of the recipient
- Type of organ needed

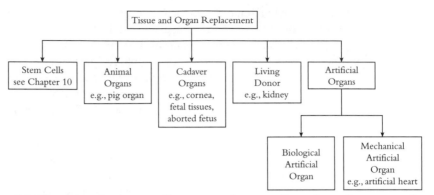

Figure 7.2 The options for those needing organ replacement. Progenitor cells or stem cells can be from the individual undergoing transplant, in which case it is called autologous (e.g., autologous bone marrow transplantation), or from an unrelated but tissue-matched donor (allogeneic transplant). When necessary, organ function can also be achieved by artificial organs, in which case the patient's blood is pumped through an animal organ or artificial organ.

Ways to Maximize the Availability of Scarce Resources

One way to better meet the need for organs is to maximize their availability through advance health care directives, which provide a place to indicate the desire to donate one's organs. The opportunity for donation can also be increased by designating this wish on a driver's license, and by raising the awareness of the need for organs through outreach to the general population. The criteria used to determine death might also be broadened (for example, to allow organ donation from patients in persistent vegetative states). The use of life support could permit organ donation at a future time from those who otherwise face imminent death, though this is still being discussed. Another controversial area is the donation of organs from executed criminals, a procedure that has been implemented in China, according to an article in the *Washington Post* (Mufson 2001). Finally, people can be encouraged to be attentive to their health (e.g., control diabetes, hypertension, alcoholism) and thereby lessen the need for transplants.

To resolve issues of equity and suspected injustices in the allocation of livers, the Institute of Medicine (IOM) investigated the situation and made recommendations to the U.S. government. The IOM panel reviewed 68,000 records relating to all patients on liver transplant waiting lists from 1995 to 1999. The report found no disparities in being accepted on a wait list or in access to transplantation for low-income patients with access to health insurance and health services, once referred for transplantation. A synopsis of the recommendations in this report is given in Appendix 7.2.

Organ transplantation is expensive. Not all health care insurance plans pay for transplants. The cost to the recipient or his family can be monumental, with a kidney transplant costing in the range of $30 to 35,000, a heart transplant around $100,000, and a liver or bone marrow transplant $150,000. These costs double or triple if complications ensue and post-transplant hospitalization is prolonged.

Ethical Considerations of Organ Transplantation

We return now to some of the ethical dilemmas raised at the beginning of this section of the chapter, dilemmas that are not easily resolved by any one moral theory such as autonomy, justice, or utilitarianism. Before the inception of UNOS, the assignment of an organ to a specific patient was determined at one time or another according to chance, lottery, first-come-first-served, worthiness, or need. While the new allocation policies are fairer, there are other questions that may still present ethical dilemmas.

Use of Fetal Tissues

A human fetus is a human in the stage of development from eight weeks after fertilization until birth (prior to that time, shortly after fertilization up to approximately eight weeks postfertilization when all adult structures can be visualized, it is referred to as an embryo). Fetal tissues are generally obtained from spontaneous abortions or from induced abortions of either unwanted pregnancies or pregnancies conceived for therapeutic or research purposes. Some have advocated that it is ethical to use organs from an aborted fetus for transplantation. While this might be a way to meet some transplant needs, the use of fetal tissues or organs might have severe consequences, depending on the health of the fetus. Common causes of abortion, other than the mother's choice, might not be apparent at the time of transplant but might compromise the recipient, such as congenital abnormalities or the presence of a metabolic abnormality. A defective fetus is often miscarried. The defect could be genetic, and if so, the use of these tissues would transfer unnatural or defective human cells that might be detrimental to the function of the transplant.

Confusion in debates on the use of fetal tissues, for either transplantation or research, stems from the different moral perspectives surrounding the distinctly different categories of a fetus. For example, the age of the fetus has a bearing on the potential for independent life. The viability of a fetus when aborted, and the reason for the abortion, are other conditions relevant to moral reasoning in the dispute over fetal tissues. The underlying moral thinking about these issues centers on the interests of society in medical advances, the need for consent, the need of individual patients for the tissues, and the autonomy of the fetus, including the imagined interest of the fetus in being left alone.

The use of fetal tissues shows promise for treating diseases such as diabetes and neurodegeneration (Alzheimer's and Parkinson's diseases). However, critics vehemently oppose the practice. The basis for the arguments against the use of fetal tissues in medical research and therapeutic trials is that since abortion for medical or social reasons (e.g., rape or congenital abnormality) is morally unacceptable, then the use of tissues from the abortion is unacceptable. Some point out that the fetus is the smallest subject, least able to protect itself, unwanted, and often without someone to protect it; others argue the fetus must be treated like any other member of the human race (Annas et al. 1977). However, while there might be fewer ethical concerns about the use of fetal tissues obtained from spontaneous abortions, one cannot rely on retrieving them in a manner timely enough to preserve tissue viability and sterility. The principal ethical problem is whether the fetus is a person with rights and whether the fetus has autonomy at the same level as the fetal tissue recipient. While the use of fetal tissues remains highly experimental, this approach will undoubtedly demand more attention in the next few years, especially if it is confirmed that these tissues are less immunogenic, faster engrafting, and/or better tolerated than more mature organs and tissues.

Transplantation Procedures for Prisoners

What if a prisoner with a life sentence for murder is acutely ill and needs a transplant? Should he get it? How can this be justly decided? There is no easy solution, so those involved must look for alternative deciding factors. One might ask the prisoner if he wants the transplant (the prisoner might decide it is not the right thing to do); an additional alternative is to contact other prisons to see how they handle similar problems. Ideally, a prison will have procedures in place regarding such problems so that a definite plan of action can be anticipated in advance.

Buying and Selling of Body Tissues and Organs

Is it ethical to buy or sell organs? What constitutes undue inducement if someone with little or no money is offered a large sum for a kidney or other organ? Is it right and just that organs go to only those who can afford to pay? In many cases, the cost of a transplant is not covered by insurance and is so expensive that accessibility is out of reach unless a patient's community rallies to provide the necessary funds. The current trafficking in organ transplantation presents ethical challenges when a person receives an organ by paying for it to bypass his or her low priority on the waiting list.

Because the sale of human organs has been deemed ethically irresponsible in most countries, including the United States, it has been widely banned. However, in some countries, including Turkey, India, China, and Russia, organs may be obtained for a price, openly and without government interference. Some surgeons have reaped large financial gains from the underground business of transplanting organs to patients who can afford to pay large sums for them. There are even brokers who locate available organs and arrange the details of the transplants.

Sometimes, deviating from conventional transplant channels offers the best option. Most cadaver transplants are taken from trauma patients in whom the kidney may have been damaged. A living-donor kidney recipient has a reasonable chance for a successful transplant lasting over 20 years, whereas with a cadaveric transplant, the expectation is about 10 years. In some countries, patients have no choice except to patronize foreign markets to obtain organs. In Israel, the hope of obtaining a donated organ is small (fewer than 3% of Israelis have signed donor cards) due to the belief in Judaism that the body must be buried intact so it will be whole at the resurrection. Since many Israelis are going elsewhere to receive living donor transplants, Israeli physicians are beginning to see the benefits of legalizing the purchase of organs (Finkel 2001a).

The financially destitute are often enticed by the prospect of financial gains from organ donation. While compensation for organs may seem large to poverty-stricken donors, they receive only a small portion of the fees charged the recipients. The majority of the money goes to the surgeons who are willing to risk their reputations to perform the surgeries, while the rest goes into transportation of donors, recipients, surgeons and nurses to and from the foreign hospital, as well as broker fees (Finkel 2001b). Most of those condemning this practice see it as a slippery slope that can lead low-income donors to be seduced by the inducement of payment for organs. Concerns have led to the establishment of an organs watch center following a 1997 report from the Bellagio Task Force on Securing Bodily Integrity for the Socially Disadvantaged in Transplant Surgery. The task force concluded that trade in human body parts was taking advantage of deprived people, paying them only a pittance for their organs and putting them at risk for medical complications of the surgery as well as for abusive, exploitive, or corrupt practices (www.centerspan.org/tnn/99111502.htm, accessed February 15, 2005).

This practice of paying for organs favors the wealthy, and the distribution of organs is no longer equitable according to need. The practice bypasses the doctrine that the only fair way of deciding who gets organs is to create a situation in which everyone has a fair chance to receive them, which is largely the case in the United States. (http://books.nap.edu/html/organ_procure/organs.pdf, accessed February 15, 2005).

Distribution of Care When Resources Are Limited

Limited availability of funds to provide medical resources for all peoples of the Earth is leading to moral problems of equity and justice. It is predicted that in the United States, the costs will reach 15% of the gross national product. Limited resources for general medical care and the details of their distribution are poised to be at the heart of major ethical and public policy debates by 2020, and the discussion of the tough decisions and policy determinations must begin now. Of the 77 million Americans who are retired, about 25 million have no retirement funds to supplement their income from Social Security. Most predict that Medicare and Social Security cannot be sustained under the current funding programs when the retiree population increases. The increase in retirees will come from those born during the Baby Boom years of the 1940s, and their increased numbers will stress an already compromised health care system. Who will decide who will receive care and who will not?

The current practice in Europe is an example of the consequences when measures to respond to these dim predictions cannot be found. An aged man with no financial support entering the hospital with a serious medical condition (e.g., stroke) is sent home to die with no further medical input. Legislators have been reluctant to take on these issues. The two major reasons for this inaction are politics and the lack of practical mechanisms for preemptively addressing the issues. In accord with the moral philosophies of this book, professional societies and appointed government officials have a duty to be proactive in implementing solutions to delivering medical care to those in need. The problem is how best to distribute limited resources.

Some solutions might include:

- Elect officials who are willing make nonpartisan decisions regarding reforms in Medicare and Social Security and who support health care research and health care industries to maintain and improve the standard of care for the aging.

- Implement a qualitative change in health care delivery, using currently available engineering and medical science innovations (e.g., biomonitoring, home care delivery systems, patient-doctor communication systems, detection of early disease).

- Seek methods that will reduce the costs of pharmaceutical development.

- Examine methods and incentives to encourage better personal health care.

- Discover the mechanisms of prevalent diseases for which there are currently ineffective therapies (e.g., heart failure, hypertension, depression), and enable treatment for diseases with known mechanisms but no cures (diabetes, HIV, other viral diseases).

- Nationalize health care. The current projections of the costs for health care show their funding will require an unacceptable percentage of the gross national product by 2020. Improvements in health care technologies are not decreasing these costs; in fact, these technologies are increasing the costs (most are very expensive). In addition, they are prolonging life, leading to an ever-increasing need for health care. Frequently, situations occur wherein lifesaving pharmaceuticals are not in sufficient supply to meet the needs of an epidemic, as illustrated in the case below.

- The problem becomes one of a global nature when we ask if equity should apply to all human beings.

Case 7.21 Limited Resources for Protection of Life

Suppose that Company A has only 50% of the amount of the new medication needed to treat all individuals exposed in an acute epidemic. The utilitarian would argue that it is better to make a majority a little bit better off or make a few very happy and give two half-doses to 50% of those exposed, whereas others might consider an even distribution to all might be more just or fair. (Note: We consider that justice and fairness are equivalent, as does Rawls [1999].) This case is an example of one of the most significant ethical problems in health care, the allocation of limited resources.

Each problem of allocation of limited resources requires consideration of the theory of justice when deciding who should receive a benefit when not all can receive it. What can we gain from applying the Four A's design strategy to the above case? First, by gathering more facts, including consultation with the international community, new approaches and alternative solutions might come to our attention. Second, by consideration of alternatives, we might find a virtuous middle road that could provide some hope for those who do not receive the curative dose. But suppose that we are left with no alternatives other than the two posed in Case 7.21? How do we select who should receive the treatment dose?

The first choice is to draw lots, that is, to use random selection; a second choice is to select those to be treated based on need. Some combination of the two alternatives may be required if, for example, many children and infants would be without parental support if parents are not also treated. Lots might be drawn for only the children, and if chosen, the parents would also be chosen. However, this might pose problems of injustice as parents with multiple children would have an advantage over parents with a single child. Yet, what are the other alternatives? Through discussion and analysis of the consequences of various lot drawing procedures, a method will finally be agreed upon. In the end, the process may involve a just distribution of lots, leaving to the community the burden of caring for infants and children whose parents are not treated. The driving force for this ethical problem is justice, and corporate involvement is only as the supplier of the drug. There were other issues in the case of the flu vaccine shortage of 2004–2005, in which there were a number of stakeholders, including the corporation manufacturing the vaccine, physicians, public policy makers, and the patients.

Case 7.22 Flu Vaccine Shortage of 2004

In the autumn of 2004, a flu vaccine shortage in the United States resulted in the need for rationing this much sought-after medical resource. The concept has similarities to triage, whereby a physician designates treatment only for those most likely to benefit when there are insufficient resources to care for all; in this case, the patients most in need of vaccination were those most likely to die from contracting the flu. Physicians evaluating patients eligible to receive the vaccine in Massachusetts were confronted with three types of patients: those pleading for the shot, those who considered others might be in greater need, and those with a high likelihood that the flu would be life-threatening. Specific national and state directives regarding the distribution of the vaccine aided physicians' decisions. The Centers for Disease Control (CDC) directed that the vaccine was to be given only to those who would benefit the most (in this case, those who would suffer the most if not protected by the shot). Additionally, the Commonwealth of Massachusetts issued a notice that those physicians not following CDC guidelines could be fined $50 to 200, imprisoned for up to six months, or both. Physician-controlled vaccine rationing was effective and without incident because the reasons for the limited vaccine availability and rationing were clearly understood. Patients accepted the rationing, even though most of those seeking the shots were denied them (Lee 2004).

General guidance for allocation of limited medical resources is given in section E-203 of the American Medical Association's Code of Medical Ethics (Appendix 7.3). The major criteria for deciding the priority for treatment are likelihood of benefit, urgency of need, potential to change the quality of life, duration of benefit, and, in some cases, the amount of resources required for successful treatment. Considerations such as ability to pay are not to be used.

Treatment in Situations of Extreme Illness

The AMA code and local policies do not address the treatment of patients whose underlying disease state is so severe that they are unlikely to benefit from the allocation of the multiple personnel and expensive resources needed to save them. There are individuals whose treatment might lead to a lifetime of unacceptable misery, such as the severely burned patient who begs to die so as not to undergo further excruciating treatments or the very small premature infant with severe congenital anomalies. The moral duties to preserve human life and to enable individuals to be healthy members of society are accepted maxims when making life and death decisions for individual patients, and when making allocation decisions that affect a population as a whole. The physician should be guided by the basic ethical principles of beneficence, non-malfeasance, autonomy, and justice. However, these principles may come into conflict.

For example, if the autonomous rights of the premature infant to receive hundreds of thousands of dollars worth of treatment to survive are in conflict with the needs of the community, how do we decide the proper allocation of resources? If beneficence for an individual patient conflicts with the needs of another patient, how can we decide who should be treated? Each case deserves individual analysis based on facts, advice, and evaluation of alternatives.

Human Experimentation and Animal Experimentation

See Chapter 8.

CHAPTER SUMMARY

The medical ethics discussed in the first part of this chapter are professional ethics involving the delivery of general health care. The underlying ethical theories of rights and duty prescribe the ethical behavior of the physician or researcher. They include the requirement to respect the patient's autonomy at all times, to treat the patient as an individual, to do no harm, to balance the possible benefits of treatment with risks of possible outcomes, and to ensure that the patient is treated justly with equal access to health care and other benefits available to other members of the population. Autonomy with respect to genetic information is an ethical aspect of the emerging technology of genomics. The conflict between the importance of genetic information of an individual being made available to society and the individual's right to privacy is currently not resolved. The end-of-life decisions for an individual made by family, associates, and professional caregivers are difficult and often complex. Allocation of scarce resources is the major ethical problem of organ transplantation for which rational procedures have been adopted, but a satisfactory solution for increasing the supply of human organs might rest on the creation of alternative methods for tissue and organ regeneration, including stem cell applications (see Chapter 10). Other serious issues involve soaring health care costs, the millions who have no health insurance, and how to allocate health care with justice to all. This chapter endeavors to show how each particular situation can be approached.

REFERENCES

American Medical Association. 1994 (June). Code of Medical Ethics, Section E-8.12. Patient Information, based on Current Opinions of the Council on Ethical and Judicial Affairs.

Annas, G. J. 2001. "Conjoined twins—the limits of law at the limits of life." *N Engl J Med* **344**:1104–1108.

Annas, G. J. 2004 (May 27). "Forcible medication for courtroom competence—the case of Charles Sell. *N Engl J Med* **350**:2297–2301.

Annas, G. J., L. H. Glantz, and B. F. Katz. 1977. "Fetal Research: The Limited Role of Informed Consent in Protecting the Unborn" (Chapter 6). In: *Informed Consent to Human Experimentation: The Subject's Dilemma*. Cambridge, MA: Ballinger Publishing Co., p. 211.

Banach, J. 1988 (September). "The Uniform Health-Care Information Act: Current Status Part II," remarks by J. McCabe. *J Am Med Rec Assoc* **59**(9):23–25.

Barnes, R. E. 2000 (January). "Reefer madness: legal and moral issues surrounding the medical prescription of marijuana." *Bioethics* **14**(1):16–41.

Beauchamp, T. L. and J. F. Childress. 2001a. *Principles of Biomedical Ethics*. Fifth edition. New York: Oxford University Press, p. 177.

Beauchamp, T. L. and J. F. Childress. 2001b, p. 178.

Cruzan (by her parents and co-guardians) v. Director, Missouri Department of Health. Supreme Court of the United States, 497 U.S. 261 (June 25, 1990, decided).

Cruzan v. Harmon, 760 S.W.2d 408 (Mo.1988).

Dahlburg, J-T. 2004 (September 24). "Florida top court tosses law keeping woman alive." *San Francisco Chronicle*, p. A4.

Eichenwald, K. 2003 (August 12). "How one hospital benefited on questionable operations." *New York Times*, pp. C3–4.

Finkel, M. 2001a (May 27). "This little kidney went to market." *New York Times Magazine*, pp. 27–33, 40, 52, and 59.

Finkel, M. 2001b, p. 30.

Goodnough, A. 2005 (June 16). "Schiavo autopsy sides with her husband. No treatment could have helped her, doctors conclude." *San Francisco Chronicle*, p. A3.

Haller, M. H. 1963. *Eugenics*. Piscataway, NJ: Rutgers University Press, p.180.

Institute of Medicine (U.S.). Committee on Organ Procurement and Transplant Policy, Division of Health Sciences Policy. 1999. *Organ Procurement and Transplantation: Assessing Current Policies and the Potential Impact of the DHHS Final Rule.* Washington, DC: National Academy Press, pp. 1–15.

Josefson, D. 2001 (May 26). "US judges rule against medical use of marijuana." *BMJ* **322**:1270.

Josefson, D. 2002 (November 9). "US government cannot revoke licenses of doctors who recommend marijuana." *BMJ* **325**:1058.

Kuczewski, M. G. 1994a. "Whose will is it, anyway? A discussion of advance directives, personal identity and consensus in medical ethics." *Bioethics* **8**(1):27–48.

Kuczewski, M.G. 1994b, p. 27.

Lee, T. H. 2004. "Rationing influenza vaccine." *N Engl J Med* **351**:2365–2366.

Lin, Z., A. B. Owen, and R. B. Altman. 2004. "Genomic research and human subject privacy." *Science* **305**:183.

Luttrell, S. and A. Sommerville. 1996 (April). "Limiting risks by curtailing rights: A response to Dr Ryan." *J Med Ethics* **22**(2):100–104.

Markel, H. 2004a. "I swear by Apollo—On taking the Hippocratic oath." *N Engl J Med* **350**:2028.

Markel, H. 2004b, p. 2026.

Mufson, S. 2001 (June 27). "Chinese doctor tells of organ removals after executions." *Washington Post*, p. A1.

National Commission for the Protection of Human Subjects of Biomedical and Behavioral Research. 1979 (April 18). *The Belmont Report*. Part B (Basic Ethical Principles), Section 2 (Beneficence), paragraph 2.

Pellegrino, E. and D. Thomasma. 1988. *For the Patient's Good: the Restoration of Beneficence in Health Care*. New York: Oxford University Press, p. 29.

Rachels, J. 2003a. *The Elements of Moral Philosophy*. Fourth Edition. New York: McGraw-Hill Publishers, p. 7.

Rachels, J. 2003b, p. 8.

Rawls, J. 1999. *A Theory of Justice*. Revised Edition. Cambridge, MA: Harvard University Press, pp. 7, 11–15.

Rosner, F. 2004. "Allocation or misallocation of limited medical resources." *Cancer Invest* **22**(5):810–812.

Ryan, C. J. 1996 (April). "Betting your life: An argument against certain advance directives." *Med Ethics* **22**(2):95–91.

Savulescu, J. 1995. "Rational non-interventional paternalism: Why doctors ought to make judgments of what is best for their patients." *J Medical Ethics* **21**:327–331.

Simpson, J. 1989. "The Final Choice (Chapter 6)." In: *Touching the Void*. New York: HarperCollins Publishers, Inc., pp. 85–106.

Steinbrook, R. 2002. "Physician-assisted suicide in Oregon—an uncertain future." *N Engl J Med* **346**(6):460–464.

United States v. Sell, 539 U.S. 166, 2003.

U.S. Department of Health and Human Services. 1996 (August 21). Health Insurance Portability and Accountability Act (HIPAA). Public Law 104-191 (104th Congress).

U.S. Department of Health and Human Services. 2002. "Standards for Privacy of Individually Identifiable Health Information" (applicable to entities covered by HIPAA). *Federal Register* **67**:53181.

Virginia Natural Death Act. 1985. Virginia Code 54-325, 8:1–13.

Vollmann, J. 2001. "Advance directives in patients with Alzheimer's disease: Ethical and clinical considerations." *Med Health Care Philos* **4**(2):161–167.

ADDITIONAL READING

Andrews, L. and D. Nelkin. *Body Bazaar: The Market for Human Tissue in the Biotechnology Age.* New York: Crown Publishers, 245 pp.

Beauchamp, T. L. and J. F. Childress. 2001. *Principles of Biomedical Ethics.* Fifth Edition. New York: Oxford University Press, Inc., 454 pp.

Buchanan, A, D. W. Brock, N. Daniels, and D. Wikler. 2000. *From Chance to Choice: Genetics and Justice.* New York: Cambridge University Press, 398 pp.

Goodman, K. (ed.). 1998. *Ethics, Computers and Medicine: Informatics and the Transformation of Healthcare.* New York: Cambridge University Press, 180 pp.

Joffe, C. 1995. *Doctors of Conscience: The Struggle to Provide Abortion Before and After Roe v. Wade.* Boston: Beacon Press, 250 pp.

Jonsen, A., M. Siegler, and W. Winslade. 1998. *Clinical Ethics: A Practical Approach to Ethical Decisions in Clinical Medicine.* Fourth Edition. New York: McGraw-Hill Companies, Inc., 206 pp.

Kuczewski, M. G. and R. L. B. Pinkus. 1999. *An Ethics Casebook for Hospitals: Practical Approaches to Everyday Cases.* Washington: Georgetown University Press, 219 pp.

Pence, G. E. 2000 *Classic Cases in Medical Ethics.* Fourth Edition. New York: McGraw-Hill Companies, Inc., 470 pp.

Shannon, T. (ed.). *Bioethics.* 1993. Fourth Edition. Mahwah, NJ: Paulist Press, 542 pp.

Shenfield, F. and C. Sureau (eds.). *Ethical Dilemmas in Reproduction.* 2002. New York: Parthenon Publishing Group, 114 pp.

APPENDICES

Appendix 7.1 California Advance Health Care Directive

(California Probate Code Section 4701)

Explanation: You have the right to give instructions about your own health care. You also have the right to name someone else to make health care decisions for you. This form lets you do either or both of these things. It also lets you express your wishes regarding donation of organs and the designation of your primary physician. If you use this form, you may complete or modify all or any part of it. You are free to use a different form.

Part 1 of this form is a power of attorney for health care. Part 1 lets you name another individual as agent to make health care decisions for you if you become incapable of making your own decisions or if you want someone else to make those decisions for you now even though you are still capable. You may also name an alternate agent to act for you if your first choice is not willing, able, or reasonably available to make decisions for you. (Your agent may not be an operator or employee of a community care facility or a residential care facility where you are receiving care, or your supervising health care provider or employee of the health care institution where you are receiving care, unless your agent is related to you or is a coworker.)

Unless the form you sign limits the authority of your agent, your agent may make all health care decisions for you. This form has a place for you to limit the authority of your agent. You need not limit the authority of your agent if you wish to rely on your agent for all health care decisions that may have to be made. If you choose not to limit the authority of your agent, your agent will have the right to:

a. Consent or refuse consent to any care, treatment, service, or procedure to maintain, diagnose, or otherwise affect a physical or mental condition.

b. Select or discharge health care providers and institutions.

c. Approve or disapprove diagnostic tests, surgical procedures, and programs of medication.

d. Direct the provision, withholding, or withdrawal of artificial nutrition and hydration and all other forms of health care, including cardiopulmonary resuscitation.

e. Make anatomical gifts, authorize an autopsy, and direct disposition of remains.

Part 2 of this form lets you give specific instructions about any aspect of your health care, whether or not you appoint an agent. Choices are provided for you to express your wishes regarding the provision, withholding, or withdrawal of treatment to keep you alive, as well as the provision of pain relief. Space is also provided for you to add to the choices you have made or for you to write out any additional wishes. If you are satisfied to allow your agent to determine what is best for you in making end-of-life decisions, you need not fill out Part 2 of this form.

Part 3 of this form lets you express an intention to donate your bodily organs and tissues following your death.

Part 4 of this form lets you designate a physician to have primary responsibility for your health care.

After completing this form, sign and date the form at the end. The form must be signed by two qualified witnesses or acknowledged before a notary public. Give a copy of the signed and completed form to your physician, to any other health care providers you may have, to any health care institution at which you are receiving care, and to any health care agents you have named. You should talk to the person you have named as agent to make sure that he or she understands your wishes and is willing to take the responsibility.

You have the right to revoke this advance health care directive or replace this form at any time.

PART 1
POWER OF ATTORNEY FOR HEALTH CARE
(1.1) **DESIGNATION OF AGENT**: I designate the following individual as my agent to make health care decisions for me:

_____

(name of individual you choose as agent)

_____

(address)   (city)   (state)   (zip code)

_____

(home phone)   (work phone)

OPTIONAL: If I revoke my agent's authority or if my agent is not willing, able, or reasonably available to make a health care decision for me, I designate as my first alternate agent:

_____

(name of individual you choose as first alternate agent)

_____

(address)   (city)   (state)   (zip code)

_____

(home phone)   (work phone)

OPTIONAL: If I revoke the authority of my agent and first alternate agent or if neither is willing, able, or reasonably available to make a health care decision for me, I designate as my second alternate agent:

_____

(name of individual you choose as second alternate agent)

_____

(address)   (city)   (state)   (zip code)

_____

(home phone)   (work phone)

(1.2) **AGENT'S AUTHORITY**: My agent is authorized to make all health care decisions for me, including decisions to provide, withhold, or withdraw artificial nutrition and hydration and all other forms of health care to keep me alive, except as I state here:

_____

_____

_____

(Add additional sheets if needed.)

(1.3) **WHEN AGENT'S AUTHORITY BECOMES EFFECTIVE**: My agent's authority becomes effective when my primary physician determines that I am unable to make my own health care decisions unless I mark the following box. If I mark this box _, my agent's authority to make health care decisions for me takes effect immediately.

(1.4) **AGENT'S OBLIGATION**: My agent shall make health care decisions for me in accordance with this power of attorney for health care, any instructions I give in Part 2 of this form, and my other wishes to the extent known to my agent. To the extent my wishes are unknown, my agent shall make health care decisions for me in accordance with what my agent determines to be in my best interest. In determining my best interest, my agent shall consider my personal values to the extent known to my agent.

(1.5) **AGENT'S POSTDEATH AUTHORITY**: My agent is authorized to make anatomical gifts, authorize an autopsy, and direct disposition of my remains, except as I state here or in Part 3 of this form:

_____

_____

_____

(Add additional sheets if needed.)

(1.6) **NOMINATION OF CONSERVATOR**: If a conservator of my person needs to be appointed for me by a court, I nominate the agent designated in this form. If that agent is not willing, able, or reasonably available to act as conservator, I nominate the alternate agents whom I have named, in the order designated.

PART 2
INSTRUCTIONS FOR HEALTH CARE
If you fill out this part of the form, you may strike any wording you do not want.

(2.1) **END-OF-LIFE DECISIONS**: I direct that my health care providers and others involved in my care provide, withhold, or withdraw treatment in accordance with the choice I have marked below:

____ (a) Choice Not To Prolong Life I do not want my life to be prolonged if (1) I have an incurable and irreversible condition that will result in my death within a relatively short time, (2) I become unconscious and, to a reasonable degree of medical certainty, I will not regain consciousness, or (3) the likely risks and burdens of treatment would outweigh the expected benefits,

OR

____ (b) Choice To Prolong Life I want my life to be prolonged as long as possible within the limits of generally accepted health care standards.

(2.2) **RELIEF FROM PAIN**: Except as I state in the following space, I direct that treatment for alleviation of pain or discomfort be provided at all times, even if it hastens my death:

_____

_____

_____

(Add additional sheets if needed.)

(2.3) **OTHER WISHES**: (If you do not agree with any of the optional choices above and wish to write your own, or if you wish to add to the instructions you have given above, you may do so here.) I direct that:

_____

_____

_____

(Add additional sheets if needed.)

PART 3
DONATION OF ORGANS AT DEATH (OPTIONAL)

(3.1) Upon my death (mark applicable box):

    ___ (a) I give any needed organs, tissues, or parts, OR

    ___ (b) I give the following organs, tissues, or parts only.

_____

(c) My gift is for the following purposes (strike any of the following you do not want):

    (1) Transplant

    (2) Therapy

    (3) Research

    (4) Education

PART 4
PRIMARY PHYSICIAN (OPTIONAL)

(4.1) I designate the following physician as my primary physician:

_____
(name of physician)

_____
(address)   (city)   (state)   (zip code)

_____
(phone)

OPTIONAL: If the physician I have designated above is not willing, able, or reasonably available to act as my primary physician, I designate the following physician as my primary physician:

_____
(name of physician)

_____
(address)   (city)   (state)   (zip code)

_____
(phone)

* * * * * * * * * * * * * * * *

PART 5

(5.1) **EFFECT OF COPY**: A copy of this form has the same effect as the original.

(5.2) **SIGNATURE**: Sign and date the form here:

_____   _____

(date)   (sign your name)

_____

(print your name)

_____

(address)

_____

(city)   (state)   (zip)

(5.3) **STATEMENT OF WITNESSES**: I declare under penalty of perjury under the laws of California (1) that the individual who signed or acknowledged this advance health care directive is personally known to me, or that the individual's identity was proven to me by convincing evidence, (2) that the individual signed or acknowledged this advance directive in my presence, (3) that the individual appears to be of sound mind and under no duress, fraud, or undue influence, (4) that I am not a person appointed as agent by this advance directive, and (5) that I am not the individual's health care provider, an employee of the individual's health care provider, the operator of a community care facility, an employee of an operator of a community care facility, the operator of a residential care facility for the elderly, nor an employee of an operator of a residential care facility for the elderly.

First witness   Second witness

_____   _____

(print name)   (print name)

_____   _____

(address)   (address)

_____   _____

(city)   (state)   (zip)   (city)   (state)   (zip)

_____   _____

(signature of witness)   (signature of witness)

_____   _____

(date)   (date)

(5.4) **ADDITIONAL STATEMENT OF WITNESSES**: At least one of the above witnesses must also sign the following declaration: I further declare under penalty of perjury under the laws of California that I am not related to the individual executing this advance health care directive by blood, marriage, or adoption, and to the best of my knowledge, I am not entitled to any part of the individual's estate upon his or her death under a will now existing or by operation of law.

_____   _____
(signature of witness)   (signature of witness)

PART 6
SPECIAL WITNESS REQUIREMENT

(6.1) The following statement is required only if you are a patient in a skilled nursing facility—a health care facility that provides the following basic services: skilled nursing care and supportive care to patients whose primary need is for availability of skilled nursing care on an extended basis. The patient advocate or ombudsman must sign the following statement:

STATEMENT OF PATIENT ADVOCATE OR OMBUDSMAN
I declare under penalty of perjury under the laws of California that I am a patient advocate or ombudsman as designated by the State Department of Aging and that I am serving as a witness as required by Section 4675 of the Probate Code.

_____   _____
(date)   (sign your name)

_____
(print your name)

_____
(address)

_____
(city)   (state)   (zip)

Appendix 7.2 Institute of Medicine. Committee on Organ Procurement and Transplant Policy, Division of Health Sciences Policy. 1999. Organ Procurement and Transplantation. Below is our synopsis and interpretation of the five recommendations of this study, which was done to determine fairness in organ allocation policies of the Department of Health and Human Services (DHHS).

- Because of data indicating that organ procurement and allocation by Organ Procurement Organizations (OPOs) serving larger population base were more effective in meeting patient needs, the committee recommended the establishment of Organ Allocation Areas (OAAs) for livers that would serve a population base of at least 9 million people, unless such an area would exceed the limits of distance and acceptable time for transport of the ischemic (without blood supply) organ to the recipient. This would provide more opportunities for the allocation of organs to those patients most in need of a transplant.

- Length of time on the transplant wait list is not the best criterion for dertemining organ allocation, especially because of the tendency to place a candidate on the wait list early in the disease process to accumulate wait time. A more appropriate method for prioritizing organ allocation might be a scoring system based on medical urgency and disease prognosis rather than wait list times. Otherwise, patients who were less ill might receive an organ before more severely ill patients, depending on the particular organ procurement organization.

- The DHHS should provide federal oversight to manage the system of organ procurement and transplantation in the public interest. This oversight should include the use of patient-centered, outcome-oriented performance measures for the transplant centers and the organ procurement and transplantation network (OPTN).

- The DHHS should establish an external, independent, multidisciplinary scientific review board responsible for insuring that organ procurement and transplantation methods are based on medical science and equity. A significant finding was that the most important predictors of equity in access to transplant services lie outside the transplantation system. The predictors are access to health insurance and to high quality health care services.

- The OPTN contractor should improve its collection of standardized and useful date regarding the system of organ procurement and transplantation and make it widely available to independent investigators and scientific reviewers in a timely manner. The DHHS should provide an independent and objective assessment of the quality and effectiveness of the data that are collected and how they are disseminated by OPTN.

Appendix 7.3 AMA Section E-2.03 Allocation of Limited Medical Resources

A physician has a duty to do all that he or she can for the benefit of the individual patient. Policies for allocating limited resources have the potential to limit the ability of physicians to fulfill this obligation to patients. Physicians have a responsibility to participate and to contribute their professional expertise in order to safeguard the interests of patients in decisions made at the societal level regarding the allocation or rationing of health resources.

Decisions regarding the allocation of limited medical resources among patients should consider only ethically appropriate criteria relating to medical need. These criteria include likelihood of benefit, urgency of need, change in quality of life, duration of benefit, and, in some cases, the amount of resources required for successful treatment. In general, only very substantial differences among patients are ethically relevant; the greater the disparities, the more justified the use of these criteria becomes. In making quality of life judgments, patients should first be prioritized so that death or extremely poor outcomes are avoided, then patients should be prioritized according to change in quality of life, but only when there are very substantial differences among patients. Non-medical criteria, such as ability to pay, age, social worth, perceived obstacles to treatment, patient contribution to illness, or past use of resources should not be considered.

Allocation decisions should respect the individuality of patients and the particulars of individual cases as much as possible. When very substantial differences do not exist among potential recipients of treatment on the basis of the appropriate criteria defined above, a "first-come-first-served" approach or some other equal opportunity mechanism should be employed to make final allocation decisions. Though there are several ethically acceptable strategies for implementing these criteria, no single strategy is ethically mandated. Acceptable approaches include a three-tiered system, a minimal threshold approach, and a weighted formula. Decision-making mechanisms should be objective, flexible, and consistent to ensure that all patients are treated equally.

The treating physician must remain a patient advocate and therefore should not make allocation decisions. Patients denied access to resources have the right to be informed of the reasoning behind the decision. The allocation procedures of institutions controlling scarce resources should be disclosed to the public as well as subject to regular peer review from the medical profession. (I,VII)

Issued March 1981; Updated June 1994 based on the report "Ethical Considerations in the Allocation of Organs and Other Scarce Medical Resources Among Patients," adopted June 1993 (Archive of Internal Medicine 1995; 155: 29–40).

PROBLEM SET

7.1 A married couple has a child whose kidneys were severely damaged from a fall out of a second story bedroom window at two years of age. The kidneys started failing two years later, and the parents began the search for a kidney donor. When a suitable donor could not be found, the parents decided to conceive another child in hopes this second child would be able to supply one of its kidneys to the sick sibling. What are the moral issues and how should the situation be dealt with by the stakeholders in the event that the new child has tissue-compatible kidneys?

7.2 A daughter is planning to marry an 18-year-old who has severe juvenile diabetes. The parents fear the offspring will also have juvenile diabetes. The parents ask a judge for an injunction to prevent the marriage. How might the judge proceed?

7.3 Why do we distinguish between those who can hasten death by having their physician cease life-sustaining treatment and those who wish to hasten death by having their physicians prescribe a lethal drug? Is it consistent to have a state law that allows a patient to refuse life-sustaining treatment but does not allow physician-assisted suicide?

7.4 Causation (assisted death) and the duty to save lives are the operative concepts in decisions to withhold treatment. What is the difference between removing a patient from a ventilator and not doing a tracheostomy in the first place? Is the difference in killing vs. letting die?

7.5 Is choosing to die a fundamental right? Should laws unduly burden personal liberty?

7.6 A convict on death row acquired a kidney infection that led to destruction of both kidneys. She is on dialysis to keep her alive until the date set for her execution six weeks away. She is a foreign alien and was convicted based on circumstantial evidence for killing an official's daughter. Her name was put on the Alameda County transplant list to await a kidney transplant as a routine because the penal policy is to give the inmates all the medical benefits of California citizens. A compatible kidney has become available for this prisoner. The transplant surgeon refuses to do the transplant, and the medical association of California has declared that this case is in the hands of the state penal authority. You are the ethics committee chair for the state penal program. How would you proceed?

7.7 Complete the advance health care directive in the appendix of this chapter (or find one on the Internet specific for the state in which you live).

7.8 This is a problem for a two-team debate and involves the true case of 17-year-old Jesica (the correct spelling of her name). She was treated at Duke University Medical Center because she needed a heart/lung transplant for longstanding heart disease, inadequate blood circulation, and poorly developed lungs. She came to the United States from Mexico about three years ago for the purpose of finding medical care for her condition. During the first week of February 2003, she received a heart-lung transplant that arrived at the operating room after her own heart and lung had been removed by the surgeons. The new heart/lung transplant was put in place, but Jesica's immune system began rejecting the organs almost immediately. Her medical condition deteriorated, including probable brain and kidney damage. She lapsed into coma and was placed on a heart-lung machine. Investigation showed that the transplanted

organs were an incompatible match (Jesica's blood type was 0 but the donor's blood type was A), an error that never should have occurred as the common practice is for the surgeon to verify the blood type compatibility. This was not done.

The error was recognized before the patient left the operating room, and a search for a compatible heart-lung donor was begun immediately. Remarkably, another heart-lung was found and she received this second set of organs 13 days later. However, she remained in critical condition, and two days later, after tests conclusively demonstrated she had irreversible brain damage, she was taken off life support.

Heart-lung transplants occur about 30 times per year, and 60% survive. Usually second transplants are less successful than first transplants. Suppose that in this case you learn that Jesica was finally put high on the transplant list in the first place because a benefactor Mr. Mahoney raised money to build a center for youths like Jesica and named it for Jesica while she was waiting for the transplant. Further, you learn she was still an alien. Suppose that you learn that the common practice for the surgeon at other hospitals is to sign a document or at least examine the papers accompanying the organ to be transplanted before removing the organ from the recipient, whether it is the heart, kidney, liver, or other organ.

Team A - Argue the entire transplant procedure was not the standard of care in the first place because the rules regarding the wait list for recipients were not followed in either the first or second transplant. Argue secondly that the second transplant was futile care in the extreme as the patient already was brain dead and had kidney failure.

Team B - Argue that the fact that Jesica was an alien was not relevant and that she did wait three years. Make an argument that the second transplant was justified, that nothing was done to move her to the top of the transplant list in either transplant situation, and that the donation of the organs was not the result of worldwide publicity surrounding the case.

7.9 Write a 20-word (or less) explanation of each of the following phrases or terms:

  Futile care

  Advance health care directive (AHCD)

  Surrogate

Physician-assisted suicide allowed by Oregon

Slippery slope

Duty of a caregiver if disconnecting life support is against his or her beliefs

Can a physician override a parent in rendering a particular procedure for a child?

What if a health care executor asks the hospital to continue care against the AHCD?

7.10   Confidentiality about the dissemination and use of genomic information is of concern. The genomic or proteomic patterns related to schizophrenia, obsessive compulsive disease, attention deficit disorders, and a host of other behavioral problems are likely in the future to be available from a blood sample taken at the time of birth or perhaps from proteomic patterns at puberty. Consider the fact that these data do not necessarily conclude that the subject will manifest these diseases, but that the probability is high and the family and caregiver will need to consider careful surveillance and possibly early treatment. The ramifications of the information on the self-image of the subject and on the educational and occupational opportunities are tremendous if this information is reported on school and job applications. How should the parents and the subject deal with information that there is a gene or protein pattern consistent with schizophrenia? Should the information not be reported on forms that require information on blood tests since the probability is that the test result is only a pattern and not a definitive diagnosis of a disease?

7.11   An eight-year-old girl has been complaining about joint pains for months. Her mother takes her to the doctor for an examination. She is diagnosed with an autoimmune disease that causes chronic kidney disease and mental deterioration, among other problems. Treatment for her illness consists of taking immunosuppressive drugs that will certainly help, but may not allow child to live out an ordinary life. The mother does not want her child to know about the illness and its severity because she fears the child will become psychologically depressed. In addition, she does not want her to receive immunosuppressive drugs because of their side effects such as loss of hair, loss of appetite, and weakness. What is the responsibility of the physician to the mother and to the child?

7.12   Racial preferences and health care: A health care provider for an elderly woman arranges for care at the woman's home. The appointed nurse is African American, and the elderly woman does not want her in the home. Considering the fact that the woman has the right to the privacy of her own home, should the health care provider comply? How should any health care institution comply with racial preferences for that matter?

7.13   With reference to duty, autonomy, justice, slippery slope, rights, utilitarianism, and virtue, which principle(s) or concept(s) fit(s) the argument *for* legalizing marijuana for medical use? Which is/are behind the argument *against* legalizing marijuana for medical use? Make a concise table with 30-word arguments for each of your choices.

7.14   A prisoner is arrested for income tax evasion and when arraigned, the prisoner appears disoriented and illogical. The judge orders him confined for medical treatment. While in prison, he becomes violent on occasion, and the prison guards request medication, which is given for their own safety. Now in a calm state, the medical confinement officer recommends the prisoner return to the courts for trial. When the prisoner learns that he is to be tried in court, he refuses medication. Though he does not appear normal to the prison authorities, he is not a threat to them and the physician decides to discontinue medications to subdue the prisoner because the prisoner refuses. There are two sides to be argued with accompanying moral theories including autonomy, duty of the courts, duty of the IRS, utilitarianism of the prison system, and execution of justice.

Write a statement to the court in defense of the rights of the prisoner, including that if the prisoner is not tried, charges must be dropped. Write a statement to the court advising the logic of bringing the prisoner to court under medication so his behavior will appear to be that of a competent person.

7.15   With regard to legalizing marijuana for medicinal use, does federal law supersede states' rights when it comes to public health and safety? Can the federal government rule that a physician cannot give advice about the appropriateness of a remedy? Present an 80-word discussion of how this situation and controversy would be viewed by philosophers and ethicists using the following theories and principles as topics: beneficence, non-malfeasance, duty, virtue, rights, utilitarianism, slippery slope, and autonomy. Argue that marijuana or its chemical ingredient should be prepared in special bottles and distributed only through a pharmacist using the "Controlled Substance" prescription for licensed physicians.

8

Ethics of Human and Animal Experimentation

The degree of risk to be taken should never exceed that determined by the humanitarian importance of the problem to be solved by the experiment. . . .

Nuremberg Code, 1949, Principle 6

ETHICS OF HUMAN EXPERIMENTATION

Background

A major responsibility of the medical profession is to conduct research on the efficacy of new drugs, instruments, and procedures developed to treat diseases. Because human subjects are used in this research, a special field of ethics known as bioethics has evolved. Although the broad definition of bioethics involves biology, medicine, public policy, law, ethics, philosophy, and theology, it was human experimentation that motivated the field of bioethics. Today, bioethical questions are raised by public health care issues such as stem cell research, human cloning, animal care, species sustainability, and genetic modification of plants and organisms. However, perhaps the most important duty that falls under the purview of bioethics is to ensure that human subjects are not exposed to any undue risks or to any risk about which they have not been informed before consenting to participate in the research and that their consent is voluntary without undue coercion.

The field of bioethics first began to evolve when it came to light during the Nuremberg trials following World War II that Nazi doctors in concentration camps had performed criminal experiments on living human subjects. The exploitation of these unwilling prisoners as research subjects was reprehensible. As a result of these injustices, the Nuremberg Code was written to set forth basic moral, ethical, and legal boundaries for medical research on human subjects (Trials of War Criminals before the Nuremberg Military Tribunals 1949). The Code specified the necessity for voluntary consent by the participant; this required that the subject have the capacity to understand the risks involved and the

277

capacity to consent. It also mandated the absence of coercion to participate and the minimization of the risk of harm to the participant. Other requirements were that the experimental study design should be appropriate and that the subject could withdraw for any reason. This code became a model for other subsequent codes, also written to ensure ethical conduct in research on human subjects.

The Declaration of Helsinki in 1964 made similar recommendations and expanded the guidelines set forth in the Nuremberg Code. The key difference between the Nuremberg Code and the Declaration of Helsinki is that the Declaration of Helsinki distinguishes between therapeutic and nontherapeutic research. The new guidelines were written in response to ethical questions raised by newly available lifesaving technologies (such as kidney dialysis, heart-lung perfusion, and surgical transplantation of the heart, liver, and kidney) for which guidelines were nonexistent as recently as 1963. These new and often highly experimental procedures raised troublesome ethical dilemmas that forced a reevaluation of such topics as the ethics of organ donation, life support, and death. A later motivation for further expanding the protections for human subjects research was the exposure in 1972 of an inhumane observational study of untreated syphilis in disadvantaged rural African American men. This unjust study was conducted over a period of years, during a time when effective treatment became available for this unrelenting disease. The men were denied penicillin in order not to interrupt the study. To the dismay of those who live in Tuskegee, the study is often referred to as the Tuskegee syphilis study but is more properly known as the U.S. Public Health Services Syphilis Study.

Case 8.1 U.S. Public Health Services Syphilis Study

This study evoked outrage when it was disclosed that 399 poor black sharecroppers in Alabama were nonconsenting subjects in the longest running observational, nontherapeutic study in history, the progression of untreated syphilis. Told they had "bad blood," the study began in 1932 and did not end until 1972. Even though it was discovered that penicillin would eradicate the disease 20 years prior to 1972, the participants received no treatment. Not only did the participants suffer from the serious consequences of this unrelenting disease, but the study also had far-reaching effects on the willingness of African Americans to participate in clinical trials, to donate organs for transplantation, to seek medical care, and to trust government officials. According to the Report of the Tuskegee Syphilis Study Legacy Committee in 1996, "In the almost 25 years since its disclosure, the Study has moved from a singular historical event to a powerful metaphor. It has come to symbolize racism in medicine, ethical misconduct in human research, paternalism by physicians, and government abuse of vulnerable people" (www.nlm.nih.gov/pubs/cbm/hum_exp.html, accessed July 30, 2004; also www.med.virginia.edu/hs-library/historical/apology/report.html, accessed July 21, 2004).

This abuse stimulated years of discussion, analysis, and debate by moral philosophers, physicians, scientists, religious leaders, and public representatives. The culmination was the Belmont Report, issued in 1979. This report synthesizes a set of moral principles to guide the policies and procedures for the protection of human subjects, including volunteers and patients, who are involved in scientific research. The historical evolution of the current Institutional Review Boards and their regulation is synopsized in Table 8.1.

Table 8.1   Evolution of Regulations Protecting Human Research Subjects

1803	First consultation with peers by a physician before proceeding with a therapeutic innovation (Percival 1803).
1949	Atrocities in Nazi concentration camps during WWII resulted in the Nuremberg Code,* which included a statement about permissible research on human subjects.
1964	The Declaration of Helsinki* was adopted by the 18th World Medical Assembly, Helsinki, Finland. Revised in 1975, 1983, 1989, and 2000, this document still serves as an international standard on rules governing therapeutic human experimentation.
1972	Discovery of the U.S. Public Health Services Syphilis Study (Tuskegee Study). The many abuses suffered by the participants led to improved guidelines in the United States for the protection of human research subjects, formalized in 1974.
1974	National Research Act created the National Commission for the Protection of Human Subjects of Biomedical and Behavioral Research. This commission evaluated and stipulated policies for the conduct of human research in the United States, and was the convening authority for the Belmont Report issued in 1979.
1979	Issuance of "The Belmont Report*: Ethical Principles and Guidelines for the Protection of Human Subjects of Research."
1983	Issuance of Department of Health and Human Services (HHS) Rules and Regulations 45 CFR (Code of Federal Regulations) Part 46, which apply to research involving human subjects funded in whole or in part by the HHS.

45 CFR 46, often referred to as the "Common Rule," was adopted by 16 federal agencies (such as NIH and FDA) involved in regulating research involving human subjects. |
| 1998 | FDA issued Guidance for Institutional Review Boards and Clinical Investigators. FDA authority is over research** involving drugs or devices regulated by the FDA [21 CFR parts 50 and 56]. FDA regulations are applicable to any individual, institution, foundation, or commercial enterprise. |

*Note: The Nuremberg Code, the Declaration of Helsinki, and the Belmont Report may be accessed at www.nlm.nih.gov/pubs/cbm/hum_exp.html #15.

**Note: When research involving products regulated by the FDA is funded, supported, or conducted by both FDA and HHS, HHS and FDA regulations apply.

Current Guidelines for Research on Human Beings

The Department of Human and Health Services oversees all research on human subjects that is supported by the federal government, and the Food & Drug Administration (FDA) oversees all research on drugs or devices under its control. Human subjects research protocols at an institution that receives federal support must be approved by the Institutional Review Board (IRB). The IRB has a mandate to ensure the preservation of the autonomy, rights, and privacy of the individual, and that the research is safe, meaningful (if risks are involved), ethical, and conducted within the law.

IRB Duties

The duty of an IRB is the protection of a research subject. It does this by ascertaining the acceptability of proposed research in light of:

- The ethical principles set forth in the Belmont Report.
- The requirements of federal regulations for the protection of human subjects (45 CFR 46).
- Applicable federal, state, and local laws.
- The standards of professional conduct and practice.

There are two situations in which a certified IRB must examine an experimental protocol:

- If the research involves professional staff or resources of a federally funded institution.
- If an experimental drug or device under the purview of the FDA is being studied at any institution or organization.

This means that all those involved in either of these types of research, including academic research institutions such as medical schools, universities, and foundations, as well as commercial enterprises, must have access to a certified IRB.

To meet its principal goal of protecting human research subjects, the IRB reviews many different types of protocols and works with the investigator to correct any deficiencies. The principles that guide the work of the IRB are outlined in the Belmont Report of 1979 (see below). Although controversial, an IRB generally refrains from evaluating the scientific merit of a protocol unless the research involves a risk to the subject that is disproportionate to the possible benefits [45CFR 46.111(a)(1)(i)]. The rationale for limiting the authority of the IRB to make judgments regarding the scientific methods and statistics of a protocol is the protection of the autonomy of the investigators who have designed the experiment. Generally, an IRB constituted to review protocols in order to protect human subjects is not always qualified to make a just decision on the scientific merit of a protocol. Substantial risk mandates the addition of qualified reviewers.

IRB Composition

Each IRB must meet exacting standards. The membership must have the professional competence necessary to review and approve specific research activities, as well as sufficient qualifications in terms of experience, expertise, and diversity of backgrounds to foster respect for its counsel. The qualifications of the members depends to some extent on the type of proposals being reviewed, because committees are required to have some expertise on the subject matter of the proposals (e.g., there should be a specialist in child research if children are involved in a proposal, 45 CFR 46.107(b)). There must be a prisoner representative on an IRB when research involving prisoners is being reviewed.

While the maximum size of IRBs differs across institutions, each IRB must have a minimum of five members. Three types of members must always be present:

- A member with expertise in the subject matter of a protocol.
- A member from outside the institution.
- A non-scientist member.

The non-scientist and outside member can be the same person. The expertise of committees also varies across institutions. For example, the composition of a committee that reviews proposals from public policy and sociology academic research departments would not be expected to have members with expertise in medical experimentation or clinical trials. An IRB may also utilize ad hoc reviewers in cases where additional or specialized expertise in protocol review would aid the board in its deliberations. However, ad hoc reviewers are not IRB members and do not vote on protocols. A quorum is 50% of the stated membership plus one person.

Guidelines for IRB Activities

The principles that guide the work of the IRB are outlined in the *Belmont Report of 1979*. This report synthesizes a set of moral principles to guide the policies and procedures for the protection of human subjects who are involved in scientific research, including volunteers and patients. These principles are:

Autonomy:

Individuals have the right to make decisions pertaining to themselves, to lead their life according to their preferences, without interference, so long as there is no serious detrimental effect on the rights or welfare of others. A consequence of this principle, and as a method for ensuring adherence to autonomy, was the institution of procedures for written and witnessed consent forms for the protection of those being asked to participate in human experimentation protocols. It should be noted that the origin of the principle of autonomy is in the ethical theory of duty. The duty to respect rights is in accord with Kantian ethical theory. This duty includes provision of information regarding the risks and benefits of a proposed procedure in the clearest possible language so that the individual can make an informed choice. The overall purpose of the activities of an IRB is to protect the individual participant in an experiment.

Beneficence and Non-Malfeasance:

The goal of medical intervention is to prevent or cure disease, to relieve symptoms, to improve or maintain functional abilities, and to do no harm. The two ethical principles guiding physicians in attaining these aims are known as beneficence and non-malfeasance. There is a distinction between the two, although both are part of doing no harm. Beneficence involves an active process of promoting welfare. It is the moral obligation to act for the benefit of others. The concept of beneficence applied to research on human subjects stems from utilitarianism, because many research studies will put the subject or patient at some risk but have the potential to benefit others. In human experimentation, the benefits of the research are not usually intended to be for the individual subject but go beyond the experimental subject to others who might benefit (Beauchamp and Childress 2001). Non-malfeasance is to do no harm and implies not actively imposing an intervention that might have adverse consequences. However, some harm might come to participants as a result of an experiment that might benefit others:

> Beneficence does not require that the researcher do any subject good. It merely requires that he have no preexisting reason to think harm is likely to the subject out of proportion to the potential for good to others (Fried 2001a).

As the results of medical intervention cannot always be predicted with certainty, the physician should offer an opinion as to the risks involved and the probability of success. If participation in a clinical trial is only one possibility for therapy, an opinion as to whether this would be in the patient's best interest may be offered, always keeping in mind that the choice or decision is the patient's.

Thus, in accord with the Belmont Report (i.e., section C.2), the physician, researcher, and Institutional Review Board must seek a proper balance between the risks to the particular patient and the probability of benefits to others if there is no clear benefit directly to the participant.

Justice:

The selection of research subjects must be fair, the benefits and burdens of research must be fairly distributed, and there must be no bias or undue burden placed upon them. To define how the burdens and benefits of the research should be distributed, the Belmont Report stipulates that:

> . . . justice demands both that these [burdens and benefits] not provide advantages only to those who can afford them and that such research should not unduly involve persons from groups unlikely to be among the beneficiaries of subsequent applications of the research. . . . Social justice requires that distinction be drawn between classes of subjects that ought, and ought not, to participate in any particular kind of research, based on the ability of members of that class to bear burdens and on the appropriateness of placing further burdens on already burdened persons.

Justice in the selection of research subjects mandates the protection of some disadvantaged or vulnerable groups of patients (for example, those who are welfare recipients, mentally incompetent, minorities, or prisoners) to ensure that they are not being selected because they are easily persuaded or easily manipulated.

Informed Consent Procedures

A major aspect of protection for the subject of experiments was the establishment of the requirement for obtaining informed consent. This ensures that the subjects understand the procedures to be used in the study, their rights, and that they can choose whether to participate or withdraw consent without compromise of their subsequent care. Informed consent requires that the subject be given sufficient information to make the decision to participate, that he or she comprehends the information, and that he or she is making a voluntary decision to participate. The required 12 elements (Appendix 8.1) of informed consent must be presented in language such that the individual can understand exactly what is being asked of him or her, and the subject must be given adequate time to understand and consider what is being asked. Those who are underage, mentally disabled, too ill, or who are comatose or otherwise cognitively impaired are consented through special procedures that usually involve a conservator.

All decisions to participate must be voluntary without undue pressure, duress, or enticement. Volunteers should not be paid so much that it becomes coercive, or that it becomes a bribe that might cause them to abandon their own best interests. They must be free agents with the freedom to assess the risks as they see them. The objective of the study must be clearly stated, with no hidden or other agendas not clearly spelled out. The subjects of clinical trials must be informed that they may be withdrawn from the study without regard to their consent, and the reasons for and any consequences of this withdrawal must be explained. In any medical study, every patient—including those of a control group, if there is one—should be assured of the best-proven care. Sources of funding for the research must be identified to fully inform the subject of this or other possible conflicts of interest.

The documents required for IRB review are the scientific protocol and all documents that will be seen by the potential participant, including the consent forms, questionnaires, scripts of oral and telephone communications, recruitment letters and advertisements, and if relevant, the video presentations to which participants will be exposed. The IRB committee does not require these materials for the purpose of evaluating the science or technical merits of the protocol and procedures unless there are some risks from the procedures. The committee is required to review the procedures and documents to be sure that the participant is not exposed to harm sometimes not anticipated by the researchers. Examples of unanticipated harms are possible loss of privacy due to poor electronic data security, failure to exclude subjects with diabetes in tests of a new food product, failure to seek permission from parents for an in-school survey of children's habits, and so on. The principal concern of the committee is the protection of the participants.

The protocol states the scientific design. If there are even minimal risks to the subject (e.g., taking blood from the subject's arm for tests), the protocol must show evidence of some benefit to society that outweigh the risks. The literature and any prior data must be thoroughly researched so as not to put human subjects at undue risk. Other means of obtaining the data should have been exhausted.

These risks must be assessed in terms of the social norms for the participants' culture, their economic status, and even their political situation because the risks change from individual to individual and culture to culture. The refusal of the patient to participate in a study must never interfere with the physician-patient relationship, nor should the research study participant continue if there is any indication it is harmful to the subject. Election by the subject to withdraw from the study should be respected without influencing future care. The well-being of the subject takes precedence over any considerations of the interests of science and society.

Two problems commonly occur when a researcher seeks consent from a potential participant to perform a research procedure. First, a participant with a disease might infer that the researcher is doing the procedure to benefit the participant's health, when in fact there are no direct benefits. Second, the participant might feel compelled to cooperate to ensure continuing preferential treatment by the caregivers. The first problem is known as therapeutic misconception (Appelbaum et al. 1987, Fried 2001b), something that can occur even if the researcher is not the participant's physician. The second problem is that a participant with a disease might believe that if he or she does not agree to participate or withdraws from participation, medical care will be compromised. This fear can occur despite reassurances from the participant's physician or the researcher otherwise. The ethical principle involved here is interference with the patient's autonomy, and both the protocol for how to obtain consent and the language of the consent form must be designed to avoid these problems.

Variety of Human Experiments

The possible experiments or clinical trials in which human beings may participate (and for which written consent is required) range from simple questionnaires that ask for opinions or personal information to complex multicenter therapy experiments in seriously ill patients. Many studies of human subjects involve no obvious risks of harm as long as confidentiality is maintained (e.g., anonymous questionnaires regarding diets, alcohol intake, sexual preferences). Yet even these studies require review by the IRB if the institution at which they are performed receives any federal funds. Thus, a graduate student at a federally funded institution who wishes to question other students anonymously about sexual preferences as a research project must make an application to the IRB for approval. In this case, the proposal might not need to go to the IRB for full review and could be found to be exempt from review. However, if videotaping were part of the project, the review might or might not go to the full IRB but would require the use of a consent form because the subjects are no longer anonymous. After review by a committee member, such a study could be expedited by a signature of an IRB committee member. Exempt protocols are exempt and need no further action. Protocols require an annual review.

Clinical trials often involve experiments to evaluate the effectiveness of new drugs or devices in the treatment of diseases. Generally, there is a group of patients who receive the treatment under study and a control group, patients who do not receive the treatment or receive a placebo (usually an inert substance made to look like the active treatment). Sometimes the participants are randomized (randomly assigned, usually by computer, in contrast to assignment by the researcher), and sometimes the researchers and the participants are "blinded" about which treatment is being given to whom (i.e., who is receiving the active treatment and who is not). The researchers are usually physicians who might or might not be the primary caregivers responsible for the participants' disease treatment, a point that should be made clear. It is also important in clinical trials to inform the participant of the financial relationships between the researcher and the commercial entity manufacturing and providing the treatment or device being studied, if there is one. For example, if a researcher might benefit financially from the analysis of the study, there is a conflict of interest. Such an arrangement should be explained to the participant as it could bias the analysis of the study. This disclosure is a requirement for the autonomy of the participant who can then decide whether to participate or not. These are but some reasons informed consent forms need to be carefully designed and discussed with the potential participant to address any unrealistic expectations or misunderstandings prior to the start of the study.

Thus, there are numerous ethical questions surrounding clinical trials. As we have seen, it is unethical to unduly sway a patient to participate in a clinical trial, although some patients are paid to participate. Further, it may be unethical to take advantage of very ill patients who might be made sicker by the treatment. Have patients been fully informed of the risks and possible benefits? In Chapter 11, "Ethics of Enhancement Technologies," in the section under gene therapy, we shall see how physician-researcher bias and undisclosed adverse events were determined to be directly responsible for the death of a patient participating in a gene therapy trial. Is it ethical to pay investigators for referrals and participation? One would think that this should not be allowed, but it has been found necessary in order to recruit patients being treated by one physician to a new or alternative treatment by a research group. Thus, these types of ethical questions and possible conflicts of interest are taken very seriously in the IRB review process.

Proposed Modifications of Regulations on Protection for Human Subjects Involved in Medical Research

As thorough as the HHS and FDA human subjects protection guidelines have been, they are now challenged by new ethical issues that have been emerging since the late 1990s. New issues include those associated with human cloning, the intergenerational risks of modifying the human germline, organ transplantation, stem cell research, and confidentiality regarding the individual's genome. Revisions will need to be incorporated into the guidelines as these technologies continue to develop. Indeed, the regulations of the two federal agencies are not always consistent, and the medical and social scientists proposing research protocols are often confused as a result. For example, when assuring a research subject that all information collected in a particular experiment will be kept strictly confidential, one must inform the subject of the exception that the records might be subject to audit by the FDA if the agent being studied is under FDA surveillance.

Improvements have been recommended in three key areas: protecting subject populations with special needs and vulnerabilities, streamlining oversight by Institutional Review Boards, and regulatory policy (Moreno et al. 1998a). The recommended modifications in these three categories are synopsized below.

Populations in Need of Special Protections:

This category involves recommendations for a number of patient populations ranging from the cognitively impaired to the need to include women, minorities, and children in biomedical research. Included is a recommendation to recognize the authority of a surrogate or advocate appointed by the advance health care directive (see Chapter 7, "Medical Ethics") of an individual to make decisions with regard to participation in human research in the event of incompetence. Regulations and guideline are needed for emergency research wherein consent needs to be waived, such as studying a new treatment of a comatose patient with a severe head injury or new methods for treating cardiac arrest.

Institutional Review Boards:

The mandate to protect human subjects needs to go beyond the IRB review of protocols and adequacy of consent forms. There should be an emphasis on the consenting process itself; that is, to specify who interacts with the participant, and what should be the qualifications of the interviewer and consenter. A recommendation to reduce paperwork and "bureaucracy" can be implemented by an experienced IRB staff. A professional staff can review the protocols based on well-defined criteria to determine their status and need for review. Those that do not require full review can receive an expedited review, which requires only the signature of an IRB member. The caveat is that at least 80% of the proposals reviewed by an IRB require some modifications (Moreno et al. 1998b). Another method to improve efficiency is to avoid overregulation of exempt protocols. These are protocols that do not put participants at risk even for invasion of privacy (e.g., questionnaires wherein the participants are anonymous). These should be identified by staff early in the review process so their review goes no further than the appropriate single member IRB signoff.

The quality of institutional review board activities as well as the need for more oversight or outside evaluation of their activities has come under scrutiny from the researchers themselves (Emanuel et al. 2004). However, requirements for more timeliness, more documentation, more evaluations of the process, and more indoctrination must be met with more funding for creating a staff to perform these activities. There also need to be inducements as well as a selection process for volunteer faculty and community citizens to serve on these committees. The inducements include relief time from teaching for academic faculty members along with some acknowledgments from the leadership of an institution that relies on the participation of board members. The responsibility for providing these resources resides with the institutions that receive funding for research involving human experimentation. As IRBs are almost entirely composed of professionals and academics on a voluntary basis and are supported by university overhead funds, one must be careful about adding more responsibilities and liabilities to IRB members who already carry a heavy load.

Policy Issues Requiring Further Study:
The current regulatory authority for the protection of human subjects resides in a federal agency that is also the leading source of research funding. As a matter of principle, this Office for Human Research Protections (OHRP) should not be located within the structure of any government funding agency. The OHRP should lead in emerging policy issues and be able to grant local institutions more flexibility to serve the spirit of the regulations. It is recommended that this goal would be ". . . improved if it were independent of any federal departments that sponsor or conduct research with human subjects" (Moreno et al. 1998b).

One of the most important recommendations is for a much more proactive empowerment of the President's Council on Bioethics, formed after expiration of the executive order for its predecessor, the National Bioethics Advisory Commission. It is important that the president and the public receive unbiased guidance based on scientific fact and not political expedience regarding emerging technologies such as stem cell research, tissue implants, and deep brain stimulation. It is also important to provide public access to medical information of importance to public health and other research that has implications for people's health and well-being. Valuable resources include members of the National Academy of Engineering and the Institute of Medicine of the National Academies, who are often asked to present timely analyses of issues regarding human experimentation.

By mid-2000, there had been significant changes in the nature of clinical science (e.g., genomic and protenomic data bases), in the financial sources of research support (e.g., international organizations), and in the institutional environment in which clinical research is conducted (e.g., human subject experimentation within nonmedical instrumentation corporations). Thus, there needs to be a mechanism for addressing new complex ethical issues, ranging from the inclusion of illiterate participants in developing countries to the cloning of embryos. These issues have become major social controversies themselves, engendering different assessments within and across IRBs (www.bioethics.gov/background/emanuelpaper.html, accessed July 14, 2004).

Noted Examples of Human Experimentation

During the 1960s, there were no ethics committees and no set guidelines for decision making in transplantation medicine or other complex clinical situations. A systematic process for deciding such issues as how to select transplant recipients, how to care for donors in the last stages of life, and even a specific definition of what constituted death of the donor before the organs could be removed for transplantation was lacking. The case below exemplifies some of the problems that arose with the first human liver transplant.

Case 8.2 First Liver Transplant (1963)

The donor for the first liver transplant was near death, comatose, and in a condition known then as brain dead. The potential donor's condition deteriorated over the weeks of his hospitalization for a terminal illness. When it appeared that the donor's death was imminent, the surgeons took the recipient to the operating room and began preparing for the transplant by disconnecting the damaged liver. Suddenly, the donor showed signs of pneumonia, for which the standard treatment is massive antibiotics. The ethical dilemma was whether to treat the pneumonia and prolong his life or whether to jeopardize the hope of the recipient for recovery. What should be the duty of the doctors supervising the first experimental liver transplant? To the donor? To the recipient? What would be the consequences of withholding treatment? What might be the consequences of treatment with antibiotics, of which one, penicillin, could cause death if either the donor or the recipient were allergic to it? After agonizing debate over whether to withhold treatment, it became clear that the duty was to treat the donor with antibiotics. He died within a few hours nonetheless, and the surgery went forward.

This next section presents examples of human experimentation selected to illustrate the ethical duty to inform volunteers of risks and to show the consequences for disrespect for the autonomy of individuals involved in research.

Case 8.3 Yellow Fever

One of first examples of the use of informed consent in human experimentation involved experiments to find the cause of yellow fever, a highly infectious and often fatal disease characterized by headache, fever, muscle aches, and jaundice. This was a disease responsible for the death of thousands of people during summer epidemics, and there was great impetus to find a way to prevent it. A four-member team, known as the Yellow Fever Commission, was dispatched to Cuba to study the problem after over 5,000 people died there of yellow fever during the war with Spain beginning in 1898, and after another epidemic occurred during the summer of 1899. Three of the four members of the commission first experimented on themselves, and subsequently on volunteers. The majority of volunteers came from a U.S. military detachment stationed in Havana, and each was offered a $100 gold piece to participate. A consent form was developed, pointing out that it was likely they would contract the disease anyway while they were there, that it was possible they might die, but that it was better to be under the supervised care of medical specialists than in the field or a remote camp where care was likely to be substandard. This concerned approach to their research earned the Yellow Fever Commission high marks for being among the first to advocate informed consent (www.med.virginia.edu/hs-library/historical/yelfev/pan11.html, accessed July 30, 2004). Their experiments showed that mosquitoes carried the infectious agent, and this led to the extermination of mosquitoes in the Panama region and a disappearance of yellow fever by 1905.

Case 8.4 Death of a Healthy Volunteer in 2001

An example of noncompliance to rules for protection of a research subject involved a healthy 24-year-old volunteer who died after participating in an asthma study at Johns Hopkins University. The drug she was given to inhale was hexamethonium, a drug removed from the market in 1972 because it was ineffective for the treatment of hypertension. The material she received was obtained from a chemical company and was labeled "For laboratory use only, not for drug, household, or other uses." The consent form referred to it as a "medication" but failed to indicate that inhaled hexamethonium was experimental and not approved by the FDA. In addition, adverse effects reported previously in the literature were not identified. Apart from these failures to provide the required elements of informed consent, the IRB erred in that it should have asked the principal investigator to obtain an Investigational New Drug (IND) permit from the FDA for inhaled hexamethonium because the drug had never been administered by the pulmonary route. Perhaps in applying for the IND, the previously reported adverse events with the drug might have been found and the treatment schedule modified. As it was, the study proceeded and a healthy volunteer died from hypoxemia (low blood oxygen levels), hypotension, and multi-organ failure resulting from the inadequate recognition of and handling of adverse events. The autopsy showed diffuse, nonspecific damage of the lung at the alveolar level (the site of the lung where oxygen is exchanged). The Johns Hopkins University report by the committee assigned to investigate the death stated that the hexamethonium was either the sole or an important contributor to her death (Steinbrook 2002).

As a result of these and other problems, the FDA and OHRP have taken an increasingly active role in stopping research studies as well as censoring investigators and IRBs that are not complying with strict research guidelines. In addition, these guidelines are continually being reassessed at local and national levels to include requirements for new developing technologies. Areas in which increasingly difficult ethical questions are arising include gene therapy, embryonic and fetal research, stem cell research, organ transplantation, and cloning. These issues are further discussed in Chapters 9, 10, and 11.

Summary of Human Experimentation

The ethical principle of individual rights or autonomy underpins the ethics of medical experimentation. The assurance that human subjects are protected from coercion or harm disproportionate to the benefits of medical experimentation is through the mechanism of informed consent. This consent process requires proper identification of the experimenters, the extent and duration of the experiment, and the risks and availability of medical treatment in the event of injury. A uniform method of evaluating a research project is provided by the establishment of the IRB, a group of medical scientists who, together with legal and local community representatives, review the procedures of the proposed experiment. Particular focus is given to ensuring that the autonomy of the subject is respected, as are the concepts of beneficence, non-malfeasance, and justice.

ETHICS OF ANIMAL EXPERIMENTATION

Cultural Attitudes toward Animals

We begin the discussion about animal experimentation ethics by tracing the attitudes of civilization toward animals in general. Controversy in this area dates back to biblical times and the story of creation:

So God created man in his own image, in the image of God created he him; male and female created he them.

And God blessed them, and God said unto them, Be fruitful, and multiply, and replenish the earth, and subdue it; and have dominion over the fish of the sea and over the fowl of the air, and over every living thing that moveth upon the earth.

The meaning of the word "dominion" has been a source of controversy in that some claim this passage commits animals to the protection of humanity to be looked after and safeguarded. However, others have interpreted this to mean that humans have been given dominance over animals and have been given license to do what they will with them, especially after God brought upon the Earth the great flood that killed all but the ark animals. Another reference to dominance in the Bible comes with this passage:

And the fear of you and dread of you shall be upon every beast of the earth and upon every fowl of the air, upon all that moveth upon the earth, and upon all the fishes of the sea; into your hand are they delivered.

These attitudes were furthered by the writings of Thomas Aquinas, who incorporated these tenets of Christianity with the ancient Greek concepts expressed by Aristotle (Singer 2000):

Plants exist for the sake of animals, and brute beasts for the sake of man—domestic animals for his use and food, wild ones (or at any rate most of them) for food and other accessories of life, such as clothing and various tools.

Since nature makes nothing purposeless or in vain, it is undeniably true that she has made all animals for the sake of man.

St. Thomas Aquinas saw no possibility of sinning against nonhuman animals or against the natural world, only against God, ourselves, or our neighbor. This thinking remained the thinking of mainstream Christianity for at least its first 18 centuries, thinking that was dominated by these views of the "natural environment" (Singer 2000):

In Western tradition, the natural world exists for the benefit of human beings. God gave human beings dominion over the natural world, and God does not care how we treat it. Human beings are the only morally important members of this world. Nature itself is of no intrinsic value, and the destruction of plants and animals cannot be sinful, unless by this destruction we harm human beings.

This human-centered ethic is the focus of many of the arguments centering on animal experimentation. Matters of experimenting on any sentient animal, of keeping animals contained in less than ideal research quarters, of inflicting pain and suffering on any living being, and the accusations that there have been inconsequential gains from animal experimentation are but some of the ethical issues surrounding animal experimentation.

Animal Malfeasance vs. Medical Benefits

The principal ethical dilemma of animal experimentation is the conflict between utilitarianism and non-malfeasance. Some medical scientists and basic biologists argue that animal experimentation is necessary to advance medical cures and basic understanding of nature; others claim that animal experimentation involves either some suffering of the animal or a sacrifice of life when the knowledge gained does not warrant this malfeasance. The intensity of the objections to the use of animals in experiments, even for the benefit of humanity, is expressed in a letter from Mark Twain to the secretary of the London Anti-Vivisection Society (1899):

> I believe I am not interested to know whether Vivisection produces results that are profitable to the human race or doesn't. To know that the results are profitable to the race would not remove my hostility to it. The pains which it inflicts upon unconsenting animals is the basis of my enmity towards it, & is to me sufficient justification of the enmity without looking further. It is so distinctly a matter of feeling, with me, & is so strong & so deeply rooted in my make & constitution that I am sure I could not even see a vivisector vivisected with anything more than a sort of qualified satisfaction. I do not say I should not go & look on; I only mean that I should almost surely fail to get out of it the degree of contentment which it ought of course to be expected to furnish. (Courtesy, The Mark Twain Project, the Bancroft Library, University of California, Berkeley.)

The prevalence of animal experimentation is increasing throughout the world, mainly because of the increased use of transgenic animal models. Transgenic animals are animals whose genome has been modified by gene manipulations adding a specific gene or genes to provide a specific function, or by knocking out specific genes to mimic the genetic aberrations of specific human genetic diseases. The resulting gene mutations allow investigators to study various aspects of gene function, expression, and regulation. Animal models of human genetic disease allow the study of treatment approaches, including gene therapy. Animals genetically modified to add a specific function, such as the secretion of a specific protein in their milk, have provided new ways to produce pharmaceuticals. Through cloning, identical animals can be produced in high numbers so that much larger and more thorough studies can be done with a high likelihood of meaningful results. Thus, transgenic animals have become a valuable addition to the researcher's armamentarium. Presently, 10 to 12 million vertebrate animals are used per year within the European Union member states (Van Zutphen 2002). An important point regarding the ethics of animal use is that the transgenic animals are bred specifically to be used as research animals in the laboratory, and had there not been a scientific goal, they would not exist in the first place. One can argue that these bred animals are in a different moral community (see Chapter 5, "Environmental Ethics") than those captured in the wild and brought back to the laboratory for study.

Opponents who object to the use of animals in medical research have presented philosophical arguments based on the lack of evidence that animal experimentation is of sufficient benefit to human beings to justify continuing it. The major point of these arguments is the utilitarian principal that if harm is done, a definite good to society must be served by this harm. However, whereas the harm is definite and known, the benefit cannot be known and is only probable. If one estimates that there is a 10% probability that a particular experiment might benefit the practice of medicine, it might be difficult to justify harming a large number of animals for this low-probability event. Another counterpoint is that the process of cloning animals has a rate of success less than 50%, and about 20% to 30% of cloned animals are born dead or malformed, although it should be noted that the survival and normalcy rates of cloned animals are now improving (Cibelli et al. 2002). A further observation by philosophers is that the creatures that suffer are not the creatures who will benefit (Lafollette and Shanks 1995), nor will the results benefit the species.

However, some find it hard to accept the statement that the utilitarian defense of animal experimentation is not valid in view of the facts known to be true. For example, the methods of organ transplantation were developed in animal studies first before exposing human subjects to experimentation. Specifically, whereas the theoretical aspects of liver transplantation were well known, it took exactly 50 attempts on dogs before the technique was sufficiently developed to permit the first successful human liver transplantation. The same is true for many other life-saving procedures (such as coronary bypass surgery and other types of tissue and organ transplantation) that are now available because of procedural and safety data gathered through animal experimentation. However, this is not to suggest animal experimentation should be allowed without adequate justification. The nature of life must be respected, and suffering must be minimized. These concepts serve as the basis for the regulations presented below. In addition to considering the current federal regulations, we will also discuss the methods of reducing, refining, and replacing animal experimentation in the context of improvements in measurement technologies.

The current U.S. regulations for animal research focus on the fact that unlike human experimentation wherein the risks infrequently include death, animal experimentation frequently carries the risk of death and usually some suffering. Thus, there must be solid evidence that the experimentation will provide some benefit, and that the benefit to medical science is great enough to accept the harm. The measure of harm is not only the amount of discomfort but also the number of animals that will be harmed. The oversight and committee approval mechanisms are very strict, not only with regard to minimizing suffering through proper anesthesia, but also in denying approval for experiments where there is not a high probability of obtaining statistically valid results useful to human health.

Federal Regulations

Federal regulations relative to the use of animals in research date from the Animal Welfare Act of 1966 (Public Law 89-544). Subsequent guidelines were added, including the Health Research Extension Act of 1985 and the *Guide for the Care and Use of Laboratory Animals* (National Academy Press 1996). Revisions and publications of guidelines and policy statements have focused on legislation for the humane care of laboratory animals through prescribed laboratory procedures and definitions of the responsibilities of researchers and oversight committees. The 1985 act was implemented and supplemented by the "U.S. Government Principles for the Utilization and Care of Vertebrate Animals Used in Testing, Research, and Training," developed and revised by the Interagency Research Animal Committee (IRAC). Though the current documents give exhaustive recommendations and expectations for researchers and institutions, the essential moral aspects are encompassed in the principles of non-malfeasance, virtue, and utilitarianism. Humane and justified treatment of animals in the service of medical science is the underlying theme.

The five essential aspects of all proposals for the care and use of animals in government-supported research stem from the 1985 Animal Welfare Act (Public Law 99-158). The five main requirements for all animal protocols that must be reviewed and approved by the institutional committee are paraphrased from the original law here:

1. The animal species and the approximate number of animals to be used must be given.
2. The rationale must be presented for the use of animals, and for the appropriateness of the particular species, as well as a justification for the number of animals. Implicit in this requirement and specified in other guidance documents is that a statistical argument must also be presented to justify the number of animals to be used in the study.
3. A complete description of the exact protocol to be used to study the animals must be given.
4. All procedures are to be designed to ensure that discomfort and injury to animals will be limited to that which is unavoidable in the conduct of scientifically valuable research, and that analgesic, anesthetic, and tranquilizing drugs will be used where indicated and appropriate to minimize discomfort and pain.
5. A description of any euthanasia method to be used must be provided.

Further guidance is given in publications from the IRAC (Public Health Service Policy on Humane Care and Use of Laboratory Animals, Health Research Extension Act of 1985, Public Law 99-158, section IV. D.1; http://grants.nih.gov/grants/olaw/references/; also see www.nal.usda.gov/awic/legislat/usdaleg1.htm.

The NIH guidelines are contained in the *Institutional Animal Care and Use Committee Guidebook*, which reviews NIH federally mandated functions, issues in proposal review, record keeping and reporting, other resources and references and sample forms for proposals (U.S. Department of Health and Human Services 1992). These regulations have encouraged the development of alternatives to the use of live animals in research and measures to ensure the safe handling and security of animals.

The federal regulations give the requirements for performance of animal research using federal funds, usually in the form of NIH grants. The law mandates requirements for proposal review, record keeping, and reporting. The law requires an institutional committee to oversee and govern animal research. This committee is known as the Institutional Animal Care and Use Committee (IACUC or ACUC). Rules governing the housing and care of animals are under a national accreditation group consisting of veterinarians and experienced animal care facility managers. Before animals are used in experiments or even purchased and shipped to an institution, the ACUC must approve the protocol for research. Many of the requirements are more stringent than those for human research because the risk to the subject (i.e., animal) is usually death, and this risk needs to be justified by the potential benefits. The requirements include an examination of the validity of the experiment, statistics, alternatives, methods of monitoring, and proof of necessity of the experiment. The arguments for the quality of the data to be obtained must be convincing to a committee of peers and non-peer concerned citizens. A review of procedures and law enforcement relative to animal experimental procedures is given by Hamm, Dell, and Van Sluyters (1995).

Case 8.5 Animal Experiment to Enable Human Exploration of Space

The exploration of interplanetary space by human beings entails a number of physiological risks ranging from bone and muscle loss to the harmful effects of radiation in space. Because most of the space radiation comes as discrete ionizing particles different from the radiation occurring in x-ray exposures, the physiological effects cannot be extrapolated from the wealth of data from x-ray exposures on Earth. To evaluate the risks from these discrete particles to astronauts during long-term space exploration, animal experiments must be conducted. Thus, the following experiment on dogs was conducted using simulated cosmic particles from a high-energy physics accelerator.

1. Eight purebred two-year-old dogs were used (four exposed, four controls).

2. The exposure was conducted at a national laboratory accelerator in 1996 using high-energy iron particles with a total dose such that each brain cell nucleus was transited by three particles on average. This is the number of hits per cell nucleus an astronaut would receive from cosmic particles during a two-year mission. (Note that though the energy per mass was 2 joules/kg, the particles deposit energy along confined tracks of about 20 nanometers diameter; thus the character of biologic damage is not similar to that from x-ray exposure at the same average energy per mass. This fact justifies this experiment with accelerator-produced particles as previous data from x-rays are not relevant.)

3. Animals were anesthetized during the exposures, which would be painless nevertheless, to ensure stable positioning.

4. During a period of eight years after the exposure, the animals were studied every few months using imaging methods such as magnetic resonance imaging and positron tomography. Each study required anesthesia and transport of the animals.

5. After eight years the animals were sacrificed in order to perform histology on the brains to determine if there were pathologic consequences of this exposure.

Did this experiment meet the requirements set out above for animal experiments? With respect to species choice, humane treatment, proper anesthesia, and methods of animal husbandry, the experiment was in compliance. The questionable aspect of this experiment and other experiments employing limited numbers of animals has to do with the statistics. The justification of an animal experiment is sound only if the results will lead to a definitive answer to the scientific question or goal of the experiment. Employing only four animals in an exposure with planned follow-up of eight years was considered risky because the animals might die of natural causes at any time. It was also possible that the cause of death could not be determined because of practical issues, such as a long weekend without surveillance. Further, if two animals showed increased pathology and two other animals showed no increase relative to controls, what conclusion could be drawn? The use of more dogs was prohibited by the lack of financial and physical resources to maintain animals in the best possible environment to assure extraneous health problems would not threaten the well-being of the animals.

In this case, the justification was based on the following scenario. Suppose that all four exposed animals and none of the control animals showed signs of pathology. Alternatively, suppose that all four exposed animals showed no pathology or imaging results to indicate harmful effects of the particular exposures. The likelihood of either of these two scenarios was just as great as the likelihood of an inconclusive experiment. Thus, this argument and the importance of the information justified the experiment. As it turned out, none of the exposed animals showed harmful effects, either by histological examination of the brain or by in vivo imaging studies. In fact, there were fewer signs of brain degeneration in the exposed animals than in the actual controls or age-matched controls from other sources.

Reduction in the Need for Animals

A substitution of new technologies for animal experiments might be a solution to the conflict between animal rights ethics and medical research needs. Three examples of the immediate benefits of new technologies that may mitigate this problem include the use of microarrays, new imaging methods, and the use of selected transgenic animals. Whereas the use of microarrays as a substitute for animal studies and imaging methods allows sequential studies without animal sacrifice, the new technology of transgenic animals might increase animal use overall because of the ease and rapidity with which it provides data.

Mircoarray and Phage Array Technologies

Mircoarrays are glass or plastic plates with a closely packed pattern of small wells each 50 to 100 µm in diameter that allow a researcher to identify sequences of unknown RNA or DNA. Adsorbed on the substrate of the wells of microarrays are sequences of DNA or RNA, which have a known relationship to a particular gene (knowns). If each of these fragments of the known gene has a fluorescent label of one color, say red, then when exposed to light, the array will show a pattern of red spots for each successful attachment of the known. The pattern could be 100 by 100 wells. These patterned arrays of knowns can be obtained commercially. A sample of tissue or body fluids to be identified is prepared so that the amount of mRNA or DNA is replicated to give enough for analysis. Each tissue sample is labeled with a fluorescent compound of a different color than the known. For example, if the known is red, then the unknown to be tested might be labeled with a dye that fluoresces green. If a sample of unknown RNA or DNA is complementary to the known RNA or DNA on the array, it will attach or hybridize. After washing off the unattached molecules, the array is illuminated with light. If an unknown fragment attaches to a particular known, green or yellow will appear. Where there is no attachment, the fluorescence will be red. The reason yellow appears is that green with red gives yellow. These tests enable a more efficient recognition of the DNA patterns of genes and gene expression. This process could replace some uses of animals because the methods are applicable to small samples from animals as well as small samples from human subjects, and obviate the need for animals for some types of experiments. This technology is also able to improve our understanding of the operation of biological systems during health and disease, or their response to other factors such as chemical insults, drugs, diet, or other experimental situations.

Although it is impossible to predict with certainty any future trends regarding animal use, microarray technology might not initially reduce animal use, as is often claimed to be the case, as argued by Jenkins et al. (2002). The accelerated pace of research as a result of the use of microarrays could increase overall animal use in basic and applied biological research by increasing the numbers of interesting genes identified for further analysis and the number of potential targets for drug development. Each new lead will require further evaluation in studies that could involve animals. In toxicity testing, microarray studies could lead to an increase in the number of animals studied if further confirmatory studies require animal verification. Before such technology can be used more extensively, several technical problems need to be overcome and the relevance of the data to biological processes needs to be assessed. Were microarray technology to be used in the manner envisaged by its protagonists, there would need to be efforts to increase the likelihood that its application would create new opportunities for reducing, refining, and replacing animal use. A critical assessment of the possible implications of the application of microarray technology on animal experimentation in various research areas is needed.

Phage array methods allow identification of chemical substances such as peptides that have an affinity for surface proteins particular to diseased tissues such as tumors. One can develop a library of different peptides on the bacteriophage and apply thousands of different bacteriophages, each with a different peptide appendage, to tissues of interest. The bacteriophages that attach to the surface proteins are extracted after the ones that do not attach are washed off. The peptide responsible for the attachment is identified using molecular biology amplification and sequencing methods. This allows a drug company to identify possible drugs without use of multiple animal models.

Animal Imaging Technologies

Animal imaging by high resolution magnetic resonance (MRI), computed tomography (CT), single photon tomography (SPECT), and positron emission tomography (PET) has the potential to drastically reduce animal sacrifice from that currently used in toxicology and scientific research. For example, the usual method to evaluate the behavior of potential drugs for treatment of a specific disease is to determine where the drug goes in the body after injection as a function of time after injection. The conventional approach is to do tissue sampling or autoradiography on groups of animals at multiple time points. A drawback is that the sampling requires animal sacrifice. Modern imaging methods allow one to make the measurements noninvasively without animal sacrifice. However, this modern technology will increase the number of experiments on a single animal, increase the use of anesthesia, and require studies on unanaesthetized animals with or without constraint because anesthesia can interfere with the chemistry being studied. Thus, no one would argue that animal imaging is more efficient and humane than previous techniques, but will the long-term consequences be an improvement in humane treatment? The evaluation of the consequences of new technologies in mitigating a current ethical problem is the subject of problem 8.7 in the Problem Set at the end of this chapter.

Transgenic Animals

As noted earlier in this section, transgenic animals are animals whose genome has been modified either by adding a specific gene or genes to provide a specific function or by knocking out specific genes to mimic the deficits of specific human genetic diseases. Transgenic animals have been developed to produce pharmaceuticals, to serve as an improved source of food (larger, leaner, with other desirable traits such as to produce more milk), as well as for research needs. The discussion of genetically modified organisms for production of commodities is in Chapter 6, "Ethics of Genetically Modified Organisms." There is a general consensus that transgenic animal development is an efficient approach to disease mechanism discovery. However, modifying animals in ways that change their essential nature (e.g., producing sedentary pigs so that they fatten more readily) is inappropriate and beyond moral acceptance to many (Banner Committee 1995).

As previously pointed out, transgenic animal evolution and new drug discovery methods have increased animal experimentation. Some of the major dilemmas for the accredited animal care committees are centered on the number of animals to be tested and the rate of testing. Review of the regulations provides little in the way of substantive guidelines. For example, the *Guide for the Care and Use of Laboratory Animals* (National Academy Press, Washington, DC, 2002) discusses optimizing the use of animals, but primarily in the context of unnecessary duplication of previous studies. The Animal Welfare Act requires that investigators consider alternatives to any procedures causing pain or distress with reduction of the number of animals being one of the possible alternatives. However, reduction in the numbers of animals studied, even in pilot studies, can lead to false positive results and repeated experiments that would not have been necessary if adequate numbers were used in the first place. Thus, because transgenic animals allow specific hypotheses to be explored with less experimental variability, there will be fewer animals required for an experiment. However, because transgenic animals provide a tool to explore new biological and medical science questions, the net result will be an increase in animal use.

Summary of Issues in Animal Experimentation

Animal experimentation presents the dilemma of medical need vs. respect for the life and painless existence of animals. The debates are as strong as those regarding right to life vs. right to choose in the abortion issue. Some solutions to the use of animals are being sought in new technologies, but there is concern that these new approaches could result in more rather than less animal use. An uncontested benefit of current regulations over experimenters and facilities is the minimization of futile experiments by demanding justification for the scientific basis for the proposed experiments. Other benefits of these regulations have been the strict requirements for cleanliness of the facilities and for the humane care of animals.

CHAPTER SUMMARY

This chapter discusses the professional ethics involved in human and animal experimentation. The underlying ethical theories of rights, duty, beneficence, and non-malfeasance prescribe ethical behavior of the physician or researcher. They include the requirement to treat the patient as an individual, to minimize or avoid harm, to balance the possible benefits of treatment with risks of possible outcomes, and to ensure that the patient is treated justly with equal access to health care and other benefits available to other members of the population. The principal focus of the procedures for human experimentation is respect for the autonomy of the subject. This autonomy requires an assurance that the subject has a full understanding of the procedures and their risks. Autonomy also requires there be no coercion by large rewards, and that there is no misconception that the experiment is a requirement for the subject's health or treatment. Being fully informed includes openness regarding possible conflicts of interest on the part of the experimenters. Protection of the subject's privacy (confidentiality) is also an important part of the requirements. Autonomy with respect to genetic information is an ethical aspect of the emerging technology of genomics. The conflict between the importance of making the genetic information of an individual available to society vs. the individual's right to privacy is currently not resolved. However, for most situations, U.S. federal guidelines give methods to assure the participant in human experimentation is not put at a greater risk than the probable benefit of the results to society. In short, the participant needs to be fully informed of all aspects of the experiment so he or she can consider participation with this knowledge.

Animal experimentation presents a difficult moral dilemma because it involves the use of a nonconsenting living thing that can suffer. The justification for animal studies has been that they benefit society. However, animal rights supporters and many scientists point to the difficulty in predicting whether an outcome will have sufficient benefit to justify the animal suffering associated with most experiments. The main goal of regulations governing the use of animals in research is to limit experiments to those that are scientifically sound with a high probability of providing valuable medical or other scientific information. A second major purpose is to insist on procedures that minimize animals' harm and suffering. Since the animal has no consenting capabilities, the animal research committee becomes the surrogate consenter.

REFERENCES

Appelbaum, P. S., L. H. Roth, C. W. Lidz, et al. 1987. "False hopes and best data: Consent to research and the therapeutic misconception." *Hastings Center Report* **17**:20–24.

Banner Committee (1995). *Report of the Committee to Consider the Ethical Implications of Emerging Technologies in the Breeding of Farm Animals*. Ministry of Agriculture, London: Fisheries and Food, HMSO.

Beauchamp, T. L. and J. F. Childress. 2001. *Principles of Biomedical Ethics*. Fifth edition. New York: Oxford University Press, pp. 166–173.

Belmont Report. 1979. See below: U.S. National Commission for the Protection of Human Subjects and Biomedical Research.

Cibelli, J. B., K. H. Campbell, G. E. Seidel, et al. 2002 (January). "The health profile of cloned animals." *Nature Biotechnology*. Nature Publishing Group, pp. 13–14, accessed July 22, 2005 via http://biotech.nature.com.

Emanuel, E. J., A. Wood, A. Fleischman, et al. 2004. "Oversight of human participants research: identifying problems to evaluate reform procedures." *Ann Intern Med* **141**:282–291.

Fried, E. 2001a. "The therapeutic misconception, beneficence, and respect." *Accountability in Research* **8**, p. 336.

Fried, E. 2001b pp. 331–348.

Hamm, T. E., Jr., R. B. Dell, and R. C. Van Sluyters. 1995. "Laboratory animal care policies and regulations: United States." *ILAR Journal* **37**(2):75–78.

Institute of Animal Resources, Commission on Life Sciences, National Research Council. 1996. *Guide for the Care and Use of Laboratory Animals*. Washington, DC: National Academy Press.

Jenkins, E. S., et al. 2002 (July–August). "The implications of microarray technology for animal use in scientific research." *Altern Lab Anim* **30**(4):459–465.

LaFollette, H. and N. Shanks. 1995. "Utilizing animals." *J Appl Philos* **12**(1):13–25.

Moreno, J., A. L. Caplan, and P. R. Wolpe. 1998a. "Updating protections for human subjects involved in research." *JAMA* **280**(22):1951–1958.

Moreno, J., A. L. Caplan, and P. R. Wolpe. 1998b, p. 1957.

Percival, T. 1803. *Medical Ethics*. Russell, London, as cited in Levine, R. J. 1981. *Ethics and Regulation of Clinical Research*. Second Edition. Baltimore, MD: Urban and Schwarzenberg, p. 208.

Singer, P. 2000. *Writings on an Ethical Life*. New York: Ecco Press/HarperCollins, p. 89.

Steinbrook, R. 2002. "Protecting research subjects—the crisis at Johns Hopkins." *N Engl J Med* **346**:716–720.

"Trials of War Criminals before the Nuremberg Military Tribunals under Control Council Law." 1949. No. 10, Vol. 2, Nuremberg, October 1946–April 1949, Washington, DC: U.S. Government Printing Office, pp. 181–182.

Twain, M. 1899 (May 26). "Letter to Sydney G. Trist, Editor of the Animals' Friend Magazine, in his capacity as Secretary of the London Anti-Vivisection Society. Location of the original unknown." A copy appeared in a pamphlet entitled "The Pains of Lowly Life" (London: The Animals' Guardian [Friend?], [circa 1900–1910?], page 130–131.

U.S. Department of Health and Human Services, Office for Protection from Research Risks, National Institutes of Health. 1992. *Institutional Animal Care and Use Committee Guidebook* (NIH publication no. 92-3415). Washington, DC: U.S. Government Printing Office. (NAL call number: HV4764.I58 1992.)

U.S. National Commission for the Protection of Human Subjects of Biomedical and Behavioral Research. 1979. *The Belmont Report: Ethical Principles and Guidelines for the Protection of Human Subjects of Research.* Washington, DC: U.S. Government Printing Office. See also http://ohsr.od.nih.gov/guidelines/belmont.html

Van Zutphen, L. F. 2002. "Use of animals in research: a science-society controversy? The European perspective." *ALTEX* **19**(3):140–144.

ADDITIONAL READING

Federman, D. D., K. E. Hanna, and L. L. Rodriquez (eds.). 2002. *Responsible Research: A Systems Approach to Protecting Research Participants.* Institute of Medicine Report. Washington, DC: National Academies Press, 290 pp.

Levine, R. J. 1986. *Ethics and Regulation of Clinical Research.* Second Edition. Baltimore, MD: Urban & Schwarzenberg, 322 pp.

Mappes, T. A. and D. DeGrazia. 1981. *Biomedical Ethics.* Fourth Edition. New York: McGraw-Hill, Inc., 645 pp.

Pence, G.E. 2000. *Classic Cases in Medical Ethics.* Fourth Edition. New York: McGraw-Hill Companies, Inc., 470 pp.

Shannon, T. (ed.). 1993. *Bioethics.* Fourth Edition. Mahwah, NJ: Paulist Press, 542 pp.

APPENDIX

Appendix 8.1 Office of Human Research Protections (OHRP) Informed Consent Checklist—Basic and Additional Elements of the Consent Form for Research Subjects

- A statement that the study involves research
- An explanation of the purposes of the research
- The expected duration of the subject's participation
- A description of the procedures to be followed

- Identification of any procedures which are experimental

- A description of any reasonably foreseeable risks or discomforts to the subject

- A description of any benefits to the subject or to others which may reasonably be expected from the research

- A disclosure of appropriate alternative procedures or courses of treatment, if any, that might be advantageous to the subject

- A statement describing the extent, if any, to which confidentiality of records identifying the subject will be maintained

- For research involving more than minimal risk, an explanation as to whether any compensation, and an explanation as to whether any medical treatments are available, if injury occurs and, if so, what they consist of, or where further information may be obtained

- An explanation of whom to contact for answers to pertinent questions about the research and research subjects' rights, and whom to contact in the event of a research-related injury to the subject

- A statement that participation is voluntary, refusal to participate will involve no penalty or loss of benefits to which the subject is otherwise entitled, and the subject may discontinue participation at any time without penalty or loss of benefits, to which the subject is otherwise entitled

Additional elements, as appropriate

- A statement that the particular treatment or procedure may involve risks to the embryo or fetus if the subject is or becomes pregnant, which are currently unforeseeable

- Anticipated circumstances under which the subject's participation may be terminated by the investigator without regard to the subject's consent

- Any additional costs to the subject that may result from participation in the research

- The consequences of a subject's decision to withdraw from the research and procedures for orderly termination of participation by the subject

- A statement that significantly new findings developed during the course of the research, which may relate to the subject's willingness to continue participation, will be provided to the subject

- The approximate number of subjects involved in the study

PROBLEM SET

8.1 What is the moral justification for research on very ill patients if this research will not benefit the patient and might make the patient more uncomfortable or possibly sicker? An example is the clinical trial of a new drug.

8.2 During the Spanish-American war beginning in 1898, over 5,000 soldiers and others died of yellow fever.

Walter Reed, after whom Walter Reed Army Medical Center is named, was commissioned to determine the cause and cure or prevention of this disease. He and his colleagues experimented on themselves with various ideas, and then on volunteers in the army. The volunteers were told they might get yellow fever anyway by virtue of their presence in the area, and if they volunteered, they would get one month's pay in compensation. The experiments showed that mosquitoes carried the infectious agent, and this led to extermination of mosquitoes in the Panama region and a disappearance of yellow fever by 1905.

Team A - Argue that it was ethically wrong to ask soldiers to volunteer and that the compensation of $100 (gold piece) was coercion, thus an immoral inducement.

Team B - Argue that the soldiers consented to participate; therefore, the soldiers knew the risks and the experiment was moral.

8.3 In 150 words or less, create an appropriate consent form with all of the required elements (see Appendix 8.1). The experiment is to withdraw 10 cc of blood from patients with Alzheimer's disease to determine the concentration of a specific protein. Make it one typed page with each element highlighted.

8.4 You are faced with the following dilemma. A researcher has formulated a protocol and consent forms for studying the relationships of DNA patterns, genes, to psychiatric disorders in patients at a psychiatric hospital. The control subjects are college students of age 18 or older. The consent form informs the persons that the data will be kept confidential and the purpose of the study is to determine what genes are associated with psychiatric symptoms. The researcher receives approval from the Institutional Review Board. During the research project, the researcher discovers an unusual cluster of gene patterns in the college students that occurred in only a few of the depressed psychiatric patients and in a somewhat higher percentage of the manic psychiatric patients. The researcher obtains the GPAs of all the volunteers in order to evaluate any correspondence between IQ and the gene pattern. Finding only a poor correlation, she goes to a nearby rival college, draws blood, and adds a $50 incentive if the volunteers provide their GPAs. Later the researcher sends her samples of extra blood to NIH for their analyses and excludes whether the student was at College A or College B, the GPA, and the anonymous data on the department in which the student is enrolled.

You are the researcher's professor, and while reading the thesis, you discover that she has gone beyond the boundaries of the protocol presented to the IRBs. How will you deal with the fact that the thesis data are extremely interesting and important and you stand to profit by joint authorship, but you are concerned that the researcher did not follow the protocol and went beyond the limits of consent?

The dilemma also is that there was no harm to any subject due to this activity and confidentiality procedures were followed strictly, so if there was no harm to the subjects, why not assume that the end justifies the means taken to get this breakthrough in science? (Your answer should be in bullet form and not more than 80 words total.)

8.5 You are running a clinical trial for a new drug aimed at treating Alzheimer's disease. Shortly before conducting the trial, you decide to visit all the patients who are to participate in the study at one of the hospitals. In hand, you have the forms that the patients have both read and signed in order to participate. However, after visiting only six of the patients, you learn that three of them have no idea what you are talking about. How should you proceed?

8.6 Four class teams will prepare an application to the Animal Care committee to do the following research. A scientific hypothesis has been made that atherosclerosis can be treated by infusion of solutions that remove cholesterol from the body. Technically these solutions consist of an engineered biomolecule known as Milano Apo A-1. This molecule is part of the high-density lipoproteins that are believed to protect against atherosclerosis. Your goal is to induce atherosclerosis in rabbits by feeding them a diet of peanut butter and mayonnaise for about six weeks. This will induce arteriosclerotic deposits in the aorta known as plaques. Next, after measuring the size of the plaques using imaging methods, you will start infusions of the new solution. You plan to measure the size of the atherosclerosis plaques in the aorta every two months using a small ultrasound device on the end of a catheter device that is inserted through the femoral artery in the groin. The ACUC has strict requirements in accord with federal guidelines regarding the written protocol, anesthesia, animal care, and related matters of scientific justification and validity of the methods to yield data that would justify the harm to the animals. Prepare a request to the ACUC on two pages maximum. This request will be presented to the class, which will act as the ACUC committee.

Four different teams will develop the application. Each team will present its application to the class using two overheads. The remaining class will act as the ACUC committee. One member of each team will be designated as the animal rightist and one member as the chair of the ACUC when reviewing the team presentations.

8.7 Determination of how a particular new drug is distributed in the body usually relies on multiple sampling of the tissues after injection of the drug. This means that to determine how the concentration changes over time in, for example, the kidney or liver, the animals will have to be sacrificed at perhaps six time points. Statistical variability is such that at least three animals are needed for each time point; thus a minimum of 18 animals is needed for these experiments. If the drug were labeled with a radioactive isotope, a fluorescent probe, or a magnetic resonance sensitive contrast agent, it would be possible to determine the information for six time points without sacrificing the animal. Thus, in principle only three animals would need to be used, a saving of 15 animals. But the isotope or other probe for noninvasive imaging might change the behavior of the drug in terms of its distribution and its toxicity. Thus, additional animal studies are needed to verify the behavior of the labeled drug. One method might be to do a few animals using the old method of sacrificing animals having received the nonradioactive drug and compare the results with the results using the labeled drug. How would you approach this situation if the ACUC questioned the scientific validity of your proposal to do the distribution study with a labeled drug, and claimed that your verification procedures for use of the labeled drug would end up sacrificing more animals in the end?

9

Ethics of Assisted Reproductive Technologies

INTRODUCTION

Various assisted reproductive technologies have been developed to help infertile couples conceive. Infertility is defined as failure to conceive after 12 months of unprotected intercourse. This problem affects approximately 10% of those of reproductive age in the United States (Mitchell 2002). Early approaches to treating infertility included artificial insemination and medication-induced ovarian stimulation. In vitro fertilization (IVF) was introduced in 1978, a procedure in which an egg fertilized by sperm in the laboratory is then placed in a woman's uterus. This procedure was first attempted in 1971 but was not successful until 1978, when Edwards and Steptoe reported the first live IVF birth in Great Britain.

Since that time, the number of clinics providing IVF has steadily increased, and the procedure is now widely accepted. The field has grown to include modifications of IVF and new methods for treating infertility, a field now known as assisted reproductive technology (ART). The method of IVF is important to understand because this technique and modifications of it are or might be used in the future for other methodologies coming to the forefront. Rather than proceeding with a pregnancy, a fertilized egg can also be incubated to the blastocyst stage from which stem cells can be obtained for use in regenerative medicine, or for the creation of animal clones that will greatly enhance the study of human diseases. The latter technologies are addressed in Chapter 6 "Ethics of Genetically Modified Organisms" and in Chapter 10 "Ethics of Stem Cell Technologies."

But first, it is necessary to understand the basic reproductive technologies. It was through in vitro fertilization and the technologies that developed around it that scientists learned the fine art of creating, incubating, and transferring an embryo. Many of these same techniques enabled the subsequent development of therapeutic and reproductive cloning. Once the facts about these procedures are in hand, we discuss the ethical questions associated with them. Below, we introduce some of the ethical issues they raise so that the reader can begin to consider them as the technologies are reviewed.

- Is unnatural fertilization morally acceptable?
- If there is a higher incidence of abnormalities in IVF-conceived infants compared to non-IVF-conceived infants, is it ethical to use this technology?
- Is it ethical to sell one's eggs or sperm?

- What are the responsibilities of an in vitro fertilization clinic to anonymous egg and sperm donors, to parents, and to spare embryos?

- Is preimplantation genetic diagnosis (PGD) and selection morally wrong because of its unnatural manipulation of nature?

- Is the use of PGD ethical as a means of embryo selection to guarantee the implantation of an embryo that will be a perfect match for an ill sibling who needs a bone marrow transplant?

We will return to these and other questions after reviewing the methodologies of the various assisted reproductive technologies.

TYPES OF ART

There are varying definitions of ART, but for the purposes of this section of the text, ART refers to any fertility treatment that involves the handling of both eggs and sperm, which are combined in the laboratory with the intention of creating an embryo. The term does not include treatment by artificial insemination (in which only the sperm are transferred to a woman desiring to become pregnant) or medical stimulation of the ovary that is done without the intention of surgically retrieving eggs. The assisted reproductive technologies that help infertile couples create a pregnancy include in vitro fertilization (IVF), gamete intrafallopian transfer (GIFT), zygote intrafallopian transfer (ZIFT), and intracytoplasmic sperm injection (ICSI). A roadmap to these techniques is found in Figure 9.1.

In Vitro Fertilization

IVF involves a series of procedures that that are often described as an "ART cycle." An ART cycle begins when the prospective mother begins medication to stimulate the ovary to produce multiple mature eggs and ends either with the transfer of an embryo to the

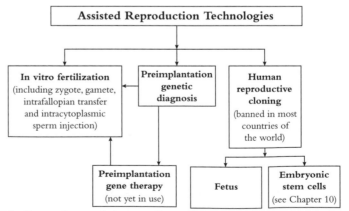

Figure 9.1 This diagram is a roadmap to understanding the various categories of assisted human reproduction. Additional and related topics of tissue and organ transplantation, as well as stem cell therapy for tissue regeneration, are also covered in Chapter 7, "Medical Ethics," Chapter 10, "Ethics of Stem Cell Technologies," and Chapter 11, "Ethics of Enhancement Technologies."

female reproductive tract for the purpose of creating a pregnancy or when the embryo is cryopreserved for future use. Not all ART cycles are successful. The duration of an ART cycle may vary, depending on individual response to therapy, but generally spans an interval of approximately two weeks (www.cdc.gov/reproductivehealth/ART02/faq.htm#5, accessed March 20, 2005). The success of an ART cycle is measured by whether or not a pregnancy results and whether the pregnancy results in a single or multiple gestation live birth(s), miscarriage, abortion, or stillbirth(s). The specific methodologies of IVF may vary from clinic to clinic, and the state of the art continues to expand and be refined, but the general procedures are reviewed in Figure 9.2.

Each ART cycle consists of the following procedures:

- Ovarian stimulation.
- Retrieval of mature eggs from the ovaries and sperm from the prospective father.
- Examination of eggs and sperm, and selection when appropriate.
- Incubation of the eggs with the sperm to initiate fertilization.
- Observation to ensure fertilization has occurred, and a short period of incubation of the fertilized egg to ensure viability.
- Transfer of the embryo to the mother's reproductive tract.

Ovarian Stimulation
Before IVF can be done, it is first necessary to obtain the eggs to be used to initiate conception. Various medications can be used to achieve this goal. The medications are potent, are usually given by injection, cause hormonal changes, and may cause serious side effects. Hormone levels are followed by serial blood tests, and serial ultrasound tests are used to visually monitor the enlarging egg follicles and the uterine lining to ascertain its readiness for implantation of an embryo. When the eggs are nearing maturity, another medication is administered to finalize the growth spurt of the egg, and approximately 36 hours later, the eggs are ready for collection.

Egg Retrieval
This is done by ultrasound-guided needle insertion through the vagina, by ultrasound-guided needle aspiration through the abdomen, or by inserting a tube (laparoscope) through a small surgical incision in the abdomen. This tube is equipped with fiber optics to allow visualization and collection of the eggs. The most commonly used procedure is through the vagina using a vaginal ultrasound probe with a hollow needle that is directed into the vicinity of the ovary under ultrasound guidance. The ovarian follicles are visualized, and the eggs are removed by suction through the needle-like tube.

Sperm Retrieval
Sperm are obtained, usually by ejaculation, and are separated from the semen (the ejaculatory fluid) in the laboratory.

Examination of the Gametes (Eggs and Sperm), Incubation to Initiate Fertilization
The harvested eggs are examined microscopically, as are the sperm. Those of the best quality are put together, usually in a small flat glass dish (petri dish) and incubated, or may be

transferred directly to the reproductive tract (a procedure known as gamete intrafallopian transfer [GIFT], in which case it cannot be immediately confirmed that fertilization has occurred). Otherwise, the combined eggs and sperm are incubated for a short period of time, after which a microscopic visualization is done to determine whether in vitro (literally meaning "in glass") fertilization has occurred. If problems with fertilization are expected, perhaps due to a low sperm count or poor sperm motility, intracytoplasmic sperm injection (ICSI, as described below) may be done, which involves injecting the sperm into the egg to increase the chances for successful fertilization.

Incubation of the Fertilized Egg

The embryo(s) are allowed to develop to the 2- to 6-cell stage (usually about 48 hours), after which one or more selected embryos (also known as zygotes) are flushed through a catheter into the woman's uterus through the cervix, or placed in a fallopian tube by laparoscopy (ZIFT). The unused fertilized eggs are frozen for possible future use, a process known as cryopreservation, and are referred to as "spare embryos." Sperm may also be frozen for a number of years for use at a later date.

In summary, GIFT, ZIFT, and ICSI are other ART techniques developed as variations of IVF to enhance reproduction. GIFT involves laparoscopy to transfer a mixture of *unfertilized*

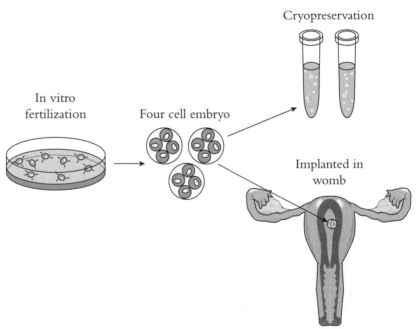

Figure 9.2 Method of in vitro fertilization. Eggs extracted from the prospective mother's ovary are fertilized in the laboratory by sperm taken from the prospective father. The embryos are incubated to the 2- to 6-cell stage of development, at which point one or more embryos are flushed into the mother's uterus or placed in a fallopian tube using a laparoscope. Implantation of the embryo takes place in the uterine lining (endometrium).

eggs and sperm to the woman's fallopian tubes, where fertilization takes place naturally. In ZIFT, the woman's eggs are *fertilized in the laboratory*, incubated, and a day later the fertilized eggs (also known as zygotes) are transferred to the fallopian tube by laparoscopy. ICSI, originally designed for use in severe male infertility possibly due to poor sperm motility or to an obstruction in the male reproductive tract preventing ejaculation, has gained wider use, perhaps because a direct injection using the best possible egg and sperm available results in a better outcome.

Preimplantation Genetic Diagnosis (PGD)

An important advance in ART has been the development of a technique called preimplantation genetic diagnosis that allows identification of genetic abnormalities in the egg prior to fertilization, or in the embryo after fertilization and before transferring to the female reproductive tract. It is known that as a woman and her eggs age, there is an increased risk of aneuploidy, a condition in which an egg has an extra or missing chromosome. The rate of aneuploidy in embryos is greater than 20% in mothers aged 35 to 39 years and is nearly 40% in mothers aged 40 years or older (www.emedicine.com/med/topic3520.htm#target3, accessed June 22, 2005). Chromosomal abnormalities can lead to miscarriage, severe disease, deformities in the fetus, a decreased fertility rate in older mothers, and an increased rate of congenital anomalies in infants of mothers over the age of 35 if the pregnancy is maintained. A number of genetic abnormalities can usually be identified prior to implantation.

Many of these abnormalities can be diagnosed using screening procedures in the second trimester of pregnancy. These tests include chorionic villus biopsy (placental tissue needle biopsy), done between 10 and 12 weeks after implantation, or amniocentesis (examination of the cells shed into the amniotic fluid that surrounds the developing fetus), done around 15 weeks after implantation. Another recently described screening test (Malone et al. 2005) can be done as early as 11 weeks after implantation. It measures fetal nuchal translucency by noninvasive ultrasound, which when combined with sequential blood tests, increased the detection rate for Down's syndrome (the most common chromosomal abnormality in older mothers) to 95%. The advantage to having this information in the first rather than the second trimester of pregnancy is that abortion is safer in early pregnancy. The situation in which a couple decides to abort a pregnancy based on the results of these tests raises difficult questions for some about the destruction of life that this represents, but it is not uncommon to take this path. Others point out that between 20 and 35% of recognizable pregnancies are aborted spontaneously anyway, many with chromosomal abnormalities (Schwartz and Palmer 1983).

However, PGD allows screening for aneuploidy and certain other genetic abnormalities *before* a pregnancy is established. Specific genetic analysis is necessary because aneuploid eggs otherwise can appear normal. At this time, not all 23 sets of chromosomes can be tested by PGD, only the chromosomes more commonly associated with abnormalities such as chromosomes 13, 16, 18, 21, and 22. The identification of these abnormalities by PGD is one way to prevent transfer of an abnormal embryo to the mother's uterus. As the ability to test for and identify additional genetic diseases evolves, there is likely to be an increased reliance on PGD to screen for them. Because not all chromosomes can be studied by PGD at the present time, there is still a small risk that a screened embryo is carrying a genetic abnormality. A current recommendation is to do either a chorionic villus biopsy or amniocentesis in addition. There are two ways of performing PGD. One is by polar body analysis of the egg; the other is by embryo biopsy.

Polar Body Analysis

The egg has two polar bodies that have the same genetic complement as the egg. One polar body is formed in the ripening egg and can be captured for analysis by a pipette inserted into the egg or retrieved after it is extruded from the egg as the egg matures. The second polar body is extruded from the egg once it has been fertilized and can be captured by micromanipulation. Thus, one polar body can be analyzed prior to fertilization for the presence of chromosomal abnormalities using DNA probes for suspect chromosomes. If no abnormalities are identified, it is inferred that the egg itself has no abnormalities. After fertilization, analysis can done a second time by evaluation of the second polar body. This repeat analysis is considered important because the analysis technique is not perfect and certain abnormalities might be detected after fertilization by the sperm. Once deemed acceptable, the embryo is incubated until of sufficient size for transfer to the maternal womb to complete the pregnancy.

Embryo Biopsy

Another option is to do preimplantation genetic diagnosis by embryo biopsy. This procedure is done soon after the embryo is formed, usually at about three days after fertilization, when it consists of a bundle of 6 to 8 rapidly dividing cells known as blastomeres. At this stage, these cells have not yet differentiated into specific cell lines for different tissues and are totipotential; that is, each can form an entire fetus or any tissue in the body. The removal of one or two cells at this stage of rapid division can be done without harm to the embryo (the risk of damage to the embryo is less than 1%, as reported by a number of leading clinics doing this procedure).

After the blastomere is gently teased away from the embryo, the embryo is incubated and continues to grow while the blastomere is evaluated microscopically for the presence of abnormal chromosomes. This testing destroys the blastomere, but the embryo is still able to continue its normal development. If the blastomere is normal, the embryo can then be transferred to the mother's uterus to continue development. In patients with recurrent miscarriages, the use of PGD reduced the number of miscarriages from 23% to 9% in one series (www.sbivf.com/miscarriage.htm, accessed July 9, 2005). Aneuploidy is not the only abnormality identified by this technique. Examples of genetic disorders that are or in the near future will be screened by PGD are listed in Table 9.1.

Table 9.1   Some Genetic Disorders Screened by PGD

- Polycystic kidney disease
- Retinitis pigmentosa
- Tay-Sachs disease
- Thalassemia
- Down's syndrome
- Duchene muscular dystrophy
- Fanconi syndrome
- Fragile X syndrome
- Gaucher's disease
- Glycogen storage disease
- Hemophilia
- Huntington's disease
- Marfan's syndrome
- Neurofibromatosis
- Sickle cell anemia
- Cystic fibrosis

The decision to perform preimplantation genetic diagnosis is based on the medical history of the parents, as listed in Table 9.2.

Table 9.2   Indications for Performing Preimplantation Genetic Diagnosis

- Advanced maternal age (35 years or older)
- Previous child with a chromosome disorder such as Down's syndrome
- One or both parents is known to have a chromosome abnormality
- In certain cases, when relatives of a child have a chromosome disorder
- Two or more spontaneous abortions
- Couples at risk for having children with X-linked disorders
- Couples wanting to choose the sex of their offspring

The increased availability of preimplantation genetic testing to screen for a wider list of diseases means that families at risk for genetic disease who have opted not to take a chance with a pregnancy are able to have children free of disease. The stress to a family will be significantly decreased knowing that a pregnancy is normal, and the financial burden to take care of a child with a genetic disease, estimated to be $2 million over the child's lifetime, could become a thing of the past. Thus, these techniques allow the birth of normal children to older women at risk for aneuploidy, and to parents who otherwise might pass on genetic disorders to their offspring. It is also hoped that the transfer of chromosomally normal embryos will lead to improved rates of embryo implantation and successful pregnancies.

As noted above, PGD techniques are not 100% accurate. At the 11th Annual Meeting of the International Working Group on Preimplantation Genetics in Vienna in 2001, it was reported that embryo or polar body PGD applied in over 3,000 clinical cycles resulted in a pregnancy rate of approximately 24%. Among children born at the time of the report in whom follow-up was available, chromosomal abnormalities were seen in 4.9%, a number comparable to the prevalence in the general population. This figure further supports the reliability of PGD (http://216.242.209.125/11m.shtml, accessed July 9, 2005).

Success Rates of ART

It is estimated that over a million IVF babies have been born worldwide in the 25 years IVF has been in use. The number of fertility clinics has been steadily increasing, and in 2002, there were over 420 fertility clinics in the United States. Each clinic is asked to provide data on its ART success rates to the Centers for Disease Control (CDC). A discouraging figure is that 65% of the ART cycles using fresh nondonor eggs or embryos (the most common procedure) did not result in a pregnancy. However, once a pregnancy was established, 83% resulted in a live birth, of which 58% were single births and 36% were multiple-gestation births (www.cdc.gov/reproductivehealth/art.htm, accessed March 20, 2005).

Successful pregnancies were highest in the under 35-year-old age group (43%) but especially poor in the >43-year-old age group (5%). Live births from these pregnancies were 91% and 40%, respectively. Table 9.3 shows this dramatic difference in the success rates when maternal age is taken into consideration.

Table 9.3 ART Pregnancy and Live Birth Rates in 2002 by Age*

Age	Pregnancy Rate (%)	Live Birth Rate (%)
<35	43	39
35–37	36	31
38–40	28	21
41–42	17	11
>42	9	4
> 43	5	2

* Using fresh nondonor eggs or embryos

Table summarizes data from the 2002 CDC report on ART found at www.cdc.gov/reproductivehealth/ART00/sect2_fig3-13.htm, accessed March 5, 2005.

Not only is there often failure to produce a live birth following the use of ART services, but they are also costly. A perusal of U.S. fertility clinic Web sites in 2005 indicated that the minimum cost for a single IVF cycle is about $6,500, two cycles might cost a minimum of about $9,000, and the cost to provide three ART cycles and stored frozen embryos for six months is about $10,000. Donor egg retrieval costs, necessary in 10% of the 100,000 cycles of oocyte collection performed annually at United States IVF clinics, can be as high as $25,000 (Okie 2005). There are additional costs for procedures such as ICSI ($2,000) and an additional $20,000 cost if genetic analysis is required. Some clinics offer risk sharing with some refund for unsuccessful pregnancies; however, whether or not this is available usually depends on the age of the prospective mother.

Risks of ART

The known risks of ART include the probability of a low to very low rate of successful implantation, a multiple gestation pregnancy, and financial peril, particularly if the procedures fail the first time and need to be repeated, or if there are multiple premature infants. ART is also known to be stressful from both medical and emotional standpoints for the mother. Another risk yet to be verified is the potential for a higher incidence of birth defects in children conceived using ART techniques. Information published to date on these risks is summarized below.

Early on, attempts were made to improve the low rates of successful implantation following ART by transferring multiple IVF-conceived embryos to the mother-to-be, particularly in the older maternal age groups. However, this often results in a multiple gestation pregnancy and medical complications for both mother and infant. Risks to the mother can include hypertension, anemia, bleeding from the placenta, premature labor, cesarean section, and even death; infants of a multiple gestation pregnancy are at risk for prematurity and all of its complications, including low birth weight, respiratory difficulties, and death (Rebar and DeCherney 2004). A recent study has cited multiple gestation pregnancy as the most important factor resulting in high neonatal mortality (Szymankiewicz et al. 2004). These complications may be costly to treat, especially since premature infants may require prolonged intensive care.

Recognizing these problems, the subject of multiple gestation pregnancies has been actively addressed by the American Society for Reproductive Medicine (ASRM). To lower the incidence of multiple gestations resulting from the use of ART, the ASRM issued guidelines for embryo transfer in 1998. These guidelines limit the maximum number of embryos to be transferred according to maternal age, reproductive history, and quality of embryos available (American Society for Reproductive Medicine 1998). Recommendations range from the transfer of no more than two embryos in women younger than 35 years of age with good-quality embryos to no more than five embryos for those with a "below average" prognosis (those who are older than age 40 or have a history of multiple failed cycles of assisted reproduction). CDC data after these guidelines went into effect showed that, except for twinning, the percentage of pregnancies with three or more fetuses declined but not the overall pregnancy rate; there was also a steady increase in the percentage of live births per cycle. These improvements may have been due to ASRM guidelines but may also have been influenced by other factors such as improved screening procedures to select genetically normal eggs, better embryo culture techniques, the use of more blastocyst transfers, or improved insurance coverage (the absence of insurance coverage led to a higher transfer of embryos to improve the chance of success for an expensive procedure likely to be performed only once) (Jain et al. 2004b). Whatever the reason for the improvements, progress is being made to minimize the risks of ART. In fact, studies have now shown that even though the risks of multiple gestation pregnancies were thought to be well established, there were few differences in the incidence of complications in age-matched populations of mothers (Kozinszky et al. 2003).

As for the risk of an increased incidence of birth defects in ART-conceived infants, there are differences of opinion as to whether or not this is an issue. The risk of major birth defects diagnosed during the first year of life was twice as high in ART infants conceived by ICSI or IVF than in naturally conceived infants (Hansen et al. 2002). In this same study, infants born through ART were more likely to have multiple major defects as well as chromosomal and musculoskeletal defects. Both findings were similar when only single births were analyzed, so the defects could not be attributed to abnormalities resulting from low birth weight of a multigestational pregnancy. However, it is important to include in these analyses the age of the mother, because this is of critical importance in comparisons of ART to natural childbirth. In the very large Hungarian study mentioned above, where matched controls were included, it was found that the perinatal outcome of singleton and twin pregnancies following ART was comparable with that of spontaneously conceived pregnancies (Kozinszky et al. 2003).

Among single infants in the United States conceived by ART, there may be a disproportionately high number of low-birth-weight ($\leq$ 2,500 grams) and very-low-birth-weight (<1,500 grams) infants compared with infants conceived naturally (Schieve et al. 2000). Their findings were unexplained by maternal age, maternal parity, or gestational age at delivery. Further study is needed to determine if the results were related to the medications used to induce ovulation or to the ART techniques themselves. It should be noted that underlying pathologies in a woman who is having difficulty conceiving could contribute to a statistic that suggests increased risk. If this is the case, then increased risk data are not valid as the control population is not controlled for the same pathologies as those seeking IVF.

It has been of concern that couples requiring ICSI, particularly the male partners, have chromosomal abnormalities as a major cause of their infertility that could be passed down to their offspring. Among 447 couples studied, there were sufficiently high rates of abnormalities found that could carry a small to moderate reproductive risk. This led to the suggestion that both partners should be routinely karyotyped before ICSI (Meschede et al. 1998, Macas et al. 2001). Using ICSI in an animal model, there was a persistence of certain structures such as the sperm acrosome and some perinuclear structures that may be responsible for the increase in chromosome abnormalities seen with this technique (Dominko et al. 2001). Concerns about ICSI continued to be published in 2004 (Aittomaki et al. 2004). Still, thousands of healthy babies have been born using this technique.

In summary, the studies cited above show there may be yet unproven health consequences for the infant conceived by the use of ART procedures. While the majority of infants conceived by these techniques are born having a normal birth weight (94%) and without a major birth defect (91%) (Mitchell 2002), the magnitude of the risks might not be significant. It will take time before the true risks of these procedures are known.

Other risks of ART not usually discussed are the mix-ups in ownership of sperm, eggs, and embryos that can occur in the clinic, in the laboratory, or at the time of embryo transfer. These situations are rare and usually arise when several patients at the clinic are undergoing procedures at the same time. The errors may or may not be discovered at the time they occur, and in some cases are discovered several years after the birth of the child. Several cases below indicate some of the problems.

Case 9.1 Birth of Mixed Race Twins to a White Couple after ICSI

This case involves a mix-up that occurred at an IVF clinic in Great Britain when the eggs of two women, one white and one black, underwent intracytoplasmic sperm injection (ICSI) on the same day. The sperm from the black donor were by error injected into the eggs of the white woman, resulting in the birth of mixed race twins to the white couple (Dyer 2002):

> The white couple, called Mr and Mrs A in court, and a black couple, Mr and Mrs B, underwent treatment by intracytoplasmic sperm injection on the same day. Mrs A conceived, but Mrs B's treatment was unsuccessful and the Bs remain childless. DNA testing showed Mr B to be the twins' biological father.
>
> An internal inquiry by the hospital failed to reach a definite conclusion but suggested that the error probably happened either when the sperm samples were placed in a centrifuge or when they were removed from a storage box immediately before injection into the eggs
>
> The Human Fertilisation and Embryology Authority, which licenses IVF clinics, has made it mandatory for all clinics to set up a double checking protocol for all stages in the procedure where a mistake could occur.

How should the question of who should have rights to raise the mixed race children be resolved? Let us consider this as a case for the step-by-step application of the Four A's design strategy described in Chapter 1, "Ethical Principles, Reasoning, and Decision Making."

Advice:
The list of possible ambiguities includes: Do the biological father and his spouse want the babies? Does the mother who carried the mixed race twins want to keep them? What are the financial and legal responsibilities of the erring clinic? Are there precedents from other clinics or previous legal decisions? Are there other cases in which newborns have been sent home with the wrong parents that might provide helpful guidance?

Alternatives:
Possibilities include giving the offspring to the sperm donor and his wife; leaving the children with the birthing family; dividing the children between the two couples; sending one or both babies for adoption; and repeat IVF with the correct gametes.

Assessment:
Considering the above alternatives as a sample of the possibilities, evaluate the dynamics of taking action with respect to the near-term and long-term consequences. For example,

in the event the children are separated, will their real identity be disclosed to them? In the event the clinic is required to compensate the parents for the error, will this compensation be fair and just for both families? And of greatest importance, will it be equally fair for both children? A successful outcome might result from the interaction between the two families, leading to new alternatives suggested by and agreed to by the families.

Action:

Possible alternatives should be exlored with the two couples. The overriding moral principle of the autonomy of the babies should be the prima facie priority for decision making on the part of the couples and the arbitrators (e.g., negotiators, legal authorities). The financial involvement of the clinic becomes an important component in the selection of alternative solutions. For example, if the best opportunity for one child is to be returned to the biologic father but he cannot afford to provide the same opportunities as Family A, then the clinic might justly provide funds for the child to receive equivalent opportunities to that of the other child. If the financial situation of one or both couples has been compromised by the costs for ICSI such that it cannot be repeated, or if educational opportunities for one or both children have been compromised, a financial supplement from the clinic can be part of a just solution. The final resolution of this case was not known as of March 2005.

Case 9.2 Woman Implanted with the Embryo of Another Couple

This case is summarized from an incident reported in 2002 (Feder 2002). An unmarried San Francisco woman, Ms. A, underwent IVF using a donated egg and donated sperm, and successfully delivered a son. In 2002, a year and a half later and soon after the Medical Board of California began investigating the clinic, she was contacted with the information that she may have received the embryo of another couple undergoing IVF at the same clinic. Later testing determined that Ms. A had indeed received the embryo of Couple B, formed after IVF of the husband's sperm and a donor egg. Further, Couple B had a daughter who was born after implantation of another embryo formed at the same time from the husband's sperm and another egg of the same donor, thereby making Couple B's daughter a sister of Ms. A's son. Couple B received temporary visitation rights until their suit for full custody of Ms. A's child was decided. Ms. A's suit against the clinic alleged that the clinic staff was aware of the mix-up the day it occurred but never informed the couples until forced to do so after the Medical Board began investigating the clinic.

- What was the responsibility of the clinic? Were they not obligated to tell the couples of the error as soon as it was discovered?
- The birth mother has refused to give up custody. To whom should the child be given? How should custody be decided?
- Has the child and/or birth mother been harmed? If so, how? What about the other couple?
- Should fertility clinics in the United States have the same stringent double-checks at every stage a possible mix-up might occur as put in place in Great Britain following the mix-up described in Case 9.1?

As can be seen, the problems are not easy to resolve as there is no right or wrong answer as to whom the child should go to or what is "best" for the child. Neither are there any guidelines as to how to address this problem. Whatever the outcome, it will not be easy for any of those concerned to live with the court decision. The student who attempts to apply the Four A's design strategy to analysis of this situation should consider all of the stakeholders. The major stakeholder is the child birthed by Ms. A. Others include the lawyers, clinic board of directors, doctors, and possibly others in addition to the couples and their relatives.

In August 2004, Ms. A was awarded a $1 million settlement from the fertility doctor who performed the implantation; a suit is still pending against the embryologist who mixed up the embryos provided for implantation (Chiang 2004). In March of 2005, the biological father, who had temporary visitation rights for two years and had sued for full custody, was awarded shared custody of the now nearly three-year-old child (Ryan 2005).

Two additional cases (Cases 9.3 and 9.4) demonstrate other consequences of errors in IVF procedures (www.kaisernetwork.org/daily_reports/print_report.cfm?DR_ID=12705&dr_cat=2, accessed April 13, 2004).

Case 9.3 Incorrect Embryo Implantation

In 1998, Donna Fasano, a New York woman, gave birth to two infants, one black and one white, after she was implanted with the embryo of another couple along with her own embryos. Although the white child is hers biologically, the black child is not. The black child's parents successfully won custody of the black child, and the birth (implanted) mother lost an appeal in which she was seeking visitation rights of the then three-year-old child.

Case 9.4 Incorrect Sperm from an Unknown Donor

A Florida couple gave birth to twins in 1995 after the woman was implanted with embryos conceived from her eggs and her husband's sperm. However, both infants were found to have B-positive blood types, an impossibility for parents who are both O-positive. It was determined that the woman's eggs had been fertilized with the wrong sperm.

Thus, the mix-ups of eggs, sperm, fertilized embryos, and erroneous embryo transfer are other risks of ART. The agonizing implications and decisions about how to handle such errors weigh heavily on everyone involved.

In summary, the risks of ART include its low rate of successful implantation, possible complications from a multiple gestation pregnancy for both mother and infants, possible risks of genetic or developmental abnormalities in the offspring, the risk of mix-ups during ART procedures, and the financial risks that result from an expensive procedure that is often not successful. Potential parents will need to evaluate whether these emerging data will affect their decision to opt for ART. Despite the risks, the use of these techniques is growing, aided by the growth in the number of clinics, the aggressive marketing of services, and programs to finance the services over time. These factors may persuade some parents to undergo ART who might have waited longer and conceived naturally. Caregivers who perform ART are ethically bound to fully inform parents of these risks and to fully inform parents immediately if and when any irregularities occur.

Regulation of ART in the United States

In the years from 1996 to 2001 in the United States, the number of clinics offering ART services increased by a factor of 1.3 from 330 to 428, and the annual number of reported ART procedures increased by a factor of 1.8 from 64,724 to 115,392 (www.cdc.gov/reproductivehealth/ART00/sect2_fig3-13.htm, accessed March 5, 2005). However, there is no evidence that the increase in the use of ART procedures is due to an increase in infertility among the general population, but rather that life-style changes and better availability have made it easier to access and utilize these kinds of services.

The Society for Assisted Reproductive Technology (SART), an affiliate of the American Society for Reproductive Medicine (ASRM), has established a registry for all facilities that provide ART services in the United States. These units must be headed by board-certified reproductive endocrinologists and must follow ASRM and SART guidelines. The Fertility Clinics Success Rate and Certification Act of 1992 (Wyden Law) was passed to implement uniformity in reporting ART success rates because misleading success rates were published by some individual clinics. Success rates for each clinic are now reported by the CDC in collaboration with ASRM and SART. The Food and Drug Administration (FDA) now requires registration of all ART programs in the United States and has developed additional regulations for donor-screening and tissue-handling practices that will be mandatory and evaluated during FDA site visits to ART facilities (*Federal Register* 2004). These regulations were published in November 2004 and went into effect in May 2005.

Issues of privacy and consent during and after IVF must be protected, and issues of ownership of the products must be addressed to ensure that rights are not violated. Principles of justice also arise to ensure that everyone, including the potential fetus, is treated with fairness and equity, and that there is equal access for all.

Ethical Concerns about ART

The modern assisted reproductive technologies raise many ethical concerns. These include whether unnatural fertilization is morally acceptable; whether preimplantation diagnosis and selection is morally wrong because of its manipulation of nature and because of its lack of respect for the embryo and its autonomy; and whether parents should have the autonomy to select ART as an option, given its possible risks to the offspring. Further concerns are that ART procedures involve the taking of potential life from blastomeres or a developing embryo, and from spare embryos that are being experimented upon, discarded, used for stem cell research, or aborted. The stakeholders are the natural donor mother or the egg donor, the surrogate mother if used, the natural donor father or semen donor, the embryo, as well as the individuals of future generations. For example:

- Who should be allowed to donate eggs, sperm, or embryos?
- Should genetic screening by preimplantation genetic diagnosis be done? Who should have access to that information? To what extent should records be generated and kept?

- To what extent does the sperm donor have rights to anonymity or to "ownership"?
- What about the rights to cryopreserve and bank eggs, sperm, or embryos?
- Who has the rights to donate embryos for research?
- What are the rights of the unborn embryo? At what point does an unborn embryo *have* these rights?
- Should parents have the right to put up for adoption their unused frozen embryos?
- How should the right to informed consent apply to an unborn fetus who cannot give consent?

ART has also raised issues of beneficence and justice in that the expense of IVF makes it available only to the few who can afford it. It also requires medical resources that might be put to better use. Other questions include the ethics of:

- Who will decide the right to be born if a fetus is found to have a genetic or other disorder?
- What licensing requirements are needed for IVF clinics? What should be the role of the FDA or other regulatory bodies in regulating IVF clinics?
- How should unused cryopreserved sperm, ova, and embryos be discarded? Should there be laws as to how long they should be stored? *Should* they be discarded? To whom do they belong? What are the legal issues? What limitations should be placed on experimental research on these entities? Who decides?
- Should the use of spare embryos be allowed for stem cell research by private institutions in the United States, even though this work is forbidden if using federal funds?

Thus, the assisted reproduction technologies raise many issues. They are complicated, often emotionally charged, and not easy to decide. We begin by looking at the ethical issues of IVF.

The Ethics of IVF

When the birth of the first baby conceived by IVF was announced in 1978, concerns were voiced that are similar to those heard today about the expanded technologies of ART. While it was understood that IVF was a new treatment for infertility and a solution for couples for whom artificial insemination was unsuccessful, the exact nature of what was done was not well understood. The procedure raised concerns from safety, societal, religious, and economic standpoints. Critics felt it reduced the act of love to a test tube or a petri dish, and objected to the medical and emotional stresses of the procedure for the mother-to-be. Many feared an IVF infant would be severely deformed or impaired due to the manipulation of the egg and the sperm outside of the body. Screening an embryo for chromosomal abnormalities was tampering with nature and the diversity of the species. Many voiced concern that the implantation of several embryos to have the best possible chance for a successful pregnancy and the increased incidence of multiple births might cause hospital nurseries to be overwhelmed with very low-birth-weight babies, and cause young families to be overwhelmed by catastrophic medical expenses for a procedure not covered by medical insurance. There was also concern about the cost of a procedure that was not available to everyone and had a relatively low probability of success.

There were also religious concerns. Even though not all religions had an official stance on the subject, the Roman Catholic Church renounced it and expressed grave misgivings about its use, even attempting to ban it. Since 1978, IVF has become increasingly well known and utilized, but the position of the Catholic Church has not changed. However, the Mormon and Jewish religions accept most aspects of ART, with the exception of reproductive cloning. The Vatican has extended its condemnation to all ART technologies. Concerns center on the treatment of embryos, the killing of spare or imperfect embryos, and their use and abuse in research after parents have donated them as a source of embryonic stem cells. The use of ART by an unmarried woman using donor sperm was also an application of IVF not anticipated in 1978, a procedure that is condemned by the Catholic Church.

Is unnatural fertilization morally acceptable?

As to the morality of the use of IVF or the techniques that are variations of IVF, some would argue that they are unethical because these procedures are unnatural. This is the position of the Roman Catholic Church. An argument against this position is that it is better to bring into the world a life supported by loving, natural parents than to create no life at all, whether the means is natural or unnatural. Though Pope John Paul's dogma is against in vitro fertilization, it is noteworthy that he did tell newsmen, "I send my most heartfelt congratulations to the English baby girl whose conception was produced artificially" (Wills 2000). With respect to discarding defective IVF embryos, one can argue that defective embryos are being discarded naturally by the uterus with far higher prevalence than the selected defective embryos discovered in the IVF clinic during preimplantation genetic diagnosis. The sides to the argument do not have to do with the timing of when life started but with the appropriateness of natural and unnatural means of selecting a person who will have a sustained life.

Those in favor of IVF point out that while it is not successful in most of the attempts, and the success declines with the age of the mother (i.e., from 39% at 30 years of age to 2% in those over 45 years of age), the procedure provides a chance for life where none existed before. Even though IVF carries some risks to mother and child, these risks are not well known and might be offset by the great benefits in family happiness that occur when the stress of infertility can be overcome and the desire to have a child can be fulfilled. Further, with guidelines limiting the number of embryos implanted and the availability of genetic screening procedures, the complications of multiple gestation pregnancies and the birth of embryos with chromosomal abnormalities can now be bypassed to a greater extent than before possible.

If there is a higher incidence of abnormalities in IVF-conceived infants compared with non-IVF-conceived infants, is it ethical to use this technology?

This is a question of autonomy and non-malfeasance. On one hand, without IVF, infertile couples who do not conceive and opt not to adopt will be deprived of a family. On the other hand, if the IVF child is at greater risk for genetic or congenital abnormalities than a naturally born child, is it fair to expose the child to a lifetime of possible disability? What exactly are the risks of these procedures? At this point in time, the extent of the risks is not known. It will take years of follow-up to determine whether there are higher risks for ART-conceived children than for the general population. As noted before, parents will need to be fully informed of the possible risks and make choices based on what they believe is right for them and their unborn child.

Is it ethical to sell eggs or sperm?

IVF also raises many questions having to do with the rights of sperm and egg donors, and the rights of the progeny to know the truth about their origin. For example, the concept of cryopreservation of sperm and of marketing eggs for use by those who want to take advantage of a particularly brilliant, talented, or otherwise gifted father or mother is a form of positive eugenics. Sperm banks are readily available for those wanting to select a sperm donor with known attributes, some of whom include Nobel laureates. It is not clear how often there is remuneration for sperm donations, although it is usual for an egg donor to be paid. Is this the first step on a slippery slope to choosing other characteristics for our offspring (e.g., designer babies selected or engineered for beauty, height, genius, resistance to specific diseases, or longevity)? Is it demeaning to society that someone would choose to sell such entities as a commodity? Should the offspring be told of the use of a donor sperm or egg? Is *not* telling children about the use of a donor an infringement of their autonomy? What are the rights of the child to have full disclosure of medical or other information by the donor?

With respect to the rights of the offspring to know the truth about their genetic origin and health history, the American Society for Reproductive Medicine Ethics Committee (2004) issued recommendations in favor of disclosure to the offspring as early in childhood as is reasonable. Disclosure fosters open communication in the family and avoids the suspicion and mistrust that may develop if disclosure is not made, or if the adoptee is accidentally informed. Some countries mandate this disclosure, or make it available on request by the offspring. Arguments against disclosure include concerns about psychological upset in the child if informed, anxiety if the child is denied access to the donor's health history, fear that the child may reject the unrelated parent, and fear of rejection of family members condemning the use of donor gametes (American Society for Reproductive Medicine Ethics Committee 2004). Withholding vital information regarding the possibility of inherited diseases is clearly morally wrong. Presumably, the sperm donor's medical history has been screened, in which case one could argue whether the child has a right to know the true father. Donor privacy is another ethical issue in gamete donation.

The market for eggs has been very active since the 1990s, as evidenced by the prevalence of newspaper advertisements to purchase female eggs. Donor eggs can be purchased by couples undergoing ART if their infertility workup reveals the need for this type of assistance. The need for donor eggs can nearly double the overall cost of the ART procedure because the egg donor must undergo ovarian stimulation and surgical retrieval of her eggs and is usually paid for her services.

What are the responsibilities of an in vitro fertilization clinic to anonymous egg and sperm donors, to parents, and to spare embryos?

The responsibilities would be expected to have as the main priority the autonomy of all the donors, the parents (including surrogate mothers), and the embryos. The details of these responsibilities should be reflected in the clinic's credo, as would be expected from any corporation (see Chapter 4, "Business Ethics"). Full informed consent and the need for confidentiality should be paramount in defining procedures and establishing policies to preempt problems in the future. It is important to define the policy for keeping confidential the identity of the sperm or egg donor. Should the policy be based on the desires of the donor, the consent of the parents, or respect for the autonomy of the person to be born? What about the need for disclosure of the donor's medical history or the need to disclose new medical information? There may also be situations in which it might not be wise for the child to know the donor father or mother. Further, the egg or sperm donor might be put at a disadvantage should his or her identity be revealed. As the reader will recognize, many of these issues and their solutions are similar to those surrounding adoption. The dilemmas regarding confidentiality can be approached using the Four A's design strategy, with care to identify the stakeholders and to be open to the changing dynamics of situations involving confidentiality (e.g., stakeholders might change their consent for disclosure of identity).

As we shall see in the next section of this chapter, similar ethical questions are evolving about reproductive and therapeutic cloning. The ethical issues surrounding these technologies are likely to be even more complex, depending on the extent to which reproductive cloning has serious consequences for society and the genome of humankind, and whether therapeutic cloning lives up to expectations for the cure of serious diseases of mankind.

Ethical Issues of PGD

The marked increase in the percent of chromosome abnormalities in women's eggs with advancing age is one of the arguments for preimplantation selection of eggs and/or embryos in older women. Although currently PGD is not used as frequently as amniocentesis, chorionic villus sampling, or ultrasound to detect the presence of a defective fetus, its use is increasing because it offers diagnosis of many diseases before the embryo is implanted. Thus, PGD bypasses the dilemma of having to decide whether to abort an established pregnancy, although having to dispose of the defective embryo also raises a difficult moral question. For most, the decision to abort an abnormal fetus postimplantation is more difficult than the procedure of discarding an abnormal embryo preimplantation. The difference lies in the perception of life or personhood in each. For those who believe the early embryo is a moral entity, the destruction of the blastomere removed from an embryo for PGD is not acceptable either.

While preimplantation diagnosis and selection can be an acceptable way to prevent genetic disease, there is the potential that its use could be viewed as unacceptable if ways were found for couples to select for cosmetic, behavioral, or other non-disease-related traits instead. Many are of the opinion that the use of these techniques for purposes of enhancement is unethical because they tamper with nature and may be a step on the slippery slope leading to the type of eugenics practiced in the past. If in the future it is possible to provide one's offspring with a means for longer life, a higher IQ, or a more attractive or muscular physique, those who can afford these options would have advantages unavailable to others unable to afford these procedures. The resulting disparities raise fears of a reemergence of financial, educational, athletic, employment, and other inequalities that could change the social structure of society. Others ask, Why not? If one can add years to a child's life or make him or her smarter and stronger, what is wrong with this? Gregory Stock and John Campbell address many of these questions in their book *Engineering the Human Germline: An Exploration of the Science and Ethics of Altering the Genes We Pass to Our Children* (Stock and Campbell 2000a). These and other questions have been the focus of ongoing debate. This topic is more fully addressed in Chapter 11, "Ethics of Enhancement Technologies."

Is preimplantation diagnosis and selection morally wrong because of its unnatural manipulation of nature?

Apart from the potential risk of damage to the embryo by this procedure, preimplantation diagnosis raises questions about diversity. Will PGD lead to a population without diversity? Will there be too much sameness that might have consequences for the future? Preimplantation diagnosis offers the option of rejecting a less than perfect embryo, but the criteria for rejecting an embryo are not well defined. Some might accept an embryo with certain defects, but how many, if given the choice, would reject any embryo considered undesirable for any reason. What effect would this have on future generations?

It has also been suggested that PGD might have unknown risks that should be reviewed and debated. However, these are speculations. The questions that need exploring with respect to concerns about interfering with natural processes are:

- Is this the first step onto the slippery slope of manipulating nature even further?
- Are we ridding society of serious diseases that bring great suffering? What *are* the intergenerational risks?
- What about the natural diversity of humanity?
- If we allow in utero surgery to correct a life-threatening anatomical problem, why would we not use PGD to bypass an embryo with a serious genetic abnormality?

Most would not argue the point that if one believes that the embryo has rights as a human being, then selecting those to be discarded presents a serious moral problem. Decisions about this problem will differ with different cultures and religions. However, the pivotal question in this debate is: At what point in time, from conception to birth, does personhood or human life exist? Some religions phrase this differently: When does individuation or ensoulment occur? At what point are actions performed upon a developing embryo or fetus perceived as having moral consequences? Many say this point in time is the moment of conception. Others argue that the point in time is during gestation when the fetus begins to develop specific organs and tissues (gastrulation). There are many answers to these questions, depending on one's religious beliefs, cultural background, and life experiences.

Is it ethical to use PGD as a means of embryo selection to guarantee the implantation of an embryo that will be a perfect match for an ill sibling in need of a bone marrow transplant?

On the one hand, creating a human being who might be harmed by a procedure to save another human being some would deem unethical. However, when the newly conceived sibling is old enough to give consent at the time of the transplant, one could argue that the end justified the means. The situation of coercion occurs for the sibling whose family has decided that he or she *must* donate bone marrow, a kidney, or part of a liver to save a sibling. A recent case is an example of the issues.

Case 9.5 Screening IVF Embryos to Ensure a Tissue Match for an Ill Sibling

A couple in Great Britain wanted to screen their IVF embryos to ensure that their next child would be a suitable donor match for a seriously ill sibling needing a bone marrow transplant and for whom no compatible match could be found. The family petitioned and won an appeal to reverse a lower court decision so that they could be granted the required license to test the embryo.

Opponents argued that such testing would lead to the slippery slope of creating designer babies for spare body parts (Ross 2003). But what parent would not do everything to try to save a sick child or relative? Since that time, further reports of similar testing of embryos in the United States have been published (Tanner 2004, also Verlinsky et al. 2004). This type of preimplantation testing of IVF embryos to ensure that the ensuing child is a compatible donor for a sibling needing bone marrow or other transplantation is becoming more common. This procedure is discussed in more detail in Chapter 10, "Ethics of Stem Cell Technologies."

CLONING

Cloning is the process whereby a copy of a living cell or organism is produced. There are two types of cloning: *reproductive* cloning and *therapeutic* cloning (see Figure 9.3). While the technique of cloning differs from ART, it takes advantage of many techniques developed for ART, such as obtaining, manipulating, and microinjecting eggs in the laboratory. Methods for producing embryonic stem cells through cloning or directly from embryos are discussed in Chapter 10, " The Ethics of Stem Cell Technologies."

Reproductive Cloning

No successful cloning of a human being has ever been verified. The cloning of animals is common, examples of which are Dolly the sheep, numerous farm animals, as well as the millions of transgenic mice and rats that are cloned for laboratory use. In animals, the most common approach to reproductive cloning has been to use cells that contribute to the structure of tissues, known as fibroblasts. These cells are easily obtained from a scrapping of the skin and grow readily in a petri dish in the laboratory. The procedure is also known as *somatic cell nuclear transfer*.

Reproductive cloning involves the substitution of the nucleus of an egg or a blastomere with the nucleus of a somatic cell, in our example, a fibroblast from the individual

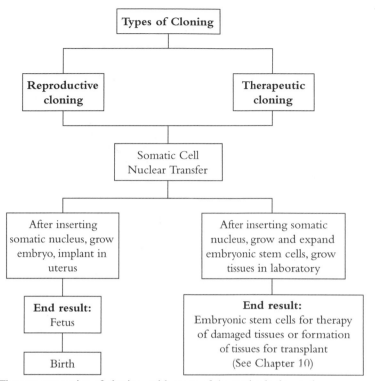

Figure 9.3 The two categories of cloning, with most of the methods denoted.

to be cloned. This is done in the following way. An egg is removed from the female and is then fertilized in vitro by the same species male, or is stimulated in such a way as to mimic fertilization (methods also used in asexual reproduction or parthenogenesis). The resulting embryo is incubated until a four-celled embryo is formed. The membrane surrounding this embryo, the zona pellucida, is removed and the four cells are separated. One of these cells is enucleated to form a blank cell into which is inserted the nucleus of a fibroblast or some other body (somatic) cell of the person or animal to be cloned. This cell is then stimulated to divide. When of sufficient size, the cloned embryo is transferred to the uterus to continue its growth to term. Figure 9.4 illustrates this procedure.

Thus, in *reproductive* cloning all of the genes of the nucleus of the clone come from a single somatic or body cell. The cell with its new genetic material is stimulated to divide such that a clonal embryo is created that is transferred to a womb with the intent of creating a fully formed living individual that would be a new entity, a clone of the person from whom the somatic cell was taken. To reiterate, to date (2006) this has only been done in animals.

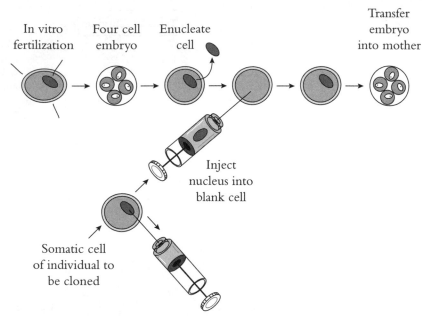

In vitro fertilization   Four cell embryo   Enucleate cell   Transfer embryo into mother

Inject nucleus into blank cell

Somatic cell of individual to be cloned

Figure 9.4 The method of in vitro fertilization is followed by nuclear substitution of the diploid nucleus of the individual to be cloned.

Reproductive and Therapeutic Cloning Differences

Reproductive cloning has the goal of giving birth to a fully formed animal or human being. Therapeutic cloning is a method of producing person-specific embryonic stem cells for cell and tissue replacement as therapy. In therapeutic cloning, the goal is to obtain embryonic stem cells to treat a patient who has supplied genes to create embryonic stem cells that will be recognized as self when used for therapy. Both types of cloning use an "empty egg" or blastomere (one from which the nucleus and therefore the chromosomes have been removed); instead of fertilization by sperm to provide paternal chromosomes, the nucleus of a somatic cell from the person to be cloned is inserted into the empty egg. This process is known as somatic cell nuclear transfer or SCNT. The cell with its new genetic material is stimulated to divide to become an embryo, but that is where the similarity ends:

- In reproductive cloning, the embryo is transferred to a womb to develop to term and results in the delivery of a live entity, a clone, that possesses the genes from only one person, the person who supplied the somatic cell and its nucleus. Only animal reproductive cloning is currently being done. The use of reproductive cloning of human beings is presently (2006) banned in the United States and many countries worldwide. This ban has been imposed because of concerns that human cloning could result in severe malformations or stillbirths, as occurs in animal cloning, although animal cloning is permitted (see Chapter 8 under animal experimentation). Other reasons to ban human cloning are interference with natural processes and concerns regarding the evolution of the species with unknown implications for society and for future generations.

The professional society known as the Association of Reproductive Health Professionals (AHRP) has taken the following position on human reproductive cloning (http://www.arhp.org/aboutarhp/positionstatements.cfm?ID=30#3, accessed March 15, 2005):

> While ARHP acknowledges the intense desire of infertile couples for a genetically related child or children, and supports reproductive choice, overwhelming scientific evidence (as summarized in the 2002 National Academies of Science report) shows that cloning is not a safe procedure for human reproduction at this time.

- In therapeutic cloning, the embryo is never implanted in a uterus and there is never an intent to give birth to a living being. Instead, the developing embryo created using an "empty egg" and patient-specific genes is incubated for about five days in the laboratory until it reaches the blastocyst stage, a hollow ball that contains approximately 200 embryonic stem cells. These cells are removed from the blastocyst and cultured to produce what is known as an embryonic stem cell line. There is never an intent or a possibility of creating a living entity from these cells. The field of therapeutic cloning is in its infancy, and scientists are only beginning to learn about its potential. This subject will be further discussed in Chapter 10, "Ethics of Stem Cell Technologies," which will give details on the kinds of stem cells, how they are obtained, and their potential uses.

The differences between reproductive and therapeutic cloning are outlined in Figure 9.3.

Experience with Reproductive Cloning in Animals

Frogs were successfully cloned over 50 years ago, and nonprimate mammals have been cloned since the birth of Dolly the sheep (Wilmut et al. 1997). The experience with mice, cattle, goats, pigs, rabbits, cats, and dogs over the past 10 years has provided the basis for recognizing the risks and for banning attempts at human coning. Less than 3 to 5 percent of cloned animal embryos survive due to spontaneous abortions, stillbirths, and failure to thrive immediately after birth. Only a few percent of the attempted reproductive cloning attempts result in a live birth; for example, the Dolly clone occurred after 277 failures. Many of the offspring develop multiple respiratory, circulatory, immune system, and growth problems (Jaenisch and Wilmut 2001, Edwards et al. 2003). Some caution is warranted in generalizing from the poor success statistics for successful cloning of a few years ago, or in extrapolations from success rates for successful pregnancy to the statistics for the health of the offspring as the efficiency of cloning has been improving as well as the health of the offspring.

Experiments to 2006 to clone primates have not been successful. Although professional societies have disapproved of cloning nonhuman primates, experiments continue with persistent failures. For example, no pregnancies resulted from the transfer of 33 embryos into

16 surrogate Rhesus monkeys (Simerly et al. 2003). It was hypothesized that the nucleus removal from the egg also removes key proteins responsible for chromosome separation during cell division. These proteins are more widely dispersed in the eggs of other species, and this is possibly the reason sheep and other species have been more easily cloned. Unpublished data from Simerly and other groups suggest that the same problem may also thwart attempts to clone humans. The failure to clone nonhuman primates is used as strong evidence that the claims by others to have cloned human beings are false. This information has not stopped cloning research because some argue that there are potential ways around this obstacle. One approach is to do cloning without initially removing the egg nucleus; that is, the nucleus to be cloned is inserted into the egg, leaving the egg's original nucleus in place until the blastomere has developed, then removing the original nucleus of the egg. It is hoped this may be helpful depending on the degree to which nuclear signaling is important to the development of the clone (Lanza et al. 2003).

ETHICAL ISSUES SURROUNDING REPRODUCTIVE AND THERAPEUTIC CLONING

Ethical Issues Surrounding Reproductive Cloning

The ethical considerations surrounding reproductive cloning can be addressed according to the rationale for the cloning, the procedure used, the source of the cells involved in the cloning, and the known and unknown consequences. The hierarchy in Figure 9.5 can guide the understanding of both reproductive and therapeutic cloning.

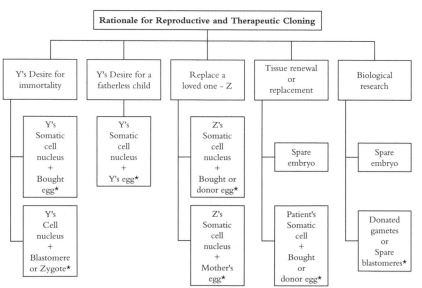

Figure 9.5 A framework for discussing the ethical issues surrounding the motives for cloning. *Denotes enucleated egg or blastomere.

There is great controversy over the moral concerns raised by reproductive cloning. The main arguments pit the autonomy of the parents against the autonomy of the clone, and the autonomy of the parents against the cultural and religious ethics surrounding reproductive cloning. Reproductive cloning creates an entirely new set of ethical issues for mankind. A clone would not have a genetic mother and father since cloning is an asexual means of reproduction. All of the genes of a clone come from a single somatic or body cell of a single individual. Thus, a clone contains genetic material from only one person instead of two, a dramatic departure in the history of reproduction. If a man cloned himself, the child would not be his son or his twin but a new category of biological relationship: his clone.

As noted previously, the safety of reproductive cloning is of great concern. However, many believe that a cloning for the purpose of forming a human being will occur in the near future. A less than perfect outcome of a clone will present ethical questions for the clone itself, for its family, for society, and for the future because the genetics of that clone might be an important perturbation of the usual human genetics. These changes carry with them unforeseen risks for the clone in later life, and for the children of the clone. Further, the essence of what is currently referred to as "family" may have new meanings for the future. Perhaps the changes will be for the best, but it is not presently clear that this will be the case.

Arguments for Reproductive Cloning

Leaving aside the current safety concerns about reproductive cloning and considering only the possible beneficial aspects, some feel that there are a number of arguments in its favor. Advocates feel parents should have the freedom to assert their reproductive autonomy in whatever way they choose, not thwarted by governmental or other restrictions. For similar reasons, scientists should not have their autonomy impinged upon, but should be free to pursue whatever research they feel is important under the guidelines presently in place for scientific research. For society to declare nonvirtuous a certain type of scientific research, or to declare the activities of scientists nonvirtuous because their research might lead to intergenerational risks, is not acceptable to most scientists.

Other arguments for reproductive cloning involve those who are infertile and for whom cloning represents another ART method that offers attractive possibilities, such as avoiding the genetic predisposition that caused their infertility, or for creating identical multiple children staged by cryopreservation to be born at appropriate times to optimize upbringing. If genetic screening or therapy could be combined with cloning, it might allow the births of children with pre-chosen characteristics, or at least without inherited diseases. Cloning allows a means of parenthood for those who are single and do not want to involve another person, or those who might be in need of an organ or tissue transplant as long as the clone will not have the defected organ. Others might want to create another genius or athlete (although there is no evidence that a clone would possess the traits for which it was bred). Some argue that reproductive cloning is going to be a thing of the future anyway, and that the best way to minimize the problems is to address them up front. Only by careful scientific research using properly designed protocols, and by encouraging open communication among investigators, ethicists, and the public, will the safety and other issues be resolved.

Arguments against Reproductive Cloning

Critics point out the profound difference between the creation of a transgenic animal and the creation of a transgenic human. In animal cloning, success is achieved only after multiple attempts. Even if implanted successfully, many animals are born dead, born so seriously impaired that they die soon after birth, or live their lives seriously deformed or die prematurely. However, the reports that cloning successes are very low and that the frequency of abnormalities is very high are misleading. Whereas the initial trials that led to cloning the first sheep required 277 attempts, the success rate after 10 years was very much greater (e.g., about 50%). The figure that is most important is the percentage of problems and abnormalities in the cases of successful cloning. Data are usually limited to single studies with few animals, but when data from all of the clone studies in the literature up to 2002 were combined (including cows, sheep, goats, mice, and pigs), 259 of 355 (77%) had no problems (Cibelli et al. 2002). These data argue that most cloned animals that are born develop normally.

Animal transgenics can be destroyed if imperfect, but a human transgenic or clone that does not measure up cannot be discarded and may have to endure, along with society, a lifetime of consequences. The scientific reasons for the deformities are simply not well enough understood at the present time to justify allowing reproductive cloning, and it remains a procedure that is banned by most countries.

Even if the safety questions of reproductive cloning can be resolved, the arguments against it on societal grounds have broader ethical implications. The process can result in a disruption of century-old definitions of family, and creates a new entity: a clone whose parentage will be differently defined. How will a clone be integrated into society? Depending on the source of its DNA, a grandmother or a sibling could be the mother of a clone, with possible societal repercussions resulting from confusion over the concepts of mother and father. Will clones be shunted to endure roles of servants or lesser beings? Or because their attributes can be chosen and their DNA enhanced, will they become dominant over others? There are other concerns having to do with the age of a clone—is it that of a new entity, or will it have the chronological age of the person supplying the DNA and thus experience rapid aging? Animal studies have not provided the answer to this question, despite the fact that the lengths of the telomeres (that shorten with each cell division) appear to be similar to those of normal embryos (Hochedlinger and Jaenisch 2003). However, Dolly and some other clones have died sooner than expected.

Further, a basic requirement for research on humans is that it be done only on the condition of informed consent, a condition that is impossible in this setting. Should parents have the right to choose these conditions for their offspring? What will be the consequences when clones wish to reproduce? Will there be unforeseen consequences for their offspring? Special concerns stem from the possibility that genetic engineering technology offers the possibility of allowing choice in physical and other attributes. New entities may be created with possible genetic anomalies that may be incorporated into the genome of generations of the future. There is also the possibility that reproductive cloning will lead to a new type of negative eugenics that could create unwanted gender, economic, social, and species gaps.

As research continues, priority should be given to considerations of the potential risks to society for severely deformed individuals, the demeaning of life, and the possibility for new classes of individuals who could be dominated by others. The sanctity of *Homo sapiens* and life as we know it today, and the unnatural acceleration of evolution of the human race to new forms, are questions for debate and for public policy.

Interestingly, these possibilities were anticipated in 1932. A popular book, Aldous Huxley's *Brave New World*, is a fantasy about the consequences of human reproductive cloning. It describes a future time in which "controllers" use reproductive technology to form populations of clones whose roles in society are preordained. The contrasts between the clones and other characters with human longings point to how precious it is to be human. Today's critics of reproductive cloning point to the book as an example of its potential to change what it means to be human. Since cloning has the potential to alter the DNA of the human species forever, they fear it will lead to a type of eugenics that generations of the future may regret. Proponents of cloning argue it is an opportunity to improve humanity in the future.

Another feature of human reproduction that will raise ethical questions in the future of human cloning has to do with the formation of an embryo using induced fusion. With this technology, it is now technically possible for two unrelated men or two unrelated women to create an embryo containing genetic material from each. The process by which this is accomplished is the opposite of identical twinning or multiple births. In the latter, the cells of the very early embryo become separated, and each goes on to develop a complete independent embryo. In induced fusion, instead of the cells separating, the cells become compressed and join. It is known that in nature, cells from two eggs ovulated separately and fertilized by the same or another father can fuse together, giving rise to a single embryo with the genetic material from each. This person is known as a chimera, and there have been a number of such people identified by genetic screening done because of the need for a transplant or for some other reason. Chimeras can appear normal, or if cells from a male and female embryo fuse, there may be ambiguous genitalia or other signs of mixed signals as the embryo develops. Embryo fusion technology would avoid this by merging two same sex embryos. This has already been done successfully in animals in the laboratory.

Thus, in the future it will be possible for the single parent to "conceive" asexually, or for parents of the same sex to conceive a child possessing genetic material from each (Stock and Campbell 2000b). This raises many of the same ethical questions and issues having to do with same-sex marriages and same-sex adoptions. It also raises concerns about how a clone formed from two cells from a single parent might fare, as in the long run there might be consequences of genetic abnormalities similar to those seen when close relatives or small populations intermarry. The same might also be true for a clone having genetic material from only one person.

Justice and Fairness in the Cloning Debate

In this ethical controversy over cloning, as well as other contemporary social and technical issues having risks and questions of morality, there is a need for open discussions and debates. All would agree that the government has a duty to respect the rights of all elements of society to express arguments and solutions. In the United States, the government usually calls on the National Academies (National Academy of Science, National

Academy of Engineering, and the Institute of Medicine) to review the technology and advise the nation on major technological and societal issues. However, this type of carefully conducted forum is not the only influence on public policy. Other strong and sometimes opposing advice comes from political influences that are not necessarily based on equal representation and virtuous dealings with appropriate representatives of science and social good. Below are two examples to substantiate the concern that the ethics of justice must underpin the open discussions of the debate on cloning, whether therapeutic or reproductive. The first example illustrates a proper forum; the second example illustrates some negatives.

Example 1: An expert panel of the National Academy of Sciences met in 2002 to discuss the scientific and medical aspects of human reproductive and therapeutic cloning. After extensive deliberations they made the following recommendations:

- Regenerative medicine had significant potential to help in the development of therapies for life-threatening and debilitating diseases and should proceed;

- Somatic cell nuclear transfer particularly has the potential for developing new ways of overcoming tissue rejection, and this information could be especially helpful in transplantation and finding new ways to regenerate tissues that are damaged;

- Ways to genetically manipulate stem cells as well as the development of a very large bank of ES cell lines should be addressed.

The panel unanimously recommended a legally enforceable ban on human reproductive cloning for the purpose of forming a human being for at least 5 years, after which time it should be reviewed. They did not recommend including in the ban therapeutic cloning or somatic cell nuclear transfer to produce stem cells as this technology has ". . . considerable potential for new therapeutic treatments for life-threatening diseases and for advancing fundamental knowledge." They further urged a national dialogue on the societal, religious, and ethical issues (National Academy of Sciences 2002).

Example 2: This is an example of political interference with public policy development on human reproductive and therapeutic cloning. It occurred in connection with President George W. Bush's Council on Bioethics, formed in 2001 and consisting of members appointed from fields including science, medicine, law, government, philosophy, theology, and other areas. Their task is to undertake ". . . fundamental inquiry into the human and moral significance of developments in biomedical and behavioral science and technology" and to ". . . advise the President on bioethical issues that may emerge as a consequence of advances in biomedical science and technology" (Creation of the President's Council on Bioethics, accessed August 20, 2004, at www.bioethics.gov/reports/executive.html).

Dr. Elizabeth Blackburn, a highly esteemed professor at the University of California San Francisco and member of the National Academy of Sciences and the Institute of Medicine, was fired from her position as a member of this council in 2004. Two other scientists were also replaced. All had opposing views to those of President Bush and the head of the panel, Dr. Leon Kass, especially with regard to therapeutic stem cell research. Not only did Dr. Blackburn believe Kass maintained a closed mind on the subject, but her and others' efforts to have their views adequately represented in the reports of the council were not honored. Further, discussions were thwarted on other issues of current importance in biomedical science. When all three were replaced by three others who more closely supported the view of Kass and the president, and who are not even medical scientists, Dr. Blackburn was invited to publish with her concerns that the council was failing to provide the best scientific information for policy decisions for the nation, and doing so for political reasons (Blackburn 2004):

> When prominent scientists must fear that descriptions of their research will be misrepresented and misused by their government to advance political ends, something is deeply wrong.

Dr. Blackburn's article is reprinted in its entirety in Appendix 9.2. Certainly President Bush's stance limiting stem cell research has raised concerns that in an area with great therapeutic potential. American scientists are being forced to stand on the sidelines while other nations are moving rapidly ahead with this research. Further concern is that the information coming to the president from the committee does not represent a balanced reporting of the latest information, and that some important facts are being either misstated or left out altogether.

While some scientists cite their concerns, it has been pointed out that the committee was formed by the president, and he is entitled to have members he chooses. But it is disheartening to think that the reporting has not been "fair and balanced" and is subject to searing criticism. Indeed, it would appear that the reports do not reflect the best interests of science due to omissions and misstatements.

CHAPTER SUMMARY

The assisted reproductive technologies, including IVF, GIFT, ZIFT, and ICSI, have been used to help millions of infertile couples conceive. The development of the technique of preimplantation genetic diagnosis for use in conjunction with IVF has expanded the ability to screen eggs and/or the embryo for genetic diseases and made it possible to bypass those that are abnormal.

There are some risks to ART procedures, but whether these are due to factors other than ART have yet to be determined. Would-be parents will have to follow developments in this area to decide the ethical implications of whether to proceed or not. There are other ethical implications of ART. There are those who feel IVF and screening procedures interfere with nature, lead to the rejection and killing of an abnormal embryo, and affect the genetic diversity of humanity. The production of spare embryos and subsequent cryopreservation of the spare embryos that are diverted to research (especially stem cell research) or are ultimately disposed of by the clinic result in the demise of potential life and should not be allowed. Heated debates have ensued because of differing opinions about when life begins, the point at which an embryo and its autonomy are to be respected, and whether parents have the right to make decisions for a being who otherwise has no advocate and whose life is at stake.

The difference between *reproductive* and *therapeutic* cloning is that reproductive cloning involves the creation of a clone from the genetic material of only one person or animal with the intent of giving birth to a new individual. Therapeutic cloning (discussed in Chapter 10, "Ethics of Stem Cell Technologies") involves the harvesting of embryonic stem cells from the blastocyst after somatic cell nuclear transfer. Reproductive cloning has been successful since 1997 only in nonprimate mammals (e.g., sheep, pigs, and other species). Therapeutic cloning has not been successful in human beings or nonhuman primates.

Reproductive cloning in human beings remains banned in most countries. One of the ethical issues surrounding human reproductive cloning is safety. The scientific facts from almost 10 years of experience up to 2006 lead to the conclusion that human reproductive cloning is likely to result in an abnormal fetus and abnormal child who might suffer from abnormalities and a limited quality of life. Other concerns are that parents who advocate cloning are to some degree violating the rights of the cloned individual in that they are creating the attributes of an offspring contrary to the traditional natural biological processes. Cloned animals have been developed with highly unique characteristics that have enhanced the study of human disease, the availability of unique pharmaceuticals, and the joy of a few animal lovers who have been successful in having their pets cloned.

Further ramifications of these new technologies are explored in discussing somatic and germline gene therapy in Chapter 10, "Ethics of Stem Cell Technologies."

REFERENCES

Aittomaki, K., U. B. Wennerholm, C. Bergh, et al. 2004 (March). "Safety issues in assisted reproduction technology: should ICSI patients have genetic testing before treatment? A practical proposition to help patient information." *Hum Reprod* **19**(3):472–476.

American Society for Reproductive Medicine Ethics Committee. 2004 (March). "Informing offspring of their conception by gamete donation." *Fertil Steril* **81**(3):528.

American Society for Reproductive Medicine Practice Committee. 1998 (January). "Guidelines on number of embryos transferred." ASRM Practice Committee Report. Birmingham, AL.

Blackburn, E. 2004. "Bioethics and the political distortion of biomedical science." *N Engl J Med* **350**:1379–1380.

Chiang, H. 2004 (August 4). "Mom awarded $1 million over embryo mix-up." *San Francisco Chronicle*, p. B3.

Cibelli, J. B., K. H. Campbell, G. E. Seidel, et al. 2002. "The health profile of cloned animals." *Nat Biotechnol* **20**:13–14.

Cowan, C. A., I. Klimanskaya, J. McMahon, et al. 2004. "Derivation of embryonic stem-cell lines from human blastocysts." *N Engl J Med* **350**:1353.

Dominko, T., L. Hewitson, C. Simerly, and G. Schatten. 2001. "Fertilization imaged in 2-, 3- and four dimensions: molecular insights for treating infertility." *Ital J Anat Embryol* **106**(Suppl 2):51–60.

Dyer, C. 2002. "White couple can keep mixed race twins after IVF blunder." *BMJ* **325**:1055b.

Edwards J. L., F. N. Schrick, M. D. McCracken, et al. 2003. "Cloning adult farm animals: a review of the possibilities and problems associated with somatic cell nuclear transfer." *Am J Reprod Immunol* **50**(2):113–123.

Feder, O. B. 2002. "Suit filed over mix-up of embryos at S.F. clinic." *San Jose Mercury*, August 2, www.mercurynews.com/mld/mercurynews/news/local/3785614.htm, accessed April 13, 2004.

Federal Register. 2004 (November 24). Department of Health and Human Services, Food and Drug Administration 21 CFR Parts 16, 1270, and 12271. Part IV. Current Good Tissue Practice for Human Cell, Tissue, and Cellular and Tissue-Based Product Establishments; Inspection and Enforcement; Final Rule. **69**(226):68612–68688.

Hansen, M., J. J. Kurinczuk, C. Bower, et al. 2002. "The risk of major birth defects after intracytoplasmic sperm injection and in vitro fertilization." *N Engl J Med* **346**(10):725.

Hochedlinger, K. and R. Jaenisch. 2003a. "Nuclear transplantation, embryonic stem cells, and the potential for cell therapy" (review article). *N Engl J Med* **349**:275-286.

Hoffman, D. I., G. L. Zellman, C. C. Fair, et al. 2003 (May). "Cryopreserved embryos in the United States and their availability for research." *Fertil Steril* **79**(5):1063.

Huxley, A. 1932. *Brave New World.* New York: HarperCollins Publishers, Inc.

Jaenisch R. and I. Wilmut. 2001. "Developmental biology. Don't clone humans!" *Science* **291**(5513):2552.

Jain, T., S. A. Missmer, and M. D. Hornstein. 2004a (April 15). "Trends in embryo-transfer practice and in outcomes of the use of assisted reproductive technology." *N Engl J Med* **350**(16):1642.

Jain, T., S. A. Missmer, and M. D. Hornstein. 2004b, pp. 1643–1644.

Kozinszky, Z., J. Zadori, H. Orvos, M. Katona, A. Pal, and L. Kovacs. 2003 (September). "Obstetric and neonatal risk of pregnancies after assisted reproductive technology: a matched control study." *Acta Obstet Gynecol Scand* **82**(9):850–856.

Lanza, L., Y. Chung, M. D. West, and K. H. S. Campbell. 2003. "Comment on molecular correlates of primate nuclear transfer failures." *Science* **301**:1482–1483.

Macas, E., B. Imthurn, and P. J. Keller. 2001 (January). "Increased incidence of numerical chromosome abnormalities in spermatozoa injected into human oocytes by ICSI." *Hum Reprod* **16**(1):115–120.

Malone, F. D., J. A. Canick, R. H. Ball, et al. 2005 (November 10). "First-trimester or second-trimester screening, or both, for Down's syndrome. *N Engl J Med* **353**:2001–2011.

Medical Record of the State of California. 1890. **37**:562.

Meschede, D., B. Lemcke, J. R. Exeler, et al. 1998 (March). "Chromosome abnormalities in 447 couples undergoing intracytoplasmic sperm injection—prevalence, types, sex distribution and reproductive relevance." *Hum Reprod* **13**(3):576.

Mitchell, A. A. 2002. "Infertility treatment—more risks and challenges" (editorial). *N Engl J Med* **346**:769.

National Academy of Sciences. 2002. "Scientific and medical aspects of human reproductive cloning." Report of the National Academy of Sciences Committee on Science, Engineering, and Public Policy. National Academy Press, Washington, DC, pages 92–100.

Newman, H. H. 1925. *Evolution, Genetics and Eugenics.* Second edition. University of Chicago Press, Chicago, IL, pp. 528–529.

Nudell, D. M. and L. I. Lipshultz. 2001 (December). "Is intracytoplasmic sperm injection safe? Current status and future concerns." *Curr Urol Rep* **2**(6):423–431.

Okie, S. 2005 (July 7). "Stem-cell research—Signposts and roadblocks (perspective). *N Engl J Med* **353**:4.

Rebar, R. W. and A. H. DeCherney. 2004. "Assisted reproductive technology in the United States." *N Eng J Med* **350**:1603.

Ross, E. 2003 (April 9). "British court permits embryo tissue-matching." *San Francisco Chronicle*, Nation section, no page number available.

Ryan, J. 2005 (April 14). "Biology, technology in conflict." *San Francisco Chronicle*, pp. B1 and B10.

San Francisco Chronicle. 2002 (April 11). "Bush tries to sway senators against cloning," p. A3.

Schieve, L. A., S. F. Meikle, C. Ferre, et al. 2000. "Low and very low birth weight in infants conceived with use of assisted reproductive technology." *N Engl J Med* **346**:731.

Schwartz S., and C. G. Palmer. 1983. "Chromosome findings in 164 couples with repeated spontaneous abortions, with special consideration to prior reproductive history." *Hum Genet* **63**:28–34.

Simerly, C., T. Dominko, C. Navara, et al. 2003. "Molecular correlates of primate nuclear transfer failures." *Science* **300**:297–299.

Steptoe, P. C. and R. G. Edwards. 1978 (August 12). Birth after the reimplantation of a human embryo. *Lancet* **2**(8085):366.

Stock, G. and J. Campbell (eds.). 2000a. *Engineering the Human Germline: An Exploration of the Science and Ethics of Altering the Genes We Pass to Our Children.* New York: Oxford University Press.

Stock, G. and J. Campbell. 2000b, pp. 64–66.

Szymankiewicz, M., P. Jedrzejczak, J. Rozycka, et al. 2004 (July–August). "Newborn outcome after assisted reproductive technology: experiences and reflections from Poland." *Int J Fertil Womens Med* **49**(4):150–154.

Tanner, L. 2004 (May 5). "Creating a baby to save a sibling: stem cells from healthy infants help ailing brothers and sisters." *San Francisco Chronicle*, p. A2.

Verlinsky, Y., S. Rechitsky, T. Sharapova, et al. 2004. "Preimplantation HLA testing." *JAMA* **291**:2079–2085.

Volpe, E. P. 1987. *Test Tube Conception: A Blend of Love and Science*. Macon, GA: Mercer University Press, p. 16.

Wills, G. 2000. *Papal Sins: Structure of Deceit*. New York: Doubleday Publishers, p. 97.

Wilmut, I., A. E. Schnieke, J. McWhir, et al. 1997. "Viable offspring derived from fetal and adult mammalian cells." *Nature* **385**(6619):810–813. Erratum in *Nature* **386**(6621):200.

ADDITIONAL READING

Andrews, L. and D. Nelkin. 2001. *Body Bazaar: The Market for Human Tissue in the Biotechnology Age*. New York: Crown Publishers, 245 pp.

Beauchamp, T. L. and J. F. Childress. 2001. *Principles of Biomedical Ethics*. Fifth Edition. New York: Oxford University Press, Inc., 454 pp.

Buchanan, A., D. W. Brock, N. Daniels, and D. Wikler. 2000. *From Chance to Choice: Genetics and Justice*. Cambridge, UK: Cambridge University Press, 398 pp.

Carlson, E. A. 2001. *The Unfit: A History of a Bad Idea*. Cold Spring Harbor, NY: Cold Spring Harbor Laboratory Press, 451 pp.

Fox, M. W. 1999. *Beyond Evolution: The Genetically Altered Future of Plants, Animals, the Earth . . . and Humans*. New York: The Lyons Press, 256 pp.

Fukuyama, F. 2002. *Our Posthuman Future: Consequences of the Biotechnology Revolution*. New York: Farrar, Straus and Giroux, 256 pp.

Goodman, K. (ed.). 1998. *Ethics, Computers and Medicine: Informatics and the Transformation of Healthcare*. Cambridge, UK: Cambridge University Press, 180 pp.

Joffe, C. 1995. *Doctors of Conscience: The Struggle to Provide Abortion Before and After Roe v. Wade*. Boston: Beacon Press, 250 pp.

Jonsen, A. R., M. Siegle, and W. J. Winslade. 1998. *Clinical Ethics: A Practical Approach to Ethical Decisions in Clinical Medicine*. Fourth Edition. New York: McGraw-Hill Health Professions Division, 206 pp.

Kass, L. R. and J. Q Wilson. 1998. *The Ethics of Human Cloning*. Washington, DC: The AEI Press, 101 pp.

Peterson, J. C. 2001. *Genetic Turning Points: The Ethics of Human Genetic Intervention*. Grand Rapids, MI: William B. Eerdmans Publishing Company, 364 pp.

Ridley, M. 2000. *Genome: The Autobiography of a Species in 23 Chapters*. First perennial edition. New York: HarperCollins Publishers, Inc., 344 pp.

Shenfield, F. and C. Sureau (eds.). 2000. *Ethical Dilemmas in Reproduction*. New York: Parthenon Press, 114 pp.

Stock, G. 2002. *Redesigning Humans: Our Inevitable Genetic Future*. Boston: Houghton Mifflin Company, 277 pp.

Walters, L. and J. G. Palmer. 1997. *The Ethics of Human Gene Therapy*. New York: Oxford University Press, 209 pp.

Wills, Gary. *Papal Sins: Structure of Deceit*. New York: Doubleday Publishers, 326 pp.

APPENDICES

Appendix 9.1

THE HUMAN CLONING PROHIBITION ACT OF 2001 AND THE CLONING PROHI-BITION ACT OF 2001 (excerpts from congressional hearings)

Today, the Subcommittee on Health will continue where the Subcommittee on Oversight and Investigations, chaired by Congressman Gene Greenwood, left off. We will examine two measures, which, in many ways, reflect the discussions of that hearing: H.R. 1644, sponsored by Congressmen Weldon and Stupak, and H.R. 2172, sponsored by Congressmen Greenwood and Deutsch.

This is a difficult issue, to say the least, and it involves many new and complex concepts. But we should all be clear, I think, about the controversies related to human cloning. The term "therapeutic cloning," which many people use to mean any type of cloning that is not intended to result in a pregnancy, is confusing.

It really includes two, distinct procedures, one of which is controversial, while the other, I think, is not. The non-controversial concept of therapeutic cloning is the cloning of human tissue that does not give rise to an embryo.

The controversial aspect involves the creation of a human embryo. This latter meaning is also the subject of both of the bills we will discuss today.

H.R. 1644 seeks to ban the creation of these cloned human embryos. H.R. 2172 seeks to prevent those who clone human embryos from implanting them in a surrogate mother.

http://energycommerce.house.gov/107/hearings/06202001Hearing291/print.htm accessed March 23, 2004

If animal cloning has taught us anything, it is that cloning has significant risk. Miscarriages, birth defects, and genetic problems are the norm when it comes to cloning. Less than 3 to 5 percent of cloned animal embryos survive. In fact, it took more than 270 tries before scientists were able to clone Dolly.

http://energycommerce.house.gov/107/hearings/06202001Hearing291/print.htm
accessed March 23, 2004

Mr. Kass. [continuing] details. I will just wind up. It seems to me, as the composition of this panel of witnesses will make clear, the issue of human cloning is not an issue of pro-life or pro-choice. It is not mainly about death and destruction. It is not about a woman's right to choose. It is not about stem cell research. It is not even about the basic freedom of scientists to inquire.

It is most emphatically about baby design and manufacture. And it is the opening skirmish in a long battle against eugenics and the post-human future.

Once the embryonic clones are produced in the laboratories, this eugenic revolution will have begun, and we will have lost our best chance to do something about it.

http://energycommerce.house.gov/107/hearings/06202001Hearing291/print.htm
accessed March 23, 2004

I am Richard M. Doerflinger, Associate Director for Policy Development at the Secretariat for Pro-Life Activities, National Conference of Catholic Bishops.

The only Catholic quote I will use is this statement from the Pontifical Academy of Life, which advises the Holy See, "In the cloning process, the basic relationships of the human person are perverted; filiation, consanguinity, kinship, parenthood. A woman can be the twin sister of her mother, lack a biological father, and be the daughter of her grandmother. In in vitro fertilization"– I am sorry, "In vitro fertilization has already led to the confusion of parentage, but cloning will mean the radical rupture of these bonds."

By reducing human reproduction to simple manufacture in the laboratory, cloning reduces the new human being to a product and then to a commodity, and obviously opens the door to these human beings, at any age, being treated as mere research fodder, as second-class human beings.

http://energycommerce.house.gov/107/hearings/06202001Hearing291/print.htm
accessed March 23, 2004

FUKUYAMA What we will soon need is a broader regulatory structure that will permit us, on a routine basis, to make decisions that distinguish between those technologies that represent positive and helpful advances for human well-being, and those that raise troubling moral and political questions. Ultimately, this regulation will have to become international in scope if it is to be more effective. We will need to think carefully about the institutional form that such a regulatory structure must take. A blanket ban on human cloning is appropriate at this time, however, because it is necessary at an early point to establish the principle that the political community has the legitimacy, authority, and power to control the direction of future biomedical research, on an issue where it is difficult to come up with compelling arguments about why there is a legitimate need for human cloning.

Appendix 9.2

The following article was written by Elizabeth Blackburn, Ph.D., who recently was dismissed as a member of President Bush's Council on Bioethics. Her concerns are about the lack of full discussion, debate, and reporting and apparent bias about negating the potential of stem cell research and other biomedical scientific advances.

Blackburn, E. 2004. "Bioethics and the political distortion of biomedical science." *N Engl J Med* **350**:1379–1380. Reproduced with permission, Massachusetts Medical Society © All rights reserved.

In late September 2001, I was asked to serve on the President's Council on Bioethics. My initial instinct was not to accept, because I was concerned that the Bush Administration would not be interested in considering fully the potential of certain controversial advances in basic biomedical research. Indeed, the administration was already on record as opposing federal funding for somatic-cell nuclear transplantation and therapeutic cloning. (Therapeutic cloning involves making early-stage preimplantation embryos for use as sources of stem cells, whereas reproductive cloning is the creation of cloned babies through the transfer of cloned embryos into a woman's uterus.) Two factors, however, tipped the balance in favor of accepting the appointment. First, as the country mourned after the tragedy of September 11, 2001, I felt that I wanted to contribute something. Second, I received strong assurance from the council's newly appointed chairman Leon Kass (and later from President George W. Bush himself) that the wisdom of a full range of experts was needed. I believed that, especially at this juncture in history, it was important to serve in this potentially critical way.

Unfortunately, my initial misgiving proved to be prescient. In a telephone call from the White House one Friday afternoon in February, I was told that my services were no longer needed. The only explanation I was offered was that "the White House has decided to make some changes in the bioethics council." Persons who are versed in such matters have since suggested that the prearranged timing of the call was not a coincidence: this administration commonly takes controversial action on Friday afternoons, when the news is expected to fall into a weekend void.

Three of the 18 members of the original bioethics council were full-time biomedical research scientists. Chairman Leon Kass, of the University of Chicago, has, in his published work, questioned modern medical and biomedical science and taken the stance of a "moral philosopher," often invoking a "wisdom of repugnance"—in other words, rejecting science, such as research involving embryonic stem cells, because it feels wrong to him. I remain convinced that this type of visceral reaction should launch, rather than end, debate. Time and time again, other members of the council, including those who were initially skeptical about the potential of stem-cell research, joined the three scientist-members in urging the chairman to account fully and fairly for the potential of this research to alleviate human suffering.

William May, an impressively thoughtful and learned theologian and medical ethicist, has also left the council, and three new members have been appointed to succeed us. These appointments portend a strong shift in the leanings of the council. Like me, May had differences with Kass on such issues as the moral value of biomedical research and the ramifications of legislating such research. May and I both voted against a ban and a moratorium on therapeutic cloning. The published views of the three new members differ sharply from mine and May's and are much closer to those espoused by Kass. Furthermore, not one of the newly appointed members is a biomedical scientist. One, a pediatric neurosurgeon, has championed religious values in public life; another, a political philosopher, has publicly praised Kass's work; the third, a political scientist, has described as "evil" any research in which embryos are destroyed.

As an experienced cell biologist and molecular biologist and a member of the Institute of Medicine and the National Academy of Science, I felt strongly that I had a duty to contribute all the knowledge and expertise I could to the council's most durable products, its written reports. Since I am not a specialist in stem-cell research, I believed, on the one hand, that I could remain above the fray in that I have no vested interest in any one branch of such research and, on the other hand, that my scientific expertise placed a particular burden on me to educate myself about this important research specialty, in preparation for the meetings and reports of the council. I therefore read and assessed the published science, attended sessions on current stem-cell research at national and international scientific conferences, and consulted with stem-cell biologists throughout the country. I was assured repeatedly by the chairman that the science would be represented clearly in our reports.

I was therefore concerned when I read the sections concerning research on embryonic stem cells in the drafts of the report and the final Report on Monitoring Stem Cell Research. Work with animal models has been indicating the potential benefits of research involving embryonic stem cells for more than two decades. More recently, research breakthroughs in the generation and differentiation of human embryonic stem cells and increased understanding of these processes have suggested that this avenue of research will eventually lead to beneficial uses in health care. Work with animal models increasingly suggests that such research may result in therapies for diabetes, Parkinson's disease, and spinal injuries, among other conditions. Yet the best possible scientific information was not incorporated and communicated clearly in the council's report, suggesting that the presentation was biased.

How might perceived bias in a federal commission such as the bioethics council affect the ability of the nation to receive the best available scientific information on which to base policy decisions? Will researchers be unwilling to provide their expert opinions regarding their field of research for fear that they will be used to promote a particular view held by the council? I am afraid that this effect is already occurring.

I was recently contacted by a world leader in research involving neural stem cells from adults; he was considering withdrawing his agreement to provide his expert opinion to the council, for fear that the potential of research involving adult stem cells would be overstated as a justification for a continued ban on federal funding for promising research on embryonic stem cells.

When prominent scientists must fear that descriptions of their research will be misrepresented and misused by their government to advance political ends, something is deeply wrong. Leading scientists are routinely called on to volunteer their expertise to the government, through study sections of the National Institutes of Health and advisory panels of the National Academy of Sciences and as advisers to departments ranging from health and human services to defense. It has been the unspoken attitude of the scientific community that it is our duty to serve our government in this manner, independent of our personal political affiliations and those of the administration in effect at the time. But something has changed. The healthy skepticism of scientists has turned to cynicism. There is a growing sense that scientific research—which, after all, is defined by the quest for truth—is being manipulated for political ends. There is evidence that such manipulation is being achieved through the stacking of the membership of advisory bodies and through the delay and misrepresentation of their reports. As a naturalized citizen of the United States, I have an immigrant's love for my country. But our country must not fail us. Scientific advice should and must be protected from the influence of politics. Will the President's Council on Bioethics be up to that challenge?

PROBLEM SET

9.1 Personnel in a hospital emergency department discover a small wooden box containing a living baby about six hours old with stubs for arms and legs and a very small head. There is no identification. The emergency department physician calls the attending pediatrician, who alerts the hospital chaplain and hospital director. The director calls an emergency meeting of the ethics committee. As chairperson of the ethics committee, how would you proceed in assessing the situation and the baby, who has now been given fluids and is in no distress?

9.2 A group of clergy present to an argument against human cloning to the Supreme Court. To some extent they rely on the logic of previous Supreme Court decisions disallowing physician-assisted suicide, namely, that the procedure is taking a life under circumstances that cannot be regulated and could easily be abused. The cloning of life similarly had difficult to regulate consequences, and in effect, a slippery slope argument is invoked. The opponents

to the anti-cloning law claim such a law is a violation of the human right to procreate, and as long as a qualified mother and father are designated, cloning should be allowed. In 60 words or less, make an argument for cloning using every moral theory that is logical. Then make an argument for the anti-cloning law, also using concepts from chapters in this text (60 words).

9.3 The use of artificial chromosomes to incorporate attributes or enhance innate capabilities has been proposed as a possible method of human enhancement. During in vitro fertilization a chromosome is injected into an engineered zygote. This chromosome has genes that can be turned on to produce traits by injection of a promoter, and the gene expression can be turned off by injection of specific chemicals. Given that there are no changes in the regulation for preimplantation modification, what are the ethical arguments for not allowing artificial chromosome methods of human enhancement? (80 words or less.)

9.4 What are the reasons for keeping confidential the identity of an egg or sperm donor or that of the surrogate mother? Give at least six reasons and then give some arguments against four of these reasons.

9.5 A 20-year-old discovers that a sperm donor is his biologic father. He is told that a condition of the sperm donation was that the donor wished to remain anonymous. However, the young man is determined to meet his real father for a number of reasons, which include learning about medical information pertinent to his family, but mainly he just wants to meet his biologic father face to face. His biologic mother and her husband are sympathetic, but feel helpless other than to inform the young man of the clinic and the date of the IVF procedure. Using the Four A's, how should the young man proceed? (120 words.) (Hint: You might start with an Internet search.)

9.6 Your 10-year-old dog has been badly injured by an automobile. You are told he is unlikely to recover from his current crippled state but is not in immediate danger of death. You decide to attempt to clone your dog using egg(s) from a donor dog and SCNT. Outline the steps that you would use to proceed, starting with an estimate of the likelihood of success and the method you would consider in view of the fact that only 5% of cloning attempts lead to a pregnancy. The answer should include a diagram of the procedure and an estimate of the time line and costs, including costs for up to 20 cloning attempts and the estimated veterinarian costs.

10

Ethics of Stem Cell Technologies

INTRODUCTION

Research to understand how stem cells are involved in the development and regeneration of the body is one of the most exciting and controversial topics in medical research. The subject has also raised numerous ethical questions, not the least of which has to do with the disaggregation of the embryo to obtain embryonic stem cells. The major moral issue is the autonomy of the embryo and the associated philosophical issue of when after conception human life begins.

In this chapter, we review the different stem cell types (e.g., embryonic stem cells, embryonic germ cells, adult stem cells, hematopoietic stem cells, and cord-blood stem cells), how they are obtained, and what is known to date about the ways in which each type of stem cell contributes to the regenerative process, and discuss the major ethical questions surrounding the subject.

BACKGROUND

Leroy Stevens at the Jackson Laboratory in Bar Harbor, Maine, did the first characterization and naming of embryonic stem cells in 1970. In 1998, James Thomson and others at the University of Wisconsin—Madison were the first to report methods for deriving and maintaining human embryonic stem cells in culture (Thomson et al. 1998). At about the same time, the derivation of embryonic germ cells obtained from the gonadal ridge of 5- to 9-week fetuses was reported (Shamblott et al. 1998). The field of human embryonic stem cell research developed rapidly thereafter.

Embryonic stem cell research and stem cell therapeutics are a subset of the general topic of tissue replacement and repair. Organ transplantation is also a subset of this topic. To date, transplantation represents the major therapeutic option for seriously ill patients whose own mechanisms of tissue replacement and repair have failed. This subject was reviewed in Chapter 7, "Medical Ethics," where issues of organ procurement, allocation, availability, and cost, and duration of function of a transplanted organ were addressed. We saw that organ transplantation procedures are complex, require the timely location of a histologically compatible donor, and often require major surgery. Allogeneic bone marrow transplants (when it is not possible to use one's own cells) and all solid organ transplants require long-term, often debilitating immunosuppressive agents to avoid rejection. Despite an active nationwide transplant network to locate potential donors and organs, many patients are not able to find a suitable donor. As there are insufficient organs for the large number of patients needing them, thousands of patients die each year as a result.

A new approach, that of the emerging stem cell technologies, has the potential to provide unique treatment options for these serious problems and for many chronic diseases. Despite their promising potential and the fact that there are sources of stem cells that are not controversial, there are widespread ethical concerns about their use. We saw in Chapter 9, "Ethics of Assisted Reproductive Technologies," that one source of stem cells is the "spare embryo," formed by IVF but unused and cryopreserved. Because many of the technologies involve embryos, moral issues are related mainly to concerns for the autonomy of the developing fetus and possible malfeasance against it. Related questions ask at what point is the embryo or fetus considered a human being whose life should be respected? At what point should it not be tampered with? Is regenerative medicine unethical because it uses embryonic stem cells taken from a developing embryo that is sacrificed to obtain them? Broader issues concern questions about the ethics of manipulation of the embryo's genome and the consequences to the family, society, and future generations. As with the debate on IVF, the concerns have to do with some of the basic moral and ethical principles of society, as exemplified below by some of the ethical questions stem cell research has raised:

- Is it morally acceptable to use donated embryos with a potential for life to obtain stem cells?

- Is it morally acceptable for parents to donate their spare embryos for stem cell research?

- Is therapeutic cloning ethical (e.g., expanding embryonic stem cells in culture to form the needed cells or tissue that can be used for restorative applications)?

- If it is shown that adult stem cells can be substituted for embryonic stem cells, is it ethical to proceed with embryonic stem cell research?

- Should the number of stem cell lines available for federally funded research be limited to those currently available?

- Should the U.S. government limit research that has the potential to benefit millions of people with serious illnesses?

- Should stem cell acquisition procedures be regulated by the state or the federal government?

We will discuss these and other ethical issues later in the chapter after introducing the basic biology, embryology, and applications of stem cell research. (See also the methodologies of the in vitro fertilization techniques discussed in Chapter 9 that are basic to understanding this section.) First, it is important to differentiate between the different types of stem cells and how they are obtained.

TYPES OF STEM CELLS AND THEIR ORIGINS

Embryonic Stem (ES) Cells

Embryonic stem cells possess unique qualities:

- Their most unusual characteristic is that they are capable of immortal self-renewal; that is, they can replicate themselves an unlimited number of times, forming a duplicate copy of self, and thereby maintain the continuation of the stem cell pool. No other human cells, even other types of stem cells we shall discuss later, have this natural capability in vitro.

- In addition to self-replication, initially undifferentiated pluripotent embryonic stem cells, uncommitted to any specific cell lineage but capable of forming all cell types, can divide to form more differentiated cells that go on to form a specific cell lineage. The resulting cells ultimately differentiate further to become an inherent part of the tissue in which they reside, and perform the function of that tissue (Weissman et al. 2001, Verfaillie 2004, Körbling and Estrov 2003a).

Thus, embryonic stem cells, present only in the embryo at the blastocyst stage, about five days after fertilization, are the cells from which all the cells of the tissues and organs of the body are formed. The hollow blastocyst contains approximately 200 embryonic stem cells that can be removed by disaggregation of the embryo. They can then be easily cultured and maintained in the laboratory. These unique pluripotent cells can be signaled by changes in chemical factors in the culture milieu in the laboratory to form specific cell lines that can be administered to patients to regenerate, renew, or repair specific types of dead, dying, or damaged tissues of the body; thus, the term *regenerative medicine*. Ultimately, the hope is that researchers will be able to engineer patient-specific ES cells and perhaps patient-specific tissues and organs for those who need them. Figure 10.1 outlines a schema of the differentiation of embryonic stem cells to form all three tissue lineages (mesodermal, endodermal, and ectodermal cells).

The early cells of the three lineages are unipotent stem cells. However, it should be noted that the literature is not always consistent with respect to definitions of types of stem cells, their pluri-, multi-, or unipotency. There are also inconsistencies about the origin of the progenitor cell responsible for tissue renewal in adult tissues. What is clear is that embryonic stem cells and the more differentiated progenitor cells that go on to form the three lineages are not capable of forming a complete fetus on their own.

While ES cells are always obtained from a blastocyst, the blastocyst itself can be obtained from several sources:

- A frozen spare embryo that is thawed, incubated, and allowed to grow to the blastocyst stage.

- An embryo produced by IVF for the sole purpose of creating an embryonic stem cell line.

- A procedure known as somatic cell nuclear transfer (SCNT) or therapeutic cloning.

The most frequently debated sources of stem cells are the spare embryos and spare blastomeres donated by parents for research. The majority of ES cell research to date

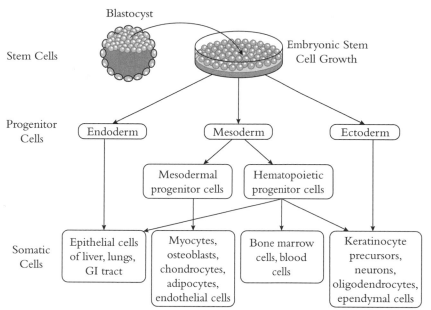

Figure 10.1 The blastocyst is the source of embryonic stem cells that are cultured in the laboratory and can be coaxed to differentiate into multiple cell lines. Stem cells are the predecessors to all tissues and organs in the body, and their plasticity is the basis for expanding their use to repair and regenerate damaged tissues. (From Körbling and Estrov 2003a, adapted with permission, © Massachusetts Medical Society 2003.)

has been done using stem cell lines developed by expanding by culture a single embryonic stem cell into many daughter cells. Embryonic stem cells can be obtained from a blastocyst-stage embryo formed by IVF and donated for research purposes (e.g., spare embryo). These spare embryos are stored in large numbers and are the major source of embryonic stem cells at the present time. In 2002, there were an estimated 400,000 spare embryos in the 430 IVF clinics in North America alone. Of these, approximately 3% are made available for medical and biologic research (Hoffman et al. 2003). Worldwide there are approximately 128 human stem cell lines (Daley 2004). There are more than 450 human ES cell lines for use in stem cell research in the United States, of which 10 are lines derived from embryos carrying genetic diseases (Verlinsky et al. 2005), 15 cell lines are those allowed by President Bush in 2001, and 17 new cell lines were developed from human blastocysts with private funding at Harvard University (Cowan et al. 2004). As discussed later under "Current Status of Embryonic Stem Cell Research," the cell lines currently in use have been found to be contaminated by murine retroviruses from prolonged physical contact with uncharacterized murine fibroblast feeder cells. It will be necessary to recreate all new stem cell lines before further ES work can go forward. New cell lines from the University of Singapore have been created since 2000 (Reubinoff et al. 2000), and these lines are free of contamination by feeder cells from another species.

Therapeutic cloning involves growing an embryo created by somatic cell nuclear transfer (SCNT) in the laboratory. SCNT is done by removing the nucleus of a stimulated or fertilized egg and replacing it with the nucleus of an adult donor cell. This can be any somatic cell of the body and is the same process used in reproductive cloning (see Figure 9.1, Chapter 9), except in therapeutic cloning the new entity is not implanted in a uterus. After the embryo is grown to the blastocyst stage in the laboratory, embryonic stem cells are harvested from its inner cell mass and are expanded in culture. These cells in culture also remain in the laboratory and are destined for use in the person who donated the somatic cell.

Embryonic stem cells are pluripotent and are incapable of generating a pregnancy if transferred to the uterus. Therefore, therapeutic cloning would not result in the birth of an infant. The intent is to utilize embryonic stem cells for their "stemness," that is, their potential to form all cell types. Tissue-specific stem cell generation is accomplished by manipulating the cellular environment to direct formation of whatever tissue (and even organs) that might be needed (see Figure 10.1).

Therapeutic cloning, or SCNT, is a procedure with the potential to offer particular advantage to a patient in need of regenerative medical treatment (e.g., transplantation with special cells, tissues, or an organ). Because the patient's own somatic cell is used in the SCNT procedure, all ES cells formed subsequently will be recognized as "self" and are not expected to provoke an immune response. Because rejection of a transplanted organ is difficult and costly to treat, and threatens the survival of the graft as well as the patient, this would be an important therapeutic advance toward providing genetically matched tissue.

Whereas the process of reproductive cloning can result in reprogramming errors and aberrant gene expression that can seriously affect clonal survival, such errors ". . . do not appear to interfere with therapeutic cloning, because the process appears to select for functional cells" (Hochedlinger and Jaenisch 2003a). The current state of knowledge regarding the differentiation of human embryonic stem cells, and of the molecular events during nuclear reprogramming, is inadequate to begin administering them to humans as medical therapy (Hochedlinger and Jaenisch 2003b). Unfortunately, it is not true that the cloned stem cells will have only the DNA of the donor of the nucleus. Mitochondrial DNA from the egg donor will also be present as has been measured in embryos from the attempts to clone endangered species using the enucleated egg from a related species (Lanza et al. 2000). However, it is expected that ES cells formed by SCNT will be better tolerated as "self" when these cells are used to regenerate tissues of a particular patient.

Many believe that stem cell therapy will offer new treatment options for millions of patients needing various types of tissue regeneration or organ transplantation. The use of somatic cell nuclear transfer to form embryonic stem cells containing DNA specific to each patient should further enable this therapy

Up to 2006, no successful primate therapeutic cloning, either human or nonhuman, has been reported. Highly publicized reports of success from South Korea (Huang et al. 2004, Huang et al. 2005) are now deemed fraudulent (Kennedy 2006). International efforts to perfect ways to perform human therapeutic cloning continue.

ES cell technologies are expected to provide other benefits as well, among them:

- A means of rapidly screening new drugs targeted for cell-specific diseases in pure populations of those cells.

- A means of testing all drugs for possible toxic effects on embryonic and fetal tissues, thereby improving safety during pregnancy.

- An opportunity to study the early development of the normal embryo in order to better understand the causes of abnormal development, the reasons for infertility, and ways to prevent or treat miscarriages and birth defects.

Alternative Sources of Embryonic Stem Cells

Several options are under study to find ways of obtaining stem cells that avoid the disaggregation of an embryo altogether. One is to reprogram somatic cells to become embryonic stem cells, bypassing the need for oocytes (Hochedlinger and Jaenisch 2003c, Perez-Pera 2003). The first success in remodeling of mammalian somatic cells was reported in July 2005 (Alberio et al. 2005).

Another option is to obtain oocytes from the germ cells of the gonadal ridge in an aborted fetus for use in nuclear transfer. Another is to grow human oocytes from cultured epithelial cells scraped from the surface of adult human ovarian tissue, as recently reported by Bukovsky et al. (2005). This option is the same as SCNT except for the origin of the egg. Another way to obtain embryonic stem cells is to use a single blastomere obtained from the four-cell stage of an early embryo, as one would do in preimplantation genetic diagnosis (see Chapter 9); instead of undergoing preimplantation genetic diagnosis, this blastomere could be cultured and allowed to develop to the blastocyst stage from which stem cells could be obtained. The first success in this approach was reported using mouse embryos in October 2005 (Chung et al. 2005). The embryo is not destroyed or even unduly harmed, and the remaining cells of the early embryo continue to develop normally. Note that the blastomere is totipotent at the four-cell stage, with the capacity to form a complete living human infant. While its use for stem cells does not involve killing a developing fetus, some may still feel that redirecting the life cycle of this cell constitutes inexorable tampering with its destiny and still constitutes interruption in the creation of a life. Others may find that the use of a blastomere for stem cells is more ethically acceptable than other options and has no risks greater than those of PGD procedures.

Another approach to embryonic stem cell harvesting involves the process of therapeutic cloning, except that before insertion of the somatic cell nucleus the genes are altered so that a developed life form cannot occur. After a blastomere is obtained, a manipulation would be performed to reverse the regulatory gene manipulation performed before SCNT (Hurlbut 2004). This procedure and its ethical implications will be discussed later.

Embryonic Germ (EG) Cells

Embryonic germ cells can be obtained from the early fetal tissue in an area known as the gonadal ridge about eight weeks after fertilization (Shamblott et al. 1998). These cells of primordial reproductive tissue can give rise to all tissue types, in addition to eggs and sperm. Currently, it is believed that although pluripotent, these types of cells have already assumed some primitive reproductive functions and may have lost some of their plasticity to differentiate into other tissue types. Currently, the only accepted source of human EG cells is the fetus after abortion. For these reasons, research efforts have focused more on embryonic and adult cells rather than on EG cells. See also the organ transplantation section of Chapter 7, "Medical Ethics," for a review of the status of fetal tissue transplantation.

Adult Stem Cells

Even though they are called adult stem cells, these are undifferentiated multipotent cells that are found in various tissues (presumably all) throughout the lifetime of the individual from birth onward. Adult stem cells are those cells that reside in and are identified within a specific tissue or organ. They are able to renew themselves by dividing to form undifferentiated daughter cells (thereby ensuring a continuing source of stem cells) and are also able to differentiate into functioning cells of a type specific to the tissue needing repair.

For example, a multipotent stem cell called a mesenchymal stem cell has been shown to produce bone, muscle, cartilage, fat, and other connective cells; bone marrow contains cells that can form all the hematopoietic cells (blood elements) of the body, including those providing immunologic functions (discussed further below). It is also known that the adult brain contains cells that can regenerate neurons, intestinal crypt cells can differentiate into cells of the intestine, and testicular cells can regenerate spermatids (the precursors of sperm). The human liver has the capacity to regenerate functioning liver cells, even when injury is extensive. If these capabilities are found to be present in other tissues in the human body, as a growing body of evidence is now suggesting, there will be even more treatment opportunities in which stem cells can be expected to regenerate or replace damaged tissue.

In addition, although originally thought to contribute to the maintenance and repair of the tissue or organ in which they reside, there are data that suggest adult stem cells may not be confined to their own tissues but may move throughout the body to form other organ-specific cells with far more plasticity than previously thought. Since the late 1960s, it has been known that cells from a transplanted organ can migrate and differentiate into cells of other organs in the recipient, and that cells from the recipient migrate into the transplanted organ. First discovered in liver transplant patients (Starzl 2004), this chimerism has also been observed in human heart (Muller et al. 2002) and other transplants. For example, in human heart transplants from female donors to male recipients, male recipient cells were found to have populated the transplanted hearts as shown by the presence of cells with XY chromosomes in the otherwise female heart (XX chromosomes).

Other evidence of this was found in humans after female recipients of an organ transplant from a male donor were found to have cells with male chromosomes in other tissues far outside the region of the original transplant. There is the possibility that these observations may have been an artifact of hematopoietic stem cells present in the donated organ at the time of transplant that made progeny that fused with cells in the target organ. However, animal data give evidence that adult tissue-derived stem cells might have the capacity to regenerate or replace tissue in distant sites. For example, mouse bone marrow cells have been reported in the scientific literature to have become integral parts of mouse skeletal muscle, liver, central nervous system, and heart muscle. Mouse muscle-derived stem cells have been shown to replenish hematopoietic elements, and some mouse hematopoietic cells can replenish cardiomyocytes (heart muscle cells) in a damaged mouse heart.

But the issue of adult stem cell plasticity is controversial, with some experimental results explained by cell fusion and others by heterogeneity of a particular stem cell population (Lakshmipathy and Verfaillie 2005). Though the literature of the early years of 2000 has shown through experimentation that adult stem cells are multipotent (Krause et al. 2001), the scientific interpretations have been challenged by experiments that indicate that the plasticity of adult stem cells is far less than many have believed (Wagers and Weissman 2004). This controversy is important for our discussions of the ethics of stem cell research, because it may be incorrect that adult stem cells will be able to function as fully as embryonic stem cells in the regenerative process.

Investigators in the fields of developmental biology and stem cell research are striving to learn how to better isolate human adult stem cells (they are not easy to obtain), and how to reprogram them to differentiate into cells for renewal and repair of specific organs. The advantages of using adult stem cells for this therapy are mainly two: 1) to avoid the ethical problems encountered with the use of embryonic stem cells, and 2) to use the patient's own adult stem cells to avoid the unwanted immune response to grafting that would be seen if allogeneic cells were used.

The best information at this time is from transplantation using adult stem cells from the blood (hematopoietic stem cells), a technique that has proven benefits described below. Other medical uses of adult stem cells are being evaluated and are discussed later in this chapter.

Hematopoietic Stem Cells

Hematopoietic stem cells are adult stem cells formed in the bone marrow, where they are obtained when bone marrow is harvested. These stem cells are also found in umbilical cord blood at the time of birth, and in the peripheral blood, where they circulate after being released from the marrow. Their main function was thought to be the formation and maintenance of blood elements (i.e., the red cells, nucleated cells, and platelets in the peripheral blood). We list them separately here because research is unfolding that shows that these cells may have a greater role in the repair and regeneration of a broader range of tissues than previously thought.

Because they move about freely in circulating blood, they can be recruited to many tissues, perhaps more easily than other stem cells. Adult mesodermal hematopoietic stem cells are able to form endodermal or ectodermal tissue cells in addition to mesodermal tissues, giving many directions for renewal and regeneration for multiple tissue types (Körbling and Estrov 2003b). Figure 10.1 shows this capability for adult hematopoietic stem cells.

Cord-Blood Stem Cells

Blood taken from the umbilical cord at the time of birth is a rich source of hematopoietic stem cells. This blood is usually discarded, but it can easily be collected in relatively large amounts after the infant is delivered, and it can be cryopreserved for future use. Studies are underway to assess how or whether umbilical cord stem cells differ from adult bone marrow stem cells. This source of stem cells may be especially valuable for young patients without a sibling donor in need of transplant because of ease and rapidity of availability, as well as success in treating the underlying disease (Kurtzberg et al. 1996).

Transplantation in adults without a related donor can also be done using cord blood from unrelated donors that is mismatched for one or two HLA antigens without compromising outcome (Laughlin et al. 2004, Rocha et al. 2004, Staba et al. 2004). Thus, umbilical cord blood is an alternative to bone marrow transplant for a number of acquired and congenital diseases in patients without an HLA-matched donor. An example of success in the use of umbilical cord blood was the treatment of Krabbe's disease in children (Escolar et al. 2005). This disease results from a lack of the enzyme galactocerebrosidase, which is essential for the development of the nervous system. Previous treatment was bone marrow transplantation from an immunologically compatible donor. Because it is sometimes impossible to find such a donor, the success of treatment using umbilical cord blood from any donor is of major importance for patients with this disease.

Diseases for which umbilical cord blood may be useful include refractory malignancies (leukemia, lymphoma, myelodysplasia, neuroblastoma), congenital and acquired bone marrow failure syndromes (aplastic anemia, Fanconi anemia), congenital immunodeficiency syndromes (severe combined immunodeficiency disease, Wiscott-Aldrich syndrome), inborn errors of metabolism, and hemoglobinopathies (sickle cell anemia and thalassemia) (http://olpa.od.nih.gov/legislation/108/pendinglegislation/stemcell.asp, accessed April 26, 2004).

The various sources of stem cells are summarized in Figure 10.2.

Current Status of Embryonic Stem Cell Research

It was hoped that wide acceptance in the United States of organ transplantation might set a precedent that would favor the use of ES cells as well. Those supporting stem cell research have used the analogy that there is little difference between donating an organ for transplantation and donating an embryo for harvesting embryonic stem cells for use in the replacement of damaged tissues and organs. Nevertheless, the early opposition to abortion, embryonic death, and IVF itself, as well as perceived similarities of methodologies of therapeutic cloning and reproductive cloning, caused opposition to the use of stem cells, with significant consequences for researchers wanting to pursue studies to enable medical therapies.

Federally funded research involving stem cells has been limited in the United States by a decision made by President Bush in 2001 that was still in effect in 2006. He made his decision to allow federally funded research only on cell lines already in possession of the researchers (e.g., no other embryos were to be used to create new stem cell lines). His address

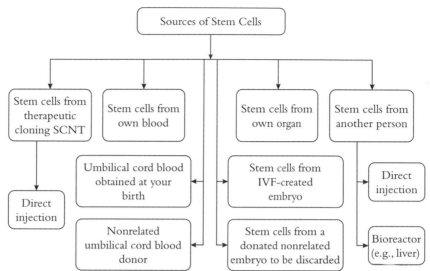

Figure 10.2 Stem cells are found in an individual's bone marrow, in circulating blood, and within organs. Some of these sources of stem cells are controversial; others are not. The ethics of cloning are discussed in Chapter 9, "Ethics of Assisted Reproductive Technologies."

to the nation in August of 2001 outlined his concerns and the rationale of his decision, a position still in place in 2006 (www.whitehouse.gov/news/releases/2001/08/20010809-2.html, accessed April 3, 2003):

> I have given this issue a great deal of thought, prayer and considerable reflection. And I have found widespread disagreement. On the first issue, are these embryos human life—well, one researcher told me he believes this five-day-old cluster of cells is not an embryo, not yet an individual, but a pre-embryo. He argued that it has the potential for life, but it is not a life because it cannot develop on its own.

> An ethicist dismissed that as a callous attempt at rationalization. Make no mistake, he told me, that cluster of cells is the same way you and I, and all the rest of us, started our lives. One goes with a heavy heart if we use these, he said, because we are dealing with the seeds of the next generation.

> And to the other crucial question, if these are going to be destroyed anyway, why not use them for good purpose—we also found different answers. Many argue these embryos are byproducts of a process that helps create life, and we should allow couples to donate them to science so they can be used for good purpose instead of wasting their potential. Others will argue there's no such thing as excess life, and the fact that a living being is going to die does not justify experimenting on it or exploiting it as a natural resource.

At its core, this issue forces us to confront fundamental questions about the beginnings of life and the ends of science. It lies at a difficult moral intersection, juxtaposing the need to protect life in all its phases with the prospect of saving and improving life in all its stages.

As the discoveries of modern science create tremendous hope, they also lay vast ethical mine fields. As the genius of science extends the horizons of what we can do, we increasingly confront complex questions about what we should do. We have arrived at that brave new world that seemed so distant in 1932, when Aldous Huxley wrote about human beings created in test tubes in what he called a "hatchery."

In recent weeks, we learned that scientists have created human embryos in test tubes solely to experiment on them. This is deeply troubling, and a warning sign that should prompt all of us to think through these issues very carefully.

Embryonic stem cell research is at the leading edge of a series of moral hazards. The initial stem cell researcher was at first reluctant to begin his research, fearing it might be used for human cloning. Scientists have already cloned a sheep. Researchers are telling us the next step could be to clone human beings to create individual designer stem cells, essentially to grow another you, to be available in case you need another heart or lung or liver.

I strongly oppose human cloning, as do most Americans. We recoil at the idea of growing human beings for spare body parts, or creating life for our convenience. And while we must devote enormous energy to conquering disease, it is equally important that we pay attention to the moral concerns raised by the new frontier of human embryo stem cell research. Even the most noble ends do not justify any means.

My position on these issues is shaped by deeply held beliefs. I'm a strong supporter of science and technology, and believe they have the potential for incredible good— to improve lives, to save life, to conquer disease. Research offers hope that millions of our loved ones may be cured of a disease and rid of their suffering. I have friends whose children suffer from juvenile diabetes. Nancy Reagan has written me about President Reagan's struggle with Alzheimer's. My own family has confronted the tragedy of childhood leukemia. And, like all Americans, I have great hope for cures.

I also believe human life is a sacred gift from our Creator. I worry about a culture that devalues life, and believe as your President I have an important obligation to foster and encourage respect for life in America and throughout the world. And while we're all hopeful about the potential of this research, no one can be certain that the science will live up to the hope it has generated.

Eight years ago, scientists believed fetal tissue research offered great hope for cures and treatments—yet, the progress to date has not lived up to its initial expectations. Embryonic stem cell research offers both great promise and great peril. So I have decided we must proceed with great care.

As a result of private research, more than 60 genetically diverse stem cell lines already exist. They were created from embryos that have already been destroyed, and they have the ability to regenerate themselves indefinitely, creating ongoing opportunities for research. I have concluded that we should allow federal funds to be used for research on these existing stem cell lines, where the life and death decision has already been made.

Leading scientists tell me research on these 60 lines has great promise that could lead to breakthrough therapies and cures. This allows us to explore the promise and potential of stem cell research without crossing a fundamental moral line, by providing taxpayer funding that would sanction or encourage further destruction of human embryos that have at least the potential for life.

I also believe that great scientific progress can be made through aggressive federal funding of research on umbilical cord placenta [blood], adult and animal stem cells which do not involve the same moral dilemma. This year, your government will spend $250 million on this important research.

I will also name a President's council to monitor stem cell research, to recommend appropriate guidelines and regulations, and to consider all of the medical and ethical ramifications of biomedical innovation. This council will consist of leading scientists, doctors, ethicists, lawyers, theologians and others, and will be chaired by Dr. Leon Kass, a leading biomedical ethicist from the University of Chicago.

This council will keep us apprised of new developments and give our nation a forum to continue to discuss and evaluate these important issues. As we go forward, I hope we will always be guided by both intellect and heart, by both our capabilities and our conscience.

I have made this decision with great care, and I pray it is the right one.

Embryonic stem cell research in the United States currently is limited because of the restriction of federal funding for new stem cell line creation and because the stem cell lines already in existence are fewer in number than previously thought and not of the best quality (Phimster and Drazen 2004). New human embryonic stem cell lines are being created by non-federally funded organizations and are made available to scientists at no cost for noncommercial research purposes (Cowan et al. 2004). Unfortunately, it was discovered in 2005 that all of the stem cell lines allowed by the president and the 17 other stem cell lines created at Harvard, and perhaps others, are contaminated with nonhuman biological proteins (Martin et al. 2005, Hampton 2005). This contamination occurred because the culture medium routinely used to nourish, expand, and maintain the ES cells in the laboratory contains nonhuman (e.g., bovine-derived) materials (Martin et al. 2005, Hampton 2005).

As the ES cells grow, foreign DNA sequences, such as uncharacterized murine retro-viruses, may become incorporated in the cells' DNA. The viral or plasmid origin of this contamination is unknown. This contamination could mean that all the cell lines and their progeny have an unknown danger of directly infecting a patient, or recombining with an endogenous retroviral sequence in the genome of a patient to produce a virulent new retro-virus. To be safe, all contaminated ES cell lines must be discarded and new ones estab-lished using completely defined conditions. In short, there can be no human applications using these contaminated lines of embryonic stem cells because their use would be reck-less and ethically unacceptable to all medical practitioners. Federal mandates will have to be changed in order to allow the establishment of new stem cell lines and their use by insti-tutions using federal support, or will need to be developed by U.S. researchers who do not receive federal funds. New cell lines from the National University of Singapore are not contaminated (Richards et al. 2004), but these cell lines cannot be used in U.S. institutions receiving federal support currently.

As of April 2002, there were almost 400,000 embryos in storage in the United States (Hoffman et al. 2003). Of these embryos, 88% were targeted for patient use and 3% were made available for use in research. It is estimated that 65% will survive the freeze-thaw cycles, but only 25% of these will be capable of developing into blastocysts. Depending on skill level of the technician, cells from these blastocysts have about a 15–30% chance of successfully developing into a stem cell line. However, only a few of the embryos donated for research purposes have been used due to the ban on federal funding for such research.

The seventeen new individual human embryonic stem cell lines recently developed with private funding at Harvard came from screening 286 frozen and thawed embryos of 6 to 12 cells each, and 58 frozen and thawed blastocysts (Cowan et al. 2004). These stem cell lines were the first to be created in the United States since President Bush made his decision in 2001 to allow federally funded embryonic stem cell research only on stem cell lines already developed.

It has been estimated that since 1978, over 200,000 embryos have been destroyed at IVF clinics in the U.S. (Hall 2001). It is the ultimate fate of these "spare embryos" around which many of the ethical debates are raised. Those in favor of embryonic stem cell research would like to see these spare embryos, which are otherwise destined to be destroyed, used instead to derive ES cell lines that can be used for the treatment of a num-ber of diseases in which cells are damaged but cannot be repaired or replaced. These con-ditions include diabetes, Parkinson's, Alzheimer's, and congestive heart failure (Gerecht-Nir and Itskovitz-Eldor 2004).

A number of state legislatures are getting around the present federal stem cell restric-tions by passing or proposing stem cell initiatives. California passed a $3 billion initiative for stem cell research in November 2004. The state is now committed to spend almost $300 million a year on stem cell research for at least 10 years through tax-free state bonds in the hope of finding cures that will save the state money on health care costs and provide income from patents. In 2004, New Jersey created a $25 million stem cell research center, and a $500 million bond measure is planned to fund embryonic stem cell research over the next 10 years. Connecticut legislators have reintroduced a bill to appropriate $10 million per year over two years for a stem cell research fund, and Wisconsin announced a $750 million plan for embryonic stem cell research in November of 2004. New York legislators have proposed a $1 billion stem cell bond measure. In 2005, Massachusetts passed a bill to facilitate embryonic stem cell research that was vetoed by the governor. His veto was promptly overturned by a two-thirds majority vote in both the Massachusetts House and the Senate (*San Francisco Chronicle* 2005).

Senator Dianne Feinstein (D-California) is concerned that the federal policy by the current administration is leading to a patchwork of state laws being proposed or enacted, when there should be instead a stronger federal stem cell policy. She would like to see federal funding for stem cell lines and federal ethical guidelines on stem cell research (http://feinstein.senate.gov/05releases/stemcell0216.htm, accessed April 7, 2005). But despite a number of bills that have made their way to the floors of the U.S. House of Representatives and the U.S. Senate, none has become law and most are stalled. A few are reviewed below to provide a sense of the direction of discussions among lawmakers.

The Brownback bill and a similar earlier Weldon bill (2001) passed in the House of Representatives would ban all forms of reproductive and therapeutic cloning, either for research on embryonic stem cells or for use for medical treatments. The bills stipulate jail time and $1 million in fines for anyone attempting to clone a human being or to create an autologous stem cell line by SCNT. While the Brownback bill has passed the House, its passage in the Senate has been stalled (as of March 2005).

The Specter-Feinstein bill was introduced in 2002. This bill, popular with scientists, would prohibit reproductive cloning but allow somatic cell nuclear transfer (SCNT) research to go forward. It would impose criminal penalties on anyone who attempts to implant the product of SCNT into a woman's uterus. Opponents to the Specter-Feinstein bill see it as the immoral destruction of a living fetus without proper regard for life. Proponents feel it is unconscionable to disallow stem cell research that has the potential to benefit 100 million people with crippling or incurable diseases (Epstein 2002).

In May of 2004, an initiative organized in the U.S. House of Representatives and signed by 206 House members was forwarded to the director of the National Institutes of Health (NIH), Elias Zerhouni, asking for a modification of the president's policy on stem cell research. A similar letter made the rounds in the Senate (Holden 2004). Zerhouni's response indicating that human embryonic stem cell research could definitely be speeded up by having access to more human embryonic stem cell lines was perceived by some as a "big deal" (Holden 2004). Scientists plan to keep ES cell research before the public's eye, and hope that rather than prohibiting this research in the United States, there can be a way to regulate it to everyone's satisfaction so that it can proceed (Drazen 2003). As of April 2005, the issue of stem cell research was again gaining attention in Congress. In a testimony to a panel of U.S. senators, Zerhouni made it clear that scientists were feeling hindered in their research efforts by the lack of cell lines. Congress is hoping to consider legislation soon that would allow federal funding for research on some of the many spare embryos slated for destruction at fertility clinics (Stolberg 2005). Other legislation is also pending, including a bill to make more viable stem cell lines available for federally funded stem cell research, and another to be introduced that will ban human cloning but allow SCNT under strict federal oversight (http://feinstein.senate.gov/05releases/stemcell0216.htm, accessed April 7, 2005).

In the meantime, private groups have increased their support for stem cell research in the private sector, donating tens of millions of dollars to fund laboratories in both the United States and abroad (Perez-Pena 2003). Examples include large donations from the

Christopher Reeve Paralysis Foundation, the Michael J. Fox Foundation for Parkinson's Research, and the Juvenile Diabetes Foundation. Private donors and large corporations are also providing financial support for this type of research to universities and teaching hospitals. None of these donations goes toward reproductive cloning, a procedure that all responsible researchers agree should remain banned. The impetus is to provide funds for research to cure those devastating diseases for which stem cell research holds the greatest promise.

Status of Adult Stem Cell Research

It has not been convincingly demonstrated that an adult stem cell from one tissue has differentiated or reprogrammed to become a fully functional tissue cell in another organ. There have been many experimental results that, while challenging previously held dogmas of biology, in fact, have been found to be flawed by phenomena that now have been identified as artifacts. A recent article summarizes the difficulties in tracing this evolution definitively (Rosenthal 2003):

> In general, the process of documenting that a stem cell from one tissue generates differentiated types of cells in another must be particularly rigorous, in order to rule out the possible contribution of cell fusion or of circulating hematopoietic stem cells. However, research to date suggests that stem-cell populations in adult mammals are not fixed entities and that after exposure to a new cellular environment, they may be able to contribute to the regeneration of multiple types of tissue.

> Although the accumulating experimental evidence of the plasticity of adult stem cells is ever more convincing, there are several hurdles to overcome. The origin of stem cells in the bone marrow is well established, but the location of stem cells in other tissues remains elusive. What are the sources of these cells? Has the body had enough forethought to set aside pluripotent stem cells from its early moments as an embryo? If not, how do adult stem cells arise in nonhematopoietic tissues, and how do they remain in suspended animation until repair is necessary? Do they need a stimulus, such as traumatic injury, to expand the population for repair of the injured tissue? If adult stem cells are truly pluripotent, why is the response to injury to tissues such as the heart or spinal cord so inefficient under normal circumstances? Do adult stem cells carry surface markers that would make them immunogenic in a foreign host, making embryonic stem cells better tolerated in the recipient? These questions may seem academic, but the future of adult stem-cell–mediated therapies depends on their resolution.

Most researchers feel it is entirely too early to draw any conclusion about whether adult stem cells can substitute for embryonic stem cells, and feel that it is important at this stage to continue to study both embryonic and adult stem cells until the question can be more definitively answered. In the meantime, research on the use of adult stem cells is proceeding in a number of directions. Before returning to the ethical questions about stem cell research posed at the start of this chapter, we review some of the clinical applications to date. It is important to note that because these applications involve the use of adult stem cells, the ethical considerations about their use are few.

Hematopoietic Stem Cell Transplantation

Hematopoietic adult stem cell therapy has been in use in bone marrow transplantation for over 30 years, making hematopoietic stem cells the most studied of all adult stem cells. Currently, there are more than 50,000 bone marrow transplants performed worldwide each year, 17,000 of which are performed in the United States. A great deal of information about this type of stem cells has been learned from the many patients undergoing this procedure since it was first introduced. Bone marrow transplantation involves the harvesting of stem cells from the bone marrow of a healthy donor who is immunologically matched to the patient needing a bone marrow transplant (allogeneic transplant). If the transplant is for a patient with cancer, treatment with radiation and chemotherapy to eradicate the cancer also destroys the patient's bone marrow where life-sustaining blood elements are formed. This function can be restored by injecting hematopoietic stem cells, obtained from a donor's bone marrow, into the peripheral blood of a cancer patient. The stem cells make their way to the patient's bone marrow, where they are able to engraft, grow, and divide, resulting in a full restoration of the cellular components of blood (red cells, nucleated cells, and platelets).

Another simpler approach to this type of transplant, first used in the late 1990s, is to isolate hematopoietic stem cells from the peripheral blood instead of taking bone marrow from within the large bone of the pelvis. The stem cells can be isolated by passing blood through a special machine. Depending on the patient's type of cancer or underlying disease, a decision is made whether to use the patient's own peripheral blood stem cells (autologous transplant) or a related donor's peripheral blood stem cells (allogeneic transplant). Obtaining these cells from the peripheral blood is a far easier procedure than obtaining bone marrow, avoids hospitalization for the donor providing bone marrow, is less painful for the donor, and is easier to process than bone marrow. In addition to these benefits, the patients receiving this type of transplant do well. A study of patients undergoing transplantation for hematologic malignancy and hematologic rescue using stem cells from either bone marrow or peripheral blood showed that the patients receiving peripheral blood stem cell transplants engrafted earlier, had fewer infections, and had a greater rate of disease-free survival at two years (Besinger et al. 2001).

Other studies of hematopoietic stem cells have assessed the transplantation of cells from older donors into young recipients and stem cells from young donors into older recipients. It was found that basal hematopoiesis (the ability to produce needed red cells, nucleated cells, and platelets) is maintained throughout life, but the ability to respond to stress decreases with age, and hematopoietic stem cells derived from older donors show some changes in their ability to differentiate (Globerson 2001). Similar experiments with careful controls need to be done with adult stem cells from other tissues to ascertain whether greater success will be achieved using adult stem cells from younger donors for therapeutic applications. Another area of study has been the administration of peripheral blood adult stem cell infusions in solid organ transplantation in an effort to diminish rejection of the transplanted organ (Salgar et al. 1999).

A new technology has evolved from IVF for a terminally ill sibling for whom no donor could be located, by which a very early embryo that has been conceived by IVF can be tested to ensure transplant compatibility prior to transfer to the womb. The procedure is known as preimplantation human leukocyte antigen (HLA) testing, a test that is routinely done to ensure immunologic compatibility between a donor and a transplant recipient. The test identifies the unique proteins present on nucleated blood cells of potential donors and compares them with those of the recipient; a match of these antigens indicates the compatibility of the tissue to be transplanted (in this case peripheral blood stem cells). Preimplantation HLA testing, when done on a blastomere of an IVF-created embryo, similarly ensures that only immunologically compatible (HLA matched) embryos are selected for implantation. This type of testing is an outgrowth of preimplantation genetic screening, except that in this case, parents are now able to provide a source of compatible hematopoietic stem cells for an ill sibling needing a transplant by conceiving an HLA-matched sibling able to donate the needed stem cells (Tanner 2004, Verlinsky et al. 2004).

Other potential clinical applications of hematopoietic stem cells for tissue repair or replacement are being studied in diseases such as hepatitis, liver cirrhosis, myocardial infarction (heart attack), impaired heart function associated with congestive heart failure, vascular disease, lung disease, retinal diseases, neurodegenerative diseases, and stroke (Körbling and Estrov 2003c). These studies should shed more light on the extent of the role of hematopoietic stem cells in regenerating multiple types of tissue.

Research in this area is being pursued in parallel in many countries, and some of the early work has shown promise that therapeutic medical applications will indeed evolve. Thus far, none has fulfilled rigorous criteria to prove that a primordial hematologic stem cell has developed into a fully functional cell in the organ to which it went to reside.

Neurogenesis

Stem cells are being studied to determine their function in neurogenesis, the regeneration of nerve tissue. Neurogenesis is known to occur in the mammalian brain, with specific examples demonstrated in the hippocampus and olfactory bulb. Blau and others have shown that transformation of adult mouse bone marrow stem cells into other tissues such as brain, heart and liver is possible. However, there is some criticism as to whether the scientific methodology used to obtain these results was sufficiently stringent, and at the present time, experiments on neurogenesis remain controversial (Blau et al. 2002, Wagers et al. 2002).

The following hypothetical case illustrates the convergence of medical hopes with the tentative and not yet validated promise of emerging tissue renewal technologies.

Case 10.1 Treatment of Alzheimer's Disease with Embryonic Stem Cells

At 50 years of age, a neurosurgeon has difficulties remembering patient's names. His work as a neurosurgeon involves inserting chemotherapy agents into brain tumors through needles that enter the brain through drill holes in the skull. A year later his forgetfulness has increased and he is diagnosed with brain neurodegeneration (i.e., Alzheimer's disease) for which there is no known cure and only temporary memory improvement from drugs.

This neurosurgeon is the major asset at the brain tumor treatment center at a major medical center. He is not demented yet, but knows he will become demented soon. He asks the hospital director, a neurologist, to treat him with embryonic stem cells from an in vitro preparation of neuronal progenitor cells. These cells can be prepared by using his fibro-blasts and a donor egg as a means of producing neuronal cells that are compatible (i.e., therapeutic cloning). The hospital director faces a number of dilemmas in making her decision. How should she proceed? The design strategy is useful in guiding the decision.

Acquire Facts:

First, she needs to determine the utility of the proposed neuronal cell replacement. In parallel, she needs information on similar situations in which an informed yet somewhat mentally compromised subject participates in an experimental procedure at the subject's own request. Consultation with peer scientists might be helpful but might violate confidentiality. She must also question whether he should still be operating on patients.

Alternative Solutions:

In this case, consideration of the alternative solutions requires understanding the risks and benefits of the proposed procedure, as well as the consequences of executing these solutions. The action to be taken includes convening two committees: one is a fact-finding scientific committee to advise about the technical appropriateness of the procedure, and the second is an ethics committee to advise regarding the protection of the subject if the proposed experiment were to proceed; another issue for this committee is to evaluate his capacity to fulfill his present and future professional obligations.

Further analysis shows that the solution of performing the proposed treatment has significant potential consequences.

- The ethics of the proposed therapeutic cloning might be questioned by the advisory committee.
- The agency of the federal government regulating therapeutic cloning might prohibit this treatment.
- The multitude of risks of injecting tissues into the brain needs to be evaluated.
- The treating physician as well as the Institutional Review Committee must be assured that the patient or his guardian understands the risks.
- Because the patient's mental status is becoming increasingly incapacitated, a guardian or family member needs to be involved.

Alternatives to the proposed therapeutic cloning should be considered. For example, there is evidence that gene therapy using viral vectors with specific genes (e.g., nerve growth factor) might help reduce the progression of disease. Another alternative is the use of progenitor cells from the neurosurgeon's blood or bone marrow, for example, identify and use progenitors of endothelial cells. Separation and use of endothelial CD34+ cells (i.e., white cells that have a surface antigen called CD34 are known to be progenitor cells for the endothelium of capillaries) from other white cells, including macrophages, might be a suitable alternative. The possible procedure is to grow these cells in a petri dish and then inject them directly into the brain.

Assessment:

The evolution of information and the analysis of the risks and potentials of the alternative therapies need to be reviewed with the neurosurgeon, the patient, and his support group or family. A respect for the patient's right to confidentiality is important, and one might anticipate that the patient may not want his family to be involved in decision making. The moral theory of rights of the ill neurosurgeon to acquire treatment for his progressing dementia will possibly conflict with three other responsibilities: 1) with professional ethics of doing no harm, 2) with FDA, possibly because these studies remain experimental with insufficient human safety data, and 3) with a duty to the major hospital stakeholders. The stakeholders include the board of trustees, participating surgeons, hospital staff, the hospital director, the treating neurologist, the afflicted neurosurgeon, and his patients.

Suppose that the alternative of injecting autologous progenitor endothelial cells seems to be the best approach, with minimum risks and some likelihood of success. This approach might be acceptable to the patient. If not, one might negotiate to do that procedure, and after six months if there is no improvement, to reconsider therapeutic cloning. Involving the FDA and the local IRB during the dynamics of this situation will be important.

Action:

In addition to the actions of the two advisory committees, preparation of appropriate stem cells might be commenced in parallel using one of the possible methods for therapeutic cloning. The institution's ethics board should be convened with appropriate consultants if necessary to review the risks and the competency of the subject to understand the risks and to practice medicine.

Myogenesis

Skeletal muscle renewal is another area in which the activities of stem cells are being studied. In late 2002, a research team at Stanford University described for the first time that by making adult mouse muscle-specific stem cells fluorescent, they were able to trace their transformation into fully functional muscle cells (LaBarge and Blau 2002). Satellite cells present in all skeletal muscles of mammals serve as the stem cells for renewal. It is now recognized that these cells respond to muscle injury and effect renewal by forming two daughter cells, one of which remains as a source of satellite cells, the other of which fuses with the injured muscle cell. At least part of the signaling pathway for this response is age-dependent in that the important signaling protein declines with age (Conboy et al. 2005).

Heart Muscle Regeneration

Multipotent hematopoietic stem cells of the adult bone marrow have been shown to play a role in heart regeneration in a series of animal studies (Orlic et al. 2001, Amado et al. 2005), and in human experiments that started in Germany in 2002. The treatment was by direct injection of the patient's own (autologous) nucleated cells through a catheter into a coronary artery which supplies the region of damaged heart muscle of human patients (Brehm et al. 2002, Strauer et al. 2003, Perin et al. 2003, Tse et al. 2003, Fuchs et al. 2003). The use of white blood or nucleated blood cells as autologous progenitor cells for heart muscle repair is illustrated in Figure 10.3. Another method is direct injection into the region of the heart damaged by coronary artery blockage. However, this method has not shown a survival of cells that can be identified as cardiomyocytes, thus raising a serious controversy regarding the efficacy of human autologous progenitor cells in myocardial regeneration (Leri et al. 2005, Balsam and Robins 2005). Nevertheless, some successes have been reported by these investigators, as evidenced by improvement in cardiac ejection fraction. The clinical results are not overwhelming, yet the promise is great.

In addition to the use of bone marrow cells, investigators are researching the possibility of encouraging local stem cells resident in the heart to participate in cardiac regeneration. The cells intrinsic to the heart, known as side population (SP) stem cells, are believed to have the potential for this regeneration. In spite of the controversy regarding the efficacy of injected blood cells in cardiac repair, the evidence that circulating cells extrinsic to the heart can also participate in heart regeneration is strong. The principal evidence comes

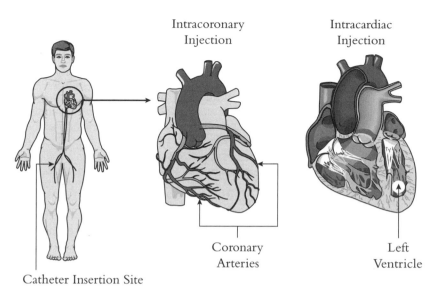

Figure 10.3 This diagram shows the procedures by which autologous progenitor cells are injected into the coronary artery (a catheter is directed to the area of injury through either the femoral or subclavian artery). The autologous progenitor cells are obtained from the peripheral blood of a patient with a heart attack. These cells are expanded in culture, from which nucleated blood cells known as progenitor monocytes are separated and injected.

from the presence of Y chromosome cells in female-to-male transplanted hearts a few months after transplant (Quaini et al. 2002). The evidence for improvement in heart function after injection of autologous as well as embryonic stem cells after a heart attack has come from clinical trials in 2001, and by 2006, this experimental practice was in use in a number of clinics throughout the world.

Below are two cases of stem cell uses in heart disease. The first is for treatment of acute heart failure associated with direct injury to the heart muscle, and the second is the treatment of chronic heart failure resulting from the consequences of uncontrolled hypertension or from previous heart attacks. The latter affects 550,000 new patients each year and is one of the major reasons stem cell research is being pursued around the world.

Case 10.2 Stem Cell Treatment of Acute Heart Failure after Heart Trauma

The first use of stem cells in human heart disease in the United States was in 2003 in a 16-year-old boy whose heart was pierced with a 3-inch nail from a nail gun. Although he survived because the nail kept his heart from bleeding out, the overall result was that one of the main pumping chambers of his heart failed to work effectively. It became apparent he would need a heart transplant, and he was placed on the transplant list. In the meantime, his doctors were considering the use of adult stem cells based on previous work in animals showing successful repair of heart tissue using bone marrow stem cells, as well as reports of trials elsewhere of the use of white cells from patients' own blood in heart repair in Germany (Brehm et al. 2002, Assmus et al. 2002, Strauer et al. 2003), the United States (Toma et al. 2002), and Tokyo (Gojo et al. 2003). This real-life situation brings up numerous medical and ethical dilemmas that are similar to other new applications of stem cell technologies, making it a good case for reviewing the use of the design strategy.

Acquire Facts:
There is evidence in the literature that white cells from rodents and human beings can aid in the return of cardiac function after a period of heart ischemia (coronary artery blockage or heart attack). However, before using methods reported in the literature, one would look to evidence of validation as the reports are on a limited number of patients (e.g., the first report from Germany involved only 10 experimental subjects and 10 control subjects). A second area of ambiguities is whether the procedure should be carried out without consulting the medical ethics committees, as well as the relevant professional conduct oversight authority that has jurisdiction in each state. Ample evidence of the potential risks and possible benefits needs to be compiled, not only so that a professional decision can be reached about whether to proceed with the infusion of stem cells into the boy's heart but also so that the boy and his parents can be advised of the risks. Exactly what were the adverse events that occurred in the other studies? Seeking advice from those who have been performing the procedure would also be prudent.

Alternatives:
The alternative *not* to do the procedure must always be considered. It is important to mention this obvious concept because when only one method is considered appropriate for a medical situation, the decision makers frequently become less objective in considering all other alternatives. For example, in this case other treatment alternatives include continuing present supportive medical treatment, human heart transplantation, or the use of an artificial ventricular assist device that connects the ventricle to the circulatory system as a bridging therapy until the heart mends itself or a transplant becomes available. All three can be used in sequence as another alternative.

Assessment:
In that there are several alternatives that might be applied in different sequences in this situation, the design strategy must include recognition of the various dynamics. For example, one alternative might be to try the stem cell approach for a given period, and if the medical situation continues to decline, to advise the use of a left ventricular assist device followed by a heart transplant. Of importance in this case is the need to place or keep the patient on the heart transplant list if the stem cell or ventricular assist device options are chosen.

Action:
The first action after gathering the facts and consulting with various oversight groups and the IRB is to consult with the patient, the parents, and the primary physician. Because the procedure under consideration is experimental, other physician consultants should be contacted as well. As noted above, there are alternative therapies.

The scientific controversy is as important here as the ethical issues, but the two are coupled. If the methods cannot yield the benefits some propose, then the use of these methods will put the patient at unwarranted risk. Whereas the reports of myocardial replication and regeneration using bone marrow stem cells have possible merit, the mechanism by which this occurs is unknown. Experimental data show that whereas bone marrow cells with specific markers identifying them as progenitor cells (i.e., antigens on the cell surface that distinguish progenitor cells from other cells) do enter the heart and remain abundant after 10 days, few of these cells show evidence of their lineages after 30 days (Balsam et al. 2004).

The bone marrow-derived cells did not express cardiac tissue-specific markers but did show markers of cells of the hematopoietic system, suggesting that transplanted progenitor cells from bone marrow do not become myocytes. Similarly, circulating nucleated blood cells were shown not to populate the injured heart. These studies done in mice are in conflict with other studies wherein male cells were shown to populate the female heart transplanted in males, both in rodents and human beings. Even more recent experiments in pigs show that there is a resident stem cell population in the heart that can be cultured to grow into little beating spheres. But the physicians treating this patient injured by the nail gun did not have all of this information, nor were they aware of the controversies regarding heart regeneration, which appeared in the literature just after they proceeded with their approach.

The action taken by the boy's doctors was to inject his peripheral blood progenitor cells into the heart vessels, which would deliver these cells to the injured region. Three months later, the ability of his heart to pump blood from the main pumping chamber of the heart had improved from 25% to 40% (normal 50 to 75%), and his physical activity capabilities improved accordingly. A year later, his heart function had returned to normal, he had returned to school, and he had no restrictions to his activities, including sports. However, the procedure was not approved by the FDA, and further treatment along these lines has been put on hold until additional safety data are available (www.beaumontpressroom.com/pls/pportal30/pportal30.story_page1?l_recent=302, accessed April 21, 2004).

We transition here from this actual case to a hypothetical case that anticipates the near future approaches using embryonic stem cells.

Case 10.3 Heart Stem Cell Transplant Using Autologous Embryonic Stem Cells

A 40-year-old woman with heart disease is dying because the scar formation in her heart muscle has decreased the ability of her heart to pump blood adequately. The woman is advised by her doctor that a new procedure is available that might substitute for a heart transplant. The new procedure is to have embryonic stem cells infused directly into her heart. Her doctor, who also heads a new clinic called the Center for Regenerative Medicine, has conducted animal studies showing that it is possible to grow cells that would aid in the regeneration of a failing heart using therapeutic cloning methods. The extraction of cells from the blastocyst grown from a donor egg and the nucleus of one of the patient's blood cells, though successful in animal studies, has never been tried in humans. From these stem cells, progenitors for heart muscle will be selected and grown into sufficient numbers for an effective regeneration of her scarred heart. These cells will be injected via heart catheterization. What are the ethical aspects of this example?

The proposed procedure is still in the research stage, and the patient needs to be informed of the alternative therapies (e.g., heart transplantation, use of a left ventricular assist device therapy, use of stem cells from more conventional sources or procedures).

In so far as the doctor is also the researcher for the new procedure, other questions arise. Is the fact that he is suggesting a potentially dangerous therapy to his patient a conflict of interest? How will the problem of a conflict of interest be resolved? Is it against the law? Is ovarian stimulation and egg retrieval overly risky to this patient, given her cardiac condition? Has the patient been informed of any problems in the research of the procedure from the past animal studies? Is the procedure approved by the FDA? Was there an Institutional Review Board evaluation made of the risks and potential benefits of this procedure?

Liver Regeneration

The most experience with the use of stem cells in organ regeneration has been in liver regeneration, where it is known that local cells can replace hepatic tissue to a degree not seen in other tissues. Less well known or accepted is the concept that circulating blood cells from the bone marrow also can regenerate liver (Dahlke et al. 2004). In situ split liver transplantation can be accomplished without complications and provides results that are superior to those obtained previously with ex vivo methods that involve exposing the liver to periods without nutrients (Goss et al 1997).

Pancreas Regeneration

Pancreas renewal has been a target for the argument that embryonic stem cells provide the only possible therapy for type I diabetes, a devastating disease affecting children and adults. This disease is caused by a reaction of the person's immune system to the insulin-generating cells of the pancreas, the result of which is a lifelong dependence on blood sugar monitoring and insulin injections. Now with the focus on adult stem cells, there is some promise that replacement of this function of the pancreas could be effected by the injection of functioning adult pancreatic stem cells that would differentiate into functioning islet (insulin-producing) cells of the individual's pancreas. The efficacy of ES cells or some other form of stem cells, such as multipotent adult progenitor cells (MAPCs, discussed below) or the individual's own stem cells, has still to be demonstrated. One of the principal concerns of this type of tissue replacement is that the immunologic process that led to the destruction of the original tissues might also lead to the destruction of the stem cells. The goal is to implant autologous stem-cell-derived islet cell implants without lifelong immunosuppressive therapy (Chaudhari et al. 2001).

Lung Renewal

The lung is another organ that must have self-renewal capabilities. This is an organ with widely varying tissue types, and in the early 2000s, evidence emerged that there are spore-like or dormant epithelial cells that can differentiate into bronchiolar tissues, the airway tubes of the lung (Bishop 2004).

Summary of Technical Status of Adult Stem Cell Research

Scientists have yet to understand whether, and if so how, cell fusion contributes to the process of adult stem cell repair or tissue renewal. Neither do they understand the extent to which hematopoietic stem cells are responsible for forming adult stem cells in other tissues. Other unknowns include the signals responsible for activating stem cells (whether hematopoietic stem cells in the bone marrow or peripheral blood, or adult stem cells in a given tissue), the nature of the response, how tissues are repaired, or how best to nudge stem cells from human donors to develop into specific cell types that can be used for therapy. It is also important to ascertain the extent to which these cells might provoke immune responses in individual recipients, and if so, whether this form of rejection by the immune system can be treated or whether the individual's own stem cells will be required for the formation of the needed tissues.

Some data in 2004 give support to the premise that adult stem cells may one day be able to substitute for ES cells for medical therapies:

- Discovery of a "master gene" or "nanog" believed to be responsible for the unique regenerative potentials in embryonic stem cells that can be used to signal somatic cells to revert into embryonic stem cells (a nanog is necessary, but not sufficient, to turn a somatic cell into an ES cell).

- Use of somatic or tumor cells has demonstrated the fact that cytoplasmic contents can reprogram cells to become cells with pluripotental capabilities.

- Conversion or reprogramming of somatic cells to embryonic stem cells has been demonstrated using frog egg nuclear substances.

- Use of nuclei from two eggs has been shown to form a diploid blastocyst.

- A woman's oocyte at metaphase II can be activated so as to maintain diploidy, and the resulting blastocyst is then disaggregated to make autologous ES cells through parthenogenesis (Cibelli et al. 2002).

Although this research is still in its early stages, there is evidence of progress in many directions. Undoubtedly, the pursuit of the answers will lead to a greater understanding of many questions, one of the most important of which is the extent to which adult stem cells can substitute for embryonic stem cells in the therapies planned for the future.

Thus, in the first five years of the 2000s, the fundamental notions of human regenerative biology have been expanded. It is now believed that regeneration in most, if not all, organs is facilitated by resident stem cells as well as circulating cells from the bone marrow. The resident stem cells can replicate and differentiate to aid in the regeneration of tissue. The evidence is growing that multipotent adult progenitor cells (MAPCs) found in the bone marrow are capable of forming numerous types of tissues of the rodent (Jiang et al. 2002). These experimental results continue to be controversial, perhaps in part due to the possibilities of artifact, since multipotent adult progenitor cells are found in only a fraction of a percent of the cell samples (Wagers et al. 2002, Wagers and Weissman 2004).

The extent to which adult stem cells will be flexible enough to perform as stem cells in tissues other than their tissue of origin is still unknown. Still to be seen is whether such cells harvested from the patient's own marrow or peripheral blood and used in the same setting as therapeutic cloning could avoid the problems with immune rejection. Experiments are proceeding to further evaluate this. It might also be that adult differentiated or specialized cells can dedifferentiate into stem cells. From an engineering control viewpoint, this possibility makes sense because a condition can exist where cell renewal requirements are greater than the supply of progenitor cells. Alternatively, the use of stem cell injection, whether adult or embryonic, might signal the target tissue, as well as circulating progenitor cells, to participate in tissue renewal. The possibility of a systemic signal having been stimulated by injection of fresh stem cells into the damaged tissue is currently a conjecture to explain why there is such a dramatic improvement in cardiac functioning after local injection of a relatively small number of progenitor cells. A mechanism might be that the injection of these stem cells leads to the recruitment of many more stem cells.

Until more information about adult stem cells is known, most researchers in the field feel that it is imperative that research continue on both ES and adult stem cells. Having more than one therapeutic option and a variety of cell types and methodologies gives one a choice to best meet the needs of an individual patient.

Ethical Issues in Embryonic Stem Cell Research and Therapy

The main ethical questions about the use of human embryonic stem cells involve concerns about the destruction or manipulation of human embryos for research purposes. For many, these procedures are contrary to the basic ethics of human decency as they involve the use of tissue obtained by disassembling living human embryos, result in the loss of a potential human life, and disregard the autonomy of the embryo. We embedded these issues in a set of questions posed earlier in this chapter and now present perspectives on each of these questions. The plan of this presentation is to weave a path from the standpoint of absolute prohibition of the use of embryonic stem cells to the possible acceptance of embryonic stem cells for therapeutic use under some conditions and with some techniques.

Is therapeutic cloning (expanding embryonic stem cells in culture and using them for restorative applications) ethical?

This question can be answered or discussed in the context of the origin of the embryonic stem cells. It helpful to review the breadth of possible sources of embryonic stem cells, all of which involve different means by which to reach the blastocyst stage:

1. Spare embryos allowed to develop to the blastocyst stage.
2. Blastomeres.
3. Already developed stem cell lines.
4. Donated human enucleated egg and the patient's somatic cell nucleus.
5. Blastocysts from the union of two human eggs from the same or different females.
6. The fusion of two blastomeres from two males or two females.

Our discussion turns to specific aspects of the moral questions surrounding the use of embryonic stem cells, usually from the perspective of spare embryo stem cell use, or partial use of a potential embryo by stealing a blastomere. Use of already developed stem cell lines might be considered the use of a spare embryo, as this was the origin of these lines. However, the next three origins of stem cells are forms of somatic cell nuclear transfer and are viewed as therapeutic cloning. Ethical issues surrounding spare embryos, as well as stem cells obtained from therapeutic cloning, are covered in the discussions of the following questions.

Is it morally acceptable to use donated embryos with a potential for life to obtain stem cells?

Most of the contemporary ethical concerns about stem cell research center on harvesting cells from embryonic tissue. Those who favor embryonic stem cell research believe that humanity has a right to benefit from this technology because it offers the hope of cure of previously incurable diseases. The concept of beneficence and that society should be able to benefit from every type of medical knowledge is at the basis of this argument. There is also the argument that scientists should have the freedom or autonomy to pursue this or any other line of research that extends human knowledge and has the potential for such broad benefits. Since medical research is bound by ethics to protect human subjects, many feel there are sufficient measures in place through informed consent and other regulations to keep the moral and other considerations in check. This is an argument to counter the slippery slope arguments.

Those who favor embryonic stem cell research argue that the debates about stem cell research have raised the degree of moral caring about the embryos and their well-being. They urge that society has the obligation to utilize embryos that would otherwise be discarded, and to honor with the highest respect these supreme gifts to science and the well-being of others. They feel that the benefits of studying and using stem cells to better mankind and treat serious diseases far outweigh the risks of not using them and wasting them. Because the decision not to implant them has been made prior to donating them for research, their fate would be to be destroyed and wasted and their benefits to humanity lost forever.

Those in opposition argue that there is no sure evidence to date that embryonic stem cell research will lead to beneficial effects. Many consider the use of cryopreserved "spare embryos" and the extraction of embryonic stem cells as immoral in that it represents the destruction of a potential life, undermines the value of life, ignores the autonomy of the embryo, and leads us onto the slippery slope toward the use of stem cell resources as commodities. Others believe the use of alternative sources of stem cells is more acceptable, and encourage research on cord blood (readily available at every birth) and adult stem cells. Still others feel stem cell research should be regulated by the government and even banned in favor of adult stem cell research. However, in the absence of data showing that adult stem cells can be manipulated to form their own or other tissues outside their cellular boundaries, most researchers feel it is imperative to continue to study all options. Many encourage the scientific pursuit of this type of research on utilitarian grounds that there will be benefits to humanity for the treatment of many diseases, but others decry this approach based on the fact that such research results in the destruction of the embryo.

A very compelling utilitarian argument can be put forward regarding the urgency of ES cell use in congestive heart failure, for which the only treatment is cardiac transplantation or artificial heart replacement. As many as 300,000 Americans die from heart failure each year, and 550,000 new cases are diagnosed each year. A promising therapy for this disease is stem-cell-assisted recovery. Animal experiments show embryonic stem cells give significantly better results than autologous stem cells in therapy after acute myocardial infarction. Heart attacks and strokes cause 17 million deaths per year in the world (World Health Organization 2004). These deaths arise from abnormalities of the arterial wall cells to stresses of the chemical and cellular environment of the individual. Embryo stem cell and adult stem cell research can also lead to a further understanding of vascular pathologies.

Stem cells can be taken from a blastocyst grown from a blastomere; obtaining the blastomere does not sacrifice the embryo. Is this ethical?

Part of the objections to this procedure centers on the question about when life begins, which indeed underlies almost all aspects of embryonic stem cell research. While usually the embryo goes on to develop fully after the removal of the blastomere, there is still a chance of its demise after removal. Some would also say that the blastomere is inexorably harmed by this procedure because it is denied its potential to become a human being. Thus, a major objection to therapeutic cloning lies in the destruction of the embryo and that a blastocytst formed either from a spare embryo or a blastomere taken from a developing embryo also has status that must be respected, just as an embryo is respected. Those supportive of therapeutic cloning would argue that the manipulation of the embryo formed without natural fertilization is manufacturing a tissue that is not entitled to be called an embryo. It is only those zygotes or blastocysts that are successfully implanted that can be considered embryos. Nevertheless, those who do not accept these ideas feel that to proceed with therapeutic cloning would lead us onto a slippery slope of also accepting reproductive cloning or other genetic manipulations. It should be noted that literally millions of embryos fail to implant and are lost every day naturally, and thousands of spontaneous abortions occur daily around the world. If the embryo is a spare from IVF, then removing a cell might be considered minimal harm and a procedure less objectionable than growing the entire embryo to the blastocyst stage and harvesting the stem cells. These considerations might have some moral merit if the principle is to do no or minimum harm to the individual in the utilitarian service to humanity.

Is it morally acceptable for parents to donate their spare embryos for research purposes?

This question differs from the previous question because it asks who is responsible for the spare embryos. Clearly, these embryos were created as part of the process of in vitro fertilization and the creation was motivated by parents who wanted a child. The extra embryos were not created for eventual use as embryo tissue for therapy but for use in case the implant was not successful. One might argue that the extra embryos should not be produced, and if the implantation or pregnancy fails, then another IVF cycle should be pursued. This would reduce the number of spare embryos and perhaps reduce the argument surrounding alternatives for spare embryos. However, the IVF cycle is a difficult procedure that may have health consequences for both the mother and infant. Since scientists and theologians have been unable to agree to the time at which the embryo is considered an individual with autonomy and other rights, a plausible argument is that the parents own and can decide the fate of spare embryos. Those who disagree argue against parental ownership, fearing that the selling of spare embryos could lead to commercialization of this resource. A counter to this slippery slope argument is to institute a law or regulations to prohibit such sales.

Is it ethical to create or use an already formed embryo for the purpose of performing experimental research?

One might immediately conclude that this procedure is morally wrong because both types of embryos have a high potential to achieve personhood if implanted. A justification for this procedure might be that the mating of a specific trait from the somatic nucleus of a donor with the genetic trait of a donor's egg would lead to an understanding of neurodegeneration or atherosclerosis. Cell lines containing genetic defects have been created for just this purpose (Verlinsky et al. 2005). Is the medical science information sufficiently important to justify the means?

What about using a frozen embryo that has died following thawing?

The definition of death for a developing embryo has not yet been formulated in the way death has been defined in a potential transplant donor. In the latter situation, organs must be harvested promptly after death to protect function. To protect the dying patient, federal guidelines have been put in place that outline the criteria that must be met before the patient's organs can be taken. However, these guidelines focus on criteria for brain death that are not applicable to the assessment of what constitutes death in a few-celled embryo. Landry and Zucker (2004) have proposed ways to refine this definition by making observations on a series of previously frozen early embryos that have failed to divide within 24 hours of thawing and would not be used for reproductive purposes. Although the embryo is failing to divide and classified as nonviable, it may still have living embryonic cells that could be made available for stem cell research. The suggested experimental protocol is to observe several hundred nondividing embryos every few hours for additional 24-hour periods. This natural history study of cleavage arrest can lead to an initial definition that can be refined with biochemical markers for irreversible arrest. Even though in irreversible arrest, this method should allow access to still-living blastomeres that could be a resource for expansion into human blastocysts and their stem cells that could be used for research and therapy. The procedure would circumvent the destruction of live human IVF spare embryos and avoid the moral dilemmas that characterize the current protocols.

There is no question that the South Korean reports of patient-specific stem cells heated up the desire to proceed with embryonic stem cell research in the United States. Congressional, state, and private endeavors to enable stem cell research are being curtailed by the current restrictive directives in place, but Bush remains adamant in his decision, the Korean report was deemed fraudulent, and the subject remains controversial.

It is now possible to do preimplantation screening and to select IVF-conceived embryos to ensure tissue compatibility with an ill sibling needing stem cell transplantation. Is this ethical?

This question stems from Case 9.5 ("Screening IVF Embryos to Ensure a Tissue Match for an Ill Sibling") in Chapter 9. If the preimplantation selection is done on spare embryos with the parents' permission, one might argue that the needs of an ill sibling should far outweigh any concern that the spare embryos might someday be implanted and are therefore potential humans. The reader might find this question brings to the stem cell and spare embryo arguments a humanitarian and utilitarian perspective that helps justify the use of spare embryo stem cells to save lives.

If it is shown that adult stem cells can be substituted for embryonic stem cells, is it ethical to proceed with embryonic stem cell research?

Embryonic stem cells are pluripotent, but tissue-specific stem cells have differentiated to the point where they are multipotent and or can become components of a specific tissue. Critics of ES cell technology argue that if it can be demonstrated that the adult stem cell is more plastic than previously thought, there may be no need for ES stem cells. Thus, if adult stem cells can be induced or reprogrammed to regenerate specific tissues, the ethical concerns about the use of embryos to obtain stem cells become moot. But this would mean that the adult stem cell would have to change from its original cell lineage to another, or change from one type of adult stem cell to another. There are also arguments about the consequences of using adult stem cells, because it appears that they do not have longevity and might evoke an immune response even in the person from whom the cells were derived because gene expression could be modified by the manipulations used to reprogram cells. It seems scientifically untenable to terminate the study of human development and embryonic stem cell research without also knowing whether therapies with adult stem cells would be successful and if there are long-term consequences.

Are the methods of the acquisition of human eggs for therapeutic cloning unethical?

Women are the source of eggs for therapeutic cloning. Although reports from South Korean scientists on the successful cloning of human embryos were fraudulent, they did draw attention to the ethical aspects of harvesting fresh eggs from volunteers and employees who may feel unjustly pressured to donate them. There are concerns that widespread use of these procedures might result in abuse of women who serve as egg donors who are unduly coerced by the possibility of financial gain, and because they may not be aware of possible adverse effects due to reported gaps in consenting procedures (Hall 2005). This is problematic because the adverse effects can be serious, one of which, though rare, is death from hyperovarian stimulation syndrome. Scientists have expressed concerns about proceeding carefully, especially noting the need for (i) ethical oversight of research collaborations between scientists working in countries with different standards, (ii) protection of oocyte donors, and (iii) avoidance of unrealistic expectations (Magnus and Cho 2005).

However, these concerns may be eased by the report that human oocytes can be grown from cultured epithelial cells scraped from the surface of adult human ovarian tissue (Bukovsky et al. 2005). If this is as potent a source of oocytes (eggs) as the authors believe, it will be possible to obtain eggs without a donor having to undergo ovarian stimulation and laparoscopic egg retrieval. In addition, this source of eggs could play an important role in the treatment of infertility, in therapeutic cloning, and in basic stem cell research. Sources include ovaries surgically removed during hysterectomy in normal women.

Is it ethical for the U.S. government to limit research that has the potential to benefit millions of people with serious illnesses?

The president's mandates have limited embryonic stem cell research that has the potential to be beneficial to millions of patients, and the U.S. House of Representatives has tried to ban research on and the use of medical treatments with cloned human stem cells. Thus far, the U.S. Congress has not been able to pass legislation to ease these restrictions, although individual states are finding ways to bypass these restrictions by funding stem cell research at the local level through private and tax dollars. There is no easy answer as to who should make these decisions and whether any person should have the authority to do so. The repercussions may come to be similar to those associated with electing a president or appointing a Supreme Court Justice who is either for or against *Roe vs. Wade.*

Should stem cell acquisition procedures be regulated by the state or the federal government, including the FDA?

This question brings the scientists' autonomy in conflict with the rights of a number of stakeholders and places the focus on the moral principles of utilitarianism and autonomy. The stakeholders include:

The embryo

The parents

A dead fetus

Patients who would benefit

Medical scientists

It is important to recognize that much of the debate revolves around the basic question of when an embryo is a person.

Is it appropriate to limit the number of stem cell lines available for research?

This question is based on the assumption that there is a wrong in having an unlimited number of stem cell lines. The majority of scientists in the United States are of the opinion that it is no longer acceptable to limit the number of stem cells available because all present stem cell lines have been contaminated with animal serum and potentially harmful viruses. They argue that if non-federally funded researchers, foundations, and other scientific or engineering enterprises are free to produce embryo cell lines, then why should there be a limitation for federally funded stem cell lines? If there is a moral wrong in federally funded stem cell lines, then this wrong should be equally applicable to non–federally funded stem cell lines.

Is it ethical to obtain embryonic stem cells from an embryo conceived for the purpose of obtaining stem cells?

A previous case involved giving birth to a child screened at the embryonic stage for immunologic compatibility with its ill sibling. In this situation, parents of a child with a disease determine that their child needs immunologically compatible stem cells for survival. They embark on creating an embryo by an in vitro fertilization cycle to create an embryo whose cells can be used to cure the afflicted child. The morality of this creation is somewhat similar to that of actually creating a living human subject for purposes of providing an immune competent organ for an older sibling who has a life-threatening disease due to absence of a kidney or a defective liver or intestine, for example. These cases illustrate the ethical dilemma of the conflict between respect for the rights or autonomy of the new child and the duty of the parents to provide the best for their first child.

The difference between the creation of an in vitro blastocyst for the treatment of their first child and the creation of a fully developed child lies in the moral question of when in the biology of human development does a person exist? One can consider the morality from the perspectives of the three stakeholders: the child with the life-threatening disease, the parents with a duty to their first child, and the embryo being created to save the life of the sick child. The dilemma in this case cannot be lessened by stating that the embryo would not even exist if there were not a need for the embryo tissues. Nor can one argue successfully that there was never an intent to create a human being, thus there is no moral wrong in this act. The considerations really must go to questions such as the following:

- Does the potential recipient of the embryo tissues consent to the method?
- When does cognate life begin?
- Could an alternate procedure, such as modification of the parent's egg or sperm, be performed before fertilization such that a living person could never develop from the in vitro fertilization?

The altering of or tampering with zygotes is similar to altering the nucleus during SCNT in order to avoid ethical concerns of therapeutic cloning proposed by Hurlbut (2004) and discussed further later. This procedure seems also to rely on the time when life or potential life is considered to begin. The assumption is that the egg and sperm are not cognate entities. The ethical issue for the alternative procedure is the safety of the resulting tissue if it has been derived from modified zygotes.

Is it ethical to obtain stem cells from a spontaneously aborted fetus?

This question is included because it is an erroneous concept that embryonic stem cells can be obtained from a fetus. The fetus from the 8–10-week stage on has developed beyond the blastocyst stage, the source of embryonic stem cells in the developing embryo. The argument that a dead fetus might have been the source of ES cells raises concerns that do not accurately reflect how ES cells are obtained. When and if it is possible to obtain fetal stem cells for therapeutic use, then these concerns will need to be revisited. Issues about the transplantation of fetal tissues were discussed in Chapter 7.

What should be society's role in promoting or negating stem cell research?

Society needs to get facts, resolve ambiguities, and encourage research into some substantial unknowns before the benefits of stem cell therapy can be appreciated or denounced. Society has not realized the full potential for the uses of either adult or embryonic stem cells. Indeed, there are new data indicating that the effects seen during injected stem cell participation in tissue renewal might not be through a direct involvement of the stem cells, but rather occur through a mechanism involving the activation of circulatory signals in the body mediated indirectly by injected stem cells (Conboy et al. 2005).

It is not presently known whether adult stem cells can be substituted for embryonic stem cells in all therapeutic settings. Because it is thought that different kinds of stem cells have different potentials for treatment of disease, there is a wide range of research still to be conducted. Scientists are now beginning to understand how to keep tissues successfully in the laboratory for long periods of time and how to signal embryonic stem cells to form specific cell lineages. The refinement of these techniques could result in new treatments for many serious diseases. However, there are numerous ethical issues of embryonic, fetal, individual, social, political, and religious importance that need to be addressed.

Whereas some elements of society would not support the use of frozen spare embryos or therapeutic cloning, most would promote the use of uncontroversial sources of stem cells, such as those harvested from the umbilical cord or adult stem cells derived from peripheral blood. To complicate matters further, it appears that the public does not clearly understand the differences between reproductive cloning (the purpose of which is the birth of a living clone, a procedure that has been banned in the United States and most countries) and therapeutic cloning (obtaining embryonic stem cells for therapeutic uses from an embryonic blastocyst). This lack of understanding, as well as other misunderstandings about stem cell research, makes it difficult to change these opinions or to make the case for other ways to obtain embryonic stem cells that might be more acceptable. Thus, the debate on moral aspects is frequently uninformed and becomes emotionally charged because not all parties understand the facts. Not all of the stem cell technologies lead to the demise of or are even harmful to the human embryo or the fetus, and not all involve genetic modification that might affect future generations. Only by understanding the issues of the various technologies, and by encouraging a broad dialogue about them, will society be able to make informed decisions about the possibilities they offer for the future.

The negative reaction to the use of stem cells for human regeneration and longevity might be in part a result of a negative reaction toward the now defunct eugenics movements of 150 years ago, which promoted improving the human race by selective mating. The principles of autonomy, non-malfeasance, and rights that applied to the creation of a superior race are applicable to issues of today such as abortion, in vitro fertilization, embryo rights, chromosome manipulation, and intergenerational risks. On the other hand, the majority might have as the basis for their objection the killing of an embryo. These two underlying objections are important to recognize and clarify when embarking on discussions regarding the ethics of stem cell research.

The decisions surrounding the ethics of embryonic stem cell research and its applications will not be easy to resolve. Because of this, the public must be kept informed about the research that is underway, and there needs to be an open, ongoing ethical debate about the issues based on scientific fact. Also needed is care in the selection of terms and language used in discussing these issues so that religious beliefs are respected and inflammatory phrases are avoided. Only in this way can these complex and controversial issues be worked through.

Is there an ethically acceptable alternative to nuclear transfer technologies?

In spring 2005, the U.S. President's Council on Bioethics proposed for consideration an approach to generating embryonic stem cells using a procedure that avoided the possible generation of a living embryo. The procedure, known as altered nuclear transfer (ANT), is to use cells that could never become a living human being by cultivating blastocysts with a defect known to interfere with the development of an organized embryo (Hurlbut 2004). For example, if during nuclear transformation the nucleus of the somatic cell is disrupted by suppressing the gene known as Cdx2, the cloned blastocyst will never develop into a living embryo because this gene is essential for the production of tissues known as the trophectoderm, needed for the development of the placenta and the viability of the embryo. The absence of this gene has allowed stem cell production in the mouse. Scientists disagree about whether the concept of tampering with the nucleus of the individual to be cloned is a good alternative to somatic cell nuclear transfer because it is not known whether the tampering will result in good embryonic stem cells without extensive experiments (Melton et al 2004). The additional concern is that a quest for developing embryonic stem cell lines through altered nuclear transformation would distract researchers from the focus on developing a medically efficacious approach to therapeutic cloning.

The proposal to use altered nuclear transfer seems to be a conciliatory effort to quiet the fierce debate between those who believe the rights of an individual start with the zygote and those who believe the benefits of society are best served by using embryos that would otherwise be destroyed. The danger of this proposal is that if accepted as the avenue for federal support in the United States or elsewhere, it would divert effort and funds toward a paradigm that might not work, thereby deflecting scientists from research that might have more promise. In effect, the ethics of the President's Committee on Bioethics can rightly be challenged on the grounds that ANT might have its own risks to the recipients of the therapy, and also on the grounds that it could divert public funding from conventional stem cell research to a path of research that might not be successful. Further, the proposed ANT approach does not address the ethical issues of egg acquisition from fertility clinics or from a volunteer donor. It does, however, make the assumption that the potential for life of an egg is not a moral issue.

If somatic cells can be reprogrammed to become embryonic stem cells without using oocytes per se, the ethical considerations change. It appears that certain genes in the nuclei of somatic cells can be manipulated to reprogram the nuclei to an embryonic state, a process known as dedifferentiation. If so, this would bypass the need for human oocytes. Could this get around the ethical and other limitations of therapeutic cloning because it would not involve the fertilization of an egg by a sperm and would not involve an embryo? Some argue that this is not merely an "activated egg" but is still morally equivalent to a human embryo and due all the rights and protections thereto (Green 2004). There will likely be a number of approaches to obtaining embryonic stem cells, and each is likely to have ethical implications that will need to be sorted out. But the avoidance of the use of an embryo, an egg and sperm, or a blastomere might quiet some of the concerns that have been voiced to date.

CHAPTER SUMMARY

The basic ethical conflict of stem cell research is between autonomy and utilitarianism. On the one hand, society stands to gain by the treatment of devastating illnesses, including Parkinson's and Alzheimer's disease, congestive heart failure, and juvenile diabetes. On the other hand, the methods to accomplish these advances involve the disaggregation of embryos that have the potential of becoming human beings. The basic conflict is how one defines the concept of "potential of becoming human beings." There are two considerations in this statement. First, if the intent is the production of stem cells without any intent to create a viable fetus, do we consider a spare embryo or a blastomere formed from an individual's somatic cell nucleus and a human egg during therapeutic cloning (Chapter 9) as truly a potential human being? If personhood exists at the zygote or blastocyst stage, then it is unethical to create a zygote or blastocyst for use as tissue replacement for another. Many argue that at this stage the embryo has moral status as a human being. The counterargument is that embryos have a lesser moral status, and there is the likelihood of significant human benefit from their use in therapy of others.

However, contrary to some beliefs, not all stem cell uses qualify as objectionable or controversial procedures, and not all stem cell research involves the destruction of the embryo from which the stem cells were taken. Stem cells taken from adults or the individual for whom the therapy is being offered (e.g., umbilical cord blood, autologous progenitor cells such as bone marrow or peripheral blood) present no ethical problems. However, the acquisition of embryonic stem cells from frozen spare embryos and the use of patient-specific stem cells for regenerative purposes or to treat specific diseases (therapeutic cloning) remain highly controversial.

Those favoring stem cell research feel a precedent has been set for the use of ES cells by the use of tissue and organ donation for transplantation. Many support research involving the use of both embryonic and adult types of stem cells for therapeutic purposes. Others support the use of adult but not embryonic stem cells. Still others favor gene therapy, arguing that to repair the cell by the installation of needed genes into the genome of the embryo to eliminate the occurrence of a serious disease is far more acceptable. Perhaps more acceptable also would be the instillation of cells with needed genes in local tissues once a disease has developed (for example, gene therapy of Parkinson's disease). However, most scientists favor the development of both embryonic stem cell and adult stem cell options in order to provide the best possible treatment choices for patients needing this therapy.

Finally, there is the need to assess the ethical problems in parallel with analysis of ongoing research and the prospects for future research:

> . . . a utilitarian perspective that allows science and medicine to trump ethics in the name of furthering medical benefits [at the cost of the life of a human embryo] is exceedingly dangerous and will eventually have grave consequences. Any serious consideration of the issues raised by research cloning must engage all basic ethical arguments against such a prospect, rather than merely dismiss them or regard them as secondary in importance.

> [The idea is] . . . not to hold back science but to encourage it to proceed in an ethical fashion. Our society must not subjugate basic ethical considerations to scientific and medical progress, lest we all become subjects mastered by our own technological prowess (Pellegrino et al. 2002).

The full potential for different kinds of stem cells to develop into useful tissues, organs, or other therapeutic remedies is not fully understood at this time. It is not known whether adult stem cells can be substituted for embryonic stem cells in the types of therapies for which, it is hoped, stem cells will help. Because of this, and because it is thought that different kinds of stem cells have different potentials for treatment of disease, there is a wide range of research still to be conducted. Scientists now are beginning to understand how to grow tissues successfully outside of the body for long periods of time, and these techniques, when refined, will undoubtedly lead to new approaches to treatments for many serious diseases. The numerous ethical issues of fetal autonomy, individual rights, social justice, political authority to regulate, and religious doctrines will continue to be debated as the reproductive technologies continue to be developed and as the benefits of stem cells become more apparent to the health of the world.

REFERENCES

Alberio R., A. D. Johnson, R. Stick, and K. H. Campbell. 2005. "Differential nuclear remodeling of mammalian somatic cells by Xenopus laevis oocyte and egg cytoplasm." *Exp Cell Res* **307**(1):131–141.

Amado, L. C., A. P. Saliaris, K. H. Schuleri, et al. 2005. "Cardiac repair with intramyocardial injection of allogeneic mesenchymal stem cells after myocardial infarction." *Proc Natl Acad Sci U S A* **102**(32):11474–11479.

Assmus, B., V. Schachinger, C. Teupe, et al. 2002. "Transplantation of progenitor cells and regeneration enhancement in acute myocardial infarction." *Circulation* **106**(24):3009–3017.

Balsam, L. B., A. J. Wagers, J. L. Christensen, et al. 2004. "Haematopoietic stem cells adopt mature haematopoietic fates in ischemic myocardium." *Nature* **428**:663–673.

Balsam, L. B., R. C. Robbins. 2005. "Haematopoietic stem cells and repair of the ischaemic heart." *Clin Sci (Lond)* **109**(6):483–492.

Besinger, W. I., P. J. Martin, B. Storer, et al. 2001 (January 18). "Transplantation of bone marrow as compared with peripheral-blood cells from HLA-identical relatives in patients with hematologic cancers." *N Engl J Med* **344**:175.

Bishop, A. E. 2004 (February). "Pulmonary epithelial stem cells." *Cell Prolif* **37**(1):89–96.

Blau, H, T. Brazelton, G. Keshet, and F. Rossi. 2002. "Something in the eye of the beholder." *Science* **298**:361.

Brehm, M., T. Zeus, and B. E. Strauer. 2002 (November). "Stem cells—clinical application and perspectives." *Herz* **27**(7):611–620.

Bukovsky, A., M. Svetlikova, and M. R. Caudle. 2005. "Oogenesis in cultures derived from adult human ovaries." *Reprod Biol Endocrinol* **3**(1):17.

Chaudhari, M., J. G. Cornelius, A. B. Peck, et al. 2001 (December). "Pancreatic stem cells: a therapeutic agent that may offer the best approach for curing type 1 diabetes." *Pediatr Diabetes* **2**(4):195–202.

Chung, Y., I. Klimanskaya, S. Becker, et al. 2006 (January 12). "Embryonic and extraembryonic stem cell lines derived from single mouse blastomeres." *Nature.* **439**:216–219.

Cibelli, J. B., K. A. Grant, K. B. Chapman, et al. 2002. "Parthenogenetic stem cells in nonhuman primates." *Science* **295**(5556):819.

Conboy I. M., M. J. Conboy, A. J. Wagers, et al. 2005. "Rejuvenation of aged progenitor cells by exposure to a young systemic environment." *Nature* **433**(7027):760–764.

Cowan, C. A., I. Klimanskaya, J. McMahon, et al. 2004. "Derivation of embryonic stem-cell lines from human blastocysts." *N Engl J Med* **350**:1353–1356.

Dahlke, M. H., F. C. Popp, S. Larsen, et al. 2004 (April). "Stem cell therapy of the liver—fusion or fiction?" *Liver Transpl* **10**(4):471.

Daley, G. Q. 2004. "Missed opportunities in embryonic stem-cell research." *N Engl J Med* **351**:627–628.

Drazen, J. M. 2003 (July 17). "Legislative myopia on stem cells." *N Engl J Med* **349**:300.

Epstein, E. 2002 (April 11). "Bush tries to sway senators against cloning." *San Francisco Chronicle*, p. A3.

Epstein, E. 2005 (May 21). "Bush vows to veto stem cell bill. He's also critical of breakthrough by Korean scientists." *San Francisco Chronicle*, p. A1.

Evans M. and M. Kaufman. 1981. "Establishment in culture of pluripotential cells from mouse embryos." *Nature* **292**:154–156.

Escolar, M. L., M. D. Poe, J. M. Provenzale, et al. 2005. "Transplantation of umbilical-cord blood in babies with infantile Krabbe's disease." *N Engl J Med* **352**(20):2069–2081.

Feng, L-X., Y. Chen, L. Dettin, et al. 2002. "Generation and in vitro differentiation of a spermatogonial cell line." *Science* **297**:392.

Fuchs, S., L. F. Satler, R. Kornowski, et al. 2003. "Catheter based autologous bone marrow myocardial injection in no option patients with advanced coronary artery disease." *J Am Col Cardiol* **41**:1721–1724.

Gerecht-Nir, S. and J. Itskovitz-Eldor. 2004. "Human embryonic stem cells: a potential source for cellular therapy." *Am J Transplant* **4**(Suppl 6):51–57.

Globerson, A. 2001. "Haematopoietic stem cell aging." *Novartis Symp* **235**:85–104.

Gojo, S., N. Gojo, Y. Takeda, et al. 2003 (August 1). "In vivo cardiovasculogenesis by direct injection of isolated adult mesenchymal stem cells." *Exp Cell Res* **288**(1):51–59.

Goss, J. A., H. Yersiz, C. R. Shackleton, et al. 1997 (September 27). "In situ splitting of the cadaveric liver for transplantation." *Transplantation* **64**(6):871–877.

Green, R. M. 2004. "Ethical Considerations." *Handbook of Stem Cells,* volume 1. R. Lanza, H. Blau, M. Moore, et al. (eds). Boston, MA: Elsevier Academic Press, pp. 759–764.

Hall, C. T. 2001 (August 20). "The forgotten embryo: Fertility clinics must store or destroy the surplus that is part of the process." *San Francisco Chronicle,* p. A1.

Hampton, T. 2005. "Human embryonic stem cells contaminated." *JAMA* **293**:789.

Hochedlinger, K. and R. Jaenisch. 2003a. "Nuclear transplantation, embryonic stem cells, and the potential for cell therapy" (review article). *N Engl J Med* **349**:275.

Hochedlinger, K. and R. Jaenisch, 2003b, p. 284.

Hochedlinger, K. and R. Jaenisch, 2003c, pp. 277–278.

Hoffman, D. I., G. L. Zellman, C. C. Fair, et al. 2003. "Cryopreserved embryos in the United States and their availability for research." *Fertil Steril* **79**(5):1063–1069.

Holden, C. 2004 (May 21). "Zerhouni's answer buoys supporters." *Science* **304**:1088.

Hurlbut, W. B. 2004 (December). "Altered nuclear transfer as a morally acceptable means for the procurement of human embryonic stem cells." The President's Council on Bioethics, www.bioethics.gov/transcripts/dec04/sessions6/htms, accessed June 13, 2005.

Hwang, W.S., Y. J. Ryu, J. H. Park, et al. 2004. "Evidence of a pluripotent human embryonic stem cell line derived from a cloned blastocyst." *Science* **303**:1669–1674.

Hwang, W. S., S. I. Roh, B. C. Lee, et al. 2005. "Patient-specific embryonic stem cells derived from human SCNT blastocysts." *Science,* **308**:1777-1783.

Jain, T., S. A. Missmer, and M. D. Hornstein. 2004. "Trends in embryo-transfer practice and in outcomes of the use of assisted reproductive technology in the United States." *N Engl J Med* **350**:1642.

Jiang, Y., B. N. Jahagirdar, R. L. Reinhardt, et al. 2002. "Pluripotency of mesenchymal stem cells derived from adult marrow." *Nature* **418**:41–49.

Kennedy, D. 2006. "Editorial expresson of concern." *Science* **311**:36.

Körbling, M. and Z. Estrov. 2003a. "Adult stem cells for tissue repair—a new therapeutic concept?" *N Engl J Med* **349**:572.

Körbling, M. and Z. Estrov. 2003b, pp. 570–573.

Körbling, M. and Z. Estrov. 2003c, pp. 574–575.

Krause, D. S., N. D. Theise, M. I. Collector, et al. 2001. "Multi-organ multi-lineage engraftment by a single bone marrow-derived stem cell." *Cell* **105**:369–377.

Kurtzberg, J, M. Laughlin, M. L. Graham, et al. 1996 (July 18). "Placental blood as a source of hematopoietic stem cells for transplantation into unrelated recipients." *N Engl J Med* **335**:157–166.

LaBarge, M. and H. M. Blau. 2002. "Biological progression from adult bone marrow to mononucleate muscle fiber in response to injury." *Cell* **111**:589–560.

Lakshmipathy, U. and C. Verfaillie. 2005 (January). "Stem cell plasticity." *Blood Rev* **19**(1):29–38.

Landry, D. W. and H. A. Zucker. 2004 (November 1). "Embryonic death and the creation of human embryonic stem cells." *J Clin Invest* **114**(9):1184–1186.

Lanza R. P., J. B. Cibelli. F. Diaz, et al. 2000. "Cloning of an endangered species (Bos gaurus) using interspecies nuclear transfer." *Cloning* **2**(2):79–90.

Laughlin, M. J., M. Eapen, P. Rubinstein, et al. 2004 (November 25). "Outcomes after transplantation of cord blood or bone marrow from unrelated donors in adults with leukemia." *N Engl J Med* **351**:2265–2275.

Leri, A., J. Kajstura, and P. Anversa. 2005. "Cardiac stem cells and mechanisms of myocardial regeneration." *Physiol Rev* **85**(4):1373–1416.

Mackay, J. and G. Mensah (eds). 2004. *Atlas of Heart Disease and Stroke.* Geneva: World Health Organization, p. 18.

Magnus, D., and K. Cho. 2005 (published online May 19). "Issues in oocyte donation for stem cell research." *Science*, accessed May 22, 2005 at www.sciencemag.org/cgi/content/abstract/1114454v1.

Martin, M. J., A. Muotri, F. Gage, and A. Varki. 2005. "Human embryonic stem cells express an immunogenic nonhuman sialic acid." *Nat Med* **11**(2):228–232.

Melton, D. A., G. Q. Daley, and C. G. Jennings. 2004 (December 30). "Altered nuclear transfer in stem-cell research—a flawed proposal." *N Engl J Med* **351**(27):2791–2792.

Muller, P., P. Pfeiffer, J. Koglin, et al. 2002. "Cardiomyocytes of noncardiac origin in myocardial biopsies of human transplanted hearts." *Circulation* **106**(1):31–35.

Nagano, M., P. Patrizio, and R. Brinster. 2000 (December). "Long-term survival of human spermatogonial stem cells in mouse testes." *Fertility and Sterility* **78**(6):1225.

National Academy of Sciences. 2002. "Scientific and Medical Aspects of Human Reproductive Cloning. Report of the National Academy of Sciences Committee on Science, Engineering, Public Policy." Washington, DC: The National Academies Press, pp. 92–100.

National Institutes of Health. 1995 (April). "Recombinant DNA Research: Actions under the Guidelines. NIH Recombinant DNA Advisory Committee Points to Consider." *Federal Register* **60**(81):20726–20737.

Orlic D., J. Kajstura, S. Chimenti, et al. 2001. "Bone marrow cells regenerate infracted myocardium." *Nature* **410**:701–705.

Pellegrino, E. D., J. F. Kilner, and K. T. FitzGerald. 2002. "Therapeutic cloning." Letter to the Editor. *N Engl J Med* **347**:1619.

Perez-Pena, R. 2003 (March 16). "Broad movement is backing embryo stem cell research." *New York Times*, National Report Section, p.18.

Perin, E. C., H. Dohmann, and R. Borojevic, et al. 2003. "Transendocardial, autologous bone marrow cell transplantation for severe chronic ischemic heart failure." *Circulation* 107:2294–2302.

Phimster, E. G. and J. M. Drazen. 2004. "Two fillips for human embryonic stem cells." *N Engl J Med* **350**:1351–1352.

Quaini, F., K. Urbanek, A. P. Beltrami, et al. 2002. "Chimerism of the transplanted heart." *N Engl J Med* **346**:5–15.

Reubinoff, B. E., M. F. Pera, C.-Y. Fong, et al. 2000. "Embryonic stem cell lines from human blastocysts: somatic differentiation in vitro." *Nature Biotechnology* **18**:399–404.

Richards, M., C. Y. Fong, S. Tan, et al. 2004. "An efficient and safe xeno-free cryopreservation method for the storage of human embryonic stem cells." *Stem Cells* **22**:779–789.

Rocha, V., M. Labopin, G. Sanz, et al. 2004 (November 25). "Transplants of umbilical-cord blood or bone marrow from unrelated donors in adults with acute leukemia." *N Engl J Med* **351**:2276–2285.

Rosenthal, N. 2003. "Prometheus's vulture and the stem-cell promise." *N Engl J Med* **349**:268, 271–272.

Salgar, S. K., R. Shapiro, F. Dodson, et al. 1999. "Infusion of donor leukocytes to induce tolerance in organ allograft recipients." *J Leukoc Biol* **66**:310–314.

San Francisco Chronicle. 2005 (June 1). "Romney veto on stem cell bill overridden." *San Francisco Chronicle,* p. A4.

Shamblott, M. J., J. Axelman, S. Wang, et al. 1998. "Derivation of pluripotent stem cells from cultured human primordial germ cells." *Proc Natl Acad Sci U S A.* **95**:13726–13731.

Staba, S. L., M. L. Escolar, M. Poe, et al. 2004 (May 6). "Cord-blood transplants from unrelated donors in patients with Hurler's syndrome." *N Engl J Med* **350**:1960–1969.

Starzl, T. E. 2004 (October 5). "Chimerism and tolerance in transplantation." *Proc Natl Acad Sci U S A.* **101**(Suppl 2):14607.

Stolberg, S. G. 2005 (April 7). "Top scientists press for more stem cells." *San Francisco Chronicle,* p. A10.

Strauer, B. E., M. Brehm, T. Zeus, et al. 2003. "Stem cell therapy in acute myocardial infarction". *Med Klin* (Munich) **98**(Suppl 2):14–18.

Tanner, L. 2004 (May 5). "Creating a baby to save a sibling: stem cells from healthy infants help ailing brothers and sisters." *San Francisco Chronicle,* p. A2.

Thomson, J. A., J. Itskovitz-Eldor, S. S. Shapiro, et al. 1998. "Embryonic stem cell lines derived from human blastocysts." *Science* **282**:1145–1147.

Toma, C., M. F. Pittenger, K. S. Cahill, et al. 2002 (January 1). "Human mesenchymal stem cells differentiate to a cardiomyocyte phenotype in the adult murine heart." *Circulation* **105**(1):93–98.

Tse, H. F., Y. L. Kwong, J. Chan, et al. 2003. "Angiogenesis in ischaemic myocardium by intramyocardial autologous bone marrow mononuclear cell implantation." *Lancet* **361**:47–48.

Verfaillie, C. M. 2004. "'Adult' stem cells: Tissue specific or not?" In: *Handbook of Stem Cells. Volume 2* R. Lanza, H. Blau, D. Melton, et al. (eds.). New York: Elsevier Press, p. 13.

Verlinsky, Y., S. Rechitsky, T. Sharapova, et al. 2004. "Preimplantation HLA testing." *JAMA* **291**:2079.

Verlinsky, Y., N. Strelchenko, V. Kukharenko, et al. 2005. "Human embryonic stem cell lines with genetic disorders." *Reprod Biomed Online* **10**(1):105–110.

Wagers, A. J., R. I. Sherwood, J. L. Christensen, and I. L. Weissman. 2002. "Little evidence for developmental plasticity of adult hematopoietic stem cells." *Science* **297**:2256–2259.

Wagers, A. J. and I. L. Weissman. 2004 (March 5). "Plasticity of adult stem cells." *Cell* **116**:639–648.

Walters, L. and J. G. Palmer. 1997. *The Ethics of Human Gene Therapy*. New York: Oxford University Press.

Weissman, I. L., D. J. Anderson, and F. Gage. 2001. "Stem and progenitor cells: origins, phenotypes, lineage commitments, and transdifferentiations." *Annu Rev Cell Dev Biol* **17**:387–403.

West, M. 2005. Personal communication.

ADDITIONAL READING

Buchanan, A., D. W. Brock, N. Daniels, and D. Wikler. 2000. *From Chance to Choice: Genetics and Justice*. New York: Cambridge University Press, 398 pp.

Fukuyama, F. 2002. *Our Posthuman Future. Consequences of the Biotechnology Revolution*. New York: Farrar, Straus and Giroux, 256 pp.

Kass, L. R. and J. Q. Wilson. 1998. *The Ethics of Human Cloning*. Washington, DC: The AEI Press, 101 pp.

Lanza, R, H. Blau, D Melton, et al. (eds). 2004. *Handbook of Stem Cells: Embryonic Stem Cells* (Vol.1), *Adult and Fetal* (Vol.2). Burlington, MA: Elsevier Academic Press,. 806 pp. and 820 pp., respectively.

National Academy of Sciences. *Scientific and Medical Aspects of Human Reproductive Cloning*. Report of the National Academy of Sciences Committee on Science, Engineering and Public Policy, 2002. Washington, DC: The National Academies Press, Washington D.C., 272 pp. See http://books.nap.edu/books/0309076374/html/index.html.

Warnock, M. *A Question of Life*. 1985. New York: Basil Blackwell Inc., 110 pp.

PROBLEM SET

10.1 A researcher wishes to establish a cell culture of heart muscle stem cells to investigate a new method of heart tissue renewal after a heart attack. Knowing the admonishments regarding embryo stem cell research, this researcher decides to clone a dog embryo using the nucleus from a fibroblast culture of dog cells and the egg of the same dog by the techniques discussed in this chapter. He cannot get permission from the animal use committee for this experiment, so he goes to Australia, where permission can be more readily obtained. There he does a successful cloning of a blastocyst and from the blastocyst is able to culture cardiac myocytes (heart muscle cells). He infuses these cells into an experimental heart attack model in the dog from which he obtained the fibroblasts and finds they help in tissue regeneration. There are a number of ethical issues, not all bad, in this example. What are they, and how would you judge the researcher's behavior?

10.2 Refer to Case 10.1 to answer this problem. The scenario given, along with alternatives and some ethical considerations, is missing very important elements that depend on some knowledge from Chapters 4 and 8. Specifically, consider all of the stakeholders associated with the clinic and the professionals in discussion of the question: What additional moral duties, rights, and virtuous considerations are important to this case?

10.3 In order to learn the extent to which nucleated blood cells from the circulation participate in tissue healing, a researcher extracts her own progenitor endothelial cells from 10 cc of her own blood. The science question is whether wound healing is from repair by local tissue cells or repair facilitated by circulating progenitor cells. She labels her cells by transfecting them with a virus which carries the gene for green fluorescent protein. After reinjecting these cells, she expects to see some green fluorescence if her hypothesis is correct. The cells she transfected will be green no matter where they distribute. They and any daughter cells will be green for their entire lifetime.

Discuss the ethics involved in the human experimentation. Recognize that the investigator is well informed regarding the experiment, so the issue of autonomy and requirements for IRB approval are not the same as for doing this experiment on a volunteer. The work is being done at a federally funded institution, and the investigator is partly funded by an NIH grant to do general stem cell research on animals.

10.4 What is the position of professional and religious societies on the uses of stem cells obtained from therapeutic cloning? (Use a Web browser or other methods to learn positions of the American Medical Association, the Association of Reproductive Health Professionals, the Christian Scientists, the Roman Catholic Church, the Baptists, etc.) Acquire the position statement for at least three societies and summarize the position of each in 30 words or less.

10.5 A business partner in a biotechnical corporation decides to set up a tissue bank for umbilical cord blood collected from a group of hospitals that would otherwise discard this valuable neonatal resource. You are asked to join in this enterprise and take on the role of marketing and licensing the product that will be used in treating human subjects and thus falls under the regulations of the Food and Drug Administration. Outline the step-by-step process you will use to obtain the umbilical cord blood. (150 words or less.)

10.6 In 2004, the state of California passed a proposition to spend 300 million dollars each year for 10 years to investigate the merits of embryonic stem cells for medical therapy. This 3 billion dollar initiative is intended to encourage research and development in an area that is forbidden to researchers who depend on federal funding. As a university researcher, you plan to use these funds by participating in research at private institutions that are not federally funded. However, you are salaried at a university that does receive federal funds, and your students are paid stipends from NIH training grants. How do you approach the dilemma of doing embryonic stem cell research and not violating the federal regulations banning research on embryonic stem cells that are newly generated and not federally approved?

10.7 A colleague in the area of human growth and development has been successful in cloning experiments wherein the ovum of a rabbit is enucleated and the nucleus of a rabbit fibroblast inserted. The resulting embryo is implanted and becomes a viable fetus. Your colleague then takes some skin cells (fibroblasts) from his arm, cultures them, and puts the nucleus of one into a freshly enucleated rabbit ovum. He proceeds to produce an embryo and grows it to the blastocyst stage. He removes stem cells from the blastocyst and cultures them to produce a stem cell line, which he intends to store for future use for "neuronal replacement" in the event he loses his cognitive powers. The dean learns of this work and decides that the experiment is a departure from acceptable practices; thus this is scientific misconduct according to the definitions.

A. How should the faculty committee approach the dean's charge of unethical behavior and reckless departure from acceptable scientific practices?

B. Would this be a violation of the prohibition for stem cell research?

C. If you were the scientist, how would you defend yourself?

D. If the committee finds this was indeed a departure from acceptable practices, does this become scientific misconduct under the definition in Chapter 2?

11

Ethics of Enhancement Technologies

INTRODUCTION

This chapter brings together the various new technologies that presently or in the future will offer new ways to enhance human health, the human genome, human performance, and human well-being. Therapy is the application of surgery, drugs, or devices to treat a disease or deformity in order to return an individual to a state of health or normalcy. *Enhancement* differs from *therapy* in that it is an intended change to improve an already healthy or average body by additions or deletions using surgery or drugs. The four major methodologies that allow these enhancements include:

- Embryo selection by preimplantation genetic diagnosis (PGD).
- Genetic manipulations by gene therapy.
- The use of drugs to enhance performance and behavior.
- Cosmetic surgery.

Ethical issues include:

- Risks to individuals.
- Equity in availability of enhancements.
- The morality of inducing human enhancements resulting from unnatural genetic manipulations that will forever be part of the DNA passed down to future generations.

We will revisit PGD not from the standpoint of screening embryos for genetic diseases (see Chapter 9, "Ethics of Assisted Reproductive Technologies") but from that of the possibilities it offers for genetic enhancement. It is already being used by parents to select embryos with desired characteristics (e.g., gender). Once the genes are identified for cosmetic and other enhancements, parents will have the option to use PGD to select these enhancements as well. As we shall see, these selections will have ethical consequences in the future. We will also study various aspects of gene therapy, including an overview of the methodologies by which new genes can be added to cells not only to restore function and but also for their potential to enhance human attributes. We will also discuss various pharmacological approaches to improving human performance and behavior, as well as surgical enhancements that are readily available and increasingly used by those who can afford them.

A multitude of enhancements are now possible because of technical advances made since the mid-1970s, some of which are reviewed below:

- In vitro fertilization (IVF) and PGD, techniques for obtaining, assessing, incubating, and manipulating eggs and sperm in the laboratory, have been further developed and refined. Similar advances have been made in techniques for assessing, incubating, and transferring embryos to the womb.

- Gene therapy became possible when information on the genetic basis of specific diseases became available. Innovative methods have been developed by which copies of a missing gene can be made and delivered to targeted body cells to restore function to individual patients with genetic diseases (*somatic gene therapy*). Eventually, it will be possible to select and administer clusters of genes to provide a broad array of selected traits to a developing embryo.

- It is now theoretically possible to perform *germline* gene therapy on reproductive cells (eggs or sperm) that carry a genetic abnormality, or on cells of the earliest embryo (either the newly formed single-celled zygote or on a totipotent blastomere). Thus, in the future, germline gene therapy will correct the genetic abnormality and restore normal function to the target cell and to all subsequent cells that are formed. Because it is the reproductive cells that carry the genetic modification, all future offspring will be free of the genetic abnormality. Early intervention will also allow the fetus to develop normally, free from the genetic abnormality that can damage it *in utero*. This is in contrast to somatic gene therapy, which restores function to only one cell type; since it does not affect the reproductive cells, the genetic additions are not passed down to offspring who may also require gene therapy.

- New drugs designed to enhance athletic performance are widely available and often escape detection in screening tests designed to identify athletes with unfair advantages over other competitors. Despite disciplinary consequences these practices contine.

- Pharmaceuticals developed for the treatment of depression and severe anxiety have been found to enhance social behavior, counteract phobias, and improve intellectual performance.

- The growing field of plastic surgery and further refinements in surgical techniques allow multiple and even radical medical and cosmetic improvements of the body.

Some of these technologies are still in their infancy and are not fully tested or available yet, but they offer profound opportunities to enhance the human condition. When considering the breadth of enhancements, it is helpful to consider whether the procedure or drug is medically indicated, not medically indicated but permissible, or ethically inappropriate (Brock 1998). At the outset, it should be pointed out that humankind is striving for betterment using a continuum of methods, many of which would not be considered unnatural or amoral. Examples are tutoring the otherwise gifted student, private schools for the affluent, individual athletic coaching, surgical facelifts, hair implants, ear tucks, nose jobs, orthodontia, and so on.

The ethical aspects of enhancement procedures also involve the duty to assess of the risks of the procedure to the individual. Equity or justice issues arise when enhancement methods are not available to all.

PREIMPLANTATION GENETIC DIAGNOSIS

In this section, we turn to an expanded discussion of preimplantation genetic diagnosis (PGD) that was introduced in Chapter 9, "Ethics of Assisted Reproductive Technologies." Recall that this technique allows genetic screening of certain chromosomes when there is a familial or other risk for genetic disease in the offspring. This is done by either polar body analysis of the egg or biopsy of the fertilized embryo, which then allows selection of a genetically normal egg for fertilization or a normal embryo for implantation. Genetic testing has not yet developed to the point that it is possible to test for all genetic diseases, but the list is growing as new genetic information becomes available. As the ability to test for and identify additional genetic diseases evolves, PGD will probably be used to screen for them.

Ethical Implications of PGD

It is likely that in the near future, PGD will allow the selection of non-disease-related genetically controlled traits for cosmetic, behavioral, or other enhancements; it will also be possible to screen for diseases for which the IVF-created embryo may be at risk over his or her lifetime. For example, a mother recently sought preimplantation genetic diagnosis because of a strong family history of a rare and severe form of early-onset Alzheimer's disease. PGD testing for the genetic mutation associated with this disorder allowed the selection of an embryo negative for the defect. Additional testing at the time of amniocentesis and again at birth indicated no evidence of predisposition to the disorder (Verlinsky et al. 2002). The birth of this infant immediately raised a number of ethical concerns, mainly that instead of just screening of embryos for disease, screening will become a shopping trip for desired traits. Another concern is that marketing pressures will lead to unnecessary and expensive genetic screening "just to be sure" that a baby is healthy. At least one bioethicist felt that the situation ". . . speaks to the need for a larger policy discussion, and regulation or some kind of oversight of assisted reproduction" (Weiss 2002). Thus, while the use of preimplantation diagnosis and selection can be acceptable for the prevention of genetic disease, societies might find it unacceptable for use to select for cosmetic, behavioral, or other nondisease traits. Others see it as a way to provide all those traits we ordinarily hope our children will possess.

In theory, it is possible to manipulate the genome in order to treat or protect against a number of abnormalities. Such "designer gene therapy" would allow the selection of a gene cocktail to enhance characteristics such as hair and eye color, height, or intelligence, or to provide protection against certain illnesses such as cancer, AIDS, or the next life-threatening virus that appears on Earth. It also offers a means for choice in gender of the infant. In China, where the one-child policy has been in effect for over 25 years, male births are highly favored and female infants suffer a high rate of abortion, adoption, or infanticide. The ratio of males to females in the population has undergone marked shifts, with cities of bachelors unable to find wives and a marked rise in the number of elderly in the population. Critics fear this policy may have other unforeseen societal effects that may be difficult to reverse (Feng 2005). Perhaps other genetic selections by PGD will also have unforeseen effects that may be detrimental to current social systems.

The procedural issues surrounding PGD are the subject of widening discussion and the focus of several groups, including the American Society for Reproductive Medicine and the Genetics and Public Policy Center. Both are involved in evaluating the extent of regulation this technology requires.

GENE THERAPY

Overview of Uses for Enhancement and Methodology

To begin the discussion of gene therapy, we present two case examples of the human enhancements predicted for the future and enabled by gene modification technology. While considering the merits of these enhancements, there are two questions that rise above the many considerations involved in proposing a genetic intervention:

- Do the benefits outweigh the risks?
- What are the consequences to other individuals and society from these enhancements?

Case 11.1 Extended Longevity

A major enhancement of the human condition will be an extension of healthy longevity far beyond the current average life span in the Western world. The search for the "Fountain of Youth" has now become a realistic quest for modern biology. The telomerase gene produces an enzyme that repairs chromosomes that are shortened with each cell division. Loss of expression of this gene is associated with aging. The return of expression of the telomerase gene was one of the first targets considered as a therapy for aging until concerns arose that a change in its expression might lead to uncontrolled growth and cancer. This caused scientists to look for aging-related genes using molecular biology methods of genomics and proteomics to identify genetic patterns and phenotypic markers in individuals over 100 years old. One of the major findings of the early 2000s was the demonstration of genome patterns related to longevity in a group of aged Ashkenazi Jews. The combination of longevity, the absence of hypertension, larger than usual high-density lipoprotein particles, and large low-density lipoprotein particles was related to a gene that produces a particular protein known as cholesteryl ester transfer protein (CETP) (Barzilai et al. 2004). Improving the expression of this gene can be considered a possible anti-aging therapy.

Another possible anti-aging therapy is modification of the functioning of a gene related to lipid metabolism and insulin regulation discovered in a study of aged Russians (Anisimov et al. 2001). CETP and the insulin regulatory gene for insulin growth factor (IGF-1) are also related to the longevity of the worm *C. elegans*, and to modifications of the insulin regulation-related gene that give mice greater longevity (Kenyon et al. 1993, Holzenberger et al. 2003). The observations that the longevity of *C. elegans* can be increased sixfold started in 1993 and have been verified in numerous experiments centering on a modification of the gene, daf-1, which regulates a host of other genes (Kenyon 2005). Further, a modification of the insulin regulatory system from normal has been associated with a significant increase in longevity in women (van Heemst et al. 2005).

These data allow us to imagine a future time when society will be challenged by the dilemma of the benefits of prolonging life by gene enhancement vs. the risks. A possible scenario such as the one described below is likely to arise in the future.

Assume that the above findings form a basis for gene therapy for longevity, and that some modification of the insulin-like growth factor gene will alter the regulatory pathway of aging to the extent that a human being's life span can be doubled. Let us further assume that research scientists and public policy are against gene therapy aimed at the embryo without a proper assessment of the safety or risks of this approach in nonhuman primates, even though studies in other mammals show only beneficial effects. However, the study of this enhancement will require many decades due to the longevity of nonhuman primates, making it impractical to evaluate the safety and risks of the modified individuals and at least one generation of offspring. On the other hand, if genes associated with longevity can be modulated by the modification of somatic cells, such as bone marrow or peripheral blood cells, then the enhancement would not entail the risks of germline modification and could generate the needed safety data in a few years in a canine model and in 10 years in non-human primates. Though modification of stem cells in bone marrow would affect a limited population of cells, the eventual population of other tissues by these cells could achieve the desired change in gene expression.

What are the ethical issues here? What are the sociological consequences of a population that lives to be 150 years old? Would they have a better quality of life than the very old of today? Could they maintain independent lives? Since one in four people over the age of 70 has some cognitive deficits consistent with Alzheimer's disease, would centenarians contribute meaningfully to their communities and to society for a longer period of time? Or would they need other therapy? What would be the impact on the age of retirement, on retirement pensions, health care, and Social Security? If people lived longer and were physically fit, they could work longer, earn salaries longer, contribute more to the work place and to advances for society. If older, more experienced workers do not retire, will the opportunities for younger citizens to find work be lessened? Is there sufficient space on the planet for the increase in population this will create over time? Is it fair to proceed with this type of embryonic gene therapy and deny the autonomy of the fetus? These and questions of safety of the procedure must be addressed before any longevity enhancement can be acceptable.

In the absence of any data on these issues, and coupled with the fact that there are no safety data in humans, proceeding with an enhancement such as this would be risky and unethical because the autonomy of the fetus is denied.

Case 11.2 Memory Improvement

It is believed that memory is closely related to the activity of a nerve cell receptor known as NMDA (N-methyl-D-aspartate). Signals are transmitted from one nerve cell to another by chemical molecules, one of which is glutamate. It is argued the efficiency of memory will improve by increasing the number or sensitivity of the NMDA receptors. Identifying the gene that enhances memory is the first step toward trials of inserting this gene into nerve cells of parts of the brain in an attempt to improve memory. The gene for the glutamate binding subunit of the NMDA receptor was discovered in 1991 (Kumar 1991).

As attractive as this prospect is, especially for patients with memory degeneration, the potential risks may be substantial but are not specifically known. For example, the NMDA receptor becomes very active after brain trauma, and this activation leads to an imbalance in the calcium concentrations in nerve cells and the supporting brain cell matrix. Thus, the manipulation of the NMDA system could make individuals susceptible to brain swelling or could result in epilepsy, a condition frequently related to a local increase in glutamate-based neuronal excitation. As is the case with any proposed new medical treatment, the risks would be assessed, but in this and the previous case, the proposed treatment is not a treatment for a disease entity but an enhancement of the performance capabilities of a normal individual. Therefore, without extensive studies of the consequences using nonhuman primates, this enhancement would put the recipient at risks beyond those currently acceptable for treatment of a disease.

Having explored some possible gene enhancements likely to be tried in the future, and the considerations that need to be evaluated before they are developed in human beings, we now turn to the methodologies involved in gene therapy. The methods for adding genes to an embryo cell for enhancement are the same as those for somatic gene modification for disease therapy. The procedures are in their infancy. Even though the first human gene therapy was given in 1990, progress has been slower than might have been anticipated at the start.

As of the year 2002, over 600 gene therapy clinical trials involving nearly 3,500 human subjects were underway in 20 countries, with an additional five clinical trials being conducted simultaneously in several countries (Roy 2003). In 2004, most of the gene therapy studies involved the use of a viral *vector* to transfer the new gene into the cells of the recipient. This vector is usually a retrovirus or adeno-associated virus chosen to take advantage of its ability to insert itself into the recipient's genome, and modified to remove its infectious capabilities. The gene is attached to this viral carrier (usually referred to as a vector), and upon incubation, the vector enters the target cell and then the cell nucleus, where the gene randomly inserts into the person's genome. The vector dissociates into its components, which are mostly proteins. Ideally, these are discarded by the cell or digested by the body's defense cells. The new genomic material takes up residence in a location on a chromosome that may or may not be near the defective genes, which remain in place. Upon integration into the genome, the gene begins to be expressed and the missing function is restored.

In most of the first 100 gene therapy studies, the genetic changes were delivered to the cells while they were outside the patient's body, after which the genetically modified cells were returned to the patients. Not all of the cells containing a malfunctioning gene will receive the new gene, but for many diseases, even a small increase in gene function can be beneficial. As the gene is randomly integrated into the genome of each cell, the genome in each cell will not be exactly the same. Further, as none of the procedures presently allows the removal of the "old" or malfunctioning genes, the long-term effects of this aspect of gene therapy are presently unknown and of some concern.

Other concerns center on the vectors themselves. The use of modified viruses as vectors is not ideal; neither is the variable uncontrolled insertion of the gene into a particular chromosome. Intense scientific work is underway to develop nonviral vectors for gene transfer, and to perfect new methods for directly injecting DNA into cells (see Chapter 12, "Ethics of Emerging Technologies," under "Synthetic Biology" for further discussion of some of these methods). Viral vectors can be toxic and may inadvertently integrate into cells other than those targeted (such as germline cells, in which case the mutation becomes intergenerational). Viral vectors can activate immune and inflammatory responses due to the foreign nature of the viral proteins, and their random insertions into the genome can cause activation of unwanted functions and further disease. While other carriers are being evaluated (such as inserting the new DNA into liposomes, which are then injected into the target cells), to date no ideal vector has been identified. Researchers are searching for safer ways to introduce the genetic changes that will not activate unintended functions.

One such approach proposed to mitigating these problems is the development of human artificial chromosomes (HACs), which have the advantage of being recognized as self and do not cause an immune response. The artificial chromosome provides a platform for the known and reproducible insertion of a gene or a cassette of genes into some well-defined location. A human artificial chromosome could be built with specific insertion sites for transgenes, tested, and then injected into a target cell. The basic technology has been used in yeast (YAC—yeast artificial chromosomes). Artificial human chromosomes should be well suited both for germline manipulations and for ex-vivo somatic manipulations. The successful addition of mini-chromosomes with centromeres (important for replication of the chromosome during cell division) and telomeres (protective elements at the ends of chromosomes) into human cell culture lines opened up this approach to gene enhancement in 1997 (Harrington et al. 1997). By 2002, the use of HACs to complement gene deficiencies had been demonstrated (Larin and Mejia 2002). This method of complementing the human genome offers potentials for germline therapies for known deficiencies that would be passed down to successive generations, as well as for enhancement of the human condition (de las Heras et al. 2003).

Somatic Gene Therapy

Between 1 and 2 percent of newborns have a single gene defect (Lubs 1977). Early gene therapy was directed toward these single-gene disorders but is now being used to treat more complex multi-gene disorders. One approach to the treatment of genetic disease is the use of somatic gene therapy. There are an estimated 4,000 diseases associated with a genetic disorder, but the cost of somatic gene therapy is estimated to be at least $100,000 per year per patient (Walters and Palmer 1997a). The reason this figure is so high is that somatic gene therapy patients require follow-up monitoring over the course of many years, and may require retreatment, depending on their response. But the potential benefits of gene therapy are great, and efforts are underway to perfect this new technology.

Somatic gene therapy involves the insertion of a new gene or cluster of genes into targeted cells of a single individual in order to restore a deficient cellular function. The desired gene is transferred, often along with other genes positioned in close proximity to the new gene (in case they somehow influence the functionality of the new gene), to as many of the patient's targeted cells as possible. The genes of the eggs or sperm (the "germ" or "germinal" cells of the individual's body) are not targeted, and it is hoped they will remain unaffected. Thus, somatic gene therapy affects only the cells of the treated individual, and the new genes are not passed down to succeeding generations. When sufficient numbers of targeted cells have incorporated the new genetic material, the missing or abnormal function can be restored. In 2003, among more than 600 somatic gene therapy clinical trials underway in the United States, only a few of which had progressed to advanced phases of study (www.ama-assn.org/ama/pub/category/2827.html, accessed April 28, 2006). Only somatic gene therapy studies are underway at the present time. As of 2005, none had been approved by FDA.

Current gene therapy studies also include the treatment of more complex genetic diseases such as cancer and cystic fibrosis and infectious diseases such as HIV and hepatitis B. In the future, it is likely that gene therapy will also be used to provide enhancements for individuals whose genetic makeup is normal but for whom parents want to provide certain desirable traits, characteristics, or abilities. However, the first diseases targeted for gene therapy were those in which a single enzyme is missing, one in which even a small change in function might be sufficient to improve the health of the patient. This approach allowed gene therapy to be done on a relatively small scale so that if something were to go wrong (production of too much of the enzyme, for example), it was hoped the consequences would be minimal.

The first disease targeted for gene therapy in the United States was a form of severe combined immunodeficiency (SCID). SCID is a rare syndrome that is due to a variety of genetic mutations that result in profound deficiencies of the immune system. Afflicted children suffer from severe life-threatening infections and usually die by age 2 unless they receive a successful bone marrow transplant. Gene therapy offers another approach to treatment of these children. Below we discuss the status of gene therapy in two forms of SCID: children with adenosine deaminase deficiency SCID (ADA-SCID) and X-linked SCID (X-SCID).

Gene Therapy for Severe Combined Immunodeficiency with Adenosine Deaminase Deficiency (ADA-SCID)

Patients with this disease lack infection-fighting cells known as T cells, B cells, and natural killer cells. The adenosine deaminase deficiency is caused by mutations in a gene on chromosome 20. The metabolism of adenosine triphosphate (ATP), which all cells use for energy, results in a turnover or recycling of the molecules that make up ATP. The backbone molecule is adensosine. Absence of the enzyme that digests excess adenosine results in a toxic accumulation of adenosine and adenosine-related compounds. The accumulation of these deoxyadenine nucleotides is directly or indirectly toxic to lymphocytes (Buckley 2000). Previous treatment by bone marrow transplantation has been successful in children receiving bone marrow from an HLA-identical sibling donor, but otherwise, these children have a generally high rate of transplant-related mortality (Hoogerbrugge et al. 1996). Restoration of function by gene therapy seemed an ideal way to treat this disease, especially since animal studies indicated that even a small amount of success in restoring ADA production resulted in significant clinical improvement.

The first official human somatic cell gene therapy in patients with ADA-SCID was done at NIH in 1990. Researchers removed blood from two children, isolated their infection-fighting cells known as T cells, and grew them in culture. Then, a normally functioning gene for ADA was inserted into the T cells, using a retrovirus as a vector to carry the genes into the cells. The transformed cells were then injected back into the patients. Both children were subsequently released from the hospital and have led relatively stable lives. The first ADA-SCID patient treated, a girl four years old at the time of treatment in 1990, has thrived. She required 11 infusions of gene-corrected T cells from September 1990 to August 1992, and has shown a circulating level of 20 to 25% gene-corrected T cells (www.frenchanderson.org/docarticles/best.html, accessed November 23, 2004). Her symptoms improved, and she has been able to lead a normal life. She attends school and has had no more infections than other children. Since the treatment of these first two patients, there have been occasional reports of significant improvement in a few patients, but the majority of patients with this disease have not been significantly helped by gene therapy. Nevertheless, treatment protocols in children with ADA-SCID are continuing, two of which are underway in the United States.

Children Undergoing Gene Therapy for X-Linked SCID

Another type of SCID, X-SCID, has also been approached using gene therapy. Patients with this disease generally have very low numbers of T cells and natural killer (NK) cells, whereas B cells are plentiful but lack specific antibody responses. X-SCID is caused by mutations of the gene known as IL2RG, the gene encoding the γ common chain (γc) of the leukocyte receptors for interleukin-2 (IL-2), IL-4, IL-7, IL-9, and IL-15. This gene participates in the delivery of growth, survival, and differentiation signals to early lymphoid progenitors (Cavazzana-Calvo et al. 2000). In one study of 136 patients with X-SCID, 95 distinct mutations were identified (Buckley 2000). As is the case in other manifestations of SCID, X-SCID patients also have multiple severe infections and die in infancy unless they receive a bone marrow transplant. Transplantation has been very successful in patients with a closely matched sibling (90% survival), less so (50 to 70% survival) in those receiving bone marrow from a parent or unrelated donor (www.fda.gov/ohrms/dockets/ac/02/slides/3902s1-07-noguchi/sld005.htm, accessed April 25, 2005). Thus, gene therapy has been an approach of great interest.

In 2002 in France, five patients with X-SCID received gene therapy that involved injecting their own bone marrow cells with a retroviral vector that had been given DNA from a normal γc chain to restore normal function to the T cells and natural killer cells. Early in this trial, there was successful reconstitution of the immune system in four of the five boys, who no longer required therapy of any kind (Hacein-Bey-Abina et al. 2002). Later data from this study showed success in 9 of 10 patients treated. However, two of the youngest of the 10 patients treated by this method later developed T cell leukemia, a rare form of cancer characterized by an uncontrolled proliferation of mature T cells. This problem may have been provoked by the retroviral vector activating oncogenes (cancer-activating genes) near the insertion site (Hacein-Bey-Abina et al. 2003). X-SCID studies were halted for a period of time in France and the United States to reassess the protocol methodologies and the benefits vs. the risks to the patients.

In January of 2005, nine months after the halted studies were resumed in May 2004, a third child in the French study developed leukemia and the studies were halted once again. An FDA advisory panel recommended in March 2005 that children with X-SCID receive gene therapy only if bone marrow transplantation has failed (Kaiser 2005). In the meantime, the two gene therapy trials for X-SCID underway in the United States remain suspended and were under FDA review in 2005. Gene therapy clinical trials for ADA-SCID underway in the United States are continuing, since the approach is different and there have been no reports of leukemia. A British trial of gene therapy for X-SCID also continues since no leukemia has been reported in seven patients treated to date, but that trial has not reached the 33-month stage, the time at which the French children developed leukemia. A number of other gene therapy studies in the United States are also continuing (www.fda.gov/ohrms/dockets/ac/05/slides/2005-409352_05.ppt#274,23, Other BRMAC Recommendations, accessed April 20, 2005).

Investigators in the field remain optimistic about gene therapy, noting that the field holds great promise for the treatment of genetic diseases. They are also optimistic that these events can be minimized by using therapeutic genes without oncogenic potential, and by improving vectors to minimize the activation of nearby genes (Berns 2004). Work continues in both areas.

History of Gene Therapy

The history of gene therapy is given in Table 11.1.

Table 11.1 History of Somatic Gene Therapy

(In part from www.dnapolicy.org/genetics/chronology.jhtml, accessed Oct. 4, 2004).

1972	First recombinant DNA molecules produced.
1976	Recombinant DNA Advisory Committee (RAC) formed in 1974 develops guidelines to prevent the safe use of genetically modified organisms or recombinant DNA.
1981	First transgenic animals (mice and fruit flies) created by adding foreign genes to their cells.
1990	First successful gene therapy performed on a patient with severe combined immunodeficiency (ADA-SCID).
1999	Gene therapy patient Jesse Gelsinger dies from a reaction to an adenovirus vector. All human gene therapy trials at the University of Pennsylvania were stopped because of departures from accepted practices of IRB approved clinical trials (see Case 11.3).
2001	Traces of DNA from the viral vector for treatment of hemophilia B are found in the seminal fluid of the first gene therapy patient, indicating the possibility of inadvertent germline genetic modification.
2002	For the second time, the FDA suspends the gene therapy trial to treat hemophilia B after DNA from the vector is discovered in the seminal fluid of another patient.
	A French child receiving gene therapy for X-SCID develops a leukemia-like disease that may have been related to the vector (carrier) used for gene transfer.
2003	The FDA suspends 27 additional trials after a second child treated in the French X-SCID trial develops a leukemia-like disease. All of the American studies employed the same vector as was used in the French trial. The NIH Recombinant DNA Advisory Committee concludes that ending all retroviral gene transfer clinical trials is not warranted but cautions that continuation of such trials is contingent upon the appropriate risk-benefit analysis, informed consent, and monitoring plans.
2004	Trials resume in May; in October one of the X-SCID leukemia patients dies.
2005	The third child in French X-SCID study develops leukemia. X-SCID studies in United States are suspended by FDA. Other gene therapy studies, including ADA-SCID, continue. Investigators remain optimistic about gene therapy despite the setbacks.

Case 11.3 Jesse Gelsinger Case: A Gene Therapy Trial Gone Awry

Another early gene therapy treatment was for a rare congenital disease called ornithine transcarbamylase (OTC) deficiency. OTC is one of several enzymes that act in the liver to metabolize protein and remove nitrogen from the body. When OTC is absent or present in insufficient amounts, nitrogen levels build up in the bloodstream in the form of ammonia, which is toxic. This case raises numerous issues about the conduct of clinical trials and the problems that can arise.

Jesse Gelsinger was an exuberant young man with a mild, well-controlled form of ornithine transcarbamylase deficiency who altruistically volunteered to participate in a gene therapy study of OTC deficiency underway at the University of Pennsylvania. He received the genes by direct injection into an artery leading to the liver in a therapy that used an adenovirus as the messenger or vector to transport the genes to their target. He died in September 1999 within four days of receiving gene therapy, most likely due to an overwhelming reaction to the viral vector (Raper et al. 2003). Subsequent investigation of this case showed, among other things, that the risks of the procedure and the potentially toxic effects of the adenovirus vector particles were not adequately explained to the patient or to his family, who were his advisors and support group. They were not informed of the deaths of monkeys during animal studies, nor did they know that previous participants in the study had suffered serious adverse events (www.fda.gov/fdac/features/2000/500_gene.html, accessed October 21, 2003). The patient and his family believed that the treatment had shown encouraging signs of efficacy when it had not, and they were not informed that the Investigator and the university had conflicts of interest with the manufacturer of the therapy and stood to benefit from the successful outcome of the trial (www.washingtonpost.com/ac2/wp-dyn/A11512-2000Nov3, accessed October 21, 2003).

This patient's death was especially tragic because he had agreed to participate in the trial even though his disease was controlled by diet and medication. Although he appeared normal, he had high ammonia levels in his bloodstream that should have excluded him from the study. The Food and Drug Administration stopped the research experiment, charging the investigators with violations of federal research regulations by not reporting prior adverse events, among other things, and failure to adequately protect the lives of patients. It was soon learned that reports of more than 600 other serious adverse events had occurred in other gene therapy trials under way since 1993, and these reports were also delayed or improperly handled. As a result, more stringent rules were put into place, but the episode delayed many trials from proceeding and forced many ethical issues of gene therapy to be readdressed (see "Ethical Issues of Somatic Gene Therapy for Enhancement" later in this chapter). The investigators received penalties in the form of heavy monetary fines and restrictions on their research for up to five years but reportedly have not admitted responsibility for Mr. Gelsinger's death (Couzin and Kaiser 2005).

Gene Therapy for Parkinson's Disease

Another targeted use of gene therapy is in patients with Parkinson's disease, where the treatment strategy is to deliver a gene, which provides the needed enzyme, dopa decarboxylase (DDC), to target sites in the brain (Bankiewicz et al. 2000). When L-dopa is administered orally to nonhuman primates, it is converted to the needed dopamine by DDC. Human trials have commenced. The method is described in Figure 11.1.

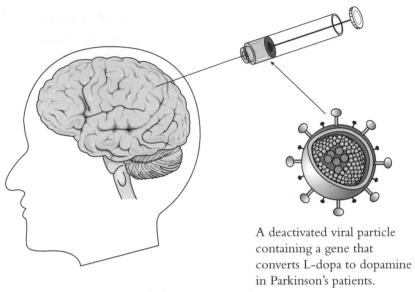

A deactivated viral particle containing a gene that converts L–dopa to dopamine in Parkinson's patients.

Figure 11.1 The method of somatic gene therapy for treatment of Parkinson's disease is illustrated here. A gene for an essential enzyme that converts therapeutic doses of L-dopa to dopamine is inserted into a viral particle and injected deep into a central region of the brain. This decarboxylase enzyme activity can be followed by imaging methods that visualize the action of the same enzyme that concentrates a particular tracer (i.e., tyrosine labeled with fluorine-18). Brain images show that the method has been effective in maintaining this enzyme activity in tissues of the nonhuman primate for over five years (Bankiewicz 2005). Human trials started in 2005.

Gene Therapy for Hemophilia

Patients with hemophilia A and B have a bleeding disorder, each due to a deficiency in a specific blood clotting factor. They are considered ideal candidates for gene therapy because they are generally healthy, have a single genetic deficiency, and because even a modest response in the production of the deficient factor VIII (hemophilia A) or factor IX (hemophilia B) corrects the bleeding disorder.

From 1999 to 2001, five clinical trials of gene therapy were done in hemophilia patients with generally minimal side effects. But at the low doses given in these early safety studies, there was also minimal efficacy in that levels of factor VIII and factor IX were insufficient to allow discontinuation of supplementary factor infusions. Gene therapy studies in hemophilia delivering factor IX into muscles using an adeno-associated virus (AAV) vector were stopped because of slightly elevated liver enzymes that occurred in two patients; another study injecting the vector carrying Factor IX into an artery in the liver was also stopped after a seventh patient developed "minor" signs of toxicity (Kaiser 2004). While the findings in the few patients were mild, there were ongoing concerns that the vectors used for gene therapy can invoke immune responses and cancers. Further, two hemophilia gene therapy patients were found to have a transgene present in their semen, leading to concerns that the transgene was affecting reproductive cells and might be passed down through their germline.

After the Gelsinger death that also involved injection of the vector into an artery in the liver, researchers have been reassessing vectors in general, and trying to minimize adverse events. The latter might be accomplished by injecting the material into local tissues rather than arteries and by keeping doses small. It is also important to determine whether the AAV vector triggered the immune response seen in trial patients and to answer questions of distribution in the body of genetic material that is administered and the extent to which it stays where it is delivered, as well as whether there is transfer of the vector to the germline (High 2003). There is also an effort underway to substitute the adenovirus or retrovirus with a synthetic virus vector that will avoid evoking a systemic immune response (Schaffer and Lauffenburger 2000). Arruda and associates have been developing a new way to administer gene therapy to achieve more extensive transduction of skeletal muscle by direct intramuscular delivery. Their studies in hemophilic dogs have shown that the new method of muscle-directed gene transfer can achieve factor IX expression in sufficiently high levels to achieve long-term (>3 years as of 2004) correction of the bleeding disorder (Arruda et al. 2004), setting the stage for studies in human patients.

Despite the problems, investigators remain generally optimistic that gene therapy represents the best chance to eliminate this disease and should succeed early in the twenty-first century (Mannucci 2002).

Somatic Gene Therapy for Human Enhancement

Suppose that somatic gene therapy was used to enhance the attributes of human beings rather than to treat diseases. The goal would be to add a selection of natural or genetically engineered material to the complement of genes that naturally occurs in a person to provide specific traits or capabilities the individual might otherwise lack. Examples might include genes that enhance muscle strength or longevity, or attributes such as height, beauty, musicality, or athletic ability.

Gene Therapy to Enhance Muscle Strength

Those who could benefit from enhanced muscle strength include patients who have muscle wasting from chronic illness, long-term bed rest, old age, and muscle-wasting disorders. These disease states are common maladies with no effective treatment options and important repercussions affecting the health and quality of life of those afflicted. The loss of muscle mass in astronauts in prolonged orbit or space flights might benefit, because methods to counteract this effect by prolonged periods of exercise are not completely satisfactory and are limited to space travelers who can endure the exercise requirements. Athletes who are normal but who would like to enhance their performance are eyeing this innovation as well.

These conditions might benefit from findings reported in an infant who at birth displayed marked muscle hypertrophy that has persisted (this child at the time his case was reported in 2004 was 4.5 years of age). The study of this child has resulted in the identification of a "loss-of-function mutation" expressed in the myostatin gene of skeletal muscles. In subsequent animal studies in mice, investigators found that when the myostatin gene was disrupted by a genetic manipulation known as a "knock-out," there was a doubling of skeletal muscle mass and evidence that myostatin may block the activation of stem cells known as satellite cells resident in skeletal muscle. If myostatin acts as a control to keep muscles from runaway buildup, perhaps inactivating myostatin could result in increased muscle bulk and strength in patients with muscle-wasting conditions (Schuelke et al. 2004).

These findings have raised concerns that athletes might use this type of untested gene therapy to gain a competitive edge by building muscle mass and strength. The procedure of injecting a viral vector with a myostatin-blocking gene would not be easy to detect and, therefore, its presence would be difficult to regulate. Since such a gene is also currently untested, there are serious safety concerns about the possibility of introducing viral vectors moving into other tissues of the body. Thus, the illicit use of this technology could be quite dangerous. Nevertheless, the lure of fame and fortune might prove more than an athlete is able to resist, although rapid advances in detecting gene expression may, in the long run, make gene therapy more easily apparent (Vogel 2004).

There have been other attempts to increase muscle strength. Studies at the University of Pennsylvania showed that the viral-vector-mediated transfer of a gene that makes an insulin-like growth factor into skeletal muscles of mice resulted in muscles that were 15 to 30% larger than normal. Rats injected with the same factor, known as IGF-1, also grew bigger muscles, which with strength training were twice as strong as uninjected muscles. Although there are concerns that this type of gene therapy may result in tendon problems, such as stress tears due to insufficient development compared with the muscles they serve, athletes will likely be tempted to use this form of enhancement (www.guardian.co.uk/genes/article/0,2763,1149823,00.html, accessed October 7, 2004). Given the rapidity with which gene therapy is developing, gene therapy to increase muscle strength could be problematic for the Olympic Games in the near future. Such therapy would be illegal (the IOC prohibits the use of any pharmacological, chemical, or physical manipulation to enhance performance). However, it is hoped that this technology will offer a new treatment for muscle disuse and disease. Enhancements for longevity or attributes such as musical or artistic abilities are in the more distant future.

In Utero *Somatic Gene Therapy*

This type of therapy is potentially useful for treating a serious disease or enhancement in a growing embryo *in utero*. Currently, surgery is the only available treatment for correctible life-threatening abnormalities in a developing fetus. The womb is opened, the affected part is surgically corrected, leaving the placenta and umbilical cord as undisturbed as possible. But now, in addition to the theoretical possibility of performing gene therapy on the germ cells and on the earliest human embryo, the technology also exists to access the developing fetus in the uterus and to transfer genetic material into somatic cells at about the 10 to 15-week stage. Animal studies to date have been encouraging, but investigators are reluctant to proceed with clinical trials until addressing concerns about the potential risks of the vector to the germline and future generations (Anderson 2000a). In order to access the embryo *in utero*, the fetus must be of sufficient size to be identifiable, and the *in utero* gene therapy must be inserted at the earliest time possible. This is so that the cells with the genetic changes can divide with the gene incorporated into the greatest number of somatic cells of the body.

NIH guidelines give the most recent policy on *in utero* gene transfer clinical research (National Institutes of Health 2001), concluding: "there is insufficient basic and preclinical data to justify the conduct of *in utero* gene transfer clinical research." A large number of critical scientific, safety, ethical, legal, and social issues must be addressed before clinical trials can be conducted. Five of these issues, listed below, reflect the essence of what risks must be evaluated from the perspectives of this text on ethics. These are (www4.od.nih.gov/oba/gtpcreport, accessed September 1, 2004):

- Possibility of generation and activation of transmissible vector or virus.
- Possibility of initiating oncogenic, immune, or degenerative processes.
- Potential risk to the fetus.
- Potential risk to the pregnant woman.
- Detection and assessment of inadvertent germline transmission.

Regulation of Somatic Gene Therapy

In the early consideration of somatic gene therapy, there was much discussion about which diseases should be treated using this approach. Were there other approaches that were sufficiently satisfactory so as not to subject a patient to the unknown risks of gene therapy? Who should receive it? Were certain conditions more amenable to this type of treatment than others? Which ones? Were there safer vectors than retro- or adenoviruses? What happens over time when the "old" malfunctioning genes remain in situ along with the new gene therapy? These and other questions were of sufficient concern that a special committee of the NIH, the Recombinant Advisory Committee (RAC), was formed in 1974 (see Table 11.1) to advise on the safety and use of recombinant DNA techniques, as well as to review all proposed gene therapy protocols.

The RAC developed "Points to Consider," ethical considerations that should be addressed in all research protocols involving gene therapy. A summary of these points follows (Walters and Palmer 1997b):

1. What is the disease to be treated?
2. What alternative interventions are available for the treatment of this disease?
3. What is the anticipated or potential harm of the experimental gene therapy procedure?
4. What is the anticipated or potential benefit of the experimental gene therapy procedure?
5. What procedure will be followed to ensure fairness in the selection of patient-subjects?
6. What steps will be taken to ensure that patients, or their parents or guardians, give informed and voluntary consent to participating in the research?
7. How will the privacy of patients and the confidentiality of their medical information be protected?

Current Status and Future Plans for Human Gene Therapy

Since 1989, the FDA has received many hundreds of requests from medical researchers and manufacturers to undertake gene therapy studies and to develop gene therapy products. In 2002, FDA was involved in overseeing approximately 210 active gene therapy studies (the latest figure on the NIH gene therapy Web site www.fda.gov/cber/infosheets/genezn.htm, accessed April 28, 2005). As discussed above, there have been FDA-mandated suspensions of various U.S. clinical trials due to adverse events. FDA, the RAC, and experts in the field continue to work closely together on the national and international levels to ensure that all precautions are observed to make this new therapy as safe as possible for the participating patients. The trial suspensions have given all involved the opportunity to reassess the protocols, to reassess the risks and benefits of the studies, to reassess the vectors, and to more fully address the causes of the problems.

Nevertheless, the fact remains that over a 15-year period from 1990 to 2005, the FDA has not yet approved the general use of any human gene therapy approach. While there have been remarkable improvements in some patients who have received gene therapy, clinical trials have not been entirely successful. They have been marked by difficulties in unanticipated medical problems, violations in following approved research protocols, and cancer induction in three patients with X-linked SCID after what had been considered one of the most successful gene therapy trials. While U.S. X-linked SCID trials were still on hold as of March 2005, investigators continue to seek ways to refine vectors currently in use, to find new vectors, and to find new ways to deliver gene therapy more safely and effectively.

It is likely that gene therapy will be successful for the treatment of a number of genetic diseases. However, it remains unknown whether any gene therapy will be entirely successful, whether this approach will lead to a permanent cure for a disease, or whether multiple treatments will be required to maintain a cure. Besides hemophilia, the diseases thought to have the best chance for success in the near term include Parkinson's disease, cystic fibrosis, and diabetes, as well as genetic diseases of the lung and liver.

In summary, gene therapy is still early in its development and is a complex procedure with risks that include:

- The development of an immune reaction to the virus vector used to deliver the gene that would predispose the cell to cancer or some other immune response.

- The overproduction of the substance the gene was designed to provide (an example would be the overproduction of Factor VIII leading to thrombosis in hemophilia patients).

- The disruption of the function of other single or multiple genes in the cell, leading to other diseases or abnormalities.

- The presence of vector genome in other cells than those to which therapy was directed.

- The development of antibodies to the new gene product.

- The gene therapy also enters the reproductive cells, affecting the offspring of the recipient.

Ethical Issues of Somatic Gene Therapy for Enhancement

Most feel there are few ethical issues when somatic gene therapy is used to treat a genetic disease, seeing little difference between it and treating a disease with a medication or by some other means. However, the ethical implications of any situation in which the human genome is altered for enhancement are numerous. The stakeholders are individuals seeking personal enhancement, health care providers, insurance providers, society, and those inheriting the effects of gene therapy if the manipulations were to enter the germline. The public stakeholders are concerned with the following questions: Who should have gene therapy? Should there be limits on the types of genetic changes allowed? How should this be decided? Who pays? Can the therapies be equally available to all? Justice dictates that patients be chosen fairly, without undue bias or incentive, and not just because they can afford it.

Another consideration that raises ethical questions is that some people will want gene therapy for enhancement or even as a preventive therapy for some future possible disease. The medical and social ethics embodied in this possible situation are similar to those when a surgical enhancement is requested though there is no medical need. Should the professional therapist agree to procedures where the medical benefit is nil, but there will be some risks from the procedure?

There are also concerns that some might take unethical shortcuts to attain enhancements before the procedures are fully tested, or by withholding important facts in an IRB-reviewed protocol, or by omissions of other steps important to the procedure. Another concern is that ". . . biotech might eventually amend the human constitution" (Abate 2002). If it becomes possible to extend human life, important moral questions may be raised about inflating an already large population of older individuals in society and meeting their health care and other needs. Societies have the right and responsibility to decide which enhancement technologies should be allowed and which should be banned (www.arhp.org/cloning, accessed March 14, 2002). Many are of the opinion that if such enhancements were available, why wouldn't one want to choose the best for one's offspring? What is wrong with this? While this might be difficult to argue against, the justice issue of limited availability or affordability of an enhancement becomes a moral problem. There will undoubtedly be those who can afford such enhancements and those who cannot. These implications for society and the creation of a separate group of the enhanced are of concern.

The principal issue in assessing the ethical aspects of gene therapy is the question of benefits versus risk. Development of a particular somatic gene therapy can be considered to be a five-stage process. Each stage demands deliberation about the benefits vs. risk to the recipient. The first stage is animal experimentation, where objective measures of efficacy are made, and adverse events are recorded and assessed. The second stage involves the institutional review board (IRB, also known as the Committee for the Protection of Human Subjects) at the research hospital or institution. IRB members are the advocates of the patient or volunteer and do their best to ensure that the patient is fully informed of the risks and can give consent without coercion. The third stage is an iterative stage between the investigator who proposes the treatment and the IRB, FDA, and, since 2005, the RAC of the NIH. The fourth stage involves the close participation of the patient during the consenting process. The fifth stage is close monitoring of the patient's status during and after therapy. This monitoring includes evaluation of the potential that gene or viral parts might inadvertently enter the patient's germ cells.

Because of the uncertainties, there must be full disclosure of all preliminary test data and a full understanding of the possible risks and their consequences. Subjects of gene therapy and investigators should not try to press ahead too fast without proper precautions.

Biotechnology companies interested in developing somatic gene therapy are likely to involve expert hospital-based physicians as consultants in the development of the methodologies. These physicians will benefit professionally and possibly financially from the success of the clinical trial. It is important to be vigilant that these interests do not cloud judgment as these products are developed and tested.

As noted previously, the use of viruses as vectors for gene therapy has definite risks such as the inadvertent introduction of the gene into a cell other than that for which it was intended. But if a specific problem could be improved by gene therapy, and there is adequate animal and other safety data indicating the potential for success, some would favor proceeding despite the risks. However, a mistake can lead to serious consequences, including death.

Germline Gene Therapy

Germline gene therapy in human beings is theoretically possible but is not currently being done. It involves the insertion of new genetic material into the germ cells, the ova or sperm, or into the earliest embryo, in order to supply new traits to an infant that neither parent possesses that will be passed into all cells of the infant and to all subsequent generations. Thus, germline engineering also involves making decisions about what might be desirable for the next generations, decisions that may or may not prove desirable in the end. The implications of this capability to alter the genome are significant. It makes it possible not only to alter which genes will and will not be passed down to one's offspring but also which genes will be passed down to all future generations. It also has the potential for the creation of new forms of human life within one generation. Preimplantation genetic screening has only a minimal effect on the human genome, even if it were widely used, because the procedure selects from the range of existing human traits. Germline engineering will have more far-reaching effects.

Methodology of Proposed Germline Gene Therapy

Theoretically, there are presently three points at which germline gene therapy could be introduced so that all subsequent cells in the individual's body contain the desired genetic changes:

1. By introducing the selected genetic material into a germ cell or gamete *prior to fertilization*. The gametes are the egg and sperm that upon fertilization form the zygote. Technically, this will not be easy, and this approach is likely to take the longest to develop.

2. By introduction of the gene into the zygote (the earliest embryo after fertilization). This approach to gene therapy must be done at the earliest possible time after fertilization, ideally at the single-cell stage so that all subsequent cell divisions will contain the change. Otherwise, some cells will already have divided and will remain unaffected by the gene therapy.

3. By gene transfer into a single totipotent blastomere removed from a four-celled developing embryo. This cell is still undifferentiated and is capable of dividing to form an entire fetus; therefore, all subsequent cells formed from the division of a blastomere after gene therapy will contain the desired changes.

Germline therapy can provide either enhancement or disease treatment. When two individuals are seeking to conceive a healthy baby where one partner has a known genetic disease that may be transmitted to the offspring. In addition, many inherited diseases are also transmitted by healthy *carriers* of a disease, neither of whom manifests the disease but whose chromosomes carry the abnormality that can result in disease in the offspring if two carriers mate. Tests need to be available so that carriers of genetic abnormalities can be identified. Such testing may become routine in the future because in order to significantly reduce the incidence of inherited diseases in the gene pool, both carriers (or in fact, *all* carriers) would need to participate in preimplantation selection or receive germline gene therapy. This may become a goal of germline gene therapy sometime in the future.

Ethical Issues of Germline Gene Therapy

The most direct germline gene transfer mechanism, as it has successfully been applied to some plants and animals, is microinjection of a gene or cassette of genes into the pronuclear (zygote just after fertilization) embryo. This method involves using a fine glass needle to inject a purified double-stranded DNA sequence into the nucleus of a fertilized mammalian oocyte. This process leads to the integration of the sequence (transgene) into the genome. As a result, the plant or animal is born with a copy of the new sequence in every cell. However, a large fraction of all animals produced by this procedure are damaged, most lethally so; even if born alive, they are frequently deformed and later die (Anderson 2000b). Animal experiments have also suggested that germline gene transfer is unsafe because its use results in a random integration of the transgene, a lack of control of the number of gene copies inserted, significant rearrangements of host genetic material, and a 5 to 10 percent frequency of insertional mutagenesis. Expression of the transgene may not be sufficiently precise to correct the deficiency that prompted the treatment. Thus, at the present time, the use of gene transfer at the gamete or embryonic stage in humans would reach far beyond the limits of acceptable risk for a medical intervention. Serious errors can occur during the gene insertion, and these errors could be lethal or might not show up until the individual is an adult. The errors could be passed on to the next generations. For these reasons, given the limits of current technology, germline gene transfer has been considered ethically impermissible.

There has been one report of inadvertent germline gene transfer that occurred in 2001 after 30 infants were born following ooplasmic transfer in their mothers. All were women with fertility problems and underwent ooplasmic transfer in an attempt to increase their ability to conceive a healthy infant. Cytoplasm containing cellular mitochondria from the egg of a healthy egg donor was injected into the egg of woman trying to conceive, a procedure that appears to increase the fertility of the egg and enhances the chances for a normal embryo. The exact mechanism for this effect, or the extent to which the mitochondria even play a role, is unknown, but the end result was that the children all had three genetic parents (the mother and the father who contributed nuclear DNA from the egg and sperm, and the egg donor who contributed cytoplasmic or mitochondrial DNA) (Barritt et al. 2001). This protocol was shut down by FDA because the proper protocol for a research study involving gene transfer had not been followed.

Since germline gene therapies will affect all succeeding generations, the ethical implications of these alterations are numerous. There are questions about who owns the DNA and whether anyone in this generation should be making decisions for subsequent generations. Further, who should decide *whether* a genetic disease should be altered? Should each one be cured or prevented? It is costly. What about accessibility? Shouldn't the treatment be equally available to all? Who should have it? How should this be decided? Who pays for it? Should there be limits on the types of genetic changes allowed? Who should decide? Other issues involve the assessment of benefit to the patient versus risk. Two approaches now used to ensure that this evaluation is appropriate include the use of written informed consent in which the patient is apprised in detail of all aspects of the trial necessary for making a decision about participating, and the review of the study by an institutional review board (both at the local level, and increasingly by FDA).

Perhaps the bigger concerns about germline gene therapy involve its use not for medical needs, but for normal individuals seeking to be "better." Do the benefits outweigh the risks sufficiently to allow a parent to choose to have a blue-eyed child when random insertion of the new genes may cause unknown adverse effects on the actions of other genes surrounding the insertion? What mechanisms need to be in place in order to be prepared for the future? While our society supports parents' rights to do all they can to enhance their children's well-being, the question is whether we have the right to make gene changes that will result in different kinds of *Homo sapiens*, gene changes that will remain forever in the human domain, or whether changes should be left to the natural evolutionary process. Since ethical debate has maintained that DNA belongs to society, what limitations should there be to protect future generations? Genetic changes seemingly appropriate at this time may not be ideal at a later time. But there is no question that new ways to prevent genetic disease are gradually coming into use. Examples include choosing a female embryo for implantation to avoid an X-linked disease that would otherwise be passed to a male offspring, implanting the embryo most genetically compatible with a sibling with leukemia needing a transplant, or selecting the embryo whose genetic makeup lacks a cancer-causing gene mutation or the gene for early-occurring Alzheimer's disease. Genetic alteration to provide protection against a lethal disease such as AIDS or the plague may be of great benefit in the future.

Currently, federal oversight for all gene therapy is done by the Recombinant DNA Advisory Committee (RAC) and the FDA. They are charged with reviewing proposals for somatic cell gene transfer but at this point in time are not considering any protocols that involve germline engineering. However, debate for and against germline engineering is beginning to heat up.

Arguments for Germline Gene Therapy

- To treat conditions for which the only hope of prevention is for genetic addition or gene replacement in a fertilized egg.
- To enable genetic changes to protect against lethal diseases such as AIDS or the plague.
- To do public good by eliminating the possibility of having a child who will suffer.
- To pave the way toward other cures of more complex inborn errors of metabolism or other genetic diseases.
- Preimplantation selection and selective abortion are only halfway approaches, which are not practical in general and are considered immoral by the pro-life public.

Arguments against Germline Gene Therapy

*   There are as yet insufficient data in human patients who have undergone somatic gene therapy to ascertain the long-term effects of this procedure, and small animal studies of germline gene therapy show alarming rates of deaths and deformities, which are too high to permit proceeding in humans and large animals at the present time.

*   Insertion of one or more genes can, in a way, be considered merely a subset of human cloning. Genetic engineering by means of germline modification would allow novel forms of human life to be created within one generation, forms which might be dysmorphic and themselves constitute ethical dilemmas.

*   There are many unknowns that influence cellular development; thus the consequences of inserting genes without proper signaling might be dire.

*   Errors may result that will be a burden to offspring, parents, and society.

*   Alternative therapies already exist for the prevention of transmission of defects (e.g., adoption, preimplantation diagnosis, and selection of unaffected embryos).

*   Procedures are highly specialized and available only to the affluent, so a situation could arise in which genetic enhancements could turn the present socioeconomic gap into a species gap.

*   A corrupt elite group controlling technology might create a malevolent superior race.

*   The procedure lacks respect for embryos; theoretically, the rights of offspring are denied by tampering with natural genes and/or tampering with an embryo/fetus lacking the means by which to grant informed consent.

*   The gene changes will remain forever in the human race, and society should decide whether and to what extent germline engineering should be allowed.

In addition to safety issues, germline modifications to improve or enhance offspring also evoke objections based on the principle of autonomy. The parent is engineering an enhancement that might not be what the offspring really wants, or might be inconsistent with other natural attributes. For example, the enhancement of muscle mass in an offspring with timid personality traits, paired with expectations from peers and parents to aggressively compete in a sport, could cause great unhappiness for the child. Should not children have the right to make their own choices?

It is clear that one must give careful consideration to the consequences of this research. What would be the consequences if everyone were interested in the same genetic alterations? What would happen to biological diversity? Would we all look the same, have the same problems, and die of the same diseases? What would be the consequences in the event genetic alterations went awry? Your and our grandchildren might be the last generation of *Homo sapiens* whose evolution has been natural and unaltered by artificial means. How should this implication be reconciled with society's wishes today and the wishes of future generations?

The current opinion is that germline engineering should be forbidden not only on ethical grounds but also on scientific grounds because the risks are too great. More research is needed before proceeding, but the prediction that human germline engineering will never occur is not warranted. Caution should be used not to become pressured by premature commercialization and marketing promotions for these procedures that have the potential for large profits for corporations. It is important to move slowly, with proper vigilance, and only after thoroughly evaluating the potential problems this technology could pose. The latter will not be easy, since the consequences are difficult to predict, and because it may take several generations after treatment for the problems to appear. Public policy discussions are important so that the public is well informed early on about the ethical implications so that they can be addressed. The time is now to decide about regulation or public oversight of these procedures; otherwise, fertility clinics may offer whatever is marketable.

Some ethicists fear that the general public is unaware of how soon these options are going to be available. Genetic engineering of human beings is proceeding, and the public debate has been minimal even though the debate among scientists has been intense. It appears that the public has not yet grasped the implications of this technology, although there is a public awareness of the possibilities of manipulating genes for some diseases, such as Parkinson's and hemophilia.

The current declarations by the FDA and NIH committees are that germline engineering is premature, but there is no focus on *the extent to which* human genetic modifications should be allowed. Will the parental selection of enhancements for babies come down to choosing ". . . the best child you could afford . . . only to find out that 3 years later when you go back for your next child that new updates make your first child a sort of Windows 95" (Burress 2004). Special focus needs to be placed on the long-term consequences of genetic manipulation for the individual, the parents, the family, for society in the near term, and especially for society in the long term. Most importantly, the rapidity with which gene therapy innovations are advancing makes their use appear inevitable, and the first genetically modified generation is likely to be that of our children's grandchildren.

Oversight for Germline Engineering

FDA is directly involved in all clinical gene transfer research studies. NIH is involved in all gene transfer research supported with NIH funds or conducted at or sponsored by institutions that receive federal funding for recombinant DNA research. Thus, the majority of somatic cell gene transfer research is subject to the NIH Guidelines. RAC will not currently consider protocols using germline gene transfer (www.dnapolicy.org/genetics/transfer. jhtml.html accessed Jan. 18, 2006.).

The RAC has denial authority for NIH-funded research and clinical testing in the area of germline and somatic gene therapy. But in the interest of furthering the research agenda, scientists argue that the policy should be changed (www.ess.ucla.edu/huge/report.htm, accessed April 5 2004). To ensure the quality of research, the FDA should extend its authority to germline engineering, as it did in 2004 to human cloning. Many feel germline investigations should be resumed, although there is overall agreement that the safety data are insufficient to allow human germline studies at this time. Further, there is currently no controlling group to represent the public interest in assuring safe methods of health care and human therapeutics when it comes to germline engineering. It is important not to wait for an international body to dictate public policy.

Similar challenges to monitor ethical considerations relative to the genome project resulted in the creation of an important group known as ELSI (Ethical, Legal and Social Implications). This group, or one similar to it, could lead an exploration of the challenges and potentials of human germline engineering. This group has not seen germline engineering to be within its purview, but this may change. The National Bioethics Advisory Commission (NBAC) is another possible resource consisting of scholarly appointees who favor regulations against therapeutic cloning, but this might not be the objective body to look after the evolving technologies of gene therapy. Some have argued that there is a need for establishing specific guidelines and possible regulations, but others argue that we are at a primitive state of knowledge and regulations could stifle research. There is much to sort through, and the answers are not presently known.

ENHANCEMENT BY DRUGS OR OTHER MEANS

The following text addresses three topics: enhancement of athletic performance by drugs, enhancement of intellectual and behavioral performance by drugs, and enhancement of physical appearance by cosmetic surgery and other means. The basic questions to be asked for all of the enhancement topics is: Where is the line between acceptable or required enhancements and unacceptable or unjust enhancements by chemical and physical procedures? The issues are frequently discussed in the framework of the difference between treatment of a disease to correct a medical problem and application of a technology to improve an already accepted norm, or the inappropriate or unjust use of a limited resource. Seven moral questions regarding drug and surgical enhancements are:

- At what point does drug use make the playing field uneven for participants?
- No matter what the morality of taking drugs might be, is there an overriding duty to follow the rules of the sport with respect to doping? The latter question seems obvious, but deserves attention because it assumes the rules are ethically just rules. But are there moral reasons that drug use in sports should be forbidden?
- Does prohibiting drug use restrict the autonomy of the athlete who wishes to use drugs?
- Does prohibiting drug use protect the health and the rights of the athlete who wants to compete without drugs?
- Is there a duty of society to institute policies to limit enhancements that carry potential risks to individuals or to societal norms?
- Will the competitive advantage of improvement in physical appearance (e.g., cosmetic surgery), behavior modification (e.g., Prozac), and intellectual performance (e.g., drugs or genetic changes of the future) be unfair if the enhancements are not available to all?
- If enhancements in education or athletic training are imposed upon a child by a parent, is there an ethical issue regarding the child's autonomy?

Drug Use to Enhance Performance in Sport

Background

Drugs were used even in ancient times to enhance sporting performance, and their use continues today, usually to provide an advantage over innate abilities. Doping is defined as the use of a drug given illegally to affect performance. Its history goes back to the ancient Greeks, who regarded sport as an important symbol of human perfection and achievement. Success in sports brought status for the competitor as well as the region, and there were great rewards and riches for the winners. Plato wrote that victory in the ancient Olympics resulted in a monetary reward approximating $500,000, with other perks as well (www. asda.org.au/dishistory.html#, accessed September 9, 2004). The high stakes made athletes rich but led to corruption (including cheating and bribery), the decline of amateurism in athletics, and the use of substances to enhance performance.

The Romans also held athletes in great esteem and elevated their status even further as spectator sports gained in popularity. The competitions in the enormous Coliseum in Rome were indicative of what was considered sport in the times of the emperors, and the use of drugs among gladiators and charioteers who drugged their horses was reported to have increased the level of violence and drama of an event (www.asda.org.au/dishistory. html, accessed October 1, 2004).

But with the coming of Christianity, the violence of Roman sport came under criticism, and in 396 A.D. all such "pagan" sports were banned by the emperor. The popularity of sport declined, especially when it became believed that intellectual growth was hindered by physical development. It remained in the background and out of vogue until the nineteenth century, when sports were transformed by the arrival of the Industrial Revolution. The rural agricultural societies became industrialized and urbanized. Migration to the cities, combined with the development of mechanized ways to do previously time-consuming jobs, and the development of better means to communicate and travel, meant more time for leisure. The loosely organized sport of rural areas became urbanized with the development of clubs, organized competitions, new sports, and professional teams. By the twentieth century, popular athletes enjoyed great acclaim and huge salaries, but the pressure to win increased the use of performance-enhancing drugs despite serious health problems that resulted. To maintain fairness in competitions, and in an effort to protect athletes from their adverse effects, the use of drugs to enhance performance has been banned in many professional sports and elite competitions. Extensive screening procedures have been put in place to detect those who use drugs.

History of Drug Use in Modern Sport

Some of the events that led to the banning of performance-enhancing drugs and methods in sport are summarized in Table 11.2 (www.asda.org.au/dishistory.html#, accessed Oct. 1, 2004).

Substances Used to Enhance Performance (Doping)

Doping refers to the use by athletes of ingested or injected substances to enhance physical performance. The International Olympic Committee (IOC) has developed a list of banned substances, which includes stimulants, beta-blockers, narcotic analgesics, diuretics, hormones (including anabolic agents, human growth hormone, and the hormone erythropoietin, which stimulates the formation of red blood cells), as well as so-called nutritional supplements such as creatine and androstenedione. The use of any pharmacological, chemical, or physical manipulation to enhance performance is prohibited.

Table 11.2 History of Drug Use in Modern Sport

1886	First recorded death from drug overdose in a cyclist.
1904	First near death in an Olympic marathon runner using brandy and strychnine.
1920s	Athletes used alcohol, strychnine, heroin, caffeine, and cocaine until the availability of heroin and cocaine was restricted.
1930s	Amphetamines were produced and quickly became favored over strychnine.
1950s	Soviets used male hormones to increase power and strength, and the Americans developed steroids as a response.
1952	Several speed skaters at the Winter Olympics needed medical attention after taking amphetamines.
1960	A cyclist collapsed and died from an amphetamine overdose at the Olympics.
1963	The International Olympic Committee (IOC) was pressured to regulate drug usage by competitors. The Council of Europe set up a Committee on Drugs but was unable to decide on a definition of doping.
1964	Muscular appearance of the athletes at the Olympics raised suspicions of drug use.
1967	The IOC established a medical commission, adopted a definition of doping, and developed a banned list of substances after a cyclist died after taking amphetamines in the Tour de France.
1968	The IOC began drug testing at the Olympics in Mexico.
1988	Ben Johnson, a gold-medal-winning Olympic track athlete, tested positive for a banned anabolic steroid, was stripped of his gold medal, and was suspended for two years.
1988–2000	Drug use continued but was more difficult to detect. Athletes who tested positive were banned from participating in Olympic competitions. Others were stripped of their medals after drug tests showed banned substances.
2004–2005	Drug use in track and field as well as in baseball gained national attention after the BALCO scandal implicated internationally known athletes using so-called "designer steroids."

Caffeine

Caffeine is commonly present in coffee and teas, soft drinks, and some medications such as Anacin and Excedrin. Its use can increase the heart rate and blood pressure (both systolic and diastolic). It can also induce excretion of fluids by the kidney (diuresis). Caffeine also increases alertness and decreases reaction time to physical stimuli but can also cause jitteriness or tremulousness that may be detrimental in those sports requiring fine motor coordination. It has been found to also affect metabolism by increasing basal metabolic rate and enhancing fat oxidation during exercise; it may also delay fatigue during exercise (Reents 2000a).

The maximum concentration of caffeine allowed by the IOC is 12 micrograms per milliliter of urine. This is equivalent to the ingestion of eight cups of coffee over a 2- to 3-hour period. The National Collegiate Athletic Association (NCAA) allows a maximum urinary concentration of 15 micrograms per milliliter. Although caffeine may increase alertness and decrease reaction times, its overall effect on performance is probably negligible at the levels specified by the IOC, although there is some effect at levels 10 times the allowed maximum. At high concentrations of caffeine achieved by direct ingestion of caffeine capsules, there is a 7% increase in work output and a 19% increase in exercise time. These effects are conjectured to be from enhanced lipolysis and free fatty acid release, leading to more available energy and some sparing of glycogen use by muscles. Another theory is that caffeine may enhance endurance through central nervous system (CNS) stimulation, perhaps by catecholamine release (www.physsportsmed.com/issues/1997/01jan/schwenk.htm, accessed Nov 8, 2004). Caffeine is not easy to detect in the urine with accuracy, but since there have been documented enhancements of performance with high-dose caffeine use, IOC guidelines stipulate that it is considered a banned substance during competitions. This stand is in keeping with the definition of doping, which is the use of substances which artificially and unfairly enhance performance.

Ephedrine, Phenylephrine, Epinephrine, Ephedra and other Alpha-Adrenergic Drugs

These drugs are commonly present in many over-the-counter medications that are used to treat upper respiratory infections and asthma. For example, phenylephrine is present in many nasal sprays, decongestants, and cough syrups. Epinephrine is often used for intervention in an acute attack of asthma or for more severe allergic reactions such as anaphylaxis (a systemic reaction to a bee sting, for example). Ephedra, also known as *ma huang*, occurs naturally in certain herbs and other botanicals and contains ephedrine as its principal active ingredient. Ephedra in its natural form is different from ephedrine, which when chemically manufactured according to specific conditions produces a drug that has been licensed by FDA as safe and effective.

These drugs stimulate the heart and have an amphetamine-like effect on the central nervous system. The usual doses cause tachycardia (rapid heart beat, palpitations), elevated blood pressure, dizziness, nausea, vomiting, anorexia (loss of appetite), headache, irritability, anxiety, mania, and, at high doses, psychosis. Ma huang was linked to the deaths of several high school students who reportedly died from CNS hemorrhage or cardiac arrhythmia after using it as a stimulant/aphrodisiac. About half of the serious adverse events that appeared to be related to the use of ephedra-containing products occurred in persons 30 years of age and younger (www.fda.gov/bbs/topics/NEWS/ephedra/summary.html, accessed April 28, 2005).

Ma huang and other naturally occurring ephedra products have been marketed widely as dietary supplements/stimulants to control weight and to enhance performance in sport. A review of several studies of ephedra, ephedrine, or ephedrine plus caffeine showed evidence of a modest effect on short-term weight loss, but insufficient data to support the use of ephedra to boost athletic performance (Shekelle et al. 2003). One study showed a modest effect of ephedrine plus caffeine on very short-term athletic performance in military recruits, but these were unusually fit subjects (www.fda.gov/bbs/topics/NEWS/ephedra/summary.html, accessed Nov. 8, 2004).

According to FDA, the use of dietary supplements containing ephedra and its by-product ephedrine has been linked temporally to serious health risks, including hypertension, tachycardia, stroke, seizures, and death. Subsequently, all ephedra-containing supplements in the United States and Canada have been removed from the market. Ephedrine-containing medications licensed by FDA remain on the market. (http://nccam.nih.gov/health/ alerts/ephedra/consumeradvisory.htm, accessed November 8, 2004). The use of ephedrine-containing products is banned by the IOC, although topical use, including phenylephrine nasal spray, is permitted. Athletes must be very careful to check with the IOC and other elite competition administrators as to what medications are and are not permitted.

Amphetamines and Methylphenidate (Ritalin)

Amphetamines are controlled substances with effects similar to cocaine on the central nervous system but longer acting. Examples are Benzedrine (bennies) and Dexedrine. Methylphenidate (Ritalin) is also a central nervous system (CNS) stimulant with effects similar to, but more potent than, caffeine and less potent than amphetamines. It has a notably calming and "focusing" effect on those with attention deficit hyperactivity disorder (ADHD), particularly children (www.nida.nih.gov/Infofax/ritalin.html, accessed April 28, 2005). High school and college students without ADHD have been using Ritalin to provide increased focus while studying and enhanced performance on exams. Research using positron emission tomography (PET) scans of the brain after giving Ritalin to healthy men demonstrated an increase in dopamine levels, and investigators hypothesized that this increased release of dopamine, a neurotransmitter, was responsible for the improvement in attention and focus (Volkow et al. 2002).

However, the stimulant properties of Ritalin, along with its tendency to suppress appetite, may also be reasons for exploiting it. Often referred to as "uppers," use of amphetamines imparts a sense of wakefulness, mood elevation, and self-confidence, along with feelings of physical energy, power, and strength. While not directly proven to enhance athletic performance, the stimulant effects may allow a longer period of anaerobic metabolism. When combined with the feelings of confidence and energy reserve, there may be a 1 to 2% increase in power in the short term, a result that at the elite level of competition may give an athlete an edge. But by impairing judgment, amphetamines have also caused athletes to underestimate the seriousness of an injury as well as to overexert. In some cases, amphetamines have caused death, as seen in the case of an Olympic cyclist in 1960. Other adverse effects include appetite loss, elevation of heart rate and blood pressure, insomnia, headaches, paranoia, and convulsions; deaths have been reported from cardiac arrhythmias, brain hemorrhage, and heatstroke (http://espn.go.com/special/s/drugsandsports/ amphet.html, accessed April 28, 2005). The amphetamines are highly addictive and are banned by the International Olympic Committee and the NCAA, although the NCAA does permit the use of methylphenidate for ADHD if this need is documented. Its permitted use in the latter condition is controversial because the diagnosis of ADHD is not an exact science and because the drug may be used inappropriately.

Growth Hormone

Human growth hormone (hGH) is used to promote growth in undersized children but has also been used by athletes who are not growth hormone deficient in an attempt to benefit athletic performance. Growth hormone supplements in male power athletes, healthy untrained males, and elderly subjects failed to improve muscle strength more that that which could be achieved by weight training alone (Reents 2000b). However, athletes and even the parents of children who are growing normally but lag behind their peers in growth at puberty push for the use of hGH. It is banned from use in competition.

The problems with parental demands for special treatment, as for example with growth hormone, are illustrated by the following analysis (summarized from Tauer 1995). Growth hormone therapy and reimbursement for this treatment are approved only for children with documented growth hormone deficiency but not for those whose predicted adult height is less than "ideal." Some argue that it is unfair to make this expensive resource ($20,000/year for an average of four years) unavailable for all children with predicted short stature, pointing out that the psychosocial and other disadvantages of short stature are the same for all, and justice requires equal treatment. Others feel there may be unknown medical adverse effects in treating otherwise normal children with growth hormone. For example, they may not achieve the expected additional growth because they already have naturally occurring growth hormone (e.g., more may not be better). Short stature has not been proven to be overly disadvantageous functionally (they can still perform the usual daily activities and drive cars without special adaptation), and further, children so treated are labeled as "abnormal" and "deficient" instead of "normal" and growing at a rate consistent with family history and their own constitution. Because only the wealthy could afford this treatment (the cost is not covered by insurance unless growth hormone deficiency has been documented), it is unfair to others.

Diuretics

It is not the purpose of this text to provide an in-depth review of the physiology of the various classes of diuretics (the reader is referred to any pharmacology book or Chapter 2 of the book by Reents 2000c). However, in general, their use results in the removal of water, often along with electrolytes, from the intra- and extravascular spaces and their excretion by the kidneys. Some diuretics more severely deplete electrolytes such as potassium and, thus, may have significant physiologic effects if not monitored. Diuretics are often used as the initial treatment in patients with high blood pressure, and they are generally effective in mild hypertension. Athletes use them to control weight and to keep within their designated weight category (e.g., wrestlers, boxers, lightweight rowers). Young athletes are also taking it upon themselves to use these products, which are often found in their parents' medicine cabinets. Coaches are encouraging their use as well, for purposes that are not entirely clear. Surprisingly, the IOC reported that diuretics ranked fourth among the drugs abused by athletes (Benzi 1994).

The results of studies of diuretics used by athletes during exercise have been varied. One of the main concerns about their use to enhance performance is that diuretics have many effects in the human body that are similar to those seen during exercise. Both promote the loss of water and electrolytes, both of which can have detrimental effects on the cardiovascular system. The acute shifts in potassium from the intracellular space can contribute to muscle fatigue, muscle cramping, cardiac arrhythmias, and decreased exercise tolerance (Reents 2000d). Thermoregulation is impaired during exercise when taking diuretics because the loss of blood volume inhibits the vasodilatation necessary to facilitate transport of heat to the skin during heavy exercise. Thus, the increased body temperature that occurs with exercise cannot be controlled. Certain diuretics (thiazides) may also cause photosensitivity if the user is exercising in the sun. Thus, diuretics used in combination with exercise can have marked effects on body physiology. Most athletes, and even their trainers, are often not aware of these exercise-related risks, nor, often, are patients who are being treated for hypertension. The use of diuretics in competition is banned by both the NCAA and the USOC.

Erythropoietin (EPO)

Another drug used to enhance performance is the bioengineered version of the natural hormone produced in the kidney called erythropoietin. Erythropoietin stimulates the production of red blood cells, the cells that carry oxygen in the bloodstream to all the tissues of the body. An increase in red cell mass means that the oxygen-carrying capacity of the blood is increased, which in turn means more oxygen is available to muscles during exercise. But EPO must have other effects as well, because not only does it increase performance during short-term intense exercise, but it also results in a marked increase in performance during longer-term exercise and in events where oxygen delivery is not so important. This has led to the speculation that EPO may act on the brain in ways similar to amphetamines, cortisone, and anabolic steroids (Noakes 2004). However, the downside is that the increase in red cells in the circulation may increase the viscosity of the blood, which can cause stroke, seizures, or thrombosis. EPO use, along with loss of water from sweating during exercise, can lead to an even further increase in viscosity that can be fatal. Sudden death has been reported with its use. EPO is banned by the NCAA, the USOC, and the IOC, as is the administration of red cells by transfusion.

The increase in red cells in the blood stream produced by EPO persists for several weeks (the average life span of a red cell is 21 days), but EPO can be detected in urine for only a few days after the last injection, making it difficult to detect by this means of screening. However, the recombinant form of EPO differs slightly from the natural erythropoietin, making it relatively easy to spot. Attempts to increase red cells by blood transfusion can be traced by analysis of a blood sample, which shows genetically different populations of red cells, unless red cells previously stored from the same individual are used. The use of these autologous cells is also banned.

Tetrahydrogestrinone (THG)

Tetrahydrogestrinone (THG) is a new chemical used by athletes. Its detection involved some sleuthing that began at the Olympic Analytical Laboratory with a suspicion that something was amiss in the urine of a cyclist and was accomplished with the help of an anonymous whistleblower. The story is worth recounting because it represents the extent to which athletes will go to gain a competitive edge and how difficult it is to detect some of the newer compounds.

Case 11.4 Discovery of a Previously Undetected Doping Drug

This is a true case of the discovery of a previously unknown doping drug. Testing underway at the Olympic Analytical Laboratory (OAL) in Los Angeles uncovered unusually low levels of natural steroids (such as testosterone, epitestosterone, and androsterone) in the urine of a female cyclist. Further testing found traces of norbolethone, an androgen developed by

Wyeth in the 1960s but shelved because of side effects. When the OAL tested this compound in animals, it was found to build muscles with relatively few masculinizing side effects. Subsequently, the U.S. Anti-Doping Agency was forwarded the residue from a used but empty syringe, with the information that a disgruntled track coach had sent it to authorities for study. Within a few weeks, the OAL identified tetrahydrogestrinone (THG). The new chemical, which had never before been described, resembles two steroids banned for use by professional athletes: gestrinone, prescribed occasionally for the treatment of endometriosis, and trenbolone, which has some uses in veterinary medicine. Both steroids have powerful anabolic effects. When authorities tested urine stored from previous competitions, they found at least a dozen THG-tainted samples, many from athletes who had connections to the Bay Area Laboratory Co-operative (BALCO) in Burlingame, California, the likely source of THG (Vogel 2004). THG was placed on the list of banned substances at the Athens Olympic Games.

Test laboratories are vigilant for new versions of new compounds that are derivatives of natural steroids. Athletes continue to search out these new derivatives, as well as nutritional supplements such as creatine and androstenedione, despite the risk of severe, even life-threatening adverse effects. This is evidence of the extent to which athletes are still willing to go to win. A moral question here is to what extent the individual athlete considers it virtuous to avoid these supplements even if there were no punishments if discovered.

Creatine and Androstenidione

Two performance drugs touted to enhance performance, known as ergogenic aids, are creatine and androstenidione. Until December 2003, they were widely sold over the counter as dietary supplements, but the FDA has issued warning letters to all manufacturers to withdraw them from the open market due to the risk of serious side effects. We include information about them because they are likely to still be available on the black market because of the perception that they enhance performance.

Creatine is an amino acid found naturally in skeletal muscle, heart, brain, testes, retina, and other tissues. When taken orally, it causes weight gain, mainly through water retention, and is thought to increase the concentration of creatine in skeletal muscles, which in turn may allow for enhanced anaerobic production of ATP during strenuous exercise (Juhn 1999). Although it has been shown in the laboratory to mildly improve performance in brief aerobic activities, there are no data from studies of athletic competitions that demonstrate that creatine improves performance (Noakes 2004).

No data are available on long-term safety. Concerns are that its potency and purity as a dietary supplement are unknown because it is not under FDA control. Known adverse events are muscle cramping, diarrhea, dehydration, fluid shifts into skeletal muscle, and renal dysfunction. The effects of creatine supplementation on the heart, brain, and reproductive organs are unknown, but it is suspected, although unproven, that it predisposes users to life-threatening health risks such as cancer and stroke (Metzl 2001). Until 2005, there were no real safeguards or checks in place for the pediatric group of athletes, and few opportunities for the types of meaningful intervention that exist for collegiate and Olympic athletes. However, guidelines have now been put into place for high school athletes as a result of the fallout from drug use in professional athletes and a congressional inquiry on the subject.

Androstenedione is another product sold over the counter as a nutritional supplement, but when metabolized in the body it functions as the anabolic steroid testosterone. Its effect is to increase muscle size and strength (the latter from 1.2% to 18.7% in trained subjects, and from −14.4% to +12.9% in untrained subjects in another study). However, anabolic steroids have serious side effects. In males these consist of diminished sperm production, shrunken testicles, enlarged breasts, hypertension, and lipid profile abnormalities. They also have adverse effects on the liver (tumors), heart (enlargement, ischemia, atherosclerosis, coronary heart disease, and premature death), and behavior (depression, euphoria, anxiety, violent rages, psychotic episodes) (Reents 2000e). Women using anabolic steroids have a lean muscular physique and an increase in muscle mass and strength, and can also become severely masculinized. Other side effects include an increase in menstrual irregularities, eating disorders, central obesity, and diabetes, in addition to the other side effects reported in males, including aggressive and hostile behavior (Elliot and Goldberg 1993; Modell, Goldstein, and Reyes 1990; Bjorntorp 1996).

Children taking androstenedione are at risk for the early onset of puberty and premature cessation of bone growth. Also at issue are the manufacturer's claims that androstenedione safely builds bigger and stronger muscles and safely enables athletes to train longer and recover more quickly. Both children and their parents are not well informed about these concerns. As of 2005, some progress was being made in organizing forums for educating children and their families about the consequences of using these drugs after publicity during congressional hearings pointed out their dangers.

FDA jurisdiction over nutritional supplements ended in 1993 after passage of the Proxmire amendment, and until December of 2003, these drugs were readily available over the counter and marketed as safe and effective even in the absence of scientific evidence for these claims. Because these products are not regulated, there is no guarantee as to their content or safe manufacture. In addition, supplements such as creatine and androstenedione have not been tested in children or adolescents. The package labeling is vague, and several have been associated with serious adverse effects. Pediatricians have been trying for years to advise their young athlete patients of the potential for problems (Centers for Disease Control and Prevention 1998, Koshy et al. 1999).

Because of these concerns, the Department of Health and Human Services announced in 2004 that it would crack down on the manufacturers of supplements by reviewing the manufacturing, labeling, and claims made for these substances. Although supplements do not have to be proven safe before going on the market, federal agencies can remove them from the market if a product's safety becomes an issue. Letters were sent by HHS's Food and Drug Administration (FDA) to 23 companies, asking them to cease distributing products sold as dietary supplements that contain androstenedione and warning them that they could face enforcement actions if they do not take appropriate actions. Another positive step is the publicity about the BALCO scandal involving professional athletes who allegedly used these and other so-called "sports designer drugs." The publicity made clear the adverse effects that can occur with their use. The U.S. Congress has also taken up this cause, with the House proposing legislation in April 2005 that would require drug testing for all drugs banned by the World Anti-Doping Agency for all professional athletes in major sports, with penalties similar to those for Olympic athletes (a two-year suspension for the first offense, a lifetime ban after the second). Former bodybuilder California governor Arnold Schwarzenegger was forced to yield to searing public pressure from parents and others to give up his lucrative associations with bodybuilding magazines that tout the use of supplements. It is hoped that younger and nonprofessional athletes will also come to more fully understand the health- and career-threatening consequences of drug use as a result of these efforts.

Reasons Athletes Use Drugs

The motives to succeed are based on rewards ranging from money to fear of disappointing family. Elite athletes who succeed receive huge salaries, lucrative contracts for product endorsements, and opportunities for future careers. Stories of minor league players who have moved up to the major leagues using steroids induce others to use drugs for even a short-lived chance at fame, despite the risks to health and career. Younger athletes cite other factors for using drugs, which include the individual's desire to enhance his or her own performance, to bulk up to look better, or to keep up with others who are using drugs. Others use drugs to respond to peer pressure from teammates to perform better, and to improve their chances to succeed. There is also pressure from coaches, family, friends, the public, or the media to "win at all costs." At times, the desire to win for one's city, state, or nation during national or international competitions can also contribute to an athlete's search for ways to gain an edge over competitors. Even knowing of the risks of ergogenic aids, when Olympians were asked if they could get a drug that would allow them to win every event in the next five years but thereafter die, over 50% said they would take the drug (Bamberger and Jaeger 1997). The same is true in many sports in which the lures of fame and fortune cause athletes to risk serious side effects with doping.

In many instances, a competitive advantage can be obtained by means other than drug use. For example, competitors can take advantage of in-depth ergonomic studies of a sport or experiment with new ways to achieve better performance. For some, improved performance resulted from the use of the Fosbury flop in high jumping, streamlined helmets for bicycle time trials, a fiberglass pole for pole vaulting, an aerodynamic wing rigger for sculling, drag-resistant fabric for swimming and speed skating, and an exaggerated lean over the tips of the skis and a V-shaped drift for ski jumping. Some would ask at what point even these measures are unfair if the entire field of competitors does not have information about the new technique. But in most instances, if the modification has benefit, it quickly catches on.

Prevalence of Doping

Buckley, Yesalis, and Friedl were among the first to document the use of performance-enhancing drugs, not only in adults but also in high school students and even in adolescents and younger children (Buckley et al.1988). They reported that about 7% of high school seniors nationwide had used anabolic steroids, and that their use was increasing in those aged 15 and younger. One-third had started steroid use by age 16, and one-third had started at age 15 or younger. Most took steroids to improve athletic performance (47%), but 27% took them to improve appearance. A study by Yesalis et al. (1990) showed that 5.9% of collegiate women competitors use anabolic steroids. More recently, there have been reports of steroid use by fifth and sixth graders as well as increasingly by seventh to twelfth graders (Yesalis and Cowart 1998).

The "win at all costs" mentality has become the norm. There are more than 30 million children and adolescents younger than 18 years playing some form of organized team sport in the United States (Landry 1992). In a recent study of over 1,000 school-age children in middle and high school athletic programs, 5.6% admitted taking creatine. The use was seen in grades 6 to 12, although use was greatest in senior athletes (44%). Participants in all sports used creatine, but most commonly it is used by football players, wrestlers, hockey players, gymnasts, and lacrosse players, usually to enhance performance (74%) or to improve appearance (61%) (Metzl 2001). Anabolic steroids have been used by an estimated 3 to 12% of male high school students and 1% of female high school students, by 14% of college athletes, and 30 to 75% of competitive bodybuilders (Reents 2000f). Hints of major league sports personalities using steroids were followed by grand jury testimonies about the use of drugs containing testosterone such as "the cream" and "the clear" by notables such as sprinter Kelli White and baseball's Jason Giambi. A central figure in the scandal, Victor Conte of BALCO, estimated in a TV interview that over 50% of professional baseball players are taking some form of anabolic steroids, and that over 80% are taking some sort of stimulant before every game (http://sfgate.com/cgi-bin/article.cgi?file=/chronicle/archive/2004/12/03/MNGK3A5L8C11.DTL, accessed December 3, 2004).

To summarize the situation in 2006, doping with performance-enhancing drugs continues to be a major problem and an ethical dilemma for athletes and coaches. Through public outcry, there are now new rules for drug testing in professional baseball (random off-season testing and 10-day suspensions for first-time offenders), although many feel these rules should be even tougher. If indeed the Olympics are fraught with illicit drug use, as has been charged, there are rumblings that the Games should be discontinued.

Because numerous studies have indicated that the use of steroids is prevalent in younger athletes, efforts to change this pattern must begin with this group. Attitudes need to change, but unfortunately, the new sport has become a competition between drug developers and drug detectives. New drugs to enhance strength, endurance, and agility will escape detection unless the analytical scientists in charge of drug monitoring can anticipate the chemical signature of the new substance. However, the public is also part of the reason doping continues. Reluctant to dismiss their heroes despite doping, many athletes continue to be revered, as if the public's need for a hero supersedes their faults.

Enhancement of Behavior through Drugs

In addition to the use of drugs to enhance athletic performance, there is an increasing use of drugs to allay social and behavioral problems. These patterns of use show a wide acceptance of and desire to take advantage of available enhancements, forecasting wide use of genetic and other enhancements when they become available.

Whereas the motivation for athletic enhancement is to win in a sports competition and the main ethical problems are a violation of duty to obey the rules and the desertion of the principles of true sport, other enhancements are motivated by a perceived need to improve personal appearance, social behavior, or intellectual capability. In many cases, a person's sense of happiness is hindered by insecurities and even sadness when self-image does not meet expectations of perfection. Perceived imperfections in skin, stature, weight, and other body features, as well as behaviors such as shyness and obsessive-compulsiveness, all serve to limit social interactions and acceptance. More and more healthy children, adolescents, and adults are turning to enhancements that are becoming more readily available. Enhancements include perfume, jewelry, tattoos, body piercing, and minor and major plastic surgery, as well as drugs to alter behavior.

There are numerous drugs developed for certain medical problems that have been found to have other desirable effects. Some are listed in Table 11.3.

Medications such Prozac and Paxil, commonly used as antidepressants, are now being requested by subjects who are not clinically depressed but want to become less shy and more self-assured. This type of psychotherapeutic use is an example of the difference between medical therapy and enhancement. The latter has been termed "cosmetic psychopharmacology" (Kramer 1993).

The enhancement of mental capacities for normal individuals has been pursued using drugs such as Ritalin. Ritalin (methylphenidate) is a stimulant for most people. Its action on the central nervous system is similar to that of amphetamines but with fewer side effects. Although classified as a stimulant, it is used to calm children diagnosed with attention deficit hyperactivity disorder (ADHD). Students without ADHD insist that Ritalin helps them focus better and longer, resulting in more productive studying and better grades. Propranolol is a beta-adrenergic blocker used by cardiac patients, but it has been found to help lessen performance anxiety in symphony and opera musicians, public speakers, test takers, and others (Slomka 1992). It also appears to be helpful in blocking the painful memories of a traumatic or stressful event (posttraumatic stress disorder), and studies are planned to confirm this effect (Stein 2004). Other cognitive performance-enhancing substances have been used by university students in an effort to improve their academic performance (Canterbury and Lloyd 1994).

Another approach to cognitive enhancers is the use of drugs developed for treating dementia, such as the anticholinesterases (Cardenas 1993). The mechanism of this action is related to enhancement of acetylcholine in the brain, which may be related to increased blood flow to the brain, but to our knowledge, there have been no studies to date (2005) to confirm the extent to which these drugs are useful for this purpose. The use of Viagra (sidenafil), which enhances nitric oxide availability, was originally intended as a coronary dilator with potential uses for angina similar to the indications for nitroglycerine. The drug

Table 11.3   Selected Drug Use for Enhancement for Nonmedical Reasons

Drug	Medical Indication	Enhancement
Prozac (fluoxetine)	Depression	Self-confidence
Ritalin (methylphenidate)	Attention deficit disorders/ADHD	Improved ability to study
Valium (diazepam)	Muscle spasm	Calming
Propranolol	Heart arrhythmia	Alleviation of phobias, posttraumatic stress
Rogaine (minoxidil)	High blood pressure	Hair growth
Viagra (sidenafil)	Heart disease (angina)	Improved erections
Provigil (modafinil)	Narcolepsy	Alertness

was patented in 1996 and approved for prescription-only sales in 1998. It was discovered that this drug enables sustained penile erections in subjects with (or without) erectile dysfunction. An estimated 23 million men and women use this drug. The drug is dangerous when used with other medications, such as "poppers," that can cause death from hypotension. The drug is reportedly responsible for cardiovascular-related deaths.

One of the most recent behavior enhancers is a drug that was designed to aid patients with daytime sleepiness known as narcolepsy. This drug, modafinil (Provigil), is now being used as a cognitive enhancer to enable wakefulness by truck drivers, shift workers, the military, and sleep-deprived health care workers (Westcott 2005). The specific site of action is unknown (Gallopin et al. 2004), though it appears to enhance the alpha-adrenergic system. Though the original indications for this drug were based on its use in narcolepsy, advertisements by the manufacturer went beyond the approved indication and made unsubstantiated claims encouraging its use for tiredness and fatigue. The FDA issued a warning letter to the company (Appendix 11.1). This issue is examined in Problem 11.10.

There are ethical questions raised by using of drugs to change one's behavior or mental abilities. For example, is there an ethical issue in the use of drugs rather than mental discipline to improve mental abilities? A guide to answering this question is to ask: Does the use of drugs cause harm to either the user or others? If there are some risks, are these risks greater than the risks of alternative approaches to achieving the same effects? If one hasn't the time, or has other duties that interfere with studying more, why not use a memory enhancer or mental stimulant such as coffee or tea instead? We should remind ourselves of the unfettered use of caffeine in coffee, tea, and sodas to achieve a mild form of mental stimulation. As is the case with other stimulants, a decision about whether or not it is ethical to use them will lie in the individual's personal values and the magnitude of the risk entailed. Students who use Ritalin while studying for the LSAT see it as an aid to studying. Others see it as cheating, plain and simple.

There is another important perspective to consider before making a judgment regarding the rationale for using therapeutic drugs to improve a person's behavioral traits. Individuals who have stage fright, shyness, or symptoms of panic on occasion (e.g., heights, airplane flights) are not considered sick but may be experiencing genuine suffering. The use of Prozac by these individuals can be argued as a necessary treatment to relieve distress (Sabin and Daniels 1994).

Often, the use of drugs to enhance behavior draws a fine line between attaining normality and altering the breadth of what it is to be human. Fukuyama (2002) points out some of the consequences of these manipulations:

> Medical technology offers us in many cases a devil's bargain: longer life but with reduced mental capacity; freedom from depression, together with freedom from creativity or spirit; therapies that blur the line between what we achieve on our own and what we achieve because of the levels of various chemicals in our brains.

Body Enhancement by Surgery

The practice of surgical modification of an individual's appearance for nonmedical means has been widely accepted and has not generally been viewed as an issue of moral concern.

Cosmetic surgery was well established throughout the world before the use of drugs for changing individual personalities became a subject of moral discussion. Ancient practices done to enhance appearance included tattooing (e.g., New Zealand Maori) as well as insertion of ornaments from the earlobes and the nose (e.g., African natives), practices that continue to this day. While many so-called plastic surgical procedures are done to enhance personal appearance, frequently there is a functional or medical reason for doing them. An example would be rhinoplasty (nose surgery), which often corrects an anatomic abnormality interfering with breathing. Contemporary orthodontia (teeth straightening) is another example of an enhancement procedure that also corrects bite and improves and preserves dental function. But what about the modern practices of "extreme" body makeover? What might be the arguments for and against these practices from ethical perspectives?

To answer this question, we examine the situation using the moral principles of justice, duty, virtue, and autonomy, keeping in mind the benefits and dangers of an individual's choice to have radical surgery in order to improve physical appearance. From this analysis, the issue of justice rises to the fore. The procedures are costly and not available to everyone. A problem in the Problem Set for this chapter offers the opportunity to apply the four A's to this situation.

Most would agree there is nothing wrong with the pursuit of excellence in terms of intellectual and physical ability, or with the use of drugs and surgical manipulations to remedy a medical defect. What appears to many as ethically questionable is the use of medical procedures to improve one's appearance. Clearly, even here there is a gradation between plastic surgery for a marked facial deformity, a minor genetic malformation, or to beautify an already normal face. The major concern is equity or justice in the distribution and availability of medical resources. As pointed out by Elliott (Caplan and Elliott 2004a), there are 46 million uninsured people who might not receive basic medical care, while the rest of the population spends a billion dollars a year on baldness.

Elliott goes on to judge that ". . . it is the fact that we have gotten so accustomed to the inequity that we do not see it as obscene." On the other side of the argument, in the same publication, Caplan argues that people will strive by whatever means available to achieve perfection, and those who oppose enhancement methods do so based on the assumption that by improving individuals, something essential about humanity will be lost (Caplan and Elliott 2004b).

Some states have passed legislation that levies a tax on those undergoing certain cosmetic surgery procedures such as facelifts, tummy tucks, and the use of Botox to relieve wrinkles. The states of New Jersey and Washington are two examples, but other states are also eyeing this opportunity to increase state coffers. It is said these monies are to be used to cover health expenses for those who have no health insurance.

Ethics of Behavioral and Physical Enhancements

The moral issues involved in the use of drugs to enhance performance in sports are duty to obey the rules of the sport, duty to not put other competitors at risk by forcing them to take harmful measures in order to be competitive, and virtue ethics of competing using only inherent capabilities. Drug use to improve behavior and surgery or other measures to improve physical appearance raise different issues. The duty to obey rules is not an issue yet, although it could be argued that the use of stimulants by students in educational tests is unfair because it puts those not using stimulants at a disadvantage. No rules or methods of monitoring drug enhancement in the classroom have been promoted. In 2005, there were no definitive studies that showed that stimulants had an advantage, though antecdotal reports claim stimulants beyond coffee do help in exam preparation, probably by keeping the student awake. But there are adverse effects from the use of stimulants, and students who believe they need them for academic achievement put their health at risk by using them.

What are the moral issues surrounding the use of antidepressants such as Prozac for individuals who are not considered clinically depressed? If there is a high probability of significant adverse effects from the use of these drugs, are there alternative treatment interventions? *Consumer Reports* studied over 3,000 subscribers with depression or anxiety, and found that talk therapy was nearly as effective as talk therapy combined with drugs (*Consumer Reports* 2004). The analogy between drug use for depression, stage fright, and agitation and drug use for physical pain and head colds will be debated as a problem in the Problem Set of this chapter.

The modification of physical appearance does not present serious moral issues, although some might question such focus and expense on self. Even practices of what some would call disfigurement of the human body (e.g., tattooing), others would call culturally motivated practices. Similarly, surgical practices undertaken to beautify the body are enhancements for which the ethical problem might be justice as these enhancements are not available to all who want them. Consider a socialized medicine culture in which all members are given equal rights to treatment. When resources are limited, those seeking elective surgical enhancements are less likely to receive them than those needing surgery for more serious conditions. But this is not as likely to restrict surgeons in private practice from providing enhancements to those who can pay for them because this would be against the ethics of autonomy, that is, individual rights. Thus, as long as the procedures are not considered unreasonably dangerous, body enhancement for those who can pay will likely continue.

There is another perspective regarding the concept of surgery for medical reasons vs. surgery for personal reasons to improve appearance. The perspective is that of relieving suffering rather than that of correcting a medical problem. Cosmetic surgery can provide relief from the severe distress felt by an individual suffering from feelings of nonacceptance and alienation from perceived imperfections in physical features. In such a case, the need for cosmetic surgery can be considered as important as the need to improve the scar from a cleft lip or for orthodontia, not only to improve mastication but also to improve personal self-image and psychological health.

Ethical Duties of the Medical Practitioner

The medical practitioner's responsibilities in delivering drugs and surgical therapies for enhancement include the following (Little 1998):

- Will prescribing drugs carry risks whose management might be beyond the control of the practitioner (e.g., excessive doses of amphetamines)?

- Does the patient understand fully the probability of the risks and the possible negative consequences of cosmetic surgery?

- Does the procedure or prescription deplete limited resources, thus depriving others with a medical need?

- By giving in to a patient's insistence for a drug or a procedure, is the practitioner guilty of complicity by supporting an ethically questionable social behavior?

Professional duties to treat medical illnesses and not to use medical treatments for other purposes is cited as the dictum leading some to question the morality of cosmetic surgery. However, insofar as cosmetic surgery is a treatment that will relieve the internal or psychological suffering one has because of disfigurements such as a large nose, pendulous breasts, or protruding ears, one can argue much of cosmetic surgery *is* the duty of the professional. One needs to consider each situation on an individual basis, and assess whether the individual's expectations of outcome are realistic. To the extent that psychological pain is being treated, the physician is performing responsibly. But the profession of plastic surgery and cosmetic surgery is not perfect, and poor judgment, particularly in not obtaining informed consent preceded by clear explanations of all the facts and consequences, is probably the most troubling aspect of current practices.

CHAPTER SUMMARY

This chapter considers the morality of enhancement. Methods range from the selection of genes for one's offspring through preimplantation selection or gene therapy, the use of drugs to enhance performance and behavior, and surgery to enhance appearance.

The technique of preimplantation genetic diagnosis brings new ethical implications when used for enhancement of the embryo instead of for genetic testing to screen out genetic diseases. For example, many parents are opting to screen and select the sex of the embryo, not for bypassing an X-chromosome-linked genetic disease, but because they can. It appears certain that once more gene sequences are identified for human physical characteristics and traits, parents will insist on being able to choose genetic characteristics for their offspring that they perceive to be "best" for them. *Germline* gene therapy on a gamete (the egg or sperm), the pronucleus of the zygote, or a totipotent blastomere, has the potential not only to restore a function that is lacking but also to make genetic changes that will affect every cell of the body, which will be passed on to all successive generations. Germline gene therapy is especially valuable because it allows a single approach to the correction of genetic abnormalities in diseases that affect multiple cell types and multiple organs, such as cystic fibrosis, whereas to cure the disease by somatic gene therapy would require multiple procedures. This is because somatic gene therapy is directed to and restores function only to those cells targeted to receive the treatment. Since somatic gene therapy changes do not affect the reproductive cells, the genes inserted to restore function are not passed down to successive generations. The ethical issues raised by the use of these techniques pertain to their safety, the lack of affordability and availability for all, the implications of eugenics, and the autonomy of yet-unborn future generations who cannot consent to the genetic changes chosen for them. Aside from the availability of PGD and plastic surgery, most of the enhancement technologies are presently in their infancy and still being studied. Although gene therapy trials have been underway for over 15 years, none has been sufficiently successful to be approved by the FDA.

Advances are occurring so rapidly that public awareness and discourse has not kept up with how genetic changes today will affect generations of the future. The unknown consequences of genetic change will affect the human race forever, and some changes may not be as desirable as hoped. It is especially important for the public to realize that the DNA of future generations will be affected by how these issues are addressed. There is a pressing need for public debate on the subject in order to better understand the issues. The public also needs to participate in the decisions about the extent of the procedures planned, when "safe" is safe enough, and whether outside regulation is needed. The ethical issues are theoretical at the present time, but will become very real as soon as the new technologies are proven safe and put into general use.

New drugs and new uses for older drugs are providing enhancements at the molecular level that can affect a wide range of performance on the athletic field, in social settings, at school, on the job, and sexually. The moral issues involved in the use of drugs to enhance performance in sports include the duty to obey the rules of the sport, the duty to not put other competitors at risk by forcing them to take harmful measures in order to be competitive, and to compete using only inherent capabilities (virtue ethics). Other enhancements are motivated by a perceived need to improve personal appearance, social behavior, or intellectual capability. Medications originally intended as antidepressants have now been found to improve stress-related disorders as well as phobias. Drugs such as Ritalin reportedly improve academic performance, though the evidence for this is weak. Guidance for the regulation of these substances is obtained by asking the question: Will the drug cause harm to either the user or others?

Another important perspective regarding the moral rationale for using therapeutic drugs is the recognition that those with stage fright, shyness, or panic disorders are not considered sick but may be experiencing genuine suffering that may be greatly alleviated by this new therapy. Thus, the use of drugs to enhance behavior draws a fine line between attaining normality and altering the breadth of what it is to be human.

A similar observation could be made about the surgical approach to enhancement, although there is a broad range of problems for which plastic surgery is more or less acceptable. Cosmetic approaches such as orthodontia, tattooing, body piercing, and face-lifts are common. But what about the more extreme makeovers gaining in popularity? Guidelines for acceptability include the moral principles of justice, duty, virtue, and autonomy, keeping in mind the benefits and dangers of an individual's choice of ways to improve physical appearance.

Many object to the use of medical procedures to improve one's appearance on the basis of equity or justice in the distribution and availability of medical resources. Some find these procedures obscene because of the expense and focus on self. Others oppose enhancement based on the assumption that by improving individuals, something essential about humanity will be lost. However, cosmetic surgery may also provide relief from the severe distress that some suffer resulting from feelings of nonacceptance and alienation because of perceived imperfections in physical features. Cosmetic surgery can be considered an important remedy. Care providers need to carefully assess the risk and benefits of the procedures, and whether the patient undergoing such procedures has realistic expectations of the positive outcome as well as the risks.

REFERENCES

Abate, T. 2002 (October 21). "Society taking steps toward 'designer babies.'" *San Francisco Chronicle*, p. E1.

Anderson, W. F. 2000a. "A new front in the battle against disease." In: *Engineering the Human Germline*. G. Stock and J. Campbell (eds.). New York: Oxford University Press, pp. 43–48.

Anderson, W. F. 2000b, p. 46.

Anisimov, S. V., M. V. Volkova, L. V. Lenskaya, et al. 2001. "Age-associated accumulation of the apolipoprotein C-III gene T-455C polymorphism C allele in a Russian population." *J Gerontol A Biol Sci Med Sci* **56**(1):B27–32.

Arruda, V. R., H. H. Stedman, T. C. Nichols, et al. 2004. "Regional intravascular delivery of AAV-2-F.IX to skeletal muscle achieves long-term correction of hemophilia B in large animal model." *Blood*. First Edition Paper, prepublished online October 12, 2004. www.bloodjournal.org/cgi/content/abstract/2004-07-2908v1, accessed October 19, 2004.

Bamberger, M. and D. Jaeger. 1997 (April 14). "Over the edge." *Sports Illustrated* **14**:62.

Bankiewicz, K. 2005. Personal communication.

Bankiewicz, K., J. Eberling, M. Kohutnicka, et al. 2000. "Convection-enhanced delivery of AAV vector in parkinsonian monkeys; in vivo detection of gene expression and restoration of dopaminergic function using pro-drug approach." *Exp Neurol* **164**:2–14.

Barritt, J. A., C. A. Brenner, H. E. Malter, and J. Cohen. 2001. "Mitochondria in human offspring derived from ooplasmic transplantation: Brief communication." *Hum Reprod* **16**:513.

Barzilai, N., G. Atzmon, C. Schechter, et al. 2004. "Unique lipoprotein phenotype and genotype associated with exceptional longevity." *JAMA* **290**(15):2030–2040.

Benzi, G. 1994. "Pharmacoepidemiology of the drugs used in sports as doping agents." *Pharmacol Res* **29**:13–26.

Berns, A. 2004. "Good news for gene therapy." *N Engl J Med* **350**:1679–1680.

Bjorntorp, P. 1996. "The android woman—a risky condition." *J Intern Med* **239**(2):105–110.

Brock D. W. 1998. "Enhancements of human function: Some distinctions for policymakers." In: *Enhancing Human Traits*. E. Parens (ed). Washington, DC: Georgetown University Press, p. 49.

Buckley, W. A., C. W. Yesalis, K. E. Friedl, et al. 1988. "Estimated prevalence of anabolic steroid use among male high school seniors." *JAMA* **260**(23):3441–3445.

Buckley, R. H. 2000 (November 2). "Primary immunodeficiency diseases due to defects in lymphocytes." *N Engl J Med* **343**(18):1313–1324.

Burress, C. 2004 (September 30). "Discussing the ethics of altering human genes." *San Francisco Chronicle*, p. B7.

Canterbury, R. J. and E. Lloyd. 1994. "Smart drugs: Implications of student use." *J Prim Prev* **14**(3):197–207.

Caplan, A. and C. Elliott. 2004a. "Is it ethical to use enhancement technologies to make us better than well?" *PLoS Med* **1**(3):172–175.

Caplan, A. and C. Elliott, 2004b, p. 175.

Cardenas, D. D. 1993. "Cognition-enhancing drugs." *J Head Trauma Rehabil* **8**(4):112–114.

Cavazzana-Calvo, M., S. Hacein-Bey, G. de Saint Basile, et al. 2000 (April 28). "Gene therapy of human severe combined immunodeficiency (SCID)-X1 disease." *Science* **288**(5466):669.

Centers for Disease Control. 1998. "Hyperthermia and dehydration-related deaths associated with rapid weight loss in three collegiate wrestlers—North Carolina, Wisconsin and Michigan." *JAMA* **279**:824–825.

Conusmer Reports. 2004 (October). "Drugs vs. talk therapy: 3,079 readers rate their care for depression and anxiety." *Consum Rep* **69**(10):22–29.

Couzin, J. 2003 (October 17). "Is long life in the blood?" *Science* **302**:373.

Couzin, J. and J. Kaiser. 2005 (February 18). "As Gelsinger case ends, gene therapy suffers another blow." *Science* **307**(5712):1028.

de las Heras, J. I., L. D'Aiuto, and H. Cooke. 2003. "Mammalian artificial chromosome formation in human cells after lipofection of a PAC precursor." In: *Mammalian Artificial Chromosomes: Methods and Protocols*. V. Sgaramella, S. Erdani, and V. Sgaramella (eds.). Totowa, NJ: Humana Press, pp. 187–206.

Elliot, D. L. and L. Goldberg. 1993. "Women and anabolic steroids." In: *Anabolic Steroids in Sports and Exercise*. C. E. Yesalis (ed.). Champaign, IL: Human Kinetics Publishers, pp. 225–239.

Feng, W. 2005 (March). "Can China afford to continue its one-child policy?" *AsiaPacific Issues*, No. 77, accessed April 16, 2005 at www.eastwestcenter.org/stored/pdfs/api077.pdf.

Fukuyama, F. 2002. *Our Posthuman Future: Consequences of the Biotechnology Revolution*. New York: Farrar, Straus and Giroux, p. 7.

Gallopin, T., P. H. Luppi, F. A. Rambert, et al. 2004. "Effect of the wake-promoting agent modafinil on sleep-promoting neurons from the ventrolateral preoptic nucleus: an in vitro pharmacologic study." *Sleep* **27**(1):19–25.

Hacein-Bey-Abina, S., F. Le Deist, F. Carlier, et al. 2002. "Sustained correction of X-linked severe combined immunodeficiency by ex vivo gene therapy." *N Engl J Med* **346**:1185–1189.

Hacein-Bey-Abina, S., C. Von Kalle, M. Schmidt, et al. 2003. "LMO2-associated clonal T cell proliferation in two patients after gene therapy for SCID-X1." *Science* **302**:415–419.

Harrington, J. J., G. Van Bokkelen, R. W. Mays, et al. 1997 (April). "Formation of de novo centromeres and construction of first-generation human artificial microchromosomes." *Nat Genet* **15**(4):345–355.

High, K. A. 2003 (March-April). "The risks of germline gene transfer." *Hastings Center Report* **33**:3.

Hoogerbrugge, P. M., V. W. van Beusechem, A. Fischer, et al. 1996 (February). "Bone marrow gene transfer in three patients with adenosine deaminase deficiency." *Gene Ther* **3**(2):179–183.

Holzenberger, M., J. Dupont, B. Ducos, et al. 2003. "IGF-1 receptor regulates lifespan and resistance to oxidative stress in mice." *Nature* **421**(6919):182–187.

Hyman, D. A. 1990. "Aesthetics and ethics: the implications of cosmetic surgery." *Perspect Biol Med* **33**(2):190–202.

Juhn, M. S. 1999 (June). "Oral creatine supplementation." *The Physician and Sportsmedicine*. **27**(5), accessed on October 20, 2004 at http://physsportsmed.com/issues/1999/05_99/juhn.htm.

Kaiser, J. 2004 (June 4). "Gene therapy: Side effects sideline hemophilia trial." *Science* **304**:1423.

Kaiser, J. 2005. "Panel urges limits on X-SCID trials." *Science* **307**(5715):1544–1545.

Kenyon, C. 2005. "The plasticity of aging: insights from long-lived mutants." *Cell* **120** (4): 449–460.

Kenyon, C., J. Chang, E. Gensch, et al. 1993. "A *C. elegans* mutant that lives twice as long as wild type." *Nature*. **366**:461–464.

Koshy, K. M., E. Griswold, and E. E. Scheenberger. 1999. "Interstitial nephritis in a patient taking creatine [letter]." *N Engl J Med* **320**:814–815.

Kramer, P. D. 1993. *Listening to Prozac*. New York: Viking Press.

Kumar, K. N. 1991. "Cloning of cDNA for the glutamate-binding subunit of an NMDA receptor complex." *Nature* **354**:70–73.

Landry, G. 1992. "Sports injuries in childhood." *Pediatr Ann* **21**:165.

Larin, Z. and J. E. Mejia. 2002. "Advances in human artificial chromosome technology." *Trends Genet* **18**:313–319.

Little, M. O. 1998. "Cosmetic surgery, suspect norms, and the ethics of complicity." In: *Enhancing Human Traits.* E. Parens (ed.). Washington, DC: Georgetown University Press, 162–176.

Lubs, H. A. 1977. "Frequency of genetic disease." In: *Genetic Counseling.* H. A. Lubs and F. de la Cruz (eds.). New York: Raven Press, pp. 1–16.

Mannucci, P. M. 2002 (January). "Hemophilia and related bleeding disorders: a story of dismay and success." *Hematology* **2002**(1):1–9.

Metzl, J. D. 2001. "Creatine use among young athletes." *Pediatrics* **108**:421.

Modell, E., D. Goldstein, and F. Reyes. 1990. "Endocrine and behavioral responses to psychological stress in hyperandrogenic women." *Fertil Steril* **53**:454–459.

National Commission on Sleep Disorders Research. 1993 (January). *Wake Up America: A National Sleep Alert.* Volume One: Executive Summary and Executive Report. Washington, DC: U.S. Government Printing Office, Doc. No. 1993-342-299/30631.

National Institutes of Health. 2001 (January 5). "Notice of action under the NIH guidelines for research involving recombinant DNA molecules (NIH Guidelines)." *Federal Register* **66**(4):1146–1147.

Noakes, T. D. 2004. "Tainted glory—doping and athletic performance." *N Engl J Med* **351**:847–849.

Raper, S. E., N. Chirmule, F. S. Lee, et al. 2003 (September-October). "Fatal systemic inflammatory response syndrome in an ornithine transcarbamylase deficient patient following adenoviral gene transfer." *Mol Genet Metab* **80**(1–2):148–158.

Reents, S. 2000a. "Caffeine." In: *Sport and Exercise Pharmacology.* Champaign, IL: Human Kinetics Publishers, pp. 277–295.

Reents, S., 2000b, p. 155.

Reents, S. 2000c. pp. 47–70.

Reents, S., 2000 d, pp. 62–64.

Reents S., 2000e, pp. 171–179.

Reents S., 2000f, p. 163.

Roy, D. J. 2003. "Gene therapy . . . its scientific and ethical complexity." *L'observatoire de la génétique*, No. 10. www.ircm.qc.ca/bioethique/obsgenetique/cadrages/cadr2003/c_no10_03/cai_no10_03_1.html, accessed December 1, 2004.

Sabin, J. and N. Daniels. 1994. "Determining 'medical necessity' in mental health practice." *Hastings Center Report* **24**(6):5–13.

Schaffer, D. V. and D. A. Lauffenburger. 2000. "Targeted synthetic gene delivery vectors." *Current Opinions in Molecular Therapeutics* **2**:155–161.

Schuelke, M., K. R. Wagner, L. E. Stolz, et al. 2004. "Myostatin mutation associated with gross muscle hypertrophy in a child." *New Engl J Med* **350**(26):2682–2688.

Shekelle, P. G., M. L. Hardy, S. C. Morton, et al. 2003. "Efficacy and safety of ephedra and ephedrine for weight loss and athletic performance." *JAMA* 289:1537–1545.

Slomka, J. 1992 (July–August). "Playing with propranolol." *Hastings Center Report* **22**(4):13–17.

Stein, R. 2004 (October 20). "Good results for pill to block painful memories may lead to wider testing." *San Francisco Chronicle*, p. A16.

Stock, G. and J. Campbell. 2000. *Engineering the Human Germline: An Exploration of the Science and Ethics of Altering the Genes We Pass to Our Children.* New York: Oxford University Press.

Tauer, C. 1995. "Human growth hormone: A case study in treatment priorities." *Hastings Center Report.* **25**(3):S18–S20.

van Heemst, D., M. Beekman, S. P. Mooijaart, et al. 2005. "Reduced insulin/IGF-1 signalling and human longevity." *Aging Cell* **4**(2):79–85.

Verlinsky, Y., S. Rechitsky, O. Verlinsky, et al. 2002. "Preimplantation diagnosis for early-onset Alzheimer disease caused by V717L mutation." *JAMA* **287**(8):1018–1021.

Vogel, G. 2004 (July 30). "A race to the starting line." *Science* **305**(5684):632–635.

Volkow, N. D., J. S. Fowler, G. Wang, et al. 2002. "Mechanism of action of methylphenidate: insights from PET imaging studies." *J Atten Disord* **6**(Suppl 1):S31–S43.

Walters, L. and J. G. Palmer. 1997a. *The Ethics of Gene Therapy.* New York: Oxford University Press, p. 53.

Walters, L. and J. G Palmer, 1997b, p. 37.

Weiss, R. 2002 (February 27). "Healthy baby born after eggs screened for Alzheimer's gene." *San Francisco Chronicle*, pp. A1, A11.

Westcott K. J. 2005. "Modafinil, sleep deprivation, and cognitive function in military and medical settings." *Mil Med.* **170**:333–335.

Yesalis, C. E., W. A. Buckley, W. A. Anderson, et al. 1990. "Athletes' projections of anabolic steroid use." *Clinical Sports Medicine* **2**:155–171.

Yesalis, C. E. and V. Cowart. 1998. *The Steroids Game.* Champaign, IL: Human Kinetics Publishers, pp. 8–11.

ADDITIONAL READING

Buchanan, A., D. W. Brock, N. Daniels, and D. Wikler. 2000. *From Chance to Choice: Genetics and Justice.* Cambridge, UK: Cambridge University Press, 398 pp.

Fox, M. W. 1999. *Beyond Evolution: The Genetically Altered Future of Plants, Animals, the Earth . . . and Humans.* New York: The Lyons Press, 256 pp.

Fukuyama, F. *Our Posthuman Future. Consequences of the Biotechnology Revolution.* New York: Farrar, Straus and Giroux, 256 pp.

Gardner, W. 1995. "Can human genetic enhancement be prohibited?" *J Med Philos* **20**(1):65–84.

Harris, J. 1992. *Wonderwoman and Superman: The Ethics of Human Biotechnology.* New York: Oxford University Press, 271 pp.

Jazwinski, S. 1996. "Longevity, genes and aging." *Science* **273**:54–59.

Kramer, P. D. 1993. *Listening to Prozac.* New York: Viking Press, 409 pp.

Larin, Z. and J. E. Mejia. 2002. "Advances in human artificial chromosome technology." *Trends Genet* **18**:313–319.

Peterson, J. C. 2001. *Genetic Turning Points: The Ethics of Human Genetic Intervention.* Grand Rapids, MI: W. B. Eerdmans Publishing Company, 364 pp.

Ridley, M. *Genome: The Autobiography of a Species in 23 Chapters.* 2000. First perennial edition, New York: HarperCollins Publishers, Inc., 344 pp.

Stock, G. 2002. *Redesigning Humans: Our Inevitable Genetic Future.* Boston: Houghton Mifflin Company, 277 pp.

Stock, G. and J. Campbell (eds.). 2000. *Engineering the Human Germline: An Exploration of the Science and Ethics of Altering the Genes We Pass to Our Children.* New York: Oxford University Press, 169 pp.

Wivel, N. A. and L. Walters. 1993. "Germ-line gene modification and disease prevention: Some medical and ethical perspectives." *Science* **262**:533–538.

APPENDIX 11.1 FDA WARNING LETTER TO PHARMA

The warning letter below is a good illustration of false advertising that could be considered dangerous to patients. The medication was intended to assist patients with sleep disorders, but it is also being used as a cognitive enhancer. The FDA found the advertisements inaccurate and dangerously misleading. However, on the basis of scientific reports in the literature since 2002, there appears to be support for some of the claims that the warning letter suggested were not true. (see Problem 11.10)

FDA Warning Letter

[Released by FDA: 1/3/02. Posted by FDA: 1/14/02]

Paul M. Kirsch
Senior Director, Regulatory Affairs
Cephalon Inc.
145 Brandywine Parkway
West Chester, PA 19380-4245

RE: NDA # 20-717
Provigil (modafinil) Tablets
MACMIS ID # 10183

Dear Mr. Kirsch:

This letter objects to Cephalon Inc's (Cephalon) dissemination of false or misleading promotional materials for Provigil (modafanil) Tablets. As a part of its routine monitoring and surveillance program, the Division of Drug Marketing, Advertising, and Communications (DDMAC) has reviewed these materials for Provigil and has concluded that they are false, lacking in fair balance, or otherwise misleading in violation of the Federal Food, Drug, and Cosmetic Act (Act), and applicable regulations. Our specific objections follow:

Promotion of Unapproved Uses

Promotional materials are false, lacking in fair balance, or otherwise misleading if they contain representations or suggestions that a drug is better, more safe, more effective, or useful in a broader range of conditions or patients than has been demonstrated by substantial evidence. Provigil is indicated in a select group of patients. Specifically, the "Indications and Usage" section of the approved product labeling (PI) for Provigil states, "Provigil is indicated to improve wakefulness in patients with excessive daytime sleepiness associated with narcolepsy."

The claims contained in your promotional materials suggest that Provigil is safe and effective for a variety of unapproved uses. For example, your journal advertisements prominently present the following misleading claims under the header "Consider PROVIGIL to improve wakefulness:"

"When patients complain of FATIGUE or TIREDNESS"
"When patients present with SLEEPINESS"
"When patients complain of SLEEPINESS"
"When patients present with FATIGUE or TIREDNESS"
"When patients complain of feeling FATIGUED or TIRED"
"When patients present with sleepiness and Decreased ACTIVITY"
"When patients complain of sleepiness and Decreased ACTIVITY"
"When patients present with Lack of ENERGY"
"When patients complain of Lack of ENERGY"

The claims are misleading because Provigil is not approved to treat such symptoms as sleepiness, tiredness, decreased activity, lack of energy, and fatigue. Therefore, the claims promote Provigil for unapproved uses.

Similarly, your sales aids prominently present the claim "[a] wake-promoting alternative for your psychiatry practice. . . " on the front cover, followed by the claim "PROVIGIL: A prescription for daytime wakefulness," on the inside front cover. These claims are misleading because they suggest that Provigil is a safe and effective treatment for anyone with daytime sleepiness. Provigil is indicated to improve wakefulness in patients with excessive daytime sleepiness associated with narcolepsy. Provigil is not approved for use as a daytime stimulant. Furthermore, presenting the indication for Provigil in small print at the bottom of the sales aids and journal advertisements does not correct the overwhelming misleading impression that Provigil can be used to improve wakefulness in all patients presenting with symptoms of daytime sleepiness, characteristic of generalized sleep disorders, whether or not they have narcolepsy.

The Provigil website also prominently presents the claim, "Provigil, a prescription for daytime wakefulness," along with a questionnaire with the headline, "Do you suffer from excessive daytime sleepiness?" Thus, the Provigil website is misleading because, like your sales aids and journal advertisements, the website does not adequately communicate the indication for Provigil. Additionally, the website promotes Provigil for unapproved uses by suggesting that Provigil is useful for anyone with excessive daytime sleepiness.

Minimization of CNS Effects and Abuse Potential

Your promotional materials present claims that "Provigil promotes wakefulness without widespread CNS stimulation in preclinical models" and "Low abuse potential" to suggest that Provigil does not have CNS properties that may lead to abuse and are common to other scheduled stimulants or stimulant-like drugs. The claim is misleading because it is inconsistent with the PI. The PI states, "[t]he abuse potential of modafinil (200, 400, and 800mg) was assessed relative to methylphenidate (45 and 90mg) in an inpatient study in individuals experienced with drugs of abuse. Results from this clinical study demonstrated that modafinil produced psychoactive and euphoric effects and feelings consistent with other scheduled CNS stimulants (methylphenidate)." Furthermore, presenting data from pre-clinical models is not considered substantial evidence to support efficacy claims.

Misleading Mechanism of Action Claims

Claims contained in your promotional materials suggest that the mechanism of action of Provigil is understood. For example, your sales aids, journal advertisements, and Provigil website present the following misleading claims:

"PROVIGIL works differently from stimulants in preclinical models."
"PROVIGIL promotes wakefulness without widespread CNS stimulation."
"PROVIGIL acts selectively in areas of the brain to regulate normal wakefulness."
"Unlike stimulants, PROVIGTL is not mediated by a dopaminergic mechanism."
"The highly selective CNS activity of PROVIGIL is distinct from amphetamine and methylphenidate in pre-clinical models."

The claims are presented with pictures that illustrate selective sites of action in the brain where Provigil is purported to have activity based on animal studies. Moreover, the claims and pictures are presented in comparison to amphetamine and methylphenidate. These presentations are misleading because they imply that the mechanism of action of Provigil is fully understood when such is not the case. The PI specifically states that "the precise mechanism(s) of action through which modafinil promotes wakefulness is unknown." Additionally, it is misleading to make claims based on data from animal studies to suggest clinical significance when, in fact, no clinical significance has been demonstrated. Furthermore, placement of statements in small pint that "the relationship of these findings in animals to the effects of Provigil in humans has not been established" or "the precise mechanism of action is unknown" does not correct the overwhelming misleading impression presented by the claims and pictures.

Misleading Switch Protocol

Your sales aids and Provigil website state or suggest that patients should be switched from traditional stimulants (e.g., methylphenidate) to Provigil, along with other claims such as "switching to Provigil is easy" and "switch to Provigil for all the right reasons." Additionally, a protocol for switching from methylphenidate to Provigil is provided in the promotional materials. The claims and switch protocol are misleading because they imply that the efficacy of Provigil and methylphenidate, for example, are equivalent when such has not been demonstrated by substantial evidence.

Unsubstantiated Superiority Claims

Your sales aids present claims that patients dissatisfied with stimulants and patients seeking a well tolerated agent are candidates for Provigil. Your sales aids also claim that patients should be switched to Provigil because Provigil has more selective activity in the brain and improves sleep latency compared to traditional stimulants. These claims are misleading because they suggest that Provigil is superior to other agents when such has not been demonstrated by substantial evidence i.e., head-to-head clinical studies. In fact, data used to support the unproved sleep latency claim was derived from a post-hoc analysis of sleep latency. Data from post-hoc analyses are not adequate evidence to support superiority or comparative efficacy claims.

Additionally, your sales aid presents the misleading claim "Provigil significantly improved daytime wakefulness in patients unsatisfactorily treated with traditional stimulants" followed by a graph entitled "Provigil improved wakefulness." The claim and accompanying graph are misleading because they suggest superiority for Provigil versus dextroamphetamine, methylphenidate, and pemoline, when such has not been demonstrated by substantial evidence.

Requested Action

We request that you immediately cease the dissemination of sales aids, journal advertisements, websites and all other promotional materials and activities for Provigil that contain the same or similar violations outlined in this letter. Your written response to the above request should be received no later than January 17, 2002. Your response should include a list of all promotional materials that are discontinued and the date that they were discontinued. If you have any questions or comments, please contact James Rogers, Pharm.D., by facsimile at (301) 594-6771, or at the Food and Drug Administration, Division of Drug Marketing, Advertising, and Communications, HFD-42, Rm 17-1320, 5600 Fishers Lane, Rockville, MD 20857. We remind you that only written communications are considered official.

In all future correspondence regarding this particular matter, please refer to MACMIS ID # 10183 in addition to the NDA number.

Sincerely,

James R. Rogers, Pharm.D.
Regulatory Review Officer
Division of Drug Marketing,
Advertising, and Communications

PROBLEM SET

11.1   Jane is competing in a contest for a National Science Foundation undergraduate fellowship. She has spent most of the last two nights studying notes taken from conversations she had with previous contestants. She has a minor head cold and generally does not take medicine for either minor or major colds. Because she feels she needs some stimulant in addition to coffee in order to keep awake, she takes two long-acting ephedrine tablets. She also takes orange juice in a much greater quantity than she has ever had in the past because she heard that higher sugar levels in the blood would enhance brain activity.

When she starts taking the exam, she notices that her hands are shaky, and two other contestants also notice this "nervousness." Because of her full bladder and possibly the medication she is taking, she leaves the exam room four times to go to the restroom. One of her competitors decides that Jane has been taking a stimulant and confers with the other observer/contestant, who believes that Jane was leaving the exam to confer with cheat sheets. The contestants confer with the exam monitor after the exam.

You are the exam monitor. How would you approach the problem knowing that Jane has probably left Washington, DC, to return to school? This problem has three parts: 1) Propose a "whistleblower" scenario, 2) propose a "let's drop it" scenario, and 3) give your arguments for or against the acceptability of taking stimulants to enhance examination performance. (60 words for each part.)

11.2   The use of artificial chromosomes to incorporate attributes or enhance innate capabilities has been proposed as a possible method of human enhancement. During in vitro fertilization, an artificial chromosome is injected into a zygote. This chromosome has genes that can be turned on to produce traits by injection of a promoter, and the gene expression can be turned off by injection of specific chemicals. Given that there are no changes in the regulation for preimplantation modification, what are the ethical arguments for not allowing artificial chromosome methods of human enhancement? (80 words or less.)

11.3   The federal government is approached by a consortium of sports medicine and pediatric organizations to pass a law that requires the establishment of a sports educational program at all pubic and private schools. The program includes a semester of lectures and demonstrations

one hour each week. The lectures give information on drugs and their effects as well as exercise training lectures and nutrition lectures. At least 50% of the education time, or a minimum of eight sessions, must be in a supervised physical education class for proper strength training.

Blue Team - Present the ethical reasons why this federal law should not be passed. You can consider arguments based on controversies regarding objections to prayer or pledges of allegiance to the flag in the classroom. Teams should know that when Olympians were asked if they could get a drug that would allow them to win every event in the next five years but thereafter die, over 50% would take the drug (Bamberger and Jaeger 1997).

Gold Team - Present the ethical reasons why it is necessary to educate and warn the public of dangers as there is no other practical opportunity to prevent young athletes from harming themselves. This is the duty of the government. You can consider the analogies to speeding laws and seat belts.

11.4   What was the root cause for legislation that controlled drug use and selling? (Hint: search the Internet.)

  a. Ancient concepts of virtue?

  b. Public health concerns?

  c. Prejudice against African Americans?

  d. Need to tax products to support government?

  e. Abnormal sexual behavior?

11.5   The government proposes regulations that all couples must undergo genetic testing for diseases that could be passed down to their offspring. You are a member of a group of individuals advocating solutions for inherited diseases such as neurofibromatosis, diabetes, cystic fibrosis, and Down's syndrome. You have witnessed the difficulties the children with these diseases and their families face, but also the happiness of these children and the meaningfulness of the child to the fulfillment of life for the parents. You decide to mount an effort to oppose the government's regulation attempt. Give arguments that you believe the government would use to oppose your arguments as well as your arguments. How would you proceed in mounting the facts for the opposition? (150 words for your answer, 80 words for the suspected government reply, 70 words for your methods.)

11.6   How should the government, the medical profession, and schools approach the problem of drug use in students without learning disorders who are taking the psychopharmaceuticals to improve test performance?

Blue Team - Defend the proposition that there should be random tests of the urine of students during exam periods in order to control unfair drug usage. You first must defend the proposition that drug use is wrong.

Gold Team - Defend the case for student rights and invasion of autonomy if a condition of attending the university or lower schools requires urine testing. You do not have to first defend the proposition that drug use is permissible.

11.7   Drugs developed for the treatment of psychiatric disorders such as depression, anxiety, and attention deficit disorders in children are now being used by "normal" individuals to cope with life or to improve performance in everyday life endeavors. Is this use morally wrong? Is there a difference between the use of these drugs and the use of coffee or aspirin?

Blue Team - Defend the assertion that the use of drugs to influence an individual's feeling of well-being or the ability of an individual to overcome fear is no different from the use of heroin or cocaine and should be prohibited except for the treatment of a diagnosed psychiatric disorder.

Gold Team - Argue that the use of stimulants or other drugs to improve performance in anxiety-provoking situations at work or in social settings is perfectly natural, and only if excessive, such as excessive use of alcohol, should their use be limited.

11.8   A remarkable aspect of the human condition is the amount of time spent in an unconscious state during sleep, the variability of sleep required by individuals, and the prevalence of sleep disorders. The individual sleep requirement varies from 4 hours to 9 hours (17 to 37% of our lives). It is generally believed this represents the time needed for biological and chemical regeneration or recuperation. In addition, many believe it is time required for dreaming, which some argue is essential for the health of the human psyche. Accidents due to lack of sleep result in billions of dollars in lost productivity (National Commission on Sleep Disorders Research 1993).

Therefore, a method that would safely reduce the need for sleep might have many benefits.

The principal controller for sleep is known as the circadian clock, tissue that is located deep in the brain in the suprachiasmic nucleus of the hypothalamus. It regulates the chemical processes that drive this mechanism. Two genes known as *c-fos* and *jun-B* stimulate a protein that is believed to have control over the circadian clock. For individuals with sleep disorders or others who wish to enhance their performance by reducing the desire for sleep, one can speculate that insertion of additional amounts of these genes into the cells of the hypothalamus might have some appeal.

Thus, by merely infusing an inactivated virus carrying the gene into cells of a specific region of the brain, an individual could in principle control the level of alertness with far more efficiency and reliability than by using drugs. There is no question of the need for remedies to treat sleep disorders and to provide ways to ensure alertness when fatigue threatens performance, yet the question of risks has not been assessed. Clearly, though, this is an example of a potential future enhancement therapy.

Blue Team - In view of the seriousness of sleep disorders, propose that the use of gene therapy, using either the insertion of one or more genes into the region of the circadian clock or therapy using deep brain stimulation by insertion of electrodes into the same region, should be tried.

Gold Team - Argue that the reason one or both of these procedures should not be undertaken is because this is too risky. Also consider the reasons that sleep is required in the first place. (Hint: search the Internet.)

11.9   Professional baseball players have been using steroids to enhance performance prior to 2005. A proposal to test these players involves random checks and a punishment only after a few violations. Through the Web, determine the latest rules and discuss why these rules should not be as severe as those for amateur athletes. (80 words or less.)

11.10   Review the FDA Warning Letter in Appendix 11.1. Are the assertions in that letter correct? What are the ethical issues? How would you respond if you were the scientific officer of the company who received the letter? (Hint: review the literature on this drug using PubMed or other search engines of the Internet.)

12

Ethics of Emerging Technologies

"Any new engineering has ethical consequences."

Norman Augustine, Professor, Princeton

"It is impossible to predict behavior of complex systems beyond a certain level of complexity."

William Wulf, NAE President

INTRODUCTION

The focus of this chapter is on the multitude of new engineering technologies emerging in the first decade of the twenty-first century. These technologies challenge the professions that are creating them to consider risks and consequences as macro-ethical problems. By macro is meant problems that go beyond the individual's virtue and behavior. These technologies lead to macro-ethical considerations because of the magnitude and varieties of stakeholders who could experience the effect of the potential consequences. The developments that require macro-ethical considerations are problems for the professions and the public as a collective group (Wulf 2004). Table 12.1 lists most of the emerging technologies of the twenty-first century.

Those that are asterisked are considered separately in this chapter because they have both apparent and hidden consequences that are especially in need of ongoing evaluation. An evaluation of the risks can lead to countermeasures that may mitigate consequences that would undermine anticipated benefits. The technologies on which this chapter focuses include:

- Brain imaging interrogation technologies
- Exploration and utilization of outerspace
- Nanotechnology
- Neuroenhancement and neuroengineering
- Synthetic biology
- Terrorism technologies

Table 12.1   Categories of Emerging Technologies

- Alternative energy sources (see Chapter 5)
- Artificial intelligence
- Automotive technology
- Brain imaging interrogation technologies*
- Communication (see Chapter 3)
- Drug enhancement of performance (see Chapter 11)
- Embedded intelligence
- Exploration and utilization of outer space*
- Genetic modification of plants and organisms (see Chapter 6)
- Health care technologies using wireless biomonitoring
- Human cloning (see Chapter 10)
- Human genetic engineering (see Chapter 11)
- Information technology (see Chapter 3)
- Life prolongation and suspended animation (see Problem 12.9)
- Nanotechnology*
- Neuroengineering and neuroenhancement*
- Population genetic and phenotype profiling (see Chapter 7)
- Robotics
- Surveillance (see Chapter 3)
- Synthetic biology*
- Terrorism technologies*
- Tissue replacement by organ transplantation (see Chapter 7)
- Tissue regeneration by stem cells (see Chapter 10)
- Transportation (see Chapter 5)
- Warfare technologies

Topics such as alternative energy sources, genetic modification of plants, enhancement of human performance, human cloning, and tissue regeneration are already in the public forum for debate and have been discussed in previous chapters, as indicated in Table 12.1. Parts of this chapter are works in progress, and the reader is encouraged to contribute by way of criticism and new ideas.

The concern that the technologies spotlighted in this chapter may have hidden risks is based on the recognition that it is impossible to predict the behavior of complex systems beyond a certain level of complexity. An example is that a 1-bit error in a digital system can have consequences that no one can predict with certainty most of the time. Whereas the digital example may entail an unpredictable software error of some consequence, the technologies to be discussed in this chapter have the potential for disastrous consequences that cannot be predicted (Wulf 2004). For example, in a simple engineering modification of part of the water system of the Everglades, which is a complex system, we cannot predict with certainty that the intended outcome of improved waterways will not be a disaster for plants and animals. As shown in Chapter 6, the modification of plants can lead to unintended consequences such as the Southern corn leaf blight, which was a consequence of inserting a single gene out of 730,000 using conventional breeding techniques. Another example is the modification of life forms such as oil-eating bacteria to enable remediation of oil spills. Another complex system is the human brain. What might be the long-term consequences of deep brain stimulation being used to treat Parkinson's disease? And what will be the consequences of medicating children and young adults with drugs to modulate mood, performance, and/or behavior?

This chapter opens a conversation on our duty to seek methods for evaluating consequences when we have no data. Risk analysis and cost-benefit analysis provide the tools for structuring inquiries into consequences of proposed technologies. However, there are no general procedures for detection of unknown dangers. While it is easy to point to the need for a proactive approach to determine with more certainty what these dangers might be, the task becomes more daunting when the consequences themselves are unknown, the probability of occurrence is unknown, and the consequence is likely to be recognized only when it occurs. This was the case when the space shuttle Columbia was lost at reentry in February of 2003, a consequence of damage from a small piece of foam broken off at launch. No one predicted the occurrence, no one predicted the outcome, and no one had time to think through a way to save the astronauts. In retrospect, some did claim that once the piece of foam insulation had dislodged, the force of the impact with the wing could have been evaluated after the event, and that the fact that damage would have occurred was predicable; however, that calculation and the associated experiment were not done until after the astronauts had died. Thus, the introduction of complex new technologies demands a more thorough analysis of the extent of the unknowns associated with them, along with weighing the risks vs. their benefits. Their introduction also requires the inclusion of the public in decision making, and consideration of the possible needs for government regulation after articulation of the risks and benefits by the professionals.

At the outset, we should caution that there is a risk also of overplanning or overreacting to a possible consequence. Recall the costs of protecting information systems when the year 2000 approached. Clearly, some of the disasters predicted for Y2K were prevented because of precautionary measures, but not all of the expensive measures were necessary. Another recent example is the debate on limiting the distribution of genetically modified plants because of feared risks discussed in Chapter 6, "Ethics of Genetically Modified Organisms." The *precautionary principle* assigns a burden of proof of safety to those who want to introduce a new technology, particularly in cases where there is some reason to suspect the risks might be greater than the benefits. However, the application of this principle without balancing the benefits, and considering safeguards, and the involvement of the public who might benefit, could stifle the introduction of important new technologies.

More so in this chapter than in previous chapters, the ethical dilemmas that arise as a consequence of new technology development can only be imagined, usually cannot be clearly articulated, and cannot be readily evaluated in the present. The ethical problems arise when rapid deployment of a new technology (e.g., high-speed transportation, nanotube-based new materials) does not wait for adequate studies of the risks. Indeed, utilitarianism, duty, non-malfeasance, and the autonomy or rights of individuals are involved, but not always simply in competition. The challenge is not to predict the unpredictable but to engage in more of an interdisciplinary investigation of the consequences of new technologies than has been practiced in the past. The professional codes of engineers and scientists point to the moral duty to avoid harm. This means that attention to possible consequences of emerging technologies must be a part of the life-long responsibilities of the innovators and technical managers. There is also a responsibility to educate the public.

Emerging technologies bring scientists and engineers to a new dimension of interdisciplinary design efforts and risk analyses. Our Four A's strategy is challenged by unprecedented situations where entirely new macro-engineering programs are being proposed, and acquisition of facts is limited because knowledge is limited. The Engineering Ethics Framework of Pinkus et al. (1997) bears repeating here. As explained in Table 1.2 of Chapter 1, this framework consists of three concepts applicable to investigating and planning new projects.

- The concept of competence extends from the individual engineer to the team of engineers where missing competency must be filled by the organization.

- The second concept is that of responsibility for the analysis of safety and failure possibilities. This responsibility extends to the organization as well.

- The third concept they call Cicero's Creed II, which is the priority of protecting public safety.

One might label the development of new technologies as "social experimentation." As espoused by Martin and Schinzinger (2005), "Engineering projects are social experiments, that generate both possibilities and risks, and engineers share responsibility for creating benefits, preventing harm and pointing out dangers." The introduction of a new technology requires an attitude of experimentation and exploration with safeguards to minimize the estimated risks. The public needs to be engaged in the early stages of introduction of new technologies. For example, introduction of a new high-speed aircraft would naturally be preceded by a multiple systems safety analysis. The first stage of engaging society with this form of transportation should be considered an experiment designed to refine the risk analyses and provide a priority for those countermeasures needed to prevent harm.

The National Academy of Engineering anticipated the need for prospective discussions and workshops in the area of emerging technologies commencing in 1999 (National Academy of Engineering 2004). It is important that scientists and engineers perform their own evaluation of safety for persons and the environment so that regulatory bodies such as the FDA and the EPA can make well-informed policies and regulations. A good example is the introduction of magnetic resonance imaging (MRI) in the early 1980s. Engineers anticipated a public concern for the possible health effects of MRI and performed their own analyses to define the safe limits for MRI. These investigations led to FDA guidelines and were instrumental in bringing the technology forward for public benefit.

Technological innovations require the early identification of potentially negative consequences. The new tools of this chapter involve methods of risk analysis, cost-benefit analysis, and the task of developing risk models. A major lesson of this chapter is that the scientist and engineer have duties to evaluate the risks or unintended consequences as they approach a new project. Using the guidance of the Four A's and applying the tools of risk assessment, the professional can embark on a responsible investigation of possible consequences. However, the responsibility does not stop there, as the principal stakeholders, the public, must become involved in discussions about risks and be willing to accept them. Involvement of the public starts by engaging the public in discussions of the benefits and risks of the new technologies. To do so effectively requires a proactive educational program so that the public is informed before factions become polarized based on poor information and fears of consequences. The current public discussions on genetic modification of organisms is an example of a complex technology that requires continuing public evaluation and surveillance of the benefits and problems resulting from their use.

RISK ANALYSIS AND COST-BENEFIT ANALYSIS

There are two methodologies appropriate to engineering new systems or modifying already developed macro-engineering systems where design and alternative pathway selection must be done in the presence of uncertainty: reliability and probabilistic risk assessment (RPRA) and cost-benefit analysis (CBA). The essence of probabilistic risk assessment and cost-benefit analyses is developed below. The cost of minimizing the risks must be balanced against the benefit, a subject that will also be discussed as it pertains to emerging technologies.

Before embarking on the ethics associated with emerging technologies, risk analysis is introduced as a tool for evaluating the consequences of these technologies. One can consider the introduction of this tool as an expansion of the design strategy wherein risk analysis becomes part of the first step of fact finding as well as part of evaluation of the assessment of alternatives. A brief description of risk analysis is given below. The reader interested in more extensive discussions can consult the recommended texts in the Additional Reading list at the end of this chapter.

The method of probabilistic risk assessment (also known as probabilistic safety assessment) can be used to estimate the magnitude of harm from current and proposed technologies to which the public is exposed (e.g., nuclear reactors, new vaccines, new hydroelectric dams, etc.). Probabilistic risk assessment asks the following questions (Kastenberg 2002): What can go wrong? How likely is it to happen? What are the consequences? An expansion on these questions will help explain the risk analysis process. To assess what can go wrong, examples of questions to ask are: What are the possible specific flaws or failures of a technology? What is the frequency of the flaw, failure, or event? What is the probability that a consequence such as injury or death would result from a particular flaw, failure, or event? Though the data available for risk analysis are limited, and, in fact, often the event that identifies the risk was not anticipated, nevertheless, risk analysis does allow one to compute relative risks given available data and best guess estimates of the probability of an event. Thus, an analysis of the risks of a new form of transportation can be made relative to the known risks of current modes of transportation. For a given category of incident (e.g., high-speed magnetic levitation train derailment), one determines the frequency (F) of the event, and then determines the probability (P) that the derailment would cause 10 deaths, 100 deaths, 200 deaths, and so on. The product of the estimated frequency that the derailment would occur, times the probability of the different consequences, gives the risk estimate.

Another example might be the risk analysis for wide deployment of hydrogen fuel cells to run automobiles. The deployment of hydrogen fuel cells requires the distribution of hydrogen gas at very high pressures. One risk estimate is the loss of life from a break in the high-pressure gas lines used for refueling. First, an estimate of the frequency (F) of such an event is made in terms of breaks per year or per 10 years. Then an estimate of the probability (P) of an injury from a single break is made. The risk estimate is the product of F and P. The general formula for risk assessment is:

$$R = \Sigma F_i \times P_i \qquad\qquad (12.1)$$

where R is the risk, F is the frequency of the event, and P is the probability of a particular consequence.

Of course, data may not be available to make these predictions, yet some analysis is required in order to arrive at the evaluation of the safety of a new enterprise. The exercise of risk assessment requires fact finding, imaginative foresight, and brainstorming, not on an individual level but through working groups consisting of experienced engineers, legal minds, and the public.

After the risk analysis is made, the analysis proceeds to a question: Are these risks acceptable? The objective is to determine if the risk is larger, smaller, or comparable to the background of commonly occurring risks, such as the incidence of cancer in the population before being exposed to a new technology. Suppose that the risk analysis is larger than the commonly occurring risks. The utilitarian point of view comes in when we try to impose on the public the value of a technology that may have undesirable consequences. Here is where the ethical aspects of risk vs. benefit need to be discussed by the informed society.

If the risks are more or less acceptable considering the benefits, the next questions are whether these risks can be reduced, and at what cost. Or is there an alternative technology with less risk? Here again, the balance of risk vs. benefit becomes important because in many technologies, the cost of improving the safety of a product will diminish its relative benefit.

One of the most important examples of risk analysis was that applied to estimating the safety of nuclear reactors sponsored by the Nuclear Regulatory Commission (1975). The goal was to estimate the likelihood of all possible accidents at a nuclear power plant and the resulting probability that the public might be harmed by reactor accidents. The method is known as fault tree analysis, a subset of probabilistic risk analysis. In short, the strategy is to follow an "event tree," starting with the probability of an initiating event (e.g., faulty pressure alarm). At the first branch, the possible events that might occur after the first failure are evaluated and the probability of each imagined event or option is calculated and recorded. Then the next branch in each of those options is noted, and so forth. If at each branch point there are two branches and the total tree has three branch points, there will be eight paths. In this case, the resultant probabilities for the eight paths are summed to give the overall risk of some consequence. This is how the risks of deaths caused by a nuclear reactor failure are calculated. Consequences are determined by previous experience, but frequently there are no historic data from which one can estimate the probability of a consequence, or even imagine what the consequence might be. The problem is not easy when dealing with a nuclear waste repository, wherein the frequency of occurrence of an ice age or an earthquake must be considered if the repositories are to be sound over 20,000 or more years.

Other examples are risks of harm from inoculations and mammograms recommended for everyone in certain populations, even though only a few are going to have problems. This is done in order to provide protection to the mass of people. One can figure a lifetime risk, and calculate over 70 years life, the risk vs. benefit. Insurance companies set a dollar figure based on the risk of dying at a particular age.

One should differentiate between the risk of some undesired outcome (defined as the probability that that outcome will occur) and the probability that an event that leads to the outcome will occur. The fraction of events leading to the outcome must be determined, as well as the probability of the event happening in the first place. Thus, if automobiles A and B have the same probability of an accident, but the consequence of death from an accident in A is 0.5 that of B, then the risk of death in A is one-half that in B. A convenient dimension that allows risk values to be compared from one situation to another is the risk per person per year. This allows a quantitative comparison of the risks of death from various modes of transportation or comparisons of the risks of death from various cancers to death from other medical problems. Below is an example of the application of risk analysis to a specific but hypothetical problem.

Case 12.1 Electromagnetic Energy Transfer from Space to Earth

Assume that a proposal is made to collect solar energy from gigantic solar panels in geosynchronous orbits at a suitable altitude over the eastern United States. The solar energy will be transferred to power stations on Earth via electromagnetic power pulses. How should the engineers go about a risk assessment? Recall the basic elements of risk assessment: What can go wrong? How likely is it that event will happen? What is the consequence of that event? First, one develops a list of unwanted but possible events:

1. Human exposure to the electromagnetic beam.
2. Animal and plant exposure.
3. Exposure of the local environment.
4. Orbit loss and beam misalignment.
5. Effect of a sudden loss of this power source on the regional power grid.
6. Component failure and risks of maintenance.
7. Cost of the enterprise is too high relative to the expected benefits.

Second, one needs an estimate of the frequency of each of the listed events. For example, by suitable fencing, warning, and detection systems, human exposure to the beam from space can be made a negligible risk. However, a flight of birds or an aircraft navigational problem will present other possible risks. If previous records of flight problems due to storms, bird migrations, or other factors show a reasonable probability of such an occurrence, this risk might be too high. These would be reasons to move the location of the beam. Other possible events, taken one at a time, will yield a frequency value with the exception of the cost (risk number 7). Having obtained these numbers, one can go on to the third step, that is, to evaluate the probability of a consequence of each listed risk should that risk occur.

For example, if the consequence of risk 5 (loss of this power source on the regional grid) is the loss of 10 lives (i.e., freezing or respirator power loss), and we have determined that the frequency of event 5 is once in 10 years, then the loss of life risk from power outage is10 lives per 10 years, or 1 life per year. However, if the population affected is 1,000,000 individuals, then the risk is 1 in a million per year.

If event 4 occurs (orbit loss and beam misalignment), the consequence could be the loss of life, depending on the probable beam misalignment that might occur before power transmission is shut down. Suppose that the probability of a serious misalignment is once in 10 years. Assume that the radius of the killing region is 10 meters and the region over which the beam might be misaligned is a radius of 100-km. The beam area is only 0.000001% (1 in 100 million) of the 100-km radius earth area. If 1 million people live in this area, then a simple calculation, assuming uniform density of inhabitants, leads to 0.001 death per 1 million per one year, or one death per thousand years per million people.

The combined risk of death from events 4 and 5 would be 1.001 per million per year. In symbols, this model of risk analysis is

$$R = \sum F_i \times Pi \qquad (12.2)$$

where P is the probability of the consequence (e.g., death) and F is the frequency of the event, as we have calculated above.

We can compare this risk to the risk of power outage using conventional energy sources to determine if this risk is too great. Alternatively, we can compare a risk calculated as shown above against the other risks individuals have in a population. The comparison must be normalized in time and numbers of individuals. Thus, the risk of cancer is a number related to the chance a person would develop cancer in one year. Recall that the risk of the outcome of death is that risk of having cancer multiplied by the probability that having a cancer will result in death, which is a much smaller number than the risk of having cancer. If the risk of the new technology is determined to be 0.1% (1/1000), or less than the combined risks for death from other causes (e.g., automobile, smoking, air travel, cancer, heart disease, etc.), then one can consider the new technology to present an acceptable risk. This arbitrary cut-off was used in establishing the safety of fission reactor installations (see Chapter 5, "Environmental Ethics").

Case 12.2 Loss of International Space Station Rescue Options

This case describes a problem for risk assessment faced by NASA in considering options for rescue from the International Space Station (ISS) in the event of a serious illness, life-threatening accident, or malfunction requiring evacuation of the station. Early on in the project, NASA had two options for rescue: a Soyuz vehicle launched from Russia with a capacity to rescue three people or the U.S. space shuttle with the capacity to bring all of the astronauts back to earth. However, events evolved that caused a reevaluation of these rescue capabilities:

- The Russians announced that the Soyuz program might be discontinued after 2006.

- NASA determined that the space shuttle was too expensive and too complicated to launch on short notice.

- NASA then designed a new vehicle called the X-38 to serve as the "lifeboat" for ISS. However, NASA substituted plans for the X-38 with plans to build a vehicle called the orbiting space ship (OSS) at a cost 10 times that of the X-38. The time delay to create the OSS would not compromise the safety aboard the ISS since Soyuz would be available through 2006, and the space shuttle would be available until completion of the OSS in 2010. But when the Columbia catastrophe jeopardized the entire space shuttle program in 2003, the ISS astronauts were left with only Soyuz rescue capabilities.

But suppose that the shuttle program is shut down and Soyuz is no longer available after 2006, is it ethical to put the astronauts at risk of death by removing this capability? This decision should have been based on an analysis of the relative risks. The engineer or technical manager faced with the problem of the ethics of removing the ambulance or "lifeboat" capability might proceed by an analysis that follows the Four A's for each of the major categories of accidents (e.g., personnel medical problems or space habitat physical failures). The example below is for medical emergencies.

Acquire Facts: The list of ambiguities and needed facts in this case is very long. Possible medical problems include internal medicine problems (e.g., infection, kidney stones, poisoning, seizures), accidents (e.g., burns, broken bones, trauma), surgical emergencies (e.g., lacerations requiring surgical repair, ruptured appendix, or other acute abdominal problem), and mental problems (e.g., severe depression, acute psychosis).

The additional facts are in the category of on-board capabilities to stabilize or treat acute illnesses and trauma. In many life-threatening situations such as burns, the capabilities depend on supplies, instruments, training of the astronauts, and access to medical advice. The critical information has to do with the length of time the sick or injured astronaut can be kept alive before a rescue mission can be launched. The other essential information is the time required to launch a rescue from Earth.

These facts are illustrative of those necessary for a **risk analysis**, which should be done in order to adequately inform the astronauts of the risks prior to their departure. Indeed, this information is essential before one can consider the consequences of removing the ambulance capability and the countermeasures to these consequences.

Below is an example of how the medical flight surgeons might proceed to determine the risks involved.

- Determine the frequency of known medical and surgical emergencies in space travel by dividing the number of cases by the astronaut flight days and separately by the number of astronaut missions. Of course, the second number for each type of emergency will be a much higher frequency and could be considered more realistic because it would include individual astronaut variability. The result of this calculation would be the expectation of, say, acute kidney problems per astronaut mission or gastrointestinal sickness per astronaut mission.

- Next, determine the likely consequence of death for each situation under the condition of no rescue for 4 days, 8 days, and 12 days as three possible scenarios. Death from dehydration from gastrointestinal diarrhea could result if there are insufficient intravenous fluids to sustain the patient for four days or more. In addition to insufficient intravenous fluids, narcotics to sustain an individual in acute kidney pain could be inadequate, or antibiotics to treat an individual for more than a few days might not be available.

- The risk analysis can proceed using the model:

Risk = sum [frequency of diarrhea × probability of death from dehydration] + [frequency of kidney infection × death from infection] + [frequency of burn × expectation of death from no fluids for n days] + . . .

This is the same formula (Eq. 12.2) introduced for the probabilistic risk assessment above. However, this estimation of risk is meaningless without considering it in the context of other risks. Other risks are calculated by the frequency of equipment failures of the space station, and whether the consequence of each failure might result in astronaut death. It is within this context that the absence of the ambulance or recovery vehicle can be explained to the astronauts as well as to policy makers and the public. If the relative risk is high, then methods for reducing the risks of having no ambulance need to be implemented.

Alternatives: It may become clear that the lack of ambulance capabilities needs to be countered by one or more of the alternatives listed here:

- Place on board equipment to sustain individuals for sufficient periods to allow rescue.
- Negotiate the continued maintenance of Soyuz or change the international prohibition regarding purchase of Soyuz from Russia.
- Engage the astronauts in the decision about whether or not to have an ambulance.
- Enable a system for earth-based rescue that can recover an astronaut in an acceptable time (i.e., reactivate the X-38 program).

Assessment: In discussing the alternatives in parallel, we learn that the excess weight of intravenous fluids and other medical supplies needed to reasonably sustain the astronauts will be 44 pounds. This added weight is not acceptable to the mission planners. This necessitates returning to step 1, which is to acquire more facts and alternatives regarding reconfiguration of the payload. Here again, the astronauts should be involved in the process.

Action: Implement alternatives in parallel. For this situation, pursuit of the alternatives should be explored in parallel. During this process, involvement of the astronaut stakeholders can be critical.

This case demonstrates how risk analysis enters into the design strategy and can give systematic guidance when working through the considerations of an ethical dilemma. We now turn to the emerging technologies themselves that have the potential for disastrous consequences that cannot be predicted and that require ongoing risk analysis now and in the future.

Cost-Benefit Analysis (CBA)

The development of CBA came as a result of the U.S. congressional act that required the U.S. Army Corps of Engineers to carry out projects for the improvement of the waterway systems in the United States. The act mandated that the total benefits of a project must exceed the costs of that project. Thus, the Corps of Engineers had to create systematic methods for measuring such benefits and costs. They did not rely on the economics professionals, and it was not until about 20 years later that economists began to develop a consistent set of methods for measuring benefits and costs to decide whether a project is worthwhile. The same type of analysis is applicable to evaluating the costs of reducing risks, and this type of analysis is known as risk cost-benefit analysis (RCBA). The principal challenge to making a CBA is the determination of the metric for cost and for benefit. If money is used for the engineering cost, then the value of the benefit needs to be estimated in money, and this estimate is not always easy to make, particularly when the benefit is avoiding the risk of death.

A simple example is the estimate of the costs vs. benefits of solar energy discussed below. Suppose that the cost of installation of a solar panel on a two-person home is $6 per watt. Assuming that a 1-kilowatt solar panel is installed, the initial cost is $6,000. The benefit is seemingly inexpensive electricity. But a cost-benefit analysis would show just how expensive this enterprise will be. Assume that the lifetime of the installation is 20 years, and the kilowatt is available for use or charging batteries for four hours each day depending on sunlight. The cost would be 20 cents per kilowatt-hour if the device survived 20 years. Assuming that unused kilowatt-hours are not sold back to the power companies, this is more than twice the cost of electricity in California, and three or more times the cost elsewhere from conventional sources in 2006. This is not the complete analysis on cost because the economist would add to this analysis the fact those dollars invested in a benefit will be unavailable for some other benefit, such as value added through investment of the $6,000. In 20 years, this value could be about $16,000. So overall in California, the cost could be about $19,000 for a benefit worth about $3,000.

The engineer faced with decisions regarding reducing risks needs some measure of the value of saving lives in many macro-engineering initiatives. Society frowns on putting dollar values on lives, as was done by the Ford Motor Co. in the example of not taking safety measures to revise the gas tank installation in the Ford Pinto (Case 4.1). However, one can certainly estimate the decrease in the probability of loss of life for extensive safety nets vs. the litigation costs for deaths resulting from accidental falls in a construction project where these problems have a high likelihood of occurring. The cost per life saved is simply calculated, given the estimates of a probability of fatal falls.

A quantitative measure for comparing the costs and benefits of two or more alternatives or calculating the risks vs. benefits is known as cost-effectiveness. It is a parameter that gives the cost in terms of a benefit, such as a decrease in hospital days, or a decrease in the number of starving children, or an increase in longevity. Suppose that the cost of a cadaver kidney transplant is $5,000 and the cost of a noncadaveric transplant is $40,000. The life-years increase for receiving a cadaver kidney is 10 and for a noncadaver kidney is 20. The cost-effectiveness is 35,000/10, or the cost effectiveness of the noncadaveric kidney transplant is $3,500/life-year to achieve the benefit of 20 years of longevity. The parameter is simply calculated as the difference in cost between two procedures divided by the difference in life years associated with patients receiving the two treatments. The operative equation is:

$$CE = (C1 - C2) / (LY1 - LY2) \qquad\qquad C1 > C2 \qquad\qquad (12.3)$$

where CE is the cost-effectiveness, $C1$ and $C2$ are the costs of treatment 1 and 2, respectively, and $LY1$ and $LY2$ are the life-year benefits for treatments 1 and 2, respectively. If the result were a negative number, one would discard treatment 1.

NANOTECHNOLOGY

Nanotechnology is one of the emerging technologies that may produce risks for our future that outweigh its benefits. It presents a range of radically new technologies that span from embedded intelligence in people and the environment to new methods of sun energy conversion to electricity and new materials configured in structures are 20 times stronger than steel. Federal funding reached $0.8 million per year in 2000 to support what is seen by military and federal research agencies as a promising investment to keep the United States competitive in the world market and able to respond to possible high-technology terrorism or warfare.

Ethical perspectives differ, depending on the way nanotechnology is defined. For example, a narrow definition such as "Nanotechnology is associated with molecular manufacturing such as nanobots that are self replicating and can assemble into complex systems, which perform useful functions" conveys a public perception of dangerous consequences. We discuss this aspect under "Synthetic Biology," below. In contrast, a definition such as "Nanotechnology is the science and engineering that focuses on manufacturing useful objects with scales less than 1 micrometer, as well as an inquiry into the new science and engineering embodied in complex systems with components less than 1 micrometer" might be more acceptable. However, this definition is so broad that it encompasses much of research before the label "nano" came into use and misses the focus the National Nanotechnology Initiative (NNI) is seeking. The NNI definition is that nanotechnology is the convergence of physics, chemistry, and biology that bridges quantum and classical science at a scale where fundamental properties emerge.

Ethical problems will emerge from the engineering applications of nano-sized objects for medical applications, nano-parts to construct self-replicating systems, nano-robots, and applications of materials that are significantly different from contemporary materials. However, as with new technologies of the past, many of the risks will not be predicted. Nanotechnology involves many more disciplines among scientists and engineers than has been the case in the past. Thus, the development and implementation of new ideas will need to stress communications between experts who in the past have not been accustomed to this interaction. The need for communications on possible risks is higher in the twenty-first century than at any time in history. The need is not only for public safety, but also for new definitions for intellectual property, ethical responsibility, and ethical journalism to portray accurately the risks and benefits.

Possibly the most important issue for nanotechnology is whether there will be a just availability of these technologies for all nations. Will nanotechnologies widen or narrow the gap between rich and poor nations? On the one hand, countries with the personnel and resources to discover, patent, and use nanotechnologies would have advantages that could widen the gap between nations. On the other hand, the potentially inexpensive nature of nanotechnology could lead to a narrowing of the gap, as is the case in communication and information technologies (e.g., India and China).

Another important issue is the conflict between maintaining the patent protection of the investment by the innovators and the desire to ursurp the patent in the event of a public health crisis that requires more widespread and cheaper availability. Questions of the priority of public health over patent protections and other restrictions have arisen with respect to pharmaceuticals for infectious diseases, including drugs for the treatment of AIDs or to increase the production of drugs needed during a worldwide pandemic.

Human flourishing, culture, and human enhancement can be a result of nanotechnologies, but without care in anticipating the consequences, the costs could outweigh the benefits. There are two examples of the new technologies that challenge autonomy. Tiny radio-based imaging systems dispersed throughout public places for safety surveillance can challenge privacy. Dispersion of engineered nano-particles through the human blood stream to detect disease can have unanticipated pathological consequences such as aggregation of the particles and clotting. There is a need for ongoing dialogue between those in science/engineering and those reflecting on the ethical and social issues. An important and well-documented resource for understanding the evolution and future likely directions of nanotechnology is the book *Our Molecular Future* by Douglas Mulhall (Mulhall 2002).

SYNTHETIC BIOLOGY

Synthetic biology is a new field, the goal of which is the synthesis of functioning life forms using parts of known biological and biochemical components, just as one would synthesize an electronic system from a combination of electronic components. Synthetic biology is the science and engineering focused on designing systems to address a host of problems that cannot presently be solved using naturally occurring nonbiological material. An example is a non-immunogenic particle that reverse transports cholesterol from diseased arteries. The components for synthesizing agents and devices are biological, and the life forms that are manufactured provide a specific function. The synthesized objects might have properties of self-replication, the potential for dissemination in a biologic system or nature in general, and even capabilities to stimulate mutations in living beings (plants, animals, humans). An example is the synthesis of a virus-like particle that would allow gene transfection or gene therapy without the serious side effect of engendering an immune reaction (Schaffer and Lauffenburger 2000). This technology could also be used for bioterrorism, an example of which would be the synthesis of the poliovirus starting with the chemical molecule of a DNA oligonucleotide known to chemically synthesize the RNA of the poliovirus (Cello et al. 2002).

The field is not limited to medical science applications. Since synthetic biology represents a convergence of nano-fabrication technologies and biomimetics, the breadth of applications includes development of a bioreactor that can convert water, carbon dioxide, and sunlight to combustible fuels. In principle, production of hydrogen from water can be accomplished by an artificial system using the same enzyme system that enables algae to produce hydrogen when deprived of oxygen and sulfur. The enzyme hydrogenase found in the vegetative cells of algae can produce hydrogen in light or dark (Mahro and Grimme 1983). Methane and other hydrocarbon fuels can be produced by different biological organisms, including trees, algae, and bacteria (Calvin and Taylor 1989).

The ethical concerns surrounding synthetic biology are mainly those of the potential hazards inherent in the synthesized life forms that are self-replicating. Synthesized materials might induce genetic changes that may present intergenerational risks or that may spread to and be toxic for unintended species. An example will help put the potential risks in perspective.

Case 12.3 Oil-Eating Bacteria

A major problem has been the contamination of costal waters from crude oil after acciden-tal spills involving oil tankers and offshore oil drilling platforms. As a result, there has been a concerted effort to create an organism that would convert crude oil to a soluble substance through a series of enzyme reactions (Erickson 1990). The general topic is known as biodegradation, the process by which compounds are naturally broken down by microbes. For example, an oil-ingesting bacterium may use the oil droplets as energy for replication during an enzymatic chemical reaction. In effect, the oil is burned by the bacterium. The enzyme responsible for the conversion is probably in the same family as the methane mono-oxygenase for methane-eating bacteria. The by-products are energy, carbon dioxide, and alcohol. Bacteria (*Pseudomonas*) were found effective in bioremediation on some of the beaches in Alaska after the catastrophic oil spill from the Exxon Valdez in 1989.

Suppose that the work underway at M.I.T. to determine the structure of the enzyme leads to a recombinant DNA type of gene production. This would allow the production of massive quantities of viral particles containing the gene for this enzyme that could be introduced to the ocean surface waters after an oil spill. The fact that the virus could replicate and protect the entire biosphere might be seen as an enhancement of its effectiveness. But what would be the consequences if this virus *infects* the biosphere after the oil spill is biodegraded? What would the virus do to birds that become infected through the waters they drink and the biological organisms they eat? What would be the consequences when the virus enters drinking water? These are the types of questions that must be answered before any attempt at bioremediation is made using the process suggested in this case. The ethics here are that the scientists and public health professionals have a duty to first answer the questions before the experiment cre-ates a disaster of its own. The above scenario is speculative, but in reality, the introduction of oil-eating *Pseudomonas* bacteria into an oil-contaminated area of Australia did result in the dis-integration of part of a nearby highway after the oil-eating bacteria attacked petroleum products in the roadbed. Unfortunately, history verifies the fact that the consequences of a technology suggested in this scenario are rarely thoroughly articulated, much less investigated. Consider the parallel situation with GMOs, in which there were inadequate investigations regarding the spread of genetically modified plants into the natural environment.

What are some compelling facts that differentiate the enterprise of synthetic biology and associated nanomedicine from contemporary recombinant DNA research, virology, and bacte-riology? A related question asks whether current regulations relative to human experimentation and dissemination of biologicals are adequate to safeguard against unanticipated consequences from synthetic biology projects. The problem is that the biological aspects of the new tech-nologies are more complex than in the past, and there is less known about possible conse-quences. Because the character of the research and the level of risks are now very much different from those of the past, there is a need to generate either rules for self-regulation or nationally promulgated guidelines for research in synthetic biology. However, premature regu-lation might not only hinder discovery and violate the autonomy of the research community but also limit our ability to create countermeasures to stop those who abuse the system.

There are a series of actions that can be undertaken to preemptively avert some of the obvious negative consequences of creating self-replicating pathogens or vectors of gene transfer. First and foremost, the scientists and engineers should come together to discuss and develop appropriate risk models. Next, a process should be agreed upon to control and inform colleagues and the FDA or other government groups with oversight responsibilities so that dangerous practices can be minimized. The information includes, for example, DNA sequences that might lead to pathogens, chemical ingredients that can lead to dangerous toxins, and biomimetic constructs that have led to pathological agents (e.g., artificial poliovirus). The scientific community can be the originators of these guidelines for safe practices. These guidelines are not motivated by regulatory goals but are intended to keep the well-meaning scientist and bioengineer from creating life forms that may be harmful to the public as well as to the investigators. In the case of genetically modified bacteria, the regulatory bodies have already taken the first action. The Environmental Protection Agency in the early 1980s assumed regulatory powers over the genetic modification of bacteria, thus overriding the authority of the NIH Recombinant DNA Advisory Committee (RAC) as a challenge to the adequacy of the RAC in matters of commercial and ecological implications of environmental releases. The authority for this is the claim that these bacteria are new chemical substances and subject to the Toxic Substances Control Act (Budiansky 1983).

ETHICS OF EMERGING THREATS OF TERRORISM

The ethical duties of a government to protect the public from the consequences of natural or man-made disasters are included here. This is because the evaluation of consequences and assessment of the duties to take preemptive action depends on the development of risk models as part of risk analysis. Several current examples of such threats to all societies include the possibility of a natural influenza pandemic, bioterrorist spread of disease vectors, or an accidental or intentional radiation release. What preparations should be undertaken ahead of time? Because of the potential for rapid global transmission of life-threatening infectious diseases that are "just an airplane ride away," there is validity to the predictions that such threats would result in major repercussions for all aspects of society. Even the limited SARS epidemic severely strained hospital, quarantine, medical, and nursing personnel resources. It has been estimated that a deliberately altered infectious agent or a pandemic of natural origins could infect one-third of the U.S. population. If (some would say when) one of these events occurs with some magnitude at some time in the future, there would likely be substantial illness, death, social disruption, and widespread panic. The last natural pandemic occurred in 1918 and resulted in the deaths of at least 20 million people globally (Crosby 2003). Estimates indicate that a pandemic in the United States could result in an estimated 18 to 42 million outpatient visits, over 700,000 hospitalizations, and over 200,000 deaths (Meltzer et al. 1999). This is a conservative death rate that could be two times higher. Similar figures are forecast for a bioterrorism event, and preparations for either a natural or an intentional event have similar requirements.

Ethical decisions in such situations would involve how to best manage scarce vaccines, drugs, laboratory, and treatment facilities, who should receive treatment (i.e., health care workers first, random draw later?), and what precautions might be necessary. The details surrounding influenza pandemic preparedness are outlined by Gensheimer et al. (2003).

It is difficult to predict and have on hand sufficient vaccine supplies for a pandemic. Nor can it be predicted which influenza strain might be involved should there be a natural event, making difficult the preparation of an efficacious vaccine ahead of time. Further, the execution of protocols specifying those who would receive treatment would be difficult to carry out in view of the expected public panic the situation would cause. The disruption of normal public health services as well as the high rates of mortality and morbidity would have wide repercussions as to who would live and who would die. Preparing for such disasters when resources are limited is probably the greatest challenge to justice and duty ethics. The principles challenged range from autonomy and utilitarianism to justice. For example, the preemptive approaches require consideration of mass immunizations and quarantines without individuals having choice, and the anticipation of a pandemic requires the diversion of public funds and priorities to prepare personnel and stockpile needed supplies. The responsibility of the government and professional societies to make preparations based on sound risk assessments is an ethical duty to which the U.S. government and the World Health Organization are responding.

Preparedness for a terrorism-based release of radioactivity into the environment, through either a nuclear explosion or detonation of a conventional device intended to spread radioactive materials is the duty of government as well as the duty of health care professionals. A risk analysis will show that the major risk for a dirty bomb is the mass hysteria that will be created such that the fear of having been exposed will lead to a saturation of care facilities, thus compromising those in need of care or decontamination. The likelihood that thousands of individuals will be put at risk from a dirty bomb is low, particularly if the public is instructed about the mode of radioactive contamination and the methods for personal decontamination. This educational program is the duty of the professional societies as well as the public health organizations.

NEUROENHANCEMENTS AND NEUROENGINEERING

A specialized field of ethics has been developed to address the rapidly evolving technologies of interfacing between the brain and machines, enhancing performance and neural therapy. The field has been dubbed neuroethics (Wolpe 2003). Table 12.2 lists the major technological developments in neuroengineering of the last few years, and each of these technologies has some ethical aspects. Deep brain stimulation can control some behaviors (e.g., tremor) and help some symptoms (e.g., depression). The stimulating electric currents are believed to interfere with the pathological nerve firing, thus preventing a signal from being transmitted from one part of the brain to another, or, in the case of tremors, from the brain to the spinal cord. Deep brain stimulation is an example of a technology with ethical implications ranging from the duty of the medical profession to relieve suffering to the stigma of brain lobotomy of the mid-1900s for the treatment of behavioral disorders.

These topics are addressed next.

Table 12.2   Neuroengineering Technologies

Deep brain stimulation
Brain-computer interfacing
Regenerative neurology
Brain imaging
Psychopharmacology

DEEP BRAIN STIMULATION

The most successful therapeutic use of electrostimulation has been in cardiac pacemakers and cardiac defibrillators. Another evolving use is electrostimulation of the brain by inducing electrical currents, either through external devices or through internal electrodes implanted in the brain. In 1985, a French surgeon named Dr. Alim-Louis Benabid learned that he could relieve a certain kind of hand tremor by electrical stimulation of a part of the brain known as the thalamus. This led to the recognition that other tremors, including those of Parkinson's disease, might be treated neurosurgically by the implantation of electrodes to deliver an electrical stimulus to some areas of the brain.

Deep brain stimulation (DBS) involves placing electrode pairs into the central parts of the brain (globus pallidus, subthalamic nuclei) and the application of small electrical currents to treat certain neurological disorders. In addition, deep brain stimulation has been proposed for treating pain, obsessive-compulsive disorders, and severe depression. Even patients with the sequelae of cerebral strokes have symptoms that might yield to deep brain stimulation. Many of the brain disorders have no known etiology, but often the cause of the disorder can be related to car accidents, falls, gunshots, and other blows to the head. In fact, 1.5 million Americans suffer from some serious head injury each year, and there are currently 5.3 million Americans who are disabled as a result of brain injury. Many who suffer from various otherwise untreatable brain disorders are hoping deep brain stimulation may bring relief. However, this is not a medically proven or widely accepted procedure for disorders other than dystonia (i.e., uncontrolled abnormal movement disorder) and the tremors of Parkinson's disease.

The ethical questions arise when we have some new technical tools that have the potential to relieve pain, tremor, aberrant moods, or other symptomatology for which no other methods work and when there is no proof that this new approach will work in individual cases. The patient and patient's family are faced with no other choices except to live with intractable pain, to use medications that make them dysfunctional, or to take a chance with an unproven treatment. Additional concerns involve the victims of brain disease or trauma who, having been saved by heroic medical measures, now require special devices or special care to maintain a quality of life. The medical profession cannot abandon them to the consequences of having been rescued. Thus, the ethical responsibilities of the professional go beyond delivery of aid to the injured.

Another ethical concern has to do with the unknown consequences of long-term deep brain stimulation. The problem is that without study data, the consequences of this procedure cannot be known. Unfortunately, the data from animals are sparse, particularly as the life span of experimental animals is short, and they are inadequate models for diseases such as depression, obsessive-compulsive disease, and thalamic pain.

How should one approach the ethics of the application of this new technology? One can turn to a careful analysis of the risks of the procedures, the known and potential physiology of brain stimulation, and the inaccuracies and problems of the technology as it is practiced today. An analysis of the current practices shows some major flaws in electrode localization, in electrode materials, in patient selection, and in overall standardization of procedures. The Four A's strategy can guide one's approach to the problems. The advice stage is facilitated by consulting with the NIH Deep Brain Stimulation Consortium, a core group of researchers funded under the National Institute for Neurological Diseases and

Stroke and the National Institute on Aging. Annual meetings (the first was in 2002) are now being held with physicians, basic scientists, patient advocates, industry representatives, and government officials to discuss issues surrounding the use and ethics of deep brain stimulation.

In order to examine how aggressive action can be taken to improve the practice of deep brain stimulation, consider the questions of where to place the electrodes, in which patients, and how to determine accurately where the electrodes should be placed. The use of magnetic resonance imaging to guide electrodes made of material that will allow imaging in a magnetic field is a technological solution to proper placement. However, not all neurosurgeons and neurologists use this method. As of 2006, there were no methods of certifying the quality of training and proficiency for those performing the procedure. The acceptance of DBS for the treatment of depression and obsessive-compulsive disorders is not widespread but is growing. Despite the absence of proof that an individual patient will benefit, there is a great temptation to proceed with the new technology anyway. Those involved must guard against false hope, but when the only alternative is unrelenting intractable pain and partially effective medications that often cause other dysfunctions, patients often opt to grasp at any chance of improvement.

BRAIN-COMPUTER INTERFACING

The concept of using the naturally generated electrical signals from the brain to control either prosthetics or other external devices was demonstrated in a baboon in 1975 (Craggs 1975). In 2000, the scattered endeavors in the field became known as Brain Computer Interface (Wolpaw et al. 2000). Since then, implanted electrodes to translate neuronal activity to computer-driven devices with robotic arms have been used successfully in animals (Carmena et al. 2003). A pilot study in human subjects is underway to give non-speaking or "locked-in" patients with severe neurological impairment the ability to control a computer cursor using their thoughts (Cyberkinetics-Neurotechnology Systems, Inc. 2005).

The goal is to use neural implants of single or multiple electrodes to record brain activity related to intended movements from the sensorimotor pathway of paralyzed patients, and to use these signals to control external devices. Research from 2000–2003 concentrated on extracting the real-time intended trajectories of the hand by recording signals primarily, but not exclusively, from the motor cortex (Serruya et al. 2002, Taylor et al. 2002).

A demonstration in monkeys in 2004 showed that the brain signals arising from intention to move a cursor on a screen can be decoded to a computer-controlled movement. This higher cognitive level associated with "thinking" about the trajectory extends the possibilities of using signals from parts of the brain other than the motor cortex (Musallam et al. 2004).

The ethical aspects of brain interfacing to computers have to do with "slippery slope" arguments concerning the possibility that brain signals could be used to control devices for the purpose of harm. However, this misuse can be averted by avoiding the use of wireless technologies and controlling the access code between the subject and the device. Other issues are whether the distribution of this technology would be just and whether there might be widespread deployment before the risks of long-term implantation are known.

REGENERATIVE AND REHABILITATIVE NEUROLOGY

Five evolving technologies for repair of the injured brain or spinal cord are:

Implantation of embryonic stem cells or fetal cells in traumatized or diseased brain or spinal cord area: The treatment of Parkinson's disease by neurosurgical implantation of fetal cells has been attempted without success in a number of countries (e.g., Sweden, England, the United States). No systematic trials with embryonic stem cells have been reported in human subjects, but successful implantation of embryonic stem cells or tissues has been observed in animal studies (Tessler 1991). The ethical issues here are those discussed for the use of embryonic stem cells or fetal tissues in general. In short, even though there is animal evidence that neuronal tissue from embryos will help neurons regenerate in the spinal cord, the clinical use of embryonic stem cells or fetal tissues is currently banned because of not yet resolved moral issues concerning their source or safety (Chapter 10).

Gene therapy to enhance brain function: In principle, the insertion of genes using viral or synthetic biology vectors to enhance brain function has the potential to restore degenerated or diseased nervous tissue. It has already been shown that replacement of the gene that produces an enzyme that is needed to convert L-dopa to dopamine can help patients with Parkinson's disease (see Chapter 11, Figure 11.1). Gene therapy is likely be useful for replacing or enhancing a multitude of genes in the central nervous system (CNS) that have been damaged by injury or disease. The ethical issues revolve around the unknown and potentially hazardous consequences when viral or synthetic agents are introduced into the body, as well as the possibility that the technique might be used inappropriately for enhancement of natural abilities (e.g., improved cognition, better hand-eye coordination, or improved motor skills).

Stimulation of nascent progenitor cells or nascent stem cells to regenerate nervous tissue: Whereas prior to 2000 most scientists and the lay public were of the opinion that brain neurons could not regenerate, it is now known that at least some parts of the brain have nascent stem cells. Methods to stimulate these cells to repair diseased tissues are being sought, particularly as it is now shown that adult somatic cells can be reprogrammed to perform the functions of stem cells (Alberio et al. 2005).

Apposition of the patient's omentum to the injured brain or spinal cord: Since 1975, spinal cord injury in animals has been known to be reversed at least in part by the application of the omentum with its vascular supply to the injured region. The omentum is a sheet of vascularized fatty and fibrous tissue that extends from the surface of the stomach to cover the intestines. It is known to participate in protection of the gut from infections and foreign bodies. It has a rich supply of growth factors and progenitor stem cells, both of which may promote vascularization and regeneration of the diseased tissues (e.g., heart, brain, and spinal cord). Human experience is limited to small groups of subjects with stroke, Alzheimer's disease, and spinal cord trauma (Goldsmith 1990). A segment of the omental fat pad with its blood supply can be brought from the abdomen to the damaged area through a tunnel created surgically under the skin. The procedure is not without controversy, however, because some studies have not shown success (Duffilla et al. 2001). However, documented success in animals and a paraplegic patient has precipitated an interest in pursuing this procedure further (Goldsmith et al. 2005). The current ethical problem is whether to allow patients to undergo this approach before understanding the mechanism, or whether to ban the method until it is better understood, thus depriving these patients of a possible return of function. Because few neurologists believe the method is an acceptable practice, it is unlikely that a proper clinical trial will be conducted with federal support or even with nonfederal support at institutions that must follow federal rules for the protection of human subjects.

This dilemma puts the precautionary principle at odds with the rights of patients and physicians. Indeed, it raises the question of how to introduce a new and dangerous technology when the practicality and expense of conducting clinical trials involving more than a handful of patients are prohibitive.

Implantation of electronic devices to assume a lost function: The use of electronic devices for hearing (cochlear implants) has had outstanding success with few ethical problems. The implantation of retinal electrodes in domestic pigs has stimulated or emulated optical nerve signals (Schanze et al. 2005). The use of implanted electrodes to interface brain signals to a computer has been shown to be possible in a few subjects. But the use of electronic chips or computers to enhance memory or cognitive function is still on the distant horizon. These technologies will have ethical problems that include equitable distribution to those in need, privacy, and long-term medical and psychological consequences.

BRAIN IMAGING

New imaging techniques have broadened our understanding of the major psychiatric illnesses. Patterns of brain metabolism and blood flow that are seen in some psychiatric illnesses and in some violent criminals have been presented in court to defend the plea of insanity or to reduce the assumption of intent in violent crimes, respectively (Fenwick 1993). Both the insanity defense and the defense of automatism depend on the demonstration of disorders of the mind, and the image-based evidence is limited only to brain physiology. Brain imaging has been proposed as a lie detector approach in criminal investigations as well as to profile potentially violent individuals. The practice of inferring intent or lack of intent from this technology is not based on good science and should be approached with great caution (Mayberg 1996).

There are three methods of brain imaging that have been used to determine intent in criminal cases. They are infrared reflection imaging, positron or single photon emission tomography, and functional magnetic resonance imaging (see Figure 12.1):

- Using infrared light-emitting diodes and miniature photodetectors, it is possible to measure the blood flow or associated blood volume response of the head (surface skin and brain) during questioning. This is an objective method to measure a subject's blushing in response to an embarrassing question. Devices placed on the forehead during questioning of a terrorist or suspected criminal have been proposed as a way of remotely sensing whether the subject is lying. The exact origin of signals from the frontal lobes of the brain vs. more superficial tissues of the head outside the brain cannot be discerned by this approach, and this, in combination with the fact that different individuals respond to questions differently whether lying or not, makes this method inaccurate and potentially unjust.

- Positron emission tomography (PET) has been used in court to try to demonstrate an association between violent behavior and an abnormal brain study. However, although a small study of two prisoners and two patients with epilepsy showed abnormal glucose metabolism, the same pattern was also seen in 8% of 20 patients with no history of violent behavior (Budinger 1993). These data negate the appropriateness of using a PET scan to ascribe a medical or natural cause of violent behavior. The ethics of scientific integrity demands carefully controlled studies before drawing conclusions in situations like this.

- Functional magnetic resonance imaging shows the pattern of blood flow changes in the brain after some form of stimulus. Some patterns of brain activation are seen when a subject is telling the truth or is lying, as evidenced by a change in blood flow in specific regions of the brain (i.e., the amygdala). The danger in using this and other imaging modalities is that the patterns are not reliable, and some individuals can control their response when lying. Thus, the use of this methodology for these purposes is premature at the present time.

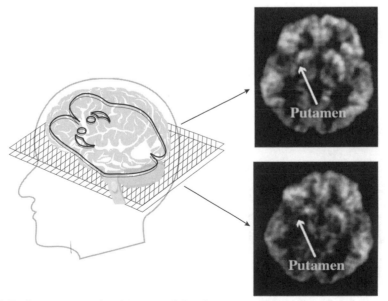

Figure 12.1 Positron tomographs give maps of the glucose metabolism of two transverse sections of the brain separated by 10 mm. The lack of uptake in the part of the brain known as the putamen could be related to the violent behavior of this individual, but this pattern is also found in subjects without a history of violent behavior.

PSYCHOPHARMACOLOGY

The emergence of drugs for behavior control is part of the ethical problems associated with neuroenhancements. These are discussed further in Chapter 11, "Ethics of Enhancement Technologies." Many of the popular drugs used to treat medical conditions, such as those listed in Table 12.3, are now also being used for other beneficial effects as well.

For example, some individuals who are timid or fearful of public speaking, giving scientific papers orally or otherwise engaging others in discussions, benefit from propranolol. On the surface, there seems to be no cause for ethical concerns if the drug is prescribed for those who have a need for symptomatic relief, in which cause the use of propranolol has a medical basis. This is not the case for Ritalin, usually prescribed for students with an attention deficit disorder but now reportedly used by students without this diagnosis because they believe it enhances alertness. Is its use analogous to the use of steroids to enhance athletic performance? To date, there are no studies documenting that Ritalin improves cognitive ability. The relatively new drug Provigil (modafinil) seems to increase alertness in fatigued individuals though it has been approved only for patients with narcolepsy.

ETHICS OF HUMAN SPACE EXPLORATION

Manned exploration of space has as its fundamental goal the discovery of knowledge. The benefits of discovery are generally considered greater than the financial costs and risks to the lives of the explorers. But in addition to minimizing risks to health and lives of explorers, the risks of the negative consequences of environment contamination and modification of the Moon and Mars need to be considered, as well as the possible risks of bringing to Earth pathogens for which we have no protection.

Case 12.4 Utility of the Hubble Space Telescope vs. Astronaut Safety

Hubble Space Telescope (HST) accomplishments include the highest resolution Earth-based images of Mars, the most convincing evidence of the existence of black holes, precise measurements of the expansion of the universe, high-resolution imaging of the births

Table 12.3   Drug Use for Enhancement for Nonmedical Reasons

Drug	Medical Indication	Enhancement
Prozac (fluoxetine)	Depression	Self-confidence
Ritalin (methylphenidate)	Attention deficit disorders/ADHD	Cognitive ability
Valium (diazepam)	Muscle spasm	Calming
Propranolol	Heart arrhythmia	Phobias, posttraumatic stress
Rogaine (minoxidil)	High blood pressure	Hair growth
Viagra (sidenafil)	Heart disease (angina)	Erectile dysfunction
Provigil (modafinil)	Narcolepsy	Alertness

and deaths of stars, and detailed imaging of the outermost planets in the solar system. But now the HST is deteriorating, and NASA has canceled future servicing missions because of perceived dangers to astronauts. NASA deemed it unethical to carry out a mission with a known lack of ability to respond to emergencies. The estimated amount of time to develop capabilities to rescue the repair team extends beyond the telescope's expected life span.

Many astronomers and other scientists believe that risk in human space exploration is unavoidable regardless of the specific mission. Astronomers argue that the scientific value of the Hubble Space Telescope is great enough to justify the inherent risk of a servicing mission. What is the most ethical decision regarding Hubble's fate? The choices are:

- Cancel the servicing mission and allow the telescope to fail.
- Reinstate the servicing mission regardless of the risk.
- Rush the development of in-orbit shuttle repair technology and reinstate the servicing mission without fully testing such technology.

In approaching the dilemma, one can employ the Four A's strategy, including the added component of risk analysis. (Assume the shuttle has not been grounded.)

Acquire Facts: The important risks are those that prevent the return of the shuttle to Earth and for which there must be a rescue mission. Once these risks are detailed, then one can consider the alternatives because as seen below, the first alternative is to obtain informed consent of those astronauts willing to accept the risks. Another consideration is whether the mission of HST can be carried out more effectively with a replacement telescope; however, that device will also require servicing. Perhaps the most important considerations are the time and costs to create an appropriate rescue vehicle.

Alternatives:

- There are currently more astronauts in NASA than there are missions. Is it ethical to conduct a servicing mission using only those astronauts who accept the risk and volunteer?
- Design a piggyback rescue device for return of the crew.
- Design the rescue vehicle.

Assessment: The concept of having astronauts involved in the risk assessment is reasonable, but should astronauts who are candidates to fly the mission be involved? They should be involved since risk analysis is not based on opinion, but it should be recognized that their inputs might be biased toward not jeopardizing their own chances to be involved in a space mission.

A rescue device might be almost as expensive as the shuttle, and might be incapacitated by the same event that incapacitated the shuttle. A rescue vehicle carried with the shuttle as a lifeboat would simplify a rescue. This option requires a cost-benefit comparison between a lifeboat configuration and a separate rescue vehicle such as the X-38.

Action: The study of the real risks should involve the principal stakeholders, the astronauts. The costs and time to build the rescue vehicle should be put to public debate as the long-term stakeholder is knowledge and the public is the steward of knowledge acquisition investments, not NASA.

ETHICS IN COLONIZATION OF THE MOON AND MARS

Earth's moon and Mars are the first candidates planned for future colonization by Earthlings. The moon, although hostile in its environment, is nearby, and its environmental drawbacks could be managed so that humans could survive there with proper protection. Exploration of the moon will provide a wealth of scientific information, and many hope that the natural resources of the moon can be exploited for the benefit of Earth. In particular, the moon is estimated to possess helium-3 isotope, an important resource for fusion processes when they are developed. With some technological advancement, helium-3 can be mined on the moon and brought back to Earth. NASA's current exploration strategy calls for returning humans to the moon by 2014. Once there, it is a likely objective to establish a method of utilizing lunar resources for the benefit of Earth and further space exploration.

- Is it ethical to pursue the mining of helium-3 on the moon?
- What ethical issues arise from mining natural resources on Earth, and how might those issues relate to the moon?
- How might such a pursuit affect the global economy?
- Does a planetary body have intrinsic value beyond its practical benefit to human life?

The planet Mars, although more distant (a year's one-way voyage away by present technology) without oxygen and inhospitable, could be transformed by currently available technology into an Earth-like planet where humans could live without the need for space suits. Many scientists and the discerning public advocate terraforming Mars to create a suitable environment for human presence. This could be done by creating an atmosphere of CO_2 and O_2 by warming the Mars South Pole ice to release CO_2, then introducing bacteria to start the process of conversion of CO_2 to O_2. The warming of the ice can be achieved by the use of giant mirrors over the South Pole to focus the sun's rays on the ice, thus vaporizing the frozen CO_2. Mars warming would require a greater greenhouse effect than that produced by releasing CO_2 from the frozen polar cap resource (Zubrin and McKay 1996). It is possible to achieve this effect by the use of man-made devices transported from Earth, which would distribute halocarbons (e.g., chlorofluorocarbon) into the Mars atmosphere. It is purported by scientists to be easier to accomplish than one might think. However, once these changes are made, they cannot be easily reversed.

- Is this an ethical method of creating suitable habitats for humans beyond Earth?
- What are the political, social, economic, and ethical ramifications of settling large numbers of humans on another planet?

A reply to these questions in good part rests on the virtue of human discovery and human enhancement of the environment beyond Earth. As long as terraforming Mars is not done with an attitude of conquering, but with the intent of colonizing and cultivating a dry, cold, radiation-intense environment to provide a habitable place for civilization, few would find an ethical argument against this endeavor.

Another argument for terraforming Mars is the potential need for a sanctuary from an uninhabitable Earth. Civilization on Earth is vulnerable to catastrophes such as an asteroid

collision, volcano eruption, overpopulation, natural resource depletion, ozone layer depletion, and nuclear or biological warfare. There is a duty to be prepared. The morally correct
thing to do is to take measures to protect the human race, and at the same time, to preserve
the planet Mars as best one can. The ethics would seem to be in the engineering methods chosen for warming Mars and providing a breathable atmosphere. A cautious start is to build
habitats or closed environments while exploring methods of converting the entire planet to
an atmosphere and temperature compatible with human physiology.

Long-term (>1 year) human presence in space is an immediate ambition for space agencies around the world. Advances in life support system technologies and a greater understanding of the space environment make it possible for humans to remain alive in space for
years. However, there is insufficient understanding of the long-term physiological consequences of extended-duration space missions. The three primary health concerns are radiation exposure due to the unprotected radiation environment that the spacecraft cannot shield
adequately, cardiovascular deconditioning, and bone strength loss due to weightlessness.

Radiation in space is unlike anything observed on Earth. Space explorers are exposed
to charged particles (i.e., the nucleus of elements such as hydrogen, helium, nitrogen, oxygen, carbon, and even iron). Short-duration space flight (6 months to 2 years) is associated
with an increased risk of cataracts, and energetic solar flares or other events have the
potential to kill individuals outside the protection of a spacecraft. Potential risks for long-
term exposure to space radiation include cancer, neurological disorders, and genetic
defects for offspring of space travelers due to random perturbations of genes from cosmic
particles. To date, there is no evidence that this has occurred.

Weightlessness leads to a number of physiological problems, but reduction in bone
strength due to calcium loss in the weightlessness of space is the most severe of the known
consequences. Reduced loading of weight-bearing bones results in an average loss of bone
mass of 1.5% per month. Similarly, postural muscles atrophy, and there is a general transition of muscle fibers from "slow twitch" (endurance muscles) to "fast twitch" (short-
duration, high-energy-output muscles). Both muscle atrophy and bone demineralization
create potential hazards in-flight and require extensive rehabilitation upon return to Earth.
It is unclear whether there is a limit to how much bones and muscles will deteriorate in
space. Exercise is the most common countermeasure to muscle atrophy.

Small sample sizes, inability to conduct comprehensive experiments on space explorers during and post-flight, and lack of fundamental knowledge about physiological mechanisms prevent us from developing a complete understanding of long-term health risks in
human space exploration. To begin developing viable countermeasures to space flight hazards, NASA has established the Bioastronautics Critical Path Roadmap (BCPR) to evaluate the relative health risks of different adaptations during human space exploration. The
National Space Biomedical Research Institute (NSBRI) is a consortium of universities
whose investigators are researching the barriers and countermeasures to enable space
exploration. Funding for this research is limited, and most research can be conducted only
through ground-based simulations of the space environment. However, these experiments
often provide only estimates of the impacts of space exploration on the human body. Much
of the research that NSBRI proposes cannot be completed in the foreseeable future. In the
year 2006, one is justified in asking the following questions:

- Based on current understanding of health risks in human space exploration, is it prudent or ethical to continue carrying out short-term space missions? How about long-term missions?

- How much should we know about the physiological risks of human space exploration before we commit to long-duration space missions?

- How valid is the argument against the expenditure of public funds for space projects when we have a duty to improve the human condition on Earth?

- What limits should be placed on the experimentation on space travelers postflight?

- Given the current state of knowledge, would you be willing to go to space if given the opportunity? For what duration?

In 2004, in spite of limited success from ground-based experiments with animals and human subjects in various simulated weightless environments, it became known that with increased exercise training while in space for many months, astronauts were able to overcome to some extent two of the most feared obstacles of long-term space occupancy: cardiovascular and musculoskeletal deconditioning. This finding was in good part due to the emergency landing of the Soyuz return vehicle from the International Space Station with three astronauts whose physical conditioning in space was maintained. Upon landing in an isolated and rugged terrain far from immediate help, they were able to exit the spacecraft and set up camp while awaiting recovery; previous astronauts could not have managed these tasks because of deconditioning. The sample size of three is small but the space community is encouraged that proper exercise protocols, including a small bicycle centrifuge (see Figure 12.2), could be the countermeasure that will enable physiologically safe missions to Mars. The U.S. space program plans to have a human being land on Mars by 2020 if all goes well.

The principal concern at present is radiation from charged particles such as protons from the sun and heavy ions that shower the space vehicle and astronauts while away from the magnetic shielding of the earth. This is a topic of intense focus of NASA research, mostly at universities around the United States (see Case 8.5).

THE ROLE OF GOVERNMENT IN EVALUATING THE RISKS OF NEW TECHNOLOGIES

Whereas this text has emphasized the prime responsibility of the engineer and scientist to consider the ethical consequences of technologies they are developing, traditionally much of the public's reliance has been on government oversight. This oversight has been accomplished through organizations such as the Nuclear Regulatory Commission and the Food and Drug Administration, of which the Center for Evaluation of Devices and Radiologic Health (CDRH) is a part. CDRH has the responsibility and the legal authority for ensuring that medical devices marketed in the United States are both reasonably safe and effective for their intended clinical use. Another major regulatory agency is the Environmental Protection Agency, which plays a role in the evaluation of and policies regarding genetic modification of plants. Its authority extends to synthetic biology through the Toxic

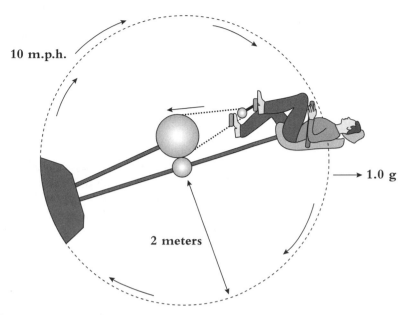

Figure 12.2 The concept of a human-powered centrifuge using a reclining bicycle ergometer. Both exercise and gravity are enabled by this device. Gravity force = (velocity)^2 × radius^{-1}. To achieve a force equivalent to earth gravity, the astronaut would need to propel the device (2-m radius) at 4.4 m/s or 10 miles/hr. The counter mass is for stability

Substance Control Act. The NIH has regulatory authority over gene therapy and matters concerning manipulation of DNA through its Recombinant Advisory Committee.

These regulatory bodies have mandates that are technologically oriented, but their evaluations are influenced by political and public perceptions of dangers and expediency except where real dangers are involved. The Nuclear Regulatory Commission has oversight ranging from reactor safety to the licensing of institutions to use radioactive materials for the treatment of diseases. Thus, the evolution of regulatory bodies in the United States as well as other countries has been to protect the public, and there is generally a reliance on the part of innovators that these regulatory groups will take responsibility for the safety of their inventions and proposed new technologies. This situation is not optimum because the precautionary principle may take precedence and stifle progress by discouraging the experiments necessary to evaluate safety and suspected consequences. One can argue that only through careful studies will we acquire data on consequences to preemptively limit catastrophes from new technologies. However, without starting the experiments, there is no way to make progress.

Two examples help clarify the above concerns. In 1957, the International Committee for the Safety of Life at Sea was approached by a number of nations to change what were considered too conservative ship speed rules in the North Atlantic shipping lanes. The lower speeds were in force in low-visibility situations such as fog in areas where there was the danger of icebergs. The International Ice Patrol was assigned the task of determining whether the then-modern post-WWII radars could detect icebergs in time to stop or change course before collision. They could not. As a result of this investigation, the strict speed rules were maintained, even when more modern radar systems became available (Budinger 1959). This was a proper exercise of the precautionary principle.

This second example describes what could have been a too conservative exercise of the precautionary principle in the regulation of the technology of magnetic resonance imaging. In 1980, there were no nonresearch magnetic resonance imaging (MRI) procedures done in the world, but by 2004, 75 million studies had been done at field strengths of 1.5 tesla (15,000 gauss, or 30,000 times the earth's magnetic field). However, in 1979 there was no experience above 0.3 tesla for magnetic field exposure, other than some stray field exposures at high-energy physics laboratories. The scientists and engineers took it upon themselves to determine the thresholds for health effects. They acquired data from genetic analyses of cells and animal and human exposures. These were used by FDA to suggest limits (not regulations), and as new data became available, the limits were changed every few years. There is now approval for clinical studies at fields of 3 tesla, and other guidance regarding other aspects of MRI procedures. The door is left open for experiments beyond the suggested limits if approved by the local institutional review boards for protection of human subjects. This is an example in which stifling regulations following the precautionary principle were avoided, and the regulatory bodies worked with the scientists to arrive at safety guidelines. The reasons the regulations kept abreast of the data were in good part that the benefits to the public were seen within a few years of the invention of MRI in 1973, and perhaps in the main that the innovators themselves directed research toward determination of the limits of safety (Budinger 1998). The current procedures for approval of gene therapy also represent the results of close interactions between peer review committees and the medical scientists.

A brief word about the future. It is likely that before 2010 some of the technologies we have discussed will lead to consequences only a few might have predicted and for which no countermeasures have been developed. Some consequences are likely to be as unpredictable as were the negative consequences of the Internet. The enormity of the possible negative consequences of gene transfection or mind control using deep brain stimulation is such that we do need to work hard to estimate risks and at the same time not stifle progress. These will be exciting challenges for the upcoming generation of scientists and engineers.

CHAPTER SUMMARY

The emerging biotechnologies associated with germline engineering, gene therapy, cloning, and methods of human enhancement, as well as other new technologies such as synthetic biology, computer-brain interfacing, brain imaging, and space exploration, carry with them some known as well as unknown risks. This chapter gives background information about these new technologies, many of which are becoming issues of heated public debate. We have added to the tools of this text the concept and general method of *risk assessment*. Although it is important to ensure a dialogue between experts, decision makers, and the public, the role of the expert becomes more important the more advanced the technology. The decision makers and public must be educated about risks; otherwise the technology and its future might become too much dependent on opinions and the precautionary principle (Pompidou 2000).

The consequences of the major new technologies will not be fully anticipated if the patterns of the past are not changed. Princeton Professor Norman Augustine captures the attention of engineers and the public to the inevitable costs that accompany new technology by the following introduction he gives to an engineering class he teaches (Augustine 2002):

> . . . a gentleman had invented a new product that virtually everyone in the world would want to have—a product that would create millions of new jobs and would greatly improve the quality of most people's lives. Furthermore, as luck would have it, he was seeking investors. When asked, most students expressed significant interest in investing in his endeavor (at least hypothetically—after all, they were students!). But when I added, "Oh yes, there is one other thing—his invention will kill a quarter of a million people each year" and asked if they would still be interested in investing, no one showed any interest in such a reprehensible product. Furthermore, most said that any such product should be banned outright. I then told them that the inventor's name was Nicholas Joseph Cugnot—and his invention was the automobile.

REFERENCES

Alberio, R., A. D. Johnson, R. Stick, and K. H. S. Campbell. 2005. "Differential nuclear remodelling of mammalian somatic cells by Xenopus laevis oocyte and egg cytoplasm." *Exp Cell Res* **307**(1):131–141.

Augustine, N. R. 2002. "Ethics and the Second Law of Thermodynamics." *The Bridge* (National Academy of Engineering) **32**(3):4–7.

Budiansky, S. 1983 (August 18). "Genetic engineering: regulation issue is resurrected by EPA." *Nature* **30**:572.

Budinger, T. F. 1959. "Iceberg detection by radar." *U.S. Coast Guard Bull No.* **45**:49–97.

Budinger, T. F. 1993. Unpublished data.

Budinger, T. F. 1998. "MR safety: past, present, and future from a historical perspective." *Magnetic Resonance Imaging Clinics of North America* **6**:701–714.

Calvin, M. and S. E. Taylor. 1989. "Fuels from algae." In: *Algal and Cyanobacterial Biotechnology*. R. C. Cresswell, T. A. Rees and N. Shah (eds.). Hoboken, NJ: John Wiley & Sons, pp. 137–157.

Carmena, J. M., M. A. Lebedev, R. E. Crist, et al. 2003. "Learning to control a brain-machine interface for reaching and grasping by primates." *PLoS Biol* **1**(2):E42.

Cello, J., A. V. Paul, and E. Wimmer. 2002 (August 9). "Chemical synthesis of poliovirus cDNA: generation of infectious virus in the absence of natural template." *Science* **297**:1016–1018.

Craggs, M. D. 1975. "Cortical control of motor prostheses: Using the cord-transected baboon as the primate model for human paraplegia." *Adv Neurol* **10**:91–101.

Crosby, A. W. 2003. *America's Forgotten Pandemic: The Influenza of 1918.* New York: Cambridge University Press.

Cyberkinetics-Neurotechnology Systems, Inc. 2005. See www.cyberkineticsinc.com, accessed January 27, 2006.

Duffilla, J., J. Buckley, D. Langa, G. Neil-Dwyera, F. McGinnc, and D. Waded. 2001. "Prospective study of omental transposition in patients with chronic spinal injury." *J Neurol Neurosurg Psychiatry* **71**:73-80.

Erickson, D. 1990. "Patent power. The oil-eating bacterium that spawned an industry." *Sci Am* **262**(6):88–93.

Fenwick P. 1993. "Brain, mind and behaviour. Some medico-legal aspects." *Br J Psychiatry* **163**:565–573.

Flanagan, R. 1996. "Engineering a cooler planet." *Earth* **5**:34–39.

Gensheimer, K. F., M. I. Meltzer, A. S. Postema, and R. A. Strikas. 2003. "Influenza pandemic preparedness." *Emerg Infect Dis* **9**:1645–1648.

Goldsmith, H. S. 1990. *The Omentum: Research and Clinical Applications.* New York: Springer-Verlag.

Goldsmith, H. S., A. Fonesca Jr., and J. Porter. 2005. "Spinal cord separation: MRI evidence of healing after omentum-collagen reconstruction." *Neurol Res* **272**:115–123.

Kastenberg, W. 2002. Bioengineering course lecture on risk analysis, University of California, Berkeley.

Lemler, J., S. B. Harris, C. Platt, and T. M. Huffman. 2004. "The arrest of biological time as a bridge to engineered negligible senescence." *Ann NY Acad Sci* **1019**:559–563.

Maguire, G. Q., Jr. and E. M. McGee. 1999 (January–February). "Implantable brain chips? Time for debate." *Hastings Center Report* **29**(1):7–13.

Marho, B. and L. H. Grimme. 1983. "Hydrogen from water—the potential role of green algae as a solar energy conversion system." In: *Biomass Utilization.* W. A. Cote (ed.). New York: Plenum Publishing Corporation, pp. 227–232.

Martin, M. W. and R. Schinzinger. 2005. *Ethics in Engineering.* Boston, MA: McGraw-Hill Publishers.

Mayberg, H. S. 1996. "Medical-legal interferences from functional neuroimaging evidence." *Semin Clinical Neurochemistry* **1**(3):195–201.

Meltzer, M. I., N. J. Cox and K. Fukuda. 1999. "Modeling the economic impact of pandemic influenza in the United States: Implications for setting priorities for intervention. *Emerg Infect Dis* **5**:159–171.

References **463**

Mulhall, D. 2002. *Our Molecular Future*. New York: Prometheus Books.

Musallam, S., B. D. Corneil, B. Greger, et al. 2004 (July). "Cognitive control signals for neural prosthetics." *Science* **305** (5681):258–262.

National Academy of Engineering 2004. *Emerging Technologies and Ethical Issues in Engineering*. Washington, DC: The National Academies Press.

Nuclear Regulatory Commission. 1975. "Reactor Safety Study". NRC report WASH-1400, NUREG 75/014 1975.

Pinkus, R. L. B., L. J. Shuman, N. P. Hummon, et al. 1997. *Engineering Ethics*. New York: Cambridge University Press.

Pompidou, A. 2000. "From bioethics to the new technology ethics." *Bull Mem Acad R Med Belg* **155**(10–12):404–414.

Reeves, D., M. J. Mills, S. B. Billick, and J. D. Brodie. 2003. "Limitations of brain imaging in forensic psychiatry." *J Am Acad Psychiatry Law* **31**(1):89–96.

Schaffer, D. V. and D. A. Lauffenburger. 2000. "Targeted synthetic gene delivery vectors." *Current Opinions in Molecular Therapeutics* **2**:155–161.

Schanze T., H. G. Sachs, C. Wiesenack, U. Brunner, and H. Sailer. 2005 (August 24). "Implantation and testing of subretinal film electrodes in domestic pigs." *Exp Eye Res* In press. Electronic version avialable at http:www.sciencedirect.com.

Serruya, M. D., N. G. Hatsopoulos, L. Paninski, et al. 2002. Brain-machine interface: Instant neural control of a movement signal. *Nature* **416**:141.

Taylor, D. M., S. I. H. Tillery and A. B. Schwartz. 2002. "Direct cortical control of 3D neuroprosthetic devices." *Science* **296**:1829.

Tessler A. 1991. "Intraspinal transplants." *Ann Neurol* **29**(2):115–123.

Ungers, L. J., P. D. Moskowitz, T. W. Owens, et al. 1982 (February). Methodology for an occupational risk assessment: an evaluation of four processes for the fabrication of photovoltaic cells." *Am Ind Hyg Assoc J* **43**(2):73–79.

Wolpaw, J. R., N. Birbaumer, W. J. Heetderks, et al. 2000. "Brain-computer interface technology: a review of the first international meeting." *IEEE Trans Rehabil Eng* **8**(2):164–173.

Wolpe, P. R. 2003. Personal communication.

Wulf, W. 2004. "Keynote Address." In: *Emerging Technologies and Ethical Issues in Engineering*. Washington, DC: National Academies Press, pp. 1–6.

Zhang, L. and A. Melis. 2002 (October 29). "Probing green algal hydrogen production." *Philos Trans Royal Soc Lond B Biol Sci* **357**:1499–1507.

Zubrin, R. and C. P. McKay. 1996. "Terraforming Mars." In: *Islands in the Sky*. S. Schmidt and R. Zubrin (eds.). New York: John Wiley & Sons, pp. 124–144.

ADDITIONAL READING

Cothran, H. (ed.). 2002. *Energy Alternatives: Opposing Viewpoints*. San Diego: Greenhaven Press, 220 pp.

Dorf, R. C. 2001. *Technology, Humans and Society: Toward a Sustainable World*. San Diego: Academic Press, 500 pp.

"Ethical Issues of New and Emerging Technologies," available via www.morst.govt.nz/uploadedfiles/Documents/ publications/research%20reports/Ethical%20issues.pdf.

"Emerging Technologies and Exponential Change: Implications for Army Transformation," available at http://carlisle-www.army.mil/usawc/Parameters/02summer/nygren.htm.

Hoffman, Peter. 2001. *Tomorrow's Energy: Hydrogen, Fuel Cells and the Prospects for a Cleaner Planet*. Cambridge, MA: The MIT Press, 289 pp.

Langton, C. G. (ed.). 1994. *Artificial Life: Proceedings of the Workshop on Artificial Life Held June, 1992 in Santa Fe, New Mexico*. Volume XVII. New York: Addison-Wesley Publishing Company, 599 pp.

National Research Council and National Academy of Engineering, 2004. *The Hydrogen Economy: Opportunities, Costs, Barriers, and R&D Needs*. Washington, DC: The National Academies Press, 240 pp.

Phimster, J. R., V. M. Bier, and H. C. Kunreuther (eds.). 2004. *Accident Precursor Analysis and Management: Reducing Technological Risk through Diligence*. Washington, DC: National Academy of Engineering, The National Academies Press, 208 pp.

Ratner, M. and D. Ratner. 2003. *Nanotechnology*. Upper Saddle River, NJ: Pearson Education, Inc./Publishing as Prentice Hall PTR, 188 pp.

PROBLEM SET

12.1 Design a land mine that will be effective for three years, after which time it will no longer be dangerous (having it self-detonate after 3 years is not an option).

12.2 Through the Internet or any other source, determine how close we are to equipping hotel rooms with micro-TV-systems (battery operated) of a size 5 mm by 10 mm that have a range for reception of 0.5 km.

12.3 Calculate the increase in surface area for a 1×1 meter panel if the panel is loaded with spheres of 20 nanometers diameter in close packed arrays vs. a panel loaded with spheres of 1 mm diameter. This problem has to do with the use of new materials as nanoparticles for conversion of the photons from the sun or even infrared from wasted heat into electricity.

12.4 The year is 2025. All countries have coalesced into a one-world democracy, referred to as UNIGOV. As a member of the UBC (UNIGOV Bioethics Council), you are discussing the legality of a new, highly advanced life-extension technology, which you strictly oppose. At the moment, it appears that the majority of the council is in favor of globally legalizing this technology. In short, their rationale consists of the following statement:

This technology possesses the prospective capability of solving all of society's problems. It may eliminate pollution, produce abundant food, and catapult other fields of technology into an intense rate of progress never before witnessed. Eliminating aging will allow our minds to grow. Hundreds of years of accumulated wisdom could prove to be the most valuable resource available to humankind. Additionally, such individuals would be more concerned about the quality of the future environment, which they themselves would expect to inhabit. Overpopulation is only a problem of the underdeveloped world. In the developed world, technology allows production to outstrip reproduction, thus this technology and economic freedom would make each member of society a net contributor to world wealth. Skyscrapers and ocean living could provide ample room for centuries of population growth. Space colonies will have access to vast energies of the sun and unlimited room.

It is now time for your rebuttal before the final votes are cast. Prepare a statement that will provide the strongest arguments against this anti-aging technology and provide it within the context of the above affirmation. Ineloquently stating the obvious is unlikely to influence the council's decision.

12.5   At the age of 50, a mathematics professor decides that he wants to be cryopreserved. He sets up the necessary funding and completes all the paperwork to ensure his transfer to cryostasis after death. However, five years later, the professor is diagnosed with brain cancer and becomes concerned that the cancer will destroy too much of his identity and memory before killing him. He goes to court and fights for his right to be preserved before legal, natural death. In the end, he loses his case. Identify reasons why the professor may have lost his case. What points would he have likely made when addressing the court?

12.6   If an automobile that weighs 1,500 kg is built of materials that are generally 20 times stronger than steel, and these materials have one-third the weight of steel, what are the possible ethical or negative engineering consequences?

12.7   A new drug is proposed for the treatment of cancer, but the measurements from animal experiments show that this drug is likely to cause a death or illness leading to death in one person in 1 million. The method of determining this number was to measure the death rate for the drug at a dose 1,000 times that planned for human treatment and divide that number by 1,000 to extrapolate to the human population. That is, in a population of 1,000 mice, one animal died from the exposure.

If the death rate is one in 100,000 from cancer, use the valuation criterion for nuclear plant safety to argue that the risk of the drug is too great. (20 words)

Is the method of determining the toxic probability of the drug valid? (20 words)

How would you determine the likely death rate from the drug if you could study only 1,000 animals? (40 words.)

12.8   From Table 12.1, select an emerging technology not analyzed in this or other chapters. Using your own imagination, discussions with others, library resources, and the Internet, present the possible long-term risks of the selected technology and show how you might engineer a precursor warning technology to prevent the event or at least mitigate the consequences. (80 words)

12.9   Cryopreservation has been proposed as a method to suspend a terminally ill person in a state from which the individual might be recovered when a medical therapy becomes available. Methods of flash freezing and techniques of tissue perfusion before freezing have been under development since successful organ perseveration for transplantation became of vital importance. Cryopreservation activities around the world are being taken seriously by for-profit businesses and by patients who will to be preserved after death. In 2005, there were at least five organizations actively cryopreserving bodies and brains in the United States: the Cryonics Institute (www.cryonics.org), BioPreservation (www.cryocare.org), American Cryonics Society, Alcor (jlemler@alcor.org), and TransTime (www.transtime.com). TransTime performed the first cryopreservation of a human in 1974 in California. The largest cryopreservation organization is the Institute for Problems of Cryobiology and Cryomedicine, located in what is now the Ukraine.

This case assumes that a method of cryopreservation has proved successful in animal studies and that human cryopreservation using intravenous infusions and a below freezing environment has been commercialized. The method involves accepting living clients who want to be preserved for a period of time sufficient for the discovery of a cure for their disease. The engineer or manager of the cryonic facilities at a foundation has been ordered to cease all operations and abandon the enterprise by the foundation's board of directors because all funding has been withdrawn. What are the ethical issues for the specific situation, and what are the issues regarding suspended animation techniques in general? Arguments for and against cryonics are to be made assuming that there will be future lifesaving treatments available for those presently cryopreserved. The arguments are to be centered on the moral theories and principles. (300 words.)

Glossary

Abortion: The premature natural or induced expulsion of an embryo or fetus from the reproductive tract.

Act utilitarianism: An action that has as its consequence a greater pleasure and lesser pain than an alternative act. See Rule utilitarianism.

Adult stem cell: An undifferentiated cell found in a differentiated tissue that can renew itself and, with certain limitations, differentiate into the specialized cell types of the tissue from which it originated.

Advance health care directive (AHCD): A legal document that communicates in writing the wishes of an individual regarding his or her medical care, usually including the appointment of an advocate to make decisions in the event a patient becomes unconscious, incompetent, or terminally ill.

Agenda 21: Arising from the 1992 Rio de Janeiro and Johannesburg conferences on preserving the environment, it is a dictum of social behavior for sustainable development and protection of the environment. Critics fear it threatens human rights and democracy by over-riding principles of freedom of choice by governmental regulation of where and how people will live.

Alternative Energy Sources: Sources other than fossil fuels and trees, such as solar conversion to electricity, gases and wind, and nuclear fission and fusion.

Altruism: Taking into consideration the interests of others in making a moral decision or in undertaking an action. Unselfish concern for the welfare of others. The opposite is egoism, the idea that actions taken in one's own self-interest are right.

Amniocentesis: A procedure done at 15 weeks of pregnancy in which a needle is inserted into the sac of fluid surrounding the developing fetus. A sample of fluid is withdrawn for genetic and other testing.

Anabolic: Relates to drugs that promote growth and metabolism. Opposite of catabolic.

Anencephaly: A condition in which an infant is born with a severe malformation of the brain that is incompatible with life.

Aneuploidy: A condition in which a human egg has an extra or missing chromosome.

Aneurysm: A ballooning of the wall of an arterial blood vessel that can rupture suddenly with serious consequences.

Anthrocentropism: The fundamental tenet is that humans alone are endowed with an intrinsic worth and are the source of all value, nature being valuable only as a means to their ends. Contrasts with biocentrism (see below).

Autologous: Pertaining to oneself. Autologous transplant is one done using one's own or genetically identical tissue or bone marrow. Contrasts with allogeneic transplant, which is done with tissue from a donor not genetically identical to the recipient.

Autonomy: A principle in ethics referring to self-determination or self-rule. People have the right to make decisions about matters that affect them as individuals, and this right should be respected as long as no one else's rights are affected and as long as no harm is done. This principle often arises in health care or human experimentation.

ART: Assisted reproductive technology; any procedure that aids in the treatment of infertility in which both eggs and sperm are manipulated outside the body.

Belmont Report: Issued on April 18, 1979, by the National Commission for the Protection of Human Subjects of Biomedical and Behavioral Research, it provides ethical principles and guidelines for the protection of human subjects participating in a research study. Among other things, it makes provisions for institutional review boards to review all protocols and establishes the necessity for written consent for any participant in a research study.

Beneficence: A principle in ethics describing the moral obligation of acting so as to do good to benefit others, to do one's best to help and protect others, to ensure their well-being, and to maximize the benefits and minimize any risks.

Biocentrism: Biocentrism holds that everything in nature has value. A basic tenet is that nature has its own intrinsic worth, independent of its utility for humans.

Biolistic method: Method for shooting genes into plant tissues using a pressure gun.

Biotechnology: The use of biology in various new technologies that benefit human beings. Examples include nanotechnology, in vitro fertilization techniques, genetically engineered food, and genetic engineering.

Blastocyst: A stage in the development of the embryo at about five days after fertilization when it consists of a hollow ball that contains a cluster of about 200 embryonic stem cells. This is the only stage in the development of the embryo where embryonic stem cells are present and can be obtained.

Blastomere: One of the cells of a 4- to 8-cell embryo; at this very early stage of development, each blastomere is totipotent, capable of forming a complete and fully functional infant.

Brain death: A condition in which the function of the brain is so impaired that the patient remains in an irreversible coma. An absence of electrical activity as sensed by the electroencephalograph is considered by some as objective evidence of brain death. This remains controversial, however, because the hibernating brain may not have detectable electrical signals.

Casuistry: The approach to an ethical decision involving contrasting previous cases that represent extremes in the possible solutions of a problem. This approach holds that the history of cases rather than fundamental principles is the foundation for ethics with a particular emphasis on bioethics.

Chimera: An individual with more than one genetically distinct population of cells.

Chimerism: The presence in an individual of more than one genetically distinct population of cells. Chimeras can occur naturally but are rare, and most likely arise from more than one zygote by cell fusion at a very early stage. Chimerism can also occur following organ transplantation with the appearance of the genome from the donor cells in cells of the recipient.

Chorionic villus: Part of the placenta nourishing the fetus *in utero*.

Chorionic villus sampling (CVS): Usually done at 10 to 12 weeks of pregnancy; a needle is used to take a biopsy of the placenta nourishing the fetus for genetic and other analyses.

Chromosome: Chromosomes carry the complete array of an organism's genetic material and are present in the nucleus of each cell. Each cell contains two sets of chromosomes, one from each parent. They consist of DNA molecules that make up the genes that pass on the inherited characteristics of the organism. Human beings have 22 pairs of chromosomes and two sex-determining chromosomes.

Clone: A genetic replica of another mammal or other living thing.

Cloning: There are two types of proposed human cloning, reproductive and therapeutic cloning. In reproductive cloning, the clonal embryo is implanted into a womb for the pregnancy to continue with the intent to create a fully formed living individual. In therapeutic cloning, the embryonic stem cells of a clonal embryo are used as stem cells to repair or create a specific tissue or organ for the person who supplied the DNA for the cloning.

Common pooled resource (CPR): An environmental resource open to all (for example, national parks). When accessed by the public, it tends to be used indiscriminately, and is less well cared for than if it were owned privately. The "tragedy of the commons" is a term that has been used to describe this process.

Conception: When sperm and ovum unite to form a zygote, the earliest embryo formed after fertilization of an egg by a sperm.

Conjoined twins: Two identical individuals in whom early embryonic cellular division fails to occur normally. The embryos continue to develop but remain joined together. Depending on the type and extent of organs shared, they may or may not be separated surgically after birth. The point of anatomic attachment varies.

Consequentialism: The doctrine that the moral value of any action always lies in its consequences. The consequence usually sought is a state of happiness or maximum utility of the individuals and avoidance of bad consequences. Utilitarianism is related to consequentialism where the approval of an act or ethical decision is based on whether the consequences will lead to the greatest happiness.

Conveyance: The transfer of information without attribution of originality to oneself or anyone else. Conveyance of information is not considered plagiarism if the information is of general knowledge or availability.

Cryopreservation: A procedure in which biological samples are frozen at very cold temperatures for prolonged periods, following which they can be thawed and used.

Cytoplasm: The living material within a cell contained in the cell membrane but exclusive of the nucleus.

Cytoplast: A cell without its nucleus.

DDC: Dopa decarboxylase, an enzyme lacking in the brain of patients with Parkinson's disease, resulting in abnormal motor function. Potentially curable by gene therapy to provide DDC to targeted sites in the brain.

Declaration of Helsinki: Adopted in 1964 soon after new ethical dilemmas in human experimentation came to light pertaining to advances in procedures in high demand and short supply, such as kidney dialysis, heart transplantation, and liver transplantation. It expanded on the Nuremberg Code, which was written following the atrocities of experimentation during World War II, and made additional recommendations and guidelines for biomedical research involving human beings.

Deep ecology: A philosophy that all life has value, and indeed, all life has equal value. Nature can be used for vital needs, but it is basically ethically wrong to waste it or to take too much. It implies that the human population now is too great for the ecosystem to maintain. It holds that we are out of balance and must look to doing things efficiently with less waste and move to a standard of living that is more responsible.

Deontology: Pertaining to a moral theory that emphasizes the need for actions according to duty and holds that some actions are required even if painful consequences result. This term is used in lieu of Kantianism by some ethicists and philosophers.

Deoxyribonucleic acid: DNA, a long chain of nucleotides each one of which consists of deoxyribose, phosphoric acid, and an organic base. DNA consists of two associated polynucleotide strands that wind together in a double helix and form the stuff of chromosomes. It is the material that carries the genetic information contained in the genes of a living organism.

Differentiation: The process by which less specialized cells in the body mature to give rise to the more specialized cells in the body. In human beings, the fertilized ovum gives rise to embryonic stem cells that have the capability to form the more than 200 cell types found in the body.

DNA: A long chain of nucleotides each one of which consists of deoxyribose, phosphoric acid, and an organic base. Double-strand DNA consists of two associated polynucleotide strands that wind together in a double helix and form chromosomes. It is the material that carries the genetic information contained in the genes of a living organism.

Ecofeminism: A concept from the 1970s that environmental problems stem from a male-dominated society that exploits nature and the environment, whereas women, if in charge, would have a much more caring approach to the environment.

Egoism: An ethical concept that actions that satisfy self-interest are right.

EG cells: Embryonic germ cells. They develop in the gonadal ridge and mature into germ cells (sperm and eggs).

Embryo: The product of conception formed by the fertilization of the egg by the sperm. The term applies to the first eight weeks of gestation, after which it is referred to as a fetus.

Embryo biopsy: A procedure done on a very early embryo at the 2- to 4-cell stage in order to detect the presence of chromosomal abnormalities or genetic diseases prior to implantation in the uterus. See also PGD.

Embryonic stem cells: In general, stem cells are important for the formation, maintenance, and repair of the tissues of the body. A stem cell has two characteristics: the ability to replicate itself, thereby maintaining the stem cell pool, and the ability to form a differentiated cell. There are two types of stem cells: embryonic stem cells that are found in the blastocyst of the developing embryo, which are pluripotent and capable of forming all the cells of the body (although not totipotent [able to form a complete fully functional fetus]), and adult stem cells, capable of replicating themselves and forming the cells of the tissue undergoing development or regeneration.

Endometrium: The innermost lining of the uterus, which receives the fertilized egg and nourishes the early embryo.

Enhancement: An alteration made in an attempt to better one's self by means of cosmetics, plastic surgery, drugs, or genetic modification, for example. Also see Genetic enhancement and Gene therapy.

Equivalence: This term is applied by regulatory agencies assessing the safety of a novel food when the food or ingredient under consideration is deemed essentially equivalent to a natural food or ingredient.

ES cells: Embryonic stem cells.

Eugenics: A term that applies to selective breeding to improve qualities of human offspring (positive eugenics) or methods to prevent the passage of traits to offspring through selection, sterilization, or castration (negative eugenics).

Euthanasia: Physician-assisted suicide; providing advice or a prescription to facilitate the patient's death. In animals, it refers to the humane destruction of an animal by a method that produces rapid unconsciousness and subsequent death without evidence of pain or distress or a method that utilizes anesthesia produced by an agent that causes painless loss of consciousness and subsequent death.

Fabrication: The construction and reporting of data not actually measured or observed.

Falsification: The modification of observed data.

Fetus: The developing human from approximately eight weeks postfertilization, when all adult tissues are identifiable, until birth.

FDA: Food and Drug Administration.

Fission: Biological fission is the splitting of biological cells. Nuclear fission is the splitting of the nucleus of an atom.

Fusion: 1) Nuclear fusion is the interaction between two atomic nuclei to form a new nucleus plus energy, as is the case for fusion of hydrogen on the sun and in the fusion bomb. 2) Biological cell fusion is the combination of two cells such as a macrophage and a heart cell wherein one cell might take on some of the genetic characteristics of the other cell, as is believed to occur when green-fluorescent-protein-tagged cells combine with host tissue in cells in stem cell research.

Futile care: The administration of medical care or treatment that is unlikely to result in improvement in disease, less pain and suffering, or better quality of life, or does more harm than good.

Gamete: A sperm or an egg.

Gene: An individual unit of DNA located on the chromosomes that determines the characteristics of an organism. Humans have approximately 30,000 pairs of genes.

Gene chip: An array of several thousand different sections of DNA which are arranged in tiny wells on a glass surface. Each strand of DNA will bind to only the DNA in the sample which is complementary to the known composition or sequence of the DNA in a specific well. The chip can be used also for identifying RNA by loading each well with known sequences to determine gene expression.

Gene therapy (somatic and germline): In general, this is the correction of a disease or condition by inserting new genetic material to correct the genetic deficit. In somatic gene therapy, DNA is inserted into the somatic cell but the change is not passed on to succeeding generations. In germline gene therapy, the germ cells (egg or sperm) are changed; thus, the changes are passed on to succeeding generations.

Genetic engineering: The manipulation of genes in an organism that produces new, permanent changes to the genome of an organism.

Genetic enhancement: To enhance the genome by selection of genes for the egg or the embryo by use of preimplantation genetic screening; currently PGD is used to screen embryos for genetic disease, but the potential exists for it to be used to enhance or select for traits such as intelligence, physical beauty, and athletic ability.

Genetic modification: The direct introduction of selected characteristics by artificial transfer of foreign or synthetic DNA (deoxyribonucleic acid, the genetic material) into an organism.

Genetically modified (GM) organisms (also known as GMOs): Genetically modified organisms are plants or animals whose genetic makeup has been intentionally changed to introduce a desired characteristic into a bacterium, plant, or animal. This is done by introducing the genes from one plant or organism of the same or another species into the genome of the recipient being genetically engineered.

Genetically modified foods: These are foods that come from plants that have had some modification of their genetic material to produce a new food product with more desirable characteristics such as drought resistance, resistance to certain pests or herbicides, faster maturation, or longer shelf life.

Genetic screening: The testing of genetic material for the purpose of identifying the presence of an abnormal gene related to the predisposition to a medical condition.

Genome: The complete array of genetic material that resides in the nucleus of each cell of an organism. It consists of two sets of chromosomes, one from each parent.

Germ cell: Pertaining to the germinal cells, either the egg or the sperm.

GIFT: Gamete intrafallopian transfer, an assisted reproduction technique in which eggs and sperm are mixed and transferred to the mother's fallopian tube, where fertilization takes place. The embryo moves from there to the uterus, where implantation takes place and the pregnancy continues to infant birth.

GMP: A genetically modified plant that differs from plants that occur naturally by having been engineered to produce more desirable characteristics, such as greater drought tolerance, faster growth, improved yield and appearance, improved nutritional characteristics, and/or greater resistance to insects or pesticides. The genetic modification is accomplished by the insertion of new genes into the genome of the plant.

Golden Mean: Espoused by Aristotle as a virtue, it is the art of finding a balance between the extremes of a situation (for example, giving sufficient information to a patient to understand an illness but not providing so much information that it is overwhelming).

Harm principle: Formulated by Mill, it was his belief that the state should be restricted to intervening in a person's private affairs only if that person's actions would lead to harm.

Hedon: A term used in an attempt to quantify the units of pleasure.

Hedonistic utilitarianism: If the utility of an act has the consequence of physical pleasure in contrast to intellectual pleasure, the act is considered hedonistic.

Hopping genes: See Transposons.

ICSI: Intracytoplasmic sperm injection. An assisted reproduction technique especially useful in cases of male infertility in which a sperm is injected directly into the egg.

IVF: In vitro fertilization. A procedure first used to help infertile couples conceive wherein egg and sperm are brought together in a glass petri dish where fertilization takes place; the embryo is then implanted into the mother's uterus, where the pregnancy continues to term. Other assisted reproductive technologies involve variations on this procedure.

Justice: A principle in ethics addressing the moral obligation to treat everyone fairly, equally, and without favoritism or bias, that there is equal burden or equal benefit, that one person or one group is not favored in a decision, that impartiality is maintained, and that the highest standard of behavior is represented in the distribution of respect, goods, property, and other resources.

Kyoto Protocol: An environmental plan that came into effect in February 2005 as a global effort to reduce greenhouse gas emissions such as CO_2 and methane to prescribed levels below the 1990 measured emissions. Not all countries have ratified it, including the United States and Australia.

Laparoscopy: A procedure in which an instrument with fiber optic capabilities is inserted through a small incision in the abdomen to view the contents or perform surgical procedures or, in the context of this book, to remove eggs from the ovary or to do intrafallopian transfer of an embryo or sperm and egg.

Living will: Describes in writing the types of procedures, care, treatments, or life support measures acceptable to the patient if rendered unconscious or unable to choose.

Malfeasance: An act that causes harm to others.

Medical futility: See Futile care.

Medical power of attorney: An advance health care directive.

Metastasis: The spread of cancer to tissues outside the tissue of origin.

Mitochondria: Intracellular apparatuses that provide energy to the cell and promote specific cellular activities. They contain their own DNA derived entirely from the ovum.

mRNA: Messenger RNA copies the code from the DNA of a gene in order to produce the specific protein associated with that gene.

Multipotential cell: A cell that has differentiated matured such that it is no longer totipotential (with the capacity to form a fully developed fetus) but now has the capacity to differentiate into different tissues of the body. Considered the same as pleuripotential.

Nanotechnology: Nanotechnology is the science and engineering that focuses on the manufacturing of useful objects with scales less than 1 micrometer. It is also an inquiry into the new science and engineering embodied in complex systems with components less than 1 micrometer. Chemical engineers, material scientists, and contemporary biotechnologists have for many years dealt with molecule assemblages and polymers of 10 to 20 nanometers, but with the successes of microfabrication technologies, an international interest in smaller and smaller devices has led to major efforts in submillimeter device development.

Non-malfeasance: A principle in ethics describing the obligation to do no harm, do no wrong, avoid injury, not put anyone at unnecessary risk or intruding on lives unnecessarily.

Nuclear transfer: See Somatic cell nuclear transfer.

Nuremberg Code: Issues regarding the protection of human subjects arose in conjunction with the Nuremberg trials following World War II after it came to light that Nazi doctors had performed atrocious experiments on living subjects. The Nuremberg Code was written to set forth basic moral, ethical, and legal boundaries for medical research on human subjects.

Oocyte: Egg or ovum.

Ooplasm: The cytoplasm of an egg.

Palliation: The treatment of a disease with the intent of providing relief from symptoms when a cure is not possible. Example: irradiating a lung tumor to reduce its size in order to relieve incessant coughing to enable the patient to sleep.

Parkinson's disease: A neurological disease in which the patient lacks the ability to produce dopamine in central sites in the brain, resulting in abnormal motor function.

PAS: Physician-assisted suicide. See Euthanasia.

Paternalism: Referred to in the relationship between physician and patient, it implies intrusion by the physician in the patient's decision-making process about medical treatment such that the patient becomes dependent upon the physician, bypassing the patient's autonomy.

Penetrance: A particular genotype might or might not produce certain phenotypes such as eye color or webbed toes or an anatomical abnormality. The penetrance is the frequency or percentage of this occurrence.

Persistent vegetative state: A condition in which the function of the brain is so severely impaired that the patient is in an altered state of consciousness for a prolonged period of time. It is a state wherein respiration, heart rate, and other bodily functions are maintained, but the person remains otherwise nonresponsive or minimally so.

Petri dish: A small flat glass dish used in the laboratory.

PGD: Preimplantation genetic diagnosis, usually done by polar body analysis of the maternal egg or analysis of a blastomere obtained by embryo biopsy.

Phenotype: The observable physical or chemical characteristics of an organism that are the result of the genetic make-up or environmental influences. Hair color, physical build, and handedness are examples of phenotypic characteristics.

Plagiarism: Stealing creative works or ideas from others and representing these as one's own.

Pleuripotential: A stem cell with the ability to form most of the tissues of the body. A pleuripotential cell is less differentiated than adult stem cells and more differentiated than totipotential stem cells.

Polar bodies: These are two bodies associated with the egg that can be used for preimplantation genetic diagnosis. One is formed in the ripening egg, which can be captured for analysis by a pipette inserted into the egg or retrieved after it is extruded from the egg as the egg matures; this polar body is not necessary for the development of the egg, but it contains the same genetic components as the egg. The second polar body is extruded from the egg once it has been fertilized. These polar bodies, known as the first and second polar bodies (IPB and IIPB), have the same genetic complement as the egg and can be analyzed for the presence of chromosomal abnormalities.

Precautionary principle: The precautionary principle assigns a burden of proof of safety to those who want to introduce a new technology, particularly in cases where there is some reason to suspect the risks might be greater than the benefits. Examples are genetically modified plants or introduction of a new vaccine.

Prima facie duty: The first intuitive notion of what one's duty should be. Frequently, a prima facie duty is in conflict with another duty. Literally, the term means "primarily appears," but the philosopher Ross, who uses this term, denies this is the meaning.

Progenitor cell: Usually refers to a cell that can differentiate to become a particular tissue type, such as a blood-circulating nucleated cell that is a progenitor cell for the endothelium. Progenitor cell designation can be confusing since it is not synonymous with stem cell because the latter is multipotent and can replicate self as well as differentiate into a specific tissue.

Promoter: Part of the gene which triggers the transcription of the DNA code to messenger RNA. The promoter is a sequence to which the enzyme RNA polymerase binds so as to start transcription.

Pronucleus: The nucleus immediately following the fusion of the sperm and the ovum after fertilization.

Proteins: An organism is made up of many different forms of proteins, each of which is composed of chains of amino acids coded for by the organism's genes.

Reproductive cloning: See Cloning.

Retrotransposons: Some transposons are retrotransposons which move about the genome by means of an RNA intermediate, a transcript of the retrotransposon DNA. To insert at another site, the RNA retrotransposon must be converted back to DNA. This is accomplished by the enzyme reverse transcriptase, which is encoded in the retrotransposon itself.

RNA: There are five types of ribonucleic acid. One type known as messenger RNA is found in the nucleus where it is synthesized. Messenger RNA carries the genetic information out of the nucleus to the cytoplasm, where it participates in protein synthesis. Transfer RNA (+RNA) and ribosomal RNA (rRNA) are also important in protein synthesis. Interface RNA (RNAi) blocks mRNA transcription. Catalytic RNAs function as enzymes.

Rule utilitarianism: An act which has a utility or consequence which gives greater pleasure or happiness and lesser pain than another act and also follows an accepted rule of behavior.

SCID: Severe combined immune deficiency, a congenital disease of the immune system in which children are born without essential infection-fighting cells, putting them at risk for severe life-threatening infections. Treatment includes monthly infusions of gamma-globulin. Bone marrow transplant has been successful in some cases.

SCNT: See Somatic cell nuclear transfer.

Slippery slope: A situation in which there is no logical disconnect from the first step (decision) and a final outcome which could be ethically unacceptable.

Somatic cell: Any cell of the body other than the sperm or egg.

Somatic cell nuclear transfer (cloning): The formation of an embryo by removing nuclear DNA from a somatic cell and transferring it to an ovum from which the nucleus has been removed, then stimulating division as if it had been fertilized. The embryo will have the same nuclear DNA as the somatic cell from which it was taken.

Stem cell: Cells that have the ability to become tissue-specific cells. These cells are generated from the embryo, or after maturation of the organism, when the nascent stem cells of mature tissues can be recruited to replace the damaged cells in most organs. See also Embryonic stem cells.

Stem cell line: The progeny of cells from a single embryonic stem cell taken from the blastocyst and expanded into many daughter cells.

Sustainable development: Meeting today's needs without compromising the ability of future generations to meet their needs. A central principle of protecting the environment and its resources. By overuse of our natural resources, we are at a point of irreversibly depleting them so they will not be available for the next generation. Sustainability is the watchword of Agenda 21, which urges the nations of the world to conserve and to use only what is needed to sustain us. It is the balance between human need and available resources.

Terminator: Part of the gene that signals the completion of transcription from the DNA code to mRNA.

Therapeutic cloning: See Cloning.

Therapeutic misconception: The situation wherein the subject of research is under the misapprehension that the research in some way will benefit him or her.

Totipotential: An embryonic stem cell still so early in development that it can form an entire, separate, fully functional fetus, or any cell in the body. The blastomere is the most differentiated cell with this potential.

Tragedy of the commons: When the public has access to an environmental resource (designated as a common pool resource), that resource is generally less well cared for than if it were owned privately. It is the tendency for people given access to a common resource to use it indiscriminately, to harm it, and ultimately to destroy it.

Transcendentalism: The concept that one needs to make a conscious decision to make things better from an environmental standpoint, that wilderness is sacred, and by itself has enormous value. Nature is a place to find one's self, and by one's relationship with nature one finds one's own humanity. The transcendentalist approach to nature is not religious but rather spiritual, and each individual's spirituality is a reflection on nature. It has similarities to Hinduism and Buddhism.

Transgenic: A plant or animal that has been genetically modified.

Transposons: Also known as transposable elements or hopping genes. These are segments of DNA that can move to a new position on the same or another chromosome, often modifying the action of neighboring genes. They were first identified in plants. While it was previously thought that these elements were excess parts of the chromosome or "junk" and had no real function in the genomes of plants and other organisms, they are now thought to have major roles in evolution, structure, and function.

Utilitarianism: The moral theory that holds that good results are results that maximize benefits and minimize harms even if this entails self-sacrifice and violates the autonomy of some individuals.

WWF: World Wildlife Fund, a global conservation organization that has as its goal to stop and ultimately reverse environmental degradation and to build a future where people live in harmony with nature.

Xenotransplantation: The insertion of organs, tumors, or tissues from a species other than that of the recipient.

ZIFT: Zygote intrafallopian transfer, a procedure in which a fertilized zygote is transferred to the fallopian tube of the mother, whence it passes to the uterus for implantation.

Zygote: The earliest embryo formed after fertilization of an egg by a sperm.

Index

477